Lecture Notes in Mathematics

Edited by A. Dold and B. Eckmann

Subseries: Instituto de Matemática Pura e Aplicada, Rio de Janeiro
Adviser: C. Camacho

1324

F. Cardoso D. G. de Figueiredo
R. Iório O. Lopes (Eds.)

Partial Differential Equations

Proceedings of ELAM VIII,
held in Rio de Janeiro, July 14–25, 1986

Springer-Verlag

Berlin Heidelberg New York London Paris Tokyo

Editors

Fernando Cardoso
Departamento de Matemática, Universidade Federal de Pernambuco
50.739 Recife Pe, Brazil

Djairo G. de Figueiredo
Departamento de Matemática, Universidade de Brasília
70.910 Brasília DF, Brazil

Rafael Iório
IMPA, Estrada Dona Castorina, 110
22.460 Rio de Janeiro RJ, Brazil

Orlando Lopes
IMECC/UNICAMP
13.081 Campinas SP, Brazil

Mathematics Subject Classification (1980): 35-XX; 35J65; 35K55; 35L70; 35P25; 35S05

ISBN 3-540-50111-8 Springer-Verlag Berlin Heidelberg New York
ISBN 0-387-50111-8 Springer-Verlag New York Berlin Heidelberg

Printed in Germany

Printing and binding: Druckhaus Beltz, Hemsbach/Bergstr.
2146/3140-543210

VIII LATIN AMERICAN SCHOOL OF MATHEMATICS

The Latin American School of Mathematics (ELAM) is one of the most important mathematical events in Latin America. It is held every other year since 1968 in a different country of the region, and its theme varies according to the areas of interest of local research groups.

The VIII ELAM took place at the Instituto de Matemática Pura e Aplicada (IMPA/CNPq), Rio de Janeiro, Brazil, during the period of July 14-25, 1986.

The subject of the school was Partial Differential Equations with emphasis on Microlocal Analysis, Scattering Theory and the applications of Nonlinear Analysis to Elliptic Equations and Hamiltonian Systems.

The School was attended by mathematicians from many countries, ARGENTINA (12), BELGIUM (2), BRAZIL (125), CHILE (8), COLOMBIA(5), FRANCE (9), GERMANY (6), ISRAEL (1), ITALY (8), JAPAN (3), MEXICO (7), PERU (2), SOVIET UNION (1), SWEDEN (2), SWITZERLAND (2), UNITED STATES (19) and VENEZUELA (5).

This volume contains most of the conference delivered at the VIII ELAM.

We would like to express our gratitude to the ELAM International Committee, whose members are Professors Jacob Palis Junior (Brasil), José Adem (México) and Carlos Segóvia (Argentina), for having given us the opportunity to organize the VIII ELAM.

We would also like to acknowledge the financial support offered to the School by the following agencies:

Organization of American States (OAS); Instituto de Matemática Pura e Aplicada (IMPA/CNPq); Financiadora de Estudos e Projetos (FINEP); Conselho Nacional de Desenvolvimento Científico e Tecnológico (CNPq); NSF (USA); GMD (Germany); CNRS (France); CONACYT (Mexico); CONICIT (Venezuela); CONICYT (Chile); Coordenação de Aperfeiçoamento de Pessoal de Nível Superior (CAPES); Fundação de Amparo à Pesquisa do Estado de São Paulo (FAPESP); IBM do Brasil; International Mathematical Union (IMU); UNESCO.

Finally we must thank the members of IMPA's staff, in particular Suely Maiato and José Manoel do Outeiro, for their priceless organizational work and Wilson Góes for typing most of the papers contained in the present volume.

The Organizing Committee,

Felix Browder (Univ. Chicago)
Fernando Cardoso (Univ.Fed. de Pernambuco)
Djairo G.Figueiredo (Univ. Brasília)
Rafael José Iório Junior (IMPA)
Jacques Louis Lions (Collège de France)
Orlando Lopes (UNICAMP,Brasil)

CONFERENCES DELIVERED DURING THE VIII ELAM

1. A.BAHRI (École Polytechnique, Paris, France) - "Critical Points at Infinity in the Variational Calculus".

2. M.S.BAOUENDI (Purdue Univ., U.S.A.) - "Holomorphic Extendability of CR Functions and Mappings".

3. M.BEN-ARTZI (Technion III, Haifa, Israel) - "The Limiting Absorption Principle for Differential Operators with Short-Range Perturbations".

4. P.BÉRARD (Univ.Savoi, France) - "On the Number of Bound States and Estimates on Some Geometric Invariants".

5. O.BOGOYAVLENSKY (Steklov Mathematical Institute, USSR) - "New Integrable Problems of Classical Mechanics".

+

6. H.BRÉZIS (Univ.Paris VI, France) - "Liquid Crystal Degree Theory and Energy Estimates for S^k Valued Maps".

++

7. F.BROWDER (Univ. of Chicago, USA) - "Strongly Non-Linear Parabolic Equations".

8. M.CRANDALL (Univ. Wisconsin, USA) - "Continuous Dependence of Solutions of a Class of Degenerated Diffusion Equations on the Equation".

9. M.ELGUETTA (Univ.Catolica de Chile) - "Large Time Behaviour of the Solutions of $U_t = (U^m)_{xx}+U$".

10. V.ENSS (Freie Universität, Germany) - "Asymptotic Time Evolutions for Strictly Outgoing Multiparticle Quantum Systems with Lon-Range Potentials".

11. J.F.ESCOBAR (Univ. of California, San Diego) - "Conformal Metrics with Prescribed Scalar Curvature".

12. D.G.FIGUEIREDO (Univ. Brasília, Brazil) - "Multiple Solutions for a Non-Linear Eigenvalue Problem".

13. D.FORTUNATO (Univ.di Bari, Italy) - "A Birkhoff-Lewis" Type Result for non Autonomous Differential Equations".

14. J.P.GOSSEZ (Univ. Libre de Bruxelles, Belgium), "Nonresonance Near the First Eigenvalue of a Second Order Elliptic Problem".

15. A.GRUNBAUM (Univ. of California, Berkeley, USA) - "Differential Equations in the Spectral Parameter and Multiphase Similarity Solutions".

16. S.HAHN-GOLDBERG (IPN, Mexico) - "An Abstract Regularity Result Applied to Quasilinear Symmetric Systems".

17. A.HARAUX (Univ. de Paris VI, France) - "Recent Results on Semi-Linear Hyperbolic Problems in a Bounded Domain".

18. L.HÖRMANDER (Univ.of Lund, Sweden) - "On the Lifespan of Solutions of Second Order Non-Linear Hyperbolic Differential Equations".

19. J.HOUNIE (Univ.Fed. Pernambuco, Brazil) - "Local Solvability with Compact Supports for Overdetermined Systems of Vector Fields".

20. R.J. IÓRIO (IMPA, Rio de Janeiro, Brazil) - "On the Cauchy Problem for the Benjamin-Ono Equation".

21. J.L.IZÉ (UNAM,Mexico) - "Degree Theory for Equivariant Maps".

22. H.JACOBOWITZ (Rutgers Univ., USA) - "Systems of Homogeneous Partial Differential Equations with Few Solutions".

23. J.KAZDAN (Univ. of Pennsylvania, USA) - "Unique Continuation for Elliptic Systems".

23. E.LAMI-DOZO (Univ. de Buenos Aires, Argentina) - "Curvatura Escalar e Produtos Ponderados".

24. M.LAPIDUS (Univ. of Georgia, USA) - "Variation of the Theme of the Schrödinger Equations: Elliptic Boundary Value Problems with Indefinite Weights and Feynman-Kac Formulae with Lebesgue-Stieltjes Measures".

25. P.LAUBIN (Univ. Liège, Belgium) - "Propagation of the Second Analytic Wave Front Set Along Gliding Rays".

26. A.LAZER (Univ. of Miami, USA) - "Introduction to Multiplicity Theory for Boundary Value Problems with Asymetric Nonlinearities".

27. O.LIESS (Univ.Bonn, Germany) - "Regularity of Solutions of Cauchy Problems with Smooth Cauchy Data".

28. H.M.MAIRE (Univ. de Genève, Switzerland) - "Necessary and Sufficient Condition for Maximal Hypoellipticity of $\bar{\partial}_b$".

29. D.MARCHESIN (PUC/Rio de Janeiro, Brazil) - "Riemann Solutions for Non-Linear Hyperbolic Conservation Laws".

30. R. MELROSE - (Massachussetts Institute of Technology,USA)- "Examples of Non-Discreteness for the Interaction Geometry of Semilinear Progressig Waves in Two Space Dimensions".

31. G. MENDOZA (IVIC, Matemáticas, Venezuela) - "Analytic Approximability of Solutions of Differential Equations".

32. S.MIZOHATA (Kyoto Univ., Japan) - "On the Cauchy Problem for Hyperbolic Equations in C^∞ and Gevrey Classes".

33. R.NAGEL (Univ. of Tubingen, Germany) - "Positivity and Stability for Cauchy Problems with Delay".

34. L.NIRENBERG (Courant Inst., New York Univ,USA) - "Subharmonic Solutions of Hamiltonian Systems".

35. G.PERLA MENZALA (LNCC and UFRJ, Rio de Janeiro, Brazil) - "On the Resonances and the Inverse Scattering Problem for Perturbed Wave Equations".

36. G.PONCE (Univ. of Chicago, USA) - "Well-Posedness of Euler and Navier-Stokes Equations in the Lebesgue Spaces $L^p_s(R^2)$".

37. P.RABINOWITZ (Univ.Wisconsin, USA) - "Periodic Solutions of Prescribed Energy of Hamiltonian Systems".

38. J.RALSTON (Univ.of California, Los Angeles, USA) - "Semi-Classical Approximation in Solid State Physics".

39. D.ROBERT (Univ.Nantes, France) - "Semiclassical Analysis in Scattering Theory" and "Semiclassical Limit and Ergodicity".

40. L.RODINO (Univ.di Torino, Italy) - "Microlocal Analysis in Gevrey Classes".

41. J.SJÖSTRAND (Univ. of Lund, Sweeden) - "Estimates on the Number of Resonances for Semiclassical Schrödinger Operators".

42. M.STRUWE (ETH-Zentrum, Switzerland) - "Heat Flow Methods for Harmonic Maps of Surfaces and Applications to Free Boundary Problems".

43. J.SYLVESTER (Duke Univ., USA) - "Inverse Boundary Value Problem".

44. C.TOMEI (PUC/Rio de Janeiro, Brazil) - "Scattering on the Line - an Overview".

45. F.TRÊVES (Rutgers Univ., USA) - "Microlocal Cohomology in Hypo-Analytic Structures" .

46. E.TRUBOWITZ (ETH-Zentrum, Switzerland) - "The Geometry of Fermi Curves and Surfaces"

47. Y.C.de VERDIÊRE (Univ.de Grenoble, France) - "Can one prescribe a Finite Part of the Spectrum (with Multiplicity of the Laplacian on a given Compact Manifold?"). Part I and "Can one prescribe a Finite Part of the Spectrum (with Multiplicity of the Laplacian on a given Compact Manifold?") Part II.

48. A.VIGNOLI (Univ. di Roma, Italy) - "Global Behaviour of the Solutions Set for Equivariant Nonlinear Equations".

49. E.ZEHNDER (Univ.Bochum, Germany) - "The V. Arnold Conjecture about Fixed Points of Simplectic Mappings".

+ 5a. J.L.BOLDRINI - "Convergence of Solutions of Capillo-Viscoelastic Perturbations of the Equations of Elasticity"

++6a. E. BRIETZKE - "Mathematical Aspects of the Minimum Mass Problem"

CONTENTS

CRITICAL POINTS AT INFINITY IN THE VARIATIONAL CALCULUS

A. BAHRI

in the honour of H. Lewy

1. ORBITS OF THE FLOW.

Let E be a space of variations (either a Hilbert space or a manifold modelled on a Hilbert space for sake of simplicity) and let :

(1) $$f : E \longrightarrow \mathbb{R}$$

be a functional which we will assume to be C^∞, for sake of simplicity also. Let :

(2) $(\, , \,)$ be the scalar product on E

and

(3) $\partial f(x)$ be the gradient of f at x in E.

We are concerned with finding a solution to the equation :

(4) $$\partial f(x) = 0 \qquad x \in E \quad .$$

For $a \in \mathbb{R}$, we introduce the level sets of the functional f at a :

(5) $$\begin{cases} f^a = \{x \in E \mid f(x) \leq a\} & \text{(sub-level set)} \\ f_a = \{x \in E \mid f(x) \geq a\} & \text{(upper-level set)} \\ {}^a f = \{x \in E \mid f(x) = a\} & \text{(level surface)} \quad . \end{cases}$$

We also consider the differential equation :

(6) $$\frac{\partial x}{\partial s} = - \partial f(x) \quad ; \quad x(0) = x_o \quad .$$

Let :

(7) $x(s,x_o)$ be the solution of (6).

We then have the following very simple principle to solve (4) :

Proposition 1 : Let $b < a$. Assume f^b is not retract by deformation of f^a (e.g. $H_*(f^a, f^b; G) \neq 0$ for a certain coefficient group G ; and H_* is a homology theory ; or $\pi_*(f^a, f^b; G) \neq 0$; π_* is homotopy ...) Then :

either (4) has a solution $\bar{x}$ with $b \leq f(\bar{x}) \leq a$

or there exists an x_o such that $b < f(x_o) \leq a$ such that

(a) $\lim_{s \to +\infty} f(x(s,x_o)) \geq b$

(b) the closure of the set $\{x(s,x_o) ; s \geq 0\}$ is non compact .

There are several directions where one can make the content of Proposition 1 more precise and in the same time more general :

1st precision : In case $\bar{x}$ exists, the basic assumption which is used to study the situation near $\bar{x}$ is that ∂f is Fredholm at $\bar{x}$. Then, if $\bar{x}$ is a non degenerate critical point of f, one computes the Morse index of f at $\bar{x}$.

We then have :

Assume the injection $f^b \to f^a$ is not a homotopy equivalence, then $\bar{x}$ cannot contribute to the difference of topology between f^b and f^a if the Morse index of f at $\bar{x}$ is infinite.

Thus a critical point $\bar{x}$ is relevant in the calculus of variations in the large if it comes with some properness assumption (Fredholm structure) and if it has a finite Morse index.

In case the Fredholm structure is available, there is a way to drop the assumption of non degeneracy (see Marino and Prodi [1]). Otherwise, very few is known.

2nd precision : In the second case of the alternative provided by Proposition 1, we have :

(8) $$\int_0^{+\infty} |\partial f(x(s,x_o))|^2 \, ds < f(x_o) - b$$

(9) $\int_0^{\tau} \partial f(x,(s,x_o))ds$ cannot be included in a compact set for $\tau \in [0,+\infty[$.

We then have a sequence (s_n) or either $(x(s_n,x_o))$ such that :

(10) $$b \leq f(x(s_n,x_o)) \leq a \quad ; \quad \partial f(x(s_n,x_o)) \longrightarrow 0 \ .$$

Classically, in the calculus of variations, there is an assumption, called the (C) condition, introduced by Palais and Smale forbidding (10). The content of this condition is the following :

(11) for any sequence (x_n) such that $b \leq f(x_n) \leq a$ and $\partial f(x_n) \to 0$, there is a convergent subsequence.

This condition forbids the second case in Proposition 1 and thus allows to find solutions to (4) via the study of the difference of topology in the level sets.
However the condition (C) forbids more tant (8)-(9). Indeed, in (8)-(9), we deal with sequences which lie on the same orbit of the flow ; while with the (C) condition, we deal with arbitrary sequences (both of them satisfying $b \leq f(x_n) \leq a$; $\partial f(x_n) \to 0$).

Thus, to do variational calculus, we need :

① either a weaker condition than (C) or (11) : namely, we can impose on the x_n's to lie on the same deformation line,

② or to study the difference of topology induced in the level sets by theses orbits of the flow which satisfy (8)-(9).

The first case ① is just an improvement of condition (C).
The second case ② is concerned with getting rid of any condition of this type, by considering the flow as a dynamical system, having possibly singularities "at infinity".
This goal of getting rid of the condition (C) has been set by S. Smale in his book : "Mathematics of the time" [2].
It is far from being concretly achieved in all variational problems of interest.

3rd precision : The invariants.

The orbits of the flow satisfying (8)-(9) are of course relevant to the flow itself , i.e. to the differential equation (6).
On the other hand, the variational calculus is concerned with the function f ; in fact, equation (4) means that it is impossible to decrease strictly with respect to f a whole neighbourhood in E of the critical point x.

From this correct interpretation of the calculus of variations, we may replace (6) by any other differential equation corresponding to a (pseudo)-gradient for f (in particular consider another scalar product on E) or even more consider any decreasing (with respect to f), globally defined deformation of the level sets.

Thus we see immediately that there is some ambiguity, if we do not make some further specification in case ②.

We discuss here this ambiguity :

1 - First of all, there is an intrinsic notion relevant to f and not ∂f (or either deformation). This notion is the difference of topology of the level sets when crossing the critical level, say c.
i.e. if this critical level c is isolated, which we will assume for sake of simplicity , then for $\varepsilon > 0$ small enough, the homotopy type of the space $f^{c+\varepsilon}/f^{c-\varepsilon}$ is an invariant.

2 - There is a second invariant which is of dynamical type.

In dynamical systems, an invariant set for a flow comes with a stable and unstable manifold, at least when it is non degenerate (or either hyperbolic), which holds generically (otherwise some perturbative argument is necessary using some kind of Fredholm structure ; these days, even degenerate cases are studied, see for instance the work of Yomdin in algebraic geometry and Cappell-Weinberger in algebraic topology).

What is invariant in a hyperbolic situation is not the stable and unstable manifolds but rather their dimension and the qualitative behaviour of the flow on the boundary of an isolating block in the sense of C.C. Conley [3].

To give the simplest picture of this invariant, the best example is the situation nearby a non degenerate critical point of a function f on a finite dimensional manifold. We then have, by Morse lemma, the following local situation (see M. Hirsch [4], for instance).

(12)

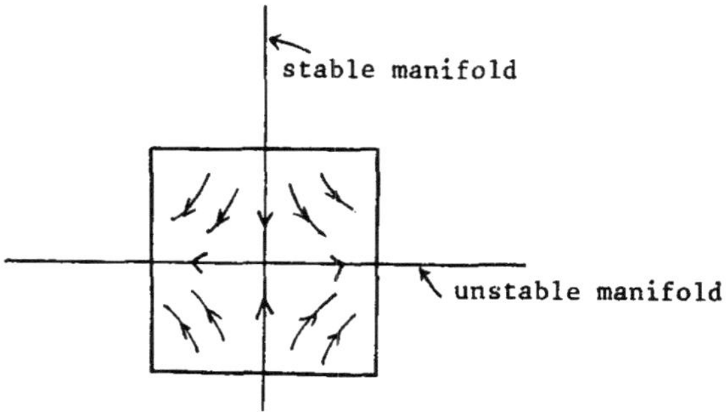

} dynamics of the flow and of any pseudo-gradient hyperbolic flow

(13) 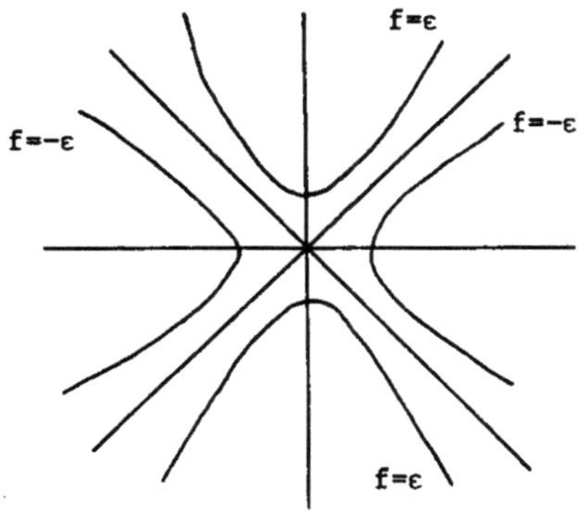

0 is the critical level.

behaviour of the level sets.

In (12), we can retain the behaviour of the flow on the boundary of the box :

(14) 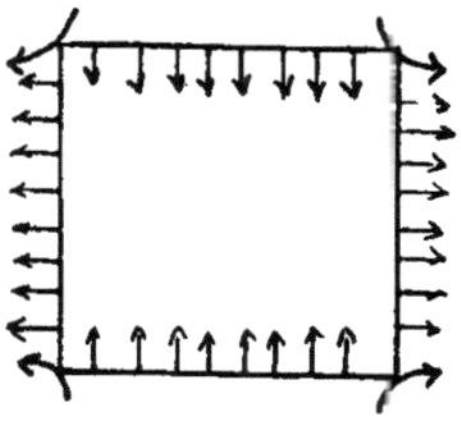

C.C Conley [3] has introduced invariants related to this Morse decomposition, namely the homology of what comes in and what comes out.
C.C. Conley introduced these invariants for a general hyperbolic flow. In such a genral situation, these invariants do not determine the invariant set and its hyperbolic structure (i.e. stable and unstable manifolds) inside the isolating blowk. However, when the flow is pseudo-gradient and the critical (or rest) point is isolated (without assumption of degeneracy), these invariants completely determine the nature of the critical point inside.

This notion of Morse decomposition and isolating block can be extended to the situation of (8)-(9), giving rise to Conley invariants related to this situation, which provide with a second set of invariants, more precise than the difference of topology in the level sets.
Qualitatively, we draw the following picture of the flow, under (8)-(9).

(15)

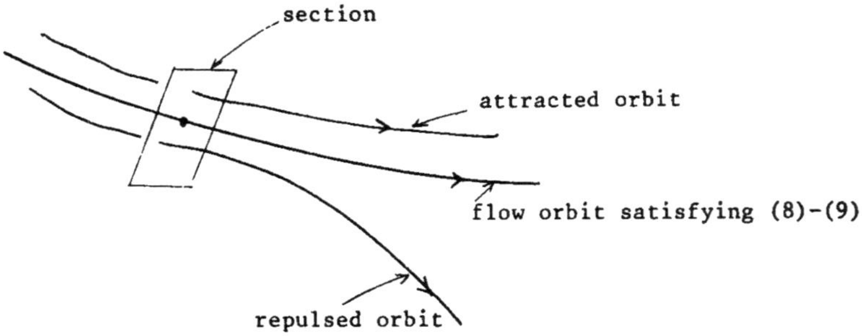

The analysis of the flow on sections provides with this second set of invariants.

4th precision : The global nature of the deformation.

With this phenomenon of the failure of the (C)-condition, there is something important which enters into account :

Consider the case of a usual non degenerate critical point for a functional f, then there is no way one can decrease with respect to the functional a whole neighbourhood of this critical point.

With a flow line going to $+\infty$, the situation is different :

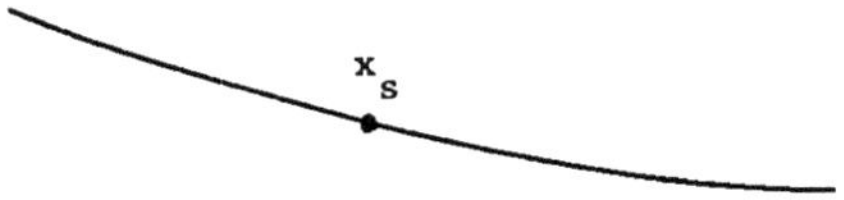

Indeed, consider a point x_s on this flow line : then there is no problem to decrease with respect to the functional a whole neighbourhood to x_s.
But, in fact, there is more : namely, one can decrease in most cases where we studied critical points at infinity a whole neighbourhood of the flow line ; in particular, the flow line itself.
What really matters is how "large" this neighbourhood can be taken, with respect to the structure of the flow in section and also (it is somewhat the same thing) how "large" it is with respect to the "end of the orbit". There is a real ambiguity here, of the same kind than one encounters with the ends of analytic functions. What is defined is a way to arrive to them (when they are "accessible") ; not really themsleves ; i.e. the way to arrive to them defines them. This is why the global nature of the

deformation, in particular the flow when necessary, is an important tool.

(For a discussion on the ends and the accessibility, see Moise, "Geometric topology in dimension 2 and 3".)

5th precision : The representation device.

Assume there are natural ends for the flow lying in some manifold V. This manifold might be given by the sequences (x_n) violating the (C)-condition. This manifold depends then only on the notion of pseudo-gradient, not on the precise pseudo-gradient chosen.

Consider then a bundle F over V with fiber a space of parameters $\Lambda = \Lambda_1 \times H$ where Λ_1 is, for sake of simplicity, finite dimensional and H is a neighbourhood of zero in a Hilbert space (or so).

Let $(\Lambda_1 \times H)_x$ be the fiber at x.

Assume now there is a way to represent the functions x of E such that $|\partial f(x)| < \varepsilon$; $|f(x) - c| < \varepsilon$ in F. Calling this representation R, we have a functional defined on F ; namely f(Rx).

Assume also that over any point u in V and for any $\lambda_1 \in \Lambda_1$, we can minimize this functional for the variations in H.

We then have a functional on Γ, a bundle over V, with fiber Λ_1 ; and we are reduced to a variational problem of finite dimensional type, which depends on the representation R.

In the simplest case, $\Lambda_1 = (\mathbb{R}^+)^p$, $p \in \mathbb{N}$ and when $\varepsilon \to 0$, $(\lambda_1,\ldots,\lambda_p) \to (0,\ldots,0)$ in $(\mathbb{R}^+)^p$.

It is then natural to look at the functional f(Rx) over V in a neighbourhood of (0,...,0) in the fiber. Then V might be thought as the space of variations at infinity (not the critical set at infinity) ; and the behaviour of the functional in a neighbourhood of Γ will select, under good conditions, the part of V which is critical "at infinity".

This relates this kind of variational problems to the study of the singularities. This is the way C.H. Taubes proceeds [13] ; and he proves that the singularities do not interfere with certain homology or homotopy classes.

After these five precisions, we introduce the notion of critical point at infinity (see Abbas Bahri [5]).

Definition 1 : A critical point at infinity is an orbit of a (pseudo)-gradient flow for the functional f, starting at a point x_0 for $s \geq 0$, such that $f(x(s,x_0)) \to c \in \mathbb{R}$, whose closure in E is non compact.

Thus a critical point at infinity is related to a (pseudo)-gradient.

If it is of hyperbolic type, or if we are dealing with a hyperbolic set of

critical points at infinity, there are invariants (in particular Conley invariants) related to such a critical point at infinity.
Finally, in case there is an appropriate extension of the variational problem nearby infinity (i.e. the bundle F) with a normal form of f, then there might exist a space of representation for these critical points at infinity.
In any case, the notion of critical point at infinity is not intrinsic to the variational problem.

2. AN ABSTRACT DEFORMATION LEMMA.

Assume we know that, with $b < a$,

(16) f^b is not retract by deformation of f^a ,

but we do not know the condition (C) to hold on $[b,a]$.
There is then a way to analyze the possible defect of compactness, which amounts to a deformation lemma we present now.
Assume also we have a function $g : E \to \mathbb{R}$ such that if

(17) (x_n) is a sequence such that $b \leq f(x_n) \leq a$; $\partial f(x_n) \to 0$ and $(g(x_n))$ is bounded, there is a convergent subsequence.

Thus, if a sequence (x_n) violates the (C)-condition, $g(x_n)$ goes to $+\infty$.
In general, there are many possible choices of g and the best one is in some sense the function g (if it exists) which measures how a sequence violating the (C)-condition leaves the compact sets of E (examples will be provided later on).

Let :

(18) $\varphi(x) = \dfrac{|\partial f|}{|\partial g|}(x)(f(x) + g(x))$ if $\partial g(x) \neq 0$; $\varphi(x) = +\infty$ if $\partial g(x) = 0$.

We assume

(19) $$\partial g(x) = 0 \Longrightarrow \partial f(x) \neq 0 \text{ if } b \leq f(x) \leq a .$$

Let

(20) $$Z(x) = (|\partial f|\ \partial g + |\partial g|\ \partial f)(x)$$

Z has the property that $(Z,\partial f) \geqslant 0$; $(Z,\partial g) \geqslant 0$; and

(21) $$(Z,\partial f)(x) = 0 \quad \text{if and only if} \quad Z(x) = 0 \quad .$$

Let $\varepsilon>0$ be given and let :

(22) $$\omega_\varepsilon : \mathbb{R}^+ \longrightarrow \mathbb{R} :$$

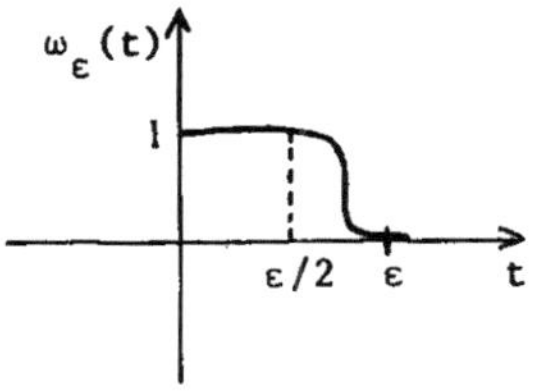

and let now :

(23) $$Z_\varepsilon(x) = (1 - \omega_\varepsilon(\varphi(x)))\quad \partial f(x) + Z(x) \quad .$$

Z_ε is locally Lipschitz.
Qualitatively, $-Z_\varepsilon$ is obtained from $-\partial f$ by adding $-Z$ which decreases f <u>and</u> g whenever φ is small.

We point out here that, in general, we may choose g such that :

(24) $$\frac{(f+g)(x)}{|\partial g(x)|} \geqslant C > 0 \quad ;$$

C uniform for any x such that $b \leqslant f(x) \leqslant a$ or even, in some situations (see section 3 below)

(25) $$\frac{(f+g)(x)}{|\partial g(x)|} \xrightarrow[g(x)\to+\infty]{} +\infty \quad \left(\text{in section 3} \quad \frac{(f+g)(x)}{|\partial g(x)|} \geqslant C\sqrt{g(x)} + C' \quad ; \quad C>0\right)$$

Thus $\varphi(x) \geqslant C\ |\partial f(x)|$ and $\varphi(x)$ goes to zero <u>implies</u> that $\partial f(x)$ goes at zero.
In this way, $-Z_\varepsilon(x)$ is a vector-field which controls the sequences violating the condition (C).
We then consider the differential equation :

(26) $$\frac{\partial x}{\partial s} = -Z_\varepsilon(x)$$

and we have :

<u>Proposition 2</u> : If for $b < a$, f^b is not retract by deformation of f^a, then, for each $\varepsilon > 0$, there exists x_ε such that :

(27) $$b \leqslant f(x_\varepsilon) \leqslant a$$

and

either $\partial f(x_\varepsilon) = 0$; if $f^b \to f^a$ is not a homotopy equivalence, x_ε has a finite Morse index if ∂f is Fredholm at x_ε and x_ε is note degenerate ;

or $Z(x_\varepsilon) = 0$; $\varphi(x_\varepsilon) \leqslant \varepsilon$. In this case, there are three relevant notions which are :

a - the set $Z(x) = 0$ around x . Generically, this is a line transverse to f ;

b - if it is a line transverse to f, the index of Z at x_ε in section to the level surface at x_ε ;

c - in the simple case when this line goes to $+\infty$ (i.e. $g(x) \to +\infty$ on this line) the unstable manifold of this line of zeros of Z.

This is summed up in the following drawing :

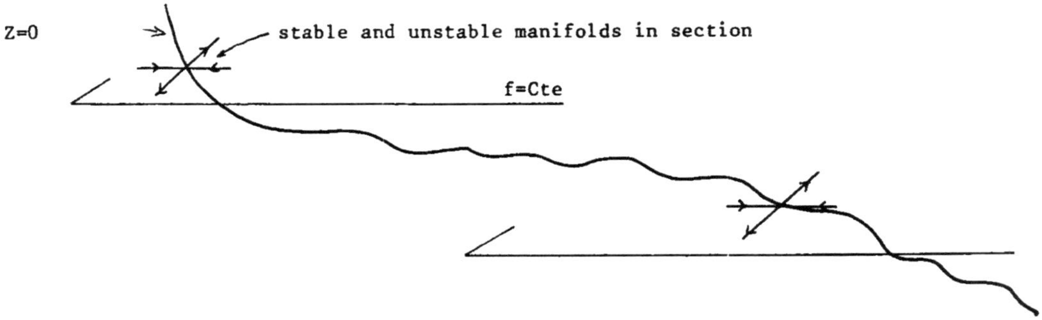

Is there a convergence process ?

In case such a line has a natural end, we end up with a <u>critical point at infinity</u> as defined in the previous section. We illustrate this by the following example (section 3).

3. PSEUDO-ORBITS OF CONTACT FORMS.

The framework is the following :

We consider a contact form α on a three dimensional compact, orientable manifold M. The assumption $n = 3$ is not essential here. There is a version of what we present in dimension 2p+1. For sake of simplicity, we restrict ourselves to this case.

Let ξ be the Reeb vector-field of α, i.e. ξ is defined by the equations

(28) $$\alpha(\xi) \equiv 1 \quad ; \quad d\alpha(\xi,.) \equiv 0 \quad .$$

Let v be a vector-field in the kernel of α which we assume to be nowhere vanishing. The existence of such a v means that the bundle in planes $\alpha = 0$ over M is trivial ; which we will assume.

Let θ_s be the one parameter group generated by v, $D\theta_s$ is differential and let θ_s be the associated map in the differential forms of M.

Let x_o be a point of M and $x_s = \theta_s(x_o)$ be the generating point of the v-orbit through x_o.

Let :

(29) $e_1(o)$ and $e_2(o)$ be two vectors tangent to M at x_o such that :

(30) $$\alpha \wedge d\alpha(v_{x_o}, e_1(o), e_2(o)) < 0 \quad .$$

Let

(31) $$e_1(s) = D\theta_s(e_1(o)) \quad ; \quad e_2(s) = D\theta_s(e_2(o)) \quad ;$$
$$w(s) = \alpha(e_1(s))e_2(s) - \alpha(e_2(s))e_1(s) \quad .$$

3.1 Some geometrical facts ; a notion of conjugacy.

The following proposition expresses along the v-orbits the fact that α is a contact form :

Proposition 3 : w(s) rotates in the direct sense of the frame $(e_1(s), e_2(s))$ when s increases.

This is expressed in the following picture :

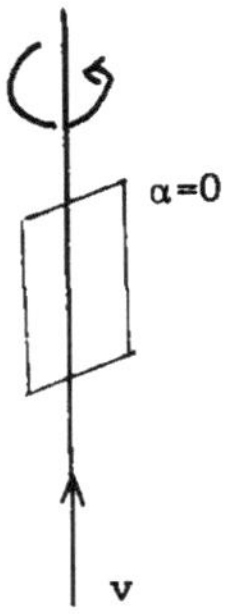

Let then :

(32) $\psi(s,x_o)$ be the angle in the moving frame $(e_1(s),e_2(s))$ which measures the rotation of $w(s)$ from 0 to s.

We define :

<u>Definition</u> 1 : We call <u>coincidence points</u> of x_o (relatively to α and v) along the v-orbit through x_o those points x_s such that $\psi(s,x_o) = 2k\pi$; $k \in \mathbb{Z}$.
At these points x_s, we have :

(33) $$(\theta_s^*\alpha)x_o = \lambda(s,x_o)\alpha_{x_o} \quad : \quad \lambda(s,x_o) > 0 \quad .$$

<u>Definition</u> 2 : We call <u>conjugate point</u> of x_o (relatively to α and v) along the v-orbit through x_o a coindence point such that :

(34) $$\lambda(s,x_o) = 1 \quad .$$

<u>Definition</u> 3 : We say that α <u>turns well</u> along v if any point x_o of M has a coincidence point distinct from itself. Let then $\gamma^i : M \to \mathbb{R}$ be the function which associates at a point x_o of M the i-th time $s = \gamma^i(x_o)$ such that x_s is a coincidence point of x_o ($i \in \mathbb{Z}$) . Let $f^i : M \to M$ be the diffeomorphism of M which sends x_o on $x_{\gamma^i(x_o)}$. We denote $\mu_i(x_o)$ the colinearity coefficient of $(\theta^*_{\gamma^i(x_o)}\alpha)_{x_o}$ and α_{x_o} :

(35) $$(\theta^*_{\gamma^i(x_o)}\alpha)_{x_o} = \mu_i(x_o)\alpha_{x_o} \quad .$$

As can be noticed, the notion of <u>conjugate point</u> is a delicate notion :

not all points of M have conjugate points distinct from themselves. At the contrary, these points live on a hypersurface of M.

If we draw the segment of v-orbit between x_o and x_{s_1}, a coincidence point :

we have a natural differential equation which comes with this piece : namely the one satisfied by α from x_o into x_{s_1} ; or either the one satisfied by w(s) in a transported frame $(e_1(s), e_2(s))$.

It has the intrinsic form :

(36)
$$\frac{\partial}{\partial s}\alpha_{x_s} = d\alpha_{x_s}(v,.)$$

$$\frac{\partial}{\partial s}(d\alpha_{x_s}(v,.) = a(x_s)\alpha_{x_s} + b(x_s)d\alpha_{x_s}(v,.)$$

thus

(37)
$$\frac{\partial^2}{\partial s^2}(\alpha_{x_s}) = a(x_s)\alpha_{x_s} + b(x_s)\frac{\partial}{\partial s}\alpha(x_s)$$

with $a(x_s) < 0$.

Hence the pendulum equation, with a periodic solution if and only of we have a conjugate point.

3.b A variational problem on a submanifold of the loop space on M.

Let

(38)
$$\beta = d\alpha(v,.) \quad .$$

We leave aside here the question of β not being a contact form (which is treated in [5]) and we study the case :

(39) $\quad \beta \wedge d\beta > 0$ with respect to $\alpha \wedge d\alpha$.

We then normalize v by multiplication by a factor λ so that :

$$\beta \wedge d\beta = \alpha \wedge d\alpha \quad . \tag{40}$$

Let then :

$$\mathcal{L}_\beta = \{x \in H^1(S^1,M) / \beta(\dot{x}) \equiv 0\} \tag{41}$$

$$C_\beta = \{x \in H^1(S^1,M) / \beta(\dot{x}) \equiv 0 \quad \text{and} \quad \alpha(\dot{x}) = Cte > 0\} \ . \tag{42}$$

We have :

Proposition 4 : C_β is a submanifold of $H^1(S^1,M)$. If α (hence β) turns well along v, then the injection of C_β in $H^1(S^1,M)$ is a weak homotopy equivalence.

Consider now a curve of C_β, x. Its tangent vector $\dot{x}$ can be split on (ξ,v) and we have :

(43) $\quad \dot{x} = a\xi + bv \qquad a$ = positive constant ; $\quad b \in L^2(S^1;\mathbb{R})$.

We then consider the functional :

$$f(x) = a \text{ on } C_\beta \quad . \tag{44}$$

As one might expect, from the first glance, the functional f does not control at all the v-component of $\dot{x}$.
We introduce :

$$g(x) = \int_0^1 b^2\, dt \quad . \tag{45}$$

Some further analysis (see [5]) shows that f does not satisfy the (C)-condition and in fact, on a sequence (x_n), with $\dot{x}_n = a_n\xi + b_n v$, the boundedness of the functional and the fact that $\partial f(x_n) \to 0$ just tells us that $\int_0^t b_n / (\int_0^1 b_n^2) \to 0$, at best.

So :

1 - f doest not satisfy (C) ;

2 - ∂f has no Fredholm structure (see [5]) ;

3 - there is a difference of topology in the level sets ; but it is by far too heavy to be due only to the critical points of f which are the periodic orbits of ξ (see also [5]).

Yet, such an ill posed variational problem has a precise meaning, of interest. For this, consider for instance the case :

(46) $M = P\mathbb{R}^3$; $\alpha = \lambda\alpha_o$ where α_o is the standard contact form on $P\mathbb{R}^3$ considered as the cotangent sphere bundle over S^2 .

Let

(47) $p : P\mathbb{R}^3 \to S^2$ be the fibration ; and v be the vector-field of the S^1-fibers.

The variational problem (44) on C_β corresponds then to the geodesic problem on S^2 (to any metric on S^2, there is a corresponding λ in (46)), in the space of immersed curves.

Solving this problem is important for the issue of counting the minimal number of geodesics on a surface Σ (then M is the S^1-fiber bundle over Σ provided by the sphere cotangent bundle).

R. Hamilton pointed out that the normal curvature flow for the projected curve from C_β along p is a pseudo-gradient for such a variational problem ; it is as well a pseudo-gradient for the area enclosed by such curves.

We thus see that this variational problem has a common pseudo-gradient with other important variational problems.

One should note here that this implies that all these problems have related critical points at infinity.

3.c The critical points at infinity of the variational problem [Note aux Comptes-Rendus, Abbas Bahri, July 1984].

(48)

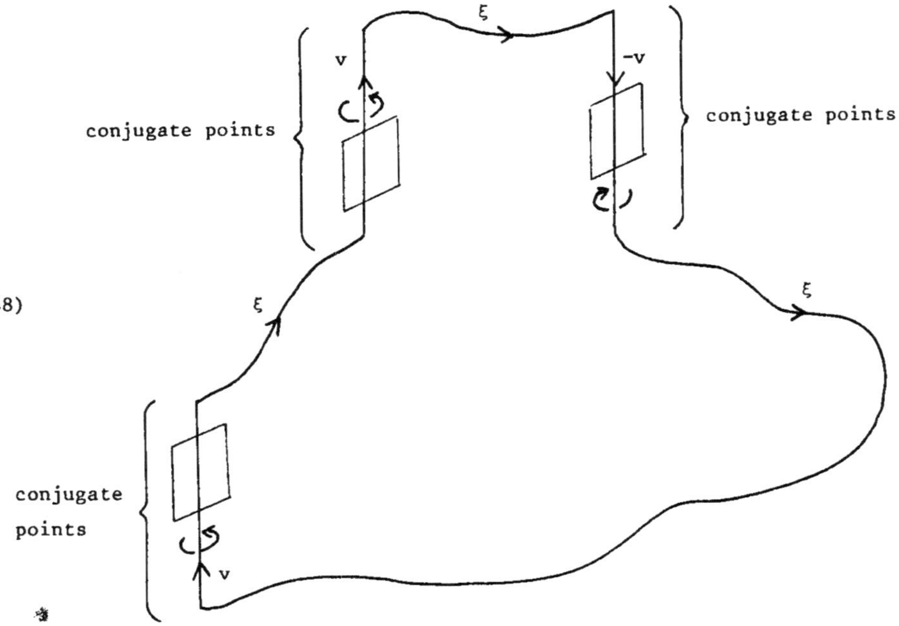

(48) is a geometrical description of the critical points at infinity.

To understand qualitatively what is going on, we have to apply a convergence process to these curves :

1 - the conjugate points

For $\varepsilon \to 0$, the curve x approximating this object (on the deformation line) forms a small bubble in section to v.
i.e. if one projects a neighbourhood of the piece rather tangent to v on x on a section to v, one finds :

(49)

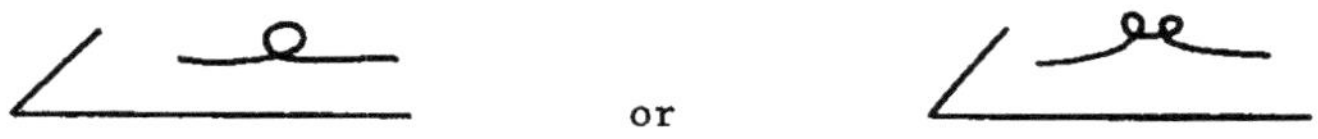

or more bubbles

These are thus points where the tangent vector to x_ε makes very rapidly an integer number of rotations, possibly growing when $\varepsilon \to 0$ by the following process :

(50)

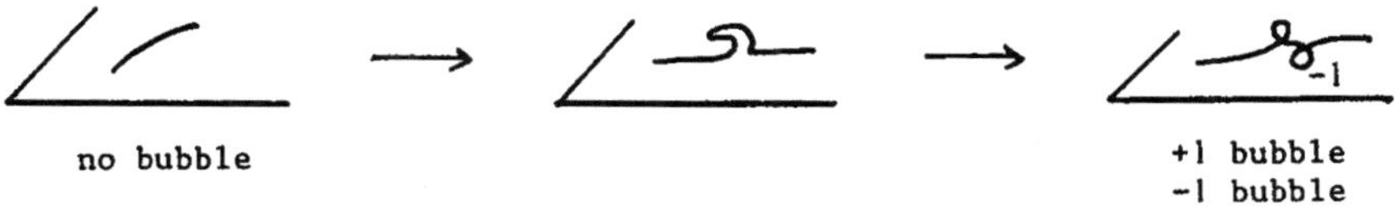

However the resulting movement is very particular, i.e. the bubble as depolyed along v will go from one point to a conjugate of this point. Thus, generically, these bubble build up at precise locations in M. At these precise locations, the singularity has a very precise and restricted normal form given by a periodic solution to (37).

To see the phenomenon, we could draw v-orbits passing through each point x of M and distinguish on these orbits the coincidence points to x :

(51)

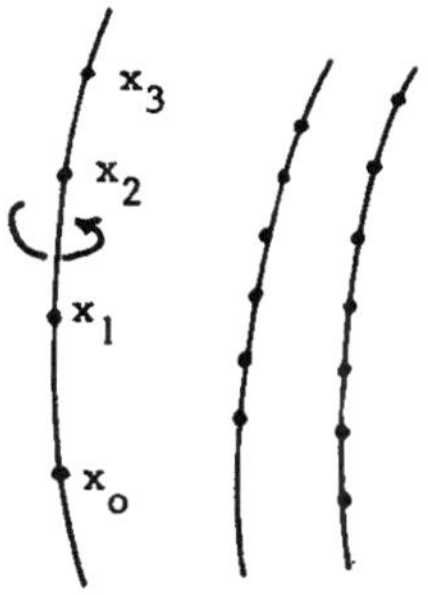

We thus have a $\mathbb{Z}$-structure along M related to α along v. For some distinguished points, we have conjugate points.

(52)

$x_o \in$ hypersurface of M

Assume x_p corresponds to time s_1 on the v-orbit starting at x_o. We have :

(53)
$$\theta^*_{s_1} \alpha = \alpha .$$

and we can compute in this situation the <u>second</u> variation of α i.e.

(54) $$\delta(\theta_s^*\alpha - \alpha)(s_1) \quad .$$

This gives rise to a <u>quadratic form</u> on tangent vectors to M at x_o, q_o ; and in the same time to a <u>quadratic form</u> on tangent vectors to M at $x_{s_1} = x_p$, q_1.

Thus, these conjugate points come out with :

1 - a <u>precise</u> location ;
2 - a <u>precise</u> normal form to the singularity ;
3 - an <u>integer</u> (the rotation of α from x_o to $x_{s_1} = x_p$) ;
4 - <u>two quadratic forms</u> q_o and q_1 ;
5 - a way to approach them by curves which <u>project on local sections on bubbles</u>.

2 - <u>the ξ-pieces</u>

These are pieces where the curve is tangent to the Reeb vector-field ; thus the curve is tangent to ξ (in this $\mathbb{Z}$-structure we introduced) <u>until it hits a point admitting a conjugate point</u>.
Then, under certain conditions stated in [5], it jumps to the conjugate point.

(55) The ξ-pieces come also with a quadratic form q_3 defined by the second variation of f along them <u>with fixed ends</u>. This quadratic form is related to a rotation of v along the ξ-piece (see [5]).

The Reeb vector-field ξ is, in the case of the cotangent unit sphere bundle of a Riemannian manifold Σ, such that its periodic orbits project on geodesics of Σ. In that case, there is no other conjugate point for a point x_o than itself and the Dirac masses describe a complete circle S^1 over a given point in Σ in $ST^*\Sigma$.
In other simple, but more complicated cases, this is what happens :

Take the case of S^3 fibering over S^2 with the Hopf fibration

(56) $$p : S^3 \longrightarrow S^2 \quad .$$

Consider $\alpha=\lambda\alpha_o$, λ a positive function on S^3 and α_o the standard contact form of S^3. Let v be the vector-field of the fibers over S^2. In this case, the Reeb vector-field ξ, when describing a fiber S^1 over a point x_o of S^2, describes in the tangent plane to S^2 at x_o the following :

(57)  i.e. two circles.

We thus have <u>two</u> choices of length on S^2, hence <u>two</u> notions of geodesics. Then, (48) projects as :

(58) 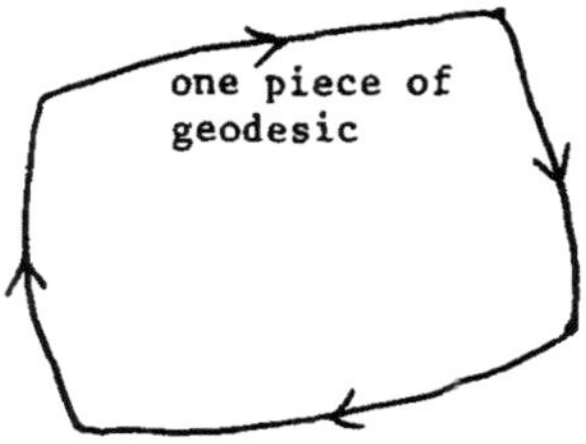

We reproduce here the theorem we announced in [6] :
Assume :

(H_1) α turns well along v ;

(H_2) v has a periodic orbit ,

(H_3) for <u>one</u> vector-field v_1, non singular and colinear to v, we have : $\exists\ k_1>0$ such that $\|D\theta_s^1\|\leq k_1 \quad \forall\ s\in\mathbb{R}$, where θ_s^1 is the one-parameter group of v_1 ;

(H_4) $\exists\ k_2$ and $k_3>0$ such that $\forall\ i\in\mathbb{Z}$, we have :

$$k_2 d(x,y)\leq d(f^i(x),f^i(y))\leq k_3 d(x,y) \qquad \forall\ x,\ y\in M \quad ;$$

(H_5) $\exists\ k_4>0$ such that $|\mu_i(x)-\mu_i(y)|\leq k_4\ d(x,y) \quad \forall\ x,y\in M$;

(H_6) $\exists\ \rho>0$ such that for any $x\in M$, the set $C_\rho(x)=\{f^i(x)/|\mu_i(x)-1|<\rho;i\in\mathbb{Z}\}$ is finite.

Then, under these hypotheses which can be considerably weakened (see [5]), we have :

<u>Theorem</u> 1 : The critical points at infinity of the variational problem are continuous and closed curves made up with pieces $[x_{2i},x_{2i+1}]$ tangent to ξ and pieces $[x_{2i+1},x_{2i+2}]$ tangent to v. x_{2i+2} is conjugate to x_{2i+1}. If the Betti numbers of the loop space are unbounded, there are unfinitely many of these curves.

Furthermore, if n is the number of v-pieces of one of these curves, we have :

(59) $$n \leqslant Ca$$

where a is the length of the curve along ξ and C is a universal constant.

As pointed out in [5] (see also [17] for application of the method developped in [5] to Yamabe type equations), the defect of compactness reveals, when analyzed along deformation lines, new geometrical structures which govern in fact the variational problem.
In turn, this variational problem somewhat expresses them through the dynamics of the flow and the difference of topology in the level sets.
This link is the important point which defines a critical point at infinity.
Out of it, one either looses the variations, hence the reason why the phenomenon exists ; or the ends of the orbits, hence the way these variations interact at infinity and are governed by these interactions.

4. THE NIRENBERG PROBLEM ; THE EQUATION $\begin{cases} \Delta u + u^{\frac{n+2}{n-2}} = 0 \\ u = 0|_{\partial\Omega} \quad u > 0 \end{cases}$; $\Omega \subset \mathbb{R}^n$ an open bounded set.

4.a The problems.

We are concerned with studying the equations :

(59) $$\begin{cases} \Delta u - \frac{n-1}{4(n-2)} u + K(x)\, u^{\frac{n+2}{n-2}} = 0 \\ \Omega \subset S^n \ ; \ u = 0_{|\partial\Omega} \ \text{ if } \ \Omega \subset\subset S^n \end{cases} \qquad u > 0 .$$

In case $\Omega = S^n$, problem (59) is the Nirenberg's problem : namely finding the functions K such that they are scalar curvature of a metric conformal to the standard one on S^n.
In case $\Omega \subset\subset S^n$, we will here mainly look at the case K(x) is a positive constant.

We wish to point out here that Lee and Jerison [7] and Chern and Hamilton [8] showed that there was a Yamabe type equation on the contact form situation. The paper by Chern and Hamilton appeared in the publication of the Berkeley MSRI in October 1984, shortly after the proof the Yamabe conjecture by R. Schoën (summer 1984).

4.b The functional and the flow.

Let

(60) $$\Sigma^+ = \{u>0 \ ; \int_\Omega |\nabla u|^2 \, dx + \frac{n-1}{4(n-2)} \int_\Omega u^2 dx = 1 \ ; \ u = 0_{|\partial\Omega} \ \text{ if } \Omega \subset\subset S^n\} \ .$$

Let

(61) $$J(u) = \frac{1}{\int_\Omega K(x) u^{2n/n-2}} \quad ; \quad u \in \Sigma^+ \ .$$

Let

(62) $$\lambda(u) = J(u)^{\frac{n-2}{4}} \ .$$

The $H^1_{(o)}$-flow of J over Σ^+ is (the parenthesis around o is meant to cover the case $\Omega = S^n$) :

(63) $$-\partial J(u) = -\lambda(u) u - \lambda(u)^{\frac{n+2}{n-2}} \, L^{-1}_{K(x)} u^{\frac{n+2}{n-2}}$$

where

(64) $L = \Delta - \frac{n-1}{4(n-2)}$ under Dirichlet boundary conditions if $\Omega \subset\subset S^n$.

We consider the differential equation :

(65) $$\frac{\partial u}{\partial s} = -\partial J(u) \ .$$

To simplify the notations, we will rather work in $\mathbb{R}^n$, transforming the problem by stereographic projection.
Then :

(66) $$\Sigma^+ = \{u>0 \ ; \ |u|^2 = \int_\Omega |\nabla u|^2 dx = 1 \quad ; \quad u = 0_{|\partial\Omega} \ \text{ if } \bar{\Omega} \text{ compact} \subset \mathbb{R}^n\} \ .$$

Let

(67) $$\delta_i(x) = \delta(x_i, \lambda_i)(x) = c_o \frac{\lambda_i^{\frac{n-2}{2}}}{(1+\lambda_i^2 \, |x-x_i|^2)^{\frac{n-2}{2}}}$$

where c_o is such that :

(68) $$\Delta \delta_i + \delta_i^{\frac{n+2}{n-2}} = 0 \ .$$

The variational problem does not satisfy the condition (C).
The sequences which violate this condition have been analyzed by J. Sacks and K. Uhlenbeck [9] for the harmonic map problem in dimension 2.
Their intuition has been extended to the Yamabe (type) equations throughout the work of Y.T. Siu and S.T. Yau [11], P.L. Lions [10], C.H. Taubes [13], M. Struwe [18] , H. Brézis and J.M. Coron [12].

We have the following proposition :

Proposition 3 : Let (u_n) be a sequence in Σ^+ such that $(J(u_n))$ is bounded and $\partial J(u_n) \to 0$. There exists $\bar{u} \geqslant 0$, $\partial J(\frac{\bar{u}}{\bar{u}}) = 0$ if $\bar{u} \neq 0$, $p \in \mathbb{N}$, such that for an extracted subsequence again denoted (u_n), we have :

(69) $$\frac{u_n - \bar{u}}{|u_n - \bar{u}|} \in V(p, \varepsilon_n) \quad ; \quad \varepsilon_n \to 0$$

where

(70) $$V(p,\varepsilon) = \Big\{ u \in \Sigma \mid \exists\, (a_1,\dots,a_p) \in \Omega^p \ ; \ \exists\, (\lambda_1,\dots,\lambda_p) \in [0,+\infty[^p$$

$$\text{s.t. } \Big|u - \frac{1}{\sqrt{p}} \sum_{i=1}^{p} \frac{1}{K(a_i)^{\frac{n-2}{4}}} \delta(a_i,\lambda_i)\Big| < \varepsilon \ ; \ \lambda_i d(a_i,\partial\Omega) > \frac{1}{\varepsilon} \quad \forall\, i \ ;$$

$$\frac{\lambda_i}{\lambda_j} + \frac{\lambda_j}{\lambda_i} + \lambda_i\lambda_j \, |a_i - a_j|^2 > \frac{1}{\varepsilon} \quad \forall\, i \neq j \Big\}$$

(71) $$V(o,\varepsilon) = \{u \in \Sigma \mid |u| < \varepsilon\} \quad .$$

From the point of view of critical points at infinity, the sets $V(p,\varepsilon)$ when $\varepsilon \to 0$ are the candidates (or potential) critical points at infinity. These are not critical points at infinity.

In fact the critical points at infinity are analyzed in the following.
Consider the problems :

(72) $$\text{Minimize } \Big|u - \sum_{i=1}^{p} \alpha_i \, \delta(x_i,\lambda_i)\Big| \quad ; \quad \alpha_i > 0 \quad ; \quad x_i \in \Omega \quad ; \quad \lambda_i > 0$$

(73) $$\text{Minimize } \Big|u - \sum_{i=1}^{p} \alpha_i \, P\delta(x_i,\lambda_i)\Big| \quad ; \quad \alpha_i > 0 \quad ; \quad x_i \in \Omega \quad ; \quad \lambda_i > 0$$

where

(74) $P\delta(x_i,\lambda_i)$ is the orthogonal projection of δ_i onto $H_o^1(\Omega)$
($P\delta_i = \delta_i$ if $\Omega = \mathbb{R}^n$).

Proposition 4 : For any p, there exists $\varepsilon_o > 0$ such that for any u in $V(p,\varepsilon_o)$, problems (72)-(73) have a unique solution up to permutation.

We consider now the differential equation (65) with a starting point

(75) $$u_o \in V(p,\varepsilon) \quad ; \quad \varepsilon < \varepsilon_o \quad .$$

We want to analyze the behaviour of the solution $u(s,u_o)$.
As long as the solution remains in $V(p,\varepsilon_o)$, we have well defined quantities :

(76) $$\alpha_i(s),\ x_i(s),\ \lambda_i(s),\ v(s) = u(s,u_o) - \sum_{i=1}^{p} \alpha_i(s)\ \delta(x_i(s),\lambda_i(s)) \quad .$$

In order for a gradient line to build up a critical point at infinity, we must have $\lambda_i(s) \to +\infty \ \forall\, i$.
In fact, the crucial behaviour is the one of the $\lambda_i(s)$.
It turns out that such a behaviour has a normal form, which we give now in the case K(x) is constant, $\bar{\Omega}$ compact.
For the general case, see A. Bahri [14].

We first introduce the regular part of the Green's function of the Laplacian on Ω.

(76) $$\begin{cases} y \longrightarrow H(x,y) \\ \Delta_y H(x,y) = 0 \quad \text{in } \Omega \\ H(x,y) = \dfrac{1}{|x-y|^{n-2}} \quad \text{on } \partial\Omega \quad . \end{cases}$$

We then define the matrix :

(77) $$M(x_1,\ldots,x_p) = (M_{ij}) \quad \text{on } \Omega^p$$

where

(78) $$M_{ij}(x_1,\ldots,x_p) = H(x_i,x_j) - \frac{1}{|x_i-x_j|^{n-2}} = G(x_i,x_j) \qquad i \neq j$$

(79) $$M_{ii}(x_1,\ldots,x_p) = H(x_i,x_i) \quad .$$

We have :

(80) $$u(s) = \sum_{i=1}^{p} \alpha_i(s)\delta(x_i(s),\lambda_i(s)) + v(s) \quad .$$

Theorem 2 : For any $\delta > 0$, there exists $\varepsilon_o > 0$ and $s_o > 0$ such that if $u(s) \in V(p,\varepsilon_o)$ for $0 \leq s \leq s_o$ and $d(x_i(s),\partial\Omega) \leq \delta$ for $0 \leq s \leq s_o$, then for all $\bar{s} \geq s_o$ such that $u(s)$ remains in $V(p,\varepsilon_o)$ for $s \in [o,\bar{s}]$, we have :

$$\begin{cases} \dfrac{\dot{\lambda}_i}{\lambda_i}(\bar{s}) = \dfrac{C_1}{\alpha_i \lambda_i^{\frac{n-2}{2}}} \left[\alpha_i^{\frac{n+2}{n-2}} \dfrac{H(x_i,x_i)}{\lambda_i^{\frac{n-2}{2}}} - \Big(\sum_{i \neq j} \dfrac{1}{\lambda_j^{n-2}} \dfrac{\alpha_i^{4/n-2}\alpha_j + \alpha_j^{\frac{n+2}{n-2}}}{|x_i - x_j|^{n-2}} - \alpha_i^{\frac{n+2}{n-2}} H(x_i,x_j)\Big) \right. \\ \qquad \left. + \sum_{j \neq i} \dfrac{\alpha_j}{\lambda_j^{\frac{n-2}{2}} |x_i - x_j|^{n+2}} \right] + o\left(\sum \dfrac{1}{\lambda_k^{n-2}} \right) \\ |\dot{x}_i|(\bar{s}) \leq \dfrac{C}{\lambda_i^{n-2}} \left(\sum \dfrac{1}{\lambda_k^{n-2}} \right) \\ \dot{\alpha}_i(\bar{s}) = C\alpha_i (1 - \alpha_i^{4/n-2} C_1 \int_{\mathbb{R}^n} \delta^{2n/n-2}) + 0 \left(\dfrac{1}{\lambda_k^{n-2}} \right) \\ |v|^2(\bar{s}) \leq K \left(\sum \dfrac{1}{{}^{n-2}_{k}} \right) \quad . \end{cases}$$

This is the dynamical behaviour of the flow nearby the singularities (i.e. in the $V(p,\varepsilon)$'s ; $\varepsilon \to 0$).

Let then :

(81) $\rho(x_1,\ldots,x_p)$ be the least eigenvalue of $M(x_1,\ldots,x_p)$.

Observe that the matrix $M(x_1,\ldots,x_p)$ is related to the equation

$$\begin{cases} \Delta G(x,x_i) = \delta_{x_i} \\ G(x,x_i) = o|_{\delta\Omega} \end{cases}$$

(δ_{x_i} = Dirac mass at x_i. Not be confounded with δ_i of (67).)

The energy interactions, as we prove it, are governed by the matrix M and ρ. In partiuclar, for two points x_i and x_j, this interaction increases along grad ρ. This gradient is related to the vector fields $D_i = \text{grad}(x,x_i)$ and $D_j = \text{grad}(x,x_j)$ which satisfy :

$$\begin{cases} \text{div } D_i = \delta_{x_i} \\ \text{div } D_j = \delta_{x_j} \end{cases} \qquad \delta_x = \text{Dirac mass at } x.$$

If there is no boundary, then D_i and D_j are $\nabla\left(\frac{1}{|x-x_i|^{n-2}}\right)$ and $\nabla\left(\frac{1}{|x-x_j|^{n-2}}\right)$.
Increasing amounts, by symmetry arguments, to move the points along $D_i(x_j) - D_j(x_i)$ which is directed by $x_i - x_j$. Hence the interaction lies along the geodesic from x_i to x_j.
Otherwise, one has to take the boundary into account ; hence the distance of the points x_i and x_j to the boundary (see [14]).
We have :

Theorem 3 : If an orbit of the flow defines a critical point at infinity then, $\lim_{s\to+\infty} \rho(x_1(s),\ldots,x_p(s)) \geqslant 0$. Conversely, if $\rho(x_1(s),\ldots,x_p(s))$ remains larger than $\delta > 0$ for a certain time interval, then the orbit will define a critical point at infinity and $\lim_{s\to+\infty} \rho(x_1(s),\ldots,x_p(s))$ will be strictly positive on such an orbit.

Then the points $x_i(s)$ converge in Ω and $\lambda_i(s) \underset{s\to+\infty}{\sim} C_i\, s^{1/n-2}$.

Note here that ρ is a "natural" extension of J at infinity (see [14]) ; its critical points in $\{x \in \Omega^p \mid \rho(x) > 0\}$ providing somewhat "more critical" points at infinity than the orbits of the flow. This is the full variational problem at infinity ; or either the variational problem with singularities.

On the other hand, we derive local expansions of J which we will give in the general case.

Using (73), we write a function u of $V(p,\varepsilon)$ in the form :

(82) $$u = \sum_{i=1}^{p} \alpha_i P\delta_i + v$$

and we have setting

(83) $$\varepsilon_{ij} = \left(\frac{\lambda_i}{\lambda_j} + \frac{\lambda_j}{\lambda_i} + \lambda_i\lambda_j |x_i - x_j|^2\right)^{-\frac{n-2}{2}} .$$

<u>Theorem</u> 4 :

$$J\left(\sum_{i=1}^{p} \alpha_i P\delta_i + v\right) = \frac{(\Sigma \alpha_i^2)^{n/n-2}}{\Sigma \alpha_i^{2n/n-2} K(x_i)} \left(\int \delta^{2n/n-2}\right) \Bigg\{ 1 - (\Sigma\alpha_j^{2n/n-2} K(x_j))^{-1} c_7 \sum_i \alpha_i^{2n/n-2} \frac{\Delta K(x_i)}{\lambda_i^2}$$

$$+ o_K \left(\Sigma \frac{1}{\lambda_i^2}\right) - c_8 \Bigg[\Sigma \frac{H(x_i,x_i)}{\lambda_i^{n-2}} \left(\frac{1}{2} \frac{\alpha_i^2}{\Sigma \alpha_j^2} - \alpha_i^2 \frac{\alpha_i^{4/n-2} K(x_i)}{\Sigma \alpha_j^2 (\alpha_j^{4/n-2} K(x_j))}\right)$$

$$+ \sum_{i\neq j} \left(\frac{H(x_i,x_j)}{(\lambda_i\lambda_j)^{n-2/2}} - \varepsilon_{ij}\right)\left(\frac{1}{2} \frac{\alpha_i\alpha_j}{\Sigma \alpha_k^2} - \alpha_i\alpha_j \frac{\alpha_i^{4/n-2} K(x_i)}{\Sigma \alpha_k^2 (\alpha_k^{4/n-2} K(x_k))}\right) \Bigg] \Bigg\}$$

$$+ \frac{c_9}{\Sigma\alpha_i^2} \left(\int |\nabla v|^2 dx - \frac{n+2}{n-2} \Sigma \alpha_j^2 \cdot \Sigma \frac{\alpha_i^{4/n-2} K(x_i)}{\Sigma \alpha_j^2 (\alpha_j^{4/n-2} K(x_j))} \int \delta_i^{4/n-2} v^2\right)$$

$$+ (f,v) + \begin{cases} + O((\int |\nabla v|^2)^{n/n-2} & \text{if } n \geq 6 \\ + O((\int |\nabla v|^2)^{3/2} & \text{if } n < 6 \end{cases} \quad + O\left(\left(\int |\nabla v|^2 dx\right)\left(\sum_{i\neq j} \varepsilon_{ij}^{2/n-2} (\log \varepsilon_{ij}^{-1})^{2/n}\right)\right.$$

$$+ O_K \left(\Sigma \frac{1}{\lambda_i}\right) + \Sigma \frac{1}{\lambda_i^2 \lambda_i^4} + \begin{cases} \Sigma \dfrac{1}{\lambda_j^2 d_j^{n-2}} & n \geq 6 \\ \Sigma \dfrac{1}{\lambda_i^{6-n/2} \lambda_j^{n-2/2}} \dfrac{1}{d_j^{n-2}} & n < 6 \end{cases} \quad + O\Bigg(\sum_{i\neq j} \varepsilon_{ij}^{n/n-2}$$

$$+ \Sigma \left(\frac{1}{\lambda_i^{n-2/2} \lambda_j^{n-2/2}} \frac{\log \lambda_i}{d_j^n} + \frac{1}{\lambda_i^{n+2/2}} \frac{1}{\lambda_j^{n-2/2}} \frac{1}{d_i^n} + \frac{1}{(\lambda_i d_i)^n} + \frac{1}{\lambda_i^2} \frac{1}{\lambda_j^{n-2}} \frac{1}{d_j^{2(n-2)}} \right)\Bigg)$$

where

$$(f,v) = \int_{\mathbb{R}^n} K(x) (\Sigma \alpha_i P\delta_i)^{\frac{n+2}{n-2}} v = O\left(\left(\int |\nabla v|^2\right)^{1/2} \left(\sum_{i\neq j} \varepsilon_{ij}^{\frac{n+2}{2(n-2)}} (\log \varepsilon_{ij}^{-1})^{\frac{n+2}{2n}}\right.\right.$$

$$\left.\left. + \Sigma \frac{1}{\lambda_i^{n+2/2} d_i^{n+2}} + \frac{1}{\lambda_i^2} \frac{1}{\lambda_j^{n-2/2}} \frac{1}{d_j^{n-2}} + O_K\left(\Sigma \frac{1}{\lambda_i}\right) + (\text{if } n < 6) \sum_{i\neq j} \varepsilon_{ij} (\log \varepsilon_{ij}^{-1})^{\frac{n-2}{n}}\right)\right)$$

and

$$o_K(\varepsilon_{ij}) \leqslant \frac{C}{\lambda_i} O(\varepsilon_{ij}(\log \varepsilon_{ij}^{-1})^{\frac{n-2}{2}} ((\frac{\log \lambda_i}{\lambda_i})^{2/n} + 1)$$

and $d_i = d(x_i, \partial\Omega)$.

The indexed (by K) quantities appear when K is non constant.

The proof of Theorem 2 is dynamical (see [14]).

The proof of Theorem 4 requires some computations (see [16]).

Theorem 4 is a kind of Morse lemma nearby infinity or more precisely the "versal deployment" of the singularities in the sense of René Thom.

4.c The defect of compactness : topological invariants.

We illustrate this problem with $K = 1$ and $\bar{\Omega}$ compact.

Setting :

(83) $$b_p = (pS)^{2/n-2}$$

we give a theorem which provides the difference of topology when crossing the energy level b_p ; i.e. for $\varepsilon > 0$ small enough a description of :

(84) $$(J^{b_p+\varepsilon} / J_p^{b_p-\varepsilon}) \cap V(p, \varepsilon_o)$$

i.e. the contribution of the critical points at infinity in the difference of topology at level b_p. For this, we need to introduce :

(85) $$I_p = \{(x_1, \ldots, x_p) \in \Omega^p / \rho(x_1, \ldots, x_p) < 0\}$$

(86) $$\Delta_{p-1} = \{(t_1, \ldots, t_p) \ ; \ t_i > 0 \ , \ \Sigma\, t_i = 1\}$$

(87) σ_p the symmetric group.

We then have :

Theorem 5 : Assume that ρ has no critical point in $\{x/\rho(x) = 0\}$. Then

$$(J^{b_p+\varepsilon} / J^{b_p-\varepsilon}) \cap V(p, \varepsilon_o) \simeq \Omega^p \times_{\sigma_p} \Delta_{p-1} / (\Omega^p \times \partial\Delta_{p-1} \times_{\sigma_p} I_p \times \Delta_{p-1}).$$

Finally, we state an existence theorem in case $K = 1$, $\bar{\Omega}$ is compact. Let Ω be connected, regular.

Theorem 6 : Assume the reduced $\mathbb{Z}_2$-homology of Ω is non zero, the equation

$$\begin{cases} \Delta u + u^{\frac{n+2}{n-2}} = 0 \\ u = 0_{|_{\partial\Omega}} \quad u > 0 \end{cases}$$

has a solution.

Proof outlined in [16].

REFERENCES

[1] A. Marino, G. Prodi, Metodi pertubattive nella teoria di Morse, B.U.M.I. 4, 11 (1975).

[2] S. Smale, The Mathematics of time, Springer 1980.

[3] C.C Conley, R.W. Easton, Isolated invariant sets and isolating blocks; (conf. on Qualitative theory of nonlinear differential and integral equations, Univ. of Wisconsin, Madison, 1978).

[4] Morris W. Hirsch, Differential topology, Springer 1976.

[5] A. Bahri, Pseudo-orbits of contact forms, preprint 1986.

[6] A. Bahri, Un problème variationnel sans compacité en géométrie de contact, Note aux Comptes Rendus de l'Académie des Sciences, Paris, Juillet 1984.

[7] D. Jerison, J. Lee, A subelliptic, non-linear eigenvalue problem and scalar curvature on CR manifolds, Microlocal Analysis, Amer. Math. Soc. Comtemporary Math. Series 27 (1984), 57-63.

[8] S.S. Chern, R. Hamilton, On Riemannian metrics adapted to three-dimensional contact manifolds, MSRI, Berkeley, October 1984.

[9] J. Sacks, K. Uhlenbeck, Ann. Math. 113 (1981), 1-24.

[10] P.L. Lions, The concentration compactness principle in the calculus of variations, the limit case, Rev. Mat. Iberoamericana 1, 1 (1985), 145-201.

[11] Y.T. Siu, S.T. Yau, Compact Kähler manifolds of positive bisectional curvature, Inv. Mathematicae 59 (1980), 189-204.

[12] H. Brezis, J.M. Coron, Convergence of solutions of H-systems or how to blow bubbles, Arch. Rat. Mech. An. 89, 1 (1985), 21-56.

[13] C.H. Taubes, Path connected Yang-Mills moduli spaces, J. Diff. Geom. 19 (1984), 337-392.

[14] A. Bahri, Critical points at infinity in some variational problems. to appear Pitman Research Notes in Mathematics, 1988.

[15] A. Bahri, J.M. Coron, Sur une équation elliptique non linéaire avec l'exposant critique de Sobolev, Note aux C.R. Acad. Sc. Paris , série I, t. 301 (1985).

[16] A. Bahri, J.M. Coron, On a non linear elliptic equation involving the critical Sobolev exponent. Communications Pure and Applied Mathematics.

[17] A. Bahri, J.M. Coron, Vers une théorie des points critiques à l'infini, Séminaire Bony-Sjöstrand-Meyer 1984-85, exposé n° VIII.

[18] M. Struwe, A global existence result for elliptic boundary value problems involving limiting nonlinearities, à paraître.

ON THE NUMBER OF BOUND STATES

AND

ESTIMATES ON SOME GEOMETRIC INVARIANTS

by

P.H. Bérard(*) and G. Besson(+)

It is a classical problem in mathematical physics to estimate the number of bound states of a given potential V in $\mathbb{R}^n$, i.e. the number of negative eigenvalues of the operator $\Delta + V$ (on $L^2(\mathbb{R}^n)$), under suitable conditions on the potential V.

For example, the Cwickel-Lieb-Rosenbljum theorem states that, for $n \geq 3$, the number of bound states in less than some (universal) constant $c(n)$ times $\int_{\mathbb{R}^n} |V|^{n/2}(x)\, dx$: [Re-Si].

In this report, we show that the problem of counting the number of bound states also appears in Riemannian geometry when one tries to give upper bounds on some geometric invariants under certain curvature assumptions. In order to avoid technicalities, we will only deal with two instances of such questions, namely: bounding the first Betti number of a closed manifold (Section I), and bounding the Morse index of a complete minimal submanifold

---

(*) Supported in part by CNPq (Brazil) at IMPA.

(+) Supported by University of Pennsylvania (Philadelphia).

(of dimension ≥ 3) in Euclidean space (Section II). In Section III, we sketch the proofs; we make some additional comments in Section IV. We refer to [Bér-Bes] for more details, and to [Bér 2] for a review on similar results.

I. EXAMPLE I: THE FIRST BETTI NUMBER.

1. Let M be a closed (i.e. compact without boundary) n-dimensional differentiable manifold. It is a well-known fact that the p-th Betti number $b_p(M)$ of M coincides with the dimension of the space $\mathcal{H}^p(M,g)$ of harmonic differential p-forms, for any Riemannian metric g on M (this is the Hodge-de Rham theorem).

An important question in Riemannian geometry is to give bounds on the Betti numbers in terms of the geometry of (M,g) - e.g. in terms of curvature assumptions, the **diameter...** - or, equivalently, to give obstructions on the existence of certain Riemannian metrics in terms of the Betti numbers (or other invariants): see [Sa] for a review.

Our first example deals with the first Betti number $b_1(M) = \dim \mathcal{H}^1(M,g)$. A theorem of S. Bochner states: if M admits a metric with positive Ricci curvature, then $b_1(M)$ vanishes; if M admits a metric with non-negative Ricci curvature, then $b_1(M) \leq n = \dim M$. See e.g. [Bér 1] Chap. III for a geometric definition of the Ricci curvature.

The proof of S. Bochner's theorem relies on the following Weitzenböck formula

$$\Delta_H \alpha = \bar{\Delta}\alpha + \mathrm{Ric}_g(\alpha^\#, \cdot), \tag{2}$$

where $\Delta_H = d^*d + dd^*$ is the Hodge Laplacian on 1-forms; $\bar{\Delta} = D^*D$ is the Bochner Laplacian on 1-forms, constructed from the covariant derivative D; $\mathrm{Ric}_g(\alpha^\#, \cdot)$ is the 1-form obtained by contracting the Ricci curvature of (M,g) with the vector field $\alpha^{\#}$ g-associated with the 1-form α (see [Bér 1] Chap. VI for more details).

Let $r_g(x)$ be the least eigenvalue of the symmetric bilinear form Ric_g on (T_xM, g_x). Applying Formula (2) to a <u>harmonic</u> 1-form α, we obtain

$$(3) \qquad \int_M |D\alpha|^2 v_g + \int_M r_g |\alpha|^2 v_g \le 0,$$

where v_g is the Riemannian measure of (M,g).

S. Bochner's theorem is an easy consequence of (3). From (3), we also conclude that the first Betti number $b_1(M)$ is less than or equal to the number of negative eigenvalues of the operator $\bar{\Delta} + r_g$ acting on 1-forms. Using ideas of E. Lieb (see [Re-Si]) and the symmetrization method of [Bér 1], one can prove

4. THEOREM. <u>Let</u> (M,g) <u>be a closed</u> n-<u>dimensional Riemannian manifold</u>. <u>Let</u> r_g <u>be the least eigenvalue of the Ricci curvature</u> Ric_g <u>on each tangent space, and decompose</u> r_g <u>into negative and positive parts</u>, $r_g = r_g^+ - r_g^-$. <u>Let</u> $D(g)$ (<u>resp.</u> $V(g) = \int_M v_g$) <u>denote the diameter of</u> (M,g) (<u>resp. the volume of</u> (M,g)). <u>Then, if</u> $r_g^+ \not\equiv 0$ <u>and</u> $n \ge 3$,

$$b_1(M) \le F(n, D(g), V(g), \|r_g^-\|_\infty, \|r_g^+\|_\infty, \|r_g^+\|_1) \int_M (r_g^-)^{n/2} v_g ,$$

<u>for some function</u> F <u>depending on the indicated arguments</u>. <u>The function</u> F <u>can be explicitely estimated</u> (by elementary numerical analysis).

Note that Theorem 4 implies that if $Ric_g \ge 0$ and $Ric_g > 0$ in some $T_{x_0}M$, then $b_1(M) = 0$: this is the first part of S. Bochner's theorem. When $r_g^+ \equiv 0$, Theorem 4 does not hold; for a generalization of the 2nd part of S. Bochner's theorem with integral bounds on r_g^-, see [Bér-Bes].

II. EXAMPLE II: THE MORSE INDEX OF A MINIMAL IMMERSION.

5. Let M^n, $n \geq 3$, be a complete non-compact minimal immersion in $\mathbb{R}^N$, with normal bundle $\nu(M)$. Minimal immersions are the critical points of the volume functional (we only consider variations with compact support). For $E \in C_o^\infty(\nu(M))$, a smooth normal vector field with compact support, the second derivative of the volume functional in the direction E is given by the quadratic form (see [Si]).

(6) $$I_M(E,E) = \int_M (|DE|^2 - \langle \mathcal{B}(E),E\rangle)\,dvol_M\,,$$

where D is the connection on $\nu(M)$, and where $\mathcal{B}$ is constructed from the second fundamental form B of the immersion. It turns out that

(7) $$\langle \mathcal{B}(E),E\rangle \leq \|B\|^2 \langle E,E\rangle,$$

where $\|B\|$ is the pointwise norm of the second fundamental form.

In this example, we are interested in the Morse index of I_M, i.e. in the number of negative eigenvalues of I_M; this number can be bounded from above by the number of negative eigenvalues of the operator $\bar{\Delta} - \|B\|^2$ acting on $C_o^\infty(\nu(M))$, where $\bar{\Delta} = D^*D$.

We have

8. THEOREM. *Let $M^n \to \mathbb{R}^N$ be a complete non-compact isometric minimal immersion,* $n \geq 3$. *Then, there exists an explicit constant* $c(n,N)$ *such that*

$$\text{Morse Index } (I_M) \leq c(n,N) \int_M \|B\|^n \,dvol_M\,.$$

III. SKETCH OF THE PROOFS OF THEOREMS 4 and 8.

9. Let us now work in the following general framework. We consider a Riemannian bundle $E \to M$ over a complete Riemannian manifold: E is equipped with a Riemannian metric $\langle \cdot,\cdot \rangle$ in each fiber and with a compatible connection D (i.e. $X.\langle s,t\rangle = \langle D_X s,t\rangle + \langle s,D_X t\rangle$ for all vector field X on M, and C^∞ sections s, t of E).

Let V be a C^∞ function on M; we decompose V into its positive and negative parts $V = V_+ - V_-$, and we write (Δ is the positive Laplacian on $C^\infty(M)$, $-d^2/dx^2$ on R)

$$(10)\quad \begin{cases} \bar{H}_o = \bar{\Delta} + V_+ , & H_o = \Delta + V_+ ; \\ \bar{H} = \bar{\Delta} + V , & H = \Delta + V ; \\ W = V_- ; & \\ \bar{H} = \bar{H}_o - W , & H = H_o - W. \end{cases}$$

Let $-e$ be a negative number and define

$N_e(\bar{H})$ = number of eigenvalues of $\bar{H}$ less than or equalt to $-e$.

The first part of the proof follows E. Lieb's ideas (see [Re-Si] p.101 ff). We only deal with the case $n=3$ (recall that the assumption in Theorems 4 and 8 is $n \geq 3$; for $n \geq 4$ use the same method as in [Re-Si]; for $n=2$ see Section IV).

11. LEMMA. <u>We have</u>

$$N_e(\bar{H}) = \#\{\lambda \in \,]0,1] : -e \text{ is an eigenvalue for } \bar{H}_o - \lambda W\}.$$

■ [Re-Si] p. 98 ■

12. LEMMA. Let $\bar{K} = W^{1/2}[(\bar{H}_o+e)^{-1} - (\bar{H}_o+W+e)^{-1}]W^{1/2}$.

Then,

$$N_e(\bar{H}) \leq \#\{\text{eigenvalues of } \bar{K} \geq \frac{1}{2}\}.$$

■ As in [Re-Si] p. 98, one observes that $-e$ is an eigenvalue of $\bar{H}_o - \lambda W$ if and only if $(\lambda(\lambda+1))^{-1}$ is an eigenvalue of $\bar{K}$. Since $\lambda \in]0,1]$, $\lambda(\lambda+1) \leq 2$ and the lemma follows ■

Since $\bar{K}$ has positive eigenvalues, one can also write

$$(13) \qquad N_e(\bar{H}) \leq 2\,\mathrm{Tr}(\bar{K}) = 2\,\mathrm{Tr}[W\{(\bar{H}_o+e)^{-1} - (\bar{H}_o+W+e)^{-1}\}].$$

The idea of E. Lieb consists in writing the left-hand side of (13) as

$$2\int_0^\infty \mathrm{Tr}\{W[\exp(-t\bar{H}_o) - \exp(-t\bar{H}_o-tW)]\}\exp(-et)dt,$$

and to use the Feynman-Kac formula. The main difficulty here is that we are dealing with operators acting on sections and not with operators acting on functions. We first reduce to the scalar case by the

14. LEMMA. The following inequality holds

$$\mathrm{Tr}\{W[(\bar{H}_o+e)^{-1}-(\bar{H}_o+W+e)^{-1}]\} \leq \mathrm{rank}(E)\mathrm{Tr}\{W[(H_o+e)^{-1}-(H_o+W+e)^{-1}]\}.$$

■ Write

$W[(\bar{H}_o+e)^{-1}-(\bar{H}_o+W+e)^{-1}] = W(\bar{H}_o+e)^{-1}W(\bar{H}_o+W+e)^{-1}$, and use the fact that $(\bar{H}_o+e)^{-1}$ (resp. $(\bar{H}_o+W+e)^{-1}$) is dominated by $(H_o+e)^{-1}$ (resp. $(H_o+W+e)^{-1}$): see [Bér 1], Appendix A ■

Finally, we have reduced our problem to estimating

$$\int_0^\infty \mathrm{Tr}\{W[\exp(-tH_o) - \exp(-tH_o-tW)]\exp(-et)dt$$

as in [Re-Si] p. 101, which gives

15. PROPOSITION. *Let* $G(u) = u(1-e^{-u})$, $u \in \mathbb{R}_+$ *and define* $\varphi(u)$ *by*

$$\varphi(u) = \begin{cases} G(u) & \text{for} \quad 0 \le u \le 2, \\ G(2) + (u-2)G'(2) & \text{for} \quad 2 \le u < \infty. \end{cases}$$

Let $k_o(t,x,y)$ *denote the kernel of* $\exp(-tH_o)$ *on* $C^\infty(M)$ *and let* ℓ *be the rank of* E. *Then*

$$N_e(\bar{H}) = \#\{\text{bound states of } \bar{\Delta}+V \le -e\},$$

$$(16) \qquad N_e(\bar{H}) \le 2\ell \int_0^\infty \left(\int_M k_o(t,x,x)\varphi(tW(x))dx\right)\exp(-et)\,\frac{dt}{t}\,.$$

17. REMARK. Recall that $n=3$. For $n \ge 4$, (16) will only give $N_e(\bar{H}) \le \infty$. To deal with the case $n \ge 3$, use the same modifications as in [Re-Si] Theorem XIII.12. For $n=2$, see Section IV.

18. PROOF OF THEOREM 8.

In this example, $E = \nu(M)$, the normal bundle to the immersion and $\ell = N-3$ (here $n=3$), and $\bar{H} = \bar{\Delta} - \|B\|^2$, i.e. $V_+ \equiv 0$, $V_- = W = \|B\|^2$. In particular, $\bar{H}_o = \bar{\Delta}$ and $H_o = \Delta$. By a theorem of S.Y. Cheng, P. Li and S.T. Yau, [C-L-Y], we have

$$k_o(t,x,x) \le (4\pi t)^{-3/2}$$

and (16) gives (take $e = 0$),

$$\text{Morse Index } (I_M) \le (N-3)c(3) \int_M \|B\|^3 \, dvol_M \,.$$

For $n \ge 3$, one obtains similarly

$$\text{Morse Index } (I_M) \le (N-n)c(n) \int_M \|B\|^n \, dvol_M \qquad \blacksquare$$

19. PROOF OF THEOREM 4

In this example, $E = T^*M$ and $\ell = n$ (here $n=3$), and $\bar{H} = \bar{\Delta} + r_g^+ - r_g^-$, $V_+ = r_g^+$, $V_- = W = r_g^-$. Since $r_g^+ \geq 0$, the function $k_o(t,x,x)$ satisfies

$$k_o(t,x,x) \leq k(t,x,x),$$

where $k(t,x,x)$ is the heat kernel for Δ on $C^\infty(M)$ (recall that $k_o(t,x,y)$ is the heat kernel for $\Delta + r_g^+$ on $C^\infty(M)$). The difficulty here is that $k(t,x,x)$ goes to $1/V(g)$ when t goes to $+\infty$, and we will have to use the fact that $k_o(t,x,x)$ decreases exponentially when t goes to $+\infty$, because $r_g^+ \not\equiv 0$ by assumption.

Indeed, one can always write

$$(20) \quad \begin{cases} k_o(t,x,x) \leq k(t,x,x) & \text{for} \quad t \leq a, \\ k_o(t,x,x) \leq \exp[\lambda_1(r_g^+)(a-t)]k(a,x,x) & \text{for} \quad t \geq a, \end{cases}$$

where $\lambda_1(r_g^+)$ is the first eigenvalue of $\Delta + r_g^+$ on $C^\infty(M)$ (notice that $\lambda_1(r_g^+) > 0$ because $r_g^+ \not\equiv 0$).

21. LEMMA. <u>There exists an explicit constant</u> $A = A(n,D(g),\|r_g^-\|_\infty)$, <u>such that</u>

$$V(g)k(t,x,x) \leq A(t/a)^{-n/2}$$

<u>for all</u> $x \in M$ <u>and</u> $t \leq a = D(g)^2$.

Proof: See [Bér 1] Chap. V ■

22. LEMMA. <u>There exists an explicit constant</u> B,

$$B = B(n,D(g),\|r_g^-\|_\infty, \lambda_1(r_g^+)),$$

<u>such that</u>

$$V(g)k_o(t,x,x) \leq B\, D(g)^n\, t^{-n/2}$$

<u>for all</u> $t > 0$ <u>and all</u> $x \in M$.

<u>Proof</u>: For $t \le a = D^2(g)$, (20) gives

$$V(g)k_o(t,x,x) \le A\, D(g)^n\, t^{-n/2} .$$

For $t \ge a = D^2(g)$, (20) gives

$$V(g)k_o(t,x,x) \le A_1 \exp(-\lambda_1(r_g^+)t),$$

where $A_1 = \exp(\lambda_1(r_g^+)D^2(g))A(n,D^2(g)\|r_g^-\|_\infty)$, and hence

$$V(g)k_o(t,x,x) \le A_1\, \lambda_1(r_g^+)^{-n/2}\, c(n)\, t^{-n/2} ,$$

where $c(n) = \sup\{e^{-u}\, u^{n/2} : u \in \mathbb{R}_+\}$. ■

From (16), we conclude that

$$\begin{aligned} b_1(M) &\le N_o(\bar\Delta + r_g) \\ &\le 2n\, B\, D(g)^n V(g)^{-1}\left(\int_M r_g^-(x)^{3/2}\, dv_g(x)\right)\int_0^\infty t^{-3/2}\,\varphi(t)\,\frac{dt}{t} \\ &\le F_1(n,D(g),V(g),\|r_g^-\|_\infty,\, \lambda_1(r_g^+)) \int_M (r_g^-)^{3/2}\, v_g . \end{aligned}$$

In order to finish the proof of Theorem 4, we need to show that $\lambda_1(r_g^+)$ can be estimated in terms of the geometry. The min-max gives

$$\lambda_1(r_g^+) \le \int_M r_g^+\, v_g \,/\, V(g)$$

and a symmetrization method gives (see [Bér-Bes] for details)

$$\lambda_1(r_g^+) \ge \Lambda(n,D(g),V(g),\|r_g^+\|_\infty,\|r_g^+\|_1) > 0.$$

Theorem 4 follows

■

V. COMMENTS.

(i) The proof of Theorem 4 generalizes to other geometric invariants. These invariants are obtained as harmonic sections for a Laplacian $\Delta_E = \bar{\Delta} + R$ acting on $C^\infty(E)$. Again, one decomposes the least eigenvalue r of R into positive and negative parts, $r = r_+ - r_-$. One needs to control $D(g)$ and $\|r_g^-\|_\infty$ on M in order to estimate the heat kernel of Δ acting on $C^\infty(M)$. The proof then goes as that of Theorem 4;

(ii) In the case of a complete non-compact Riemannian manifold, one needs an extra information (e.g. the minimallity condition of Theorem 8), to estimate the heat kernel on M. For the other applications we refer to [Bér-Bes].

(iii) In the 2-dimensional case, the situation is quite different. The Cwickel-Lieb-Rosenbljum theorem does not hold; indeed, for any $V \in C_o^\infty(\mathbb{R}^2)$, $V \le 0$ and $V \not\equiv 0$, and for any $\epsilon > 0$, the operator $\Delta + \epsilon V$ always has at least one negative eigenvalue. A simple way of proving this fact is to notice that the Dirichlet integral is a conformal invariant; one then studies a different but similar problem, namely $\Delta + \epsilon\tilde{V}a$, on S^2 [for some positive function a; $\tilde{V}$ is the pull-back of V on S^2 by stereographic projection]. In this latter case, $\Delta + \epsilon\tilde{V}a$ always has at least one negative eigenvalue because the spectrum of Δ is discrete and begins at 0. See [Re-Si] for a different proof.

(iv) Another method in order to bound geometric invariants uses Sobolev inequalities (see [Bér 2]). However, the (general) results involve the $L^{n/2+\epsilon}$-norm of the curvature-potential instead of the $L^{n/2}$-norm; they also give estimating theorems rather than vanishing theorems. We refer to [Bér-Bes] for more details.

REFERENCES

[Bér 1] BÉRARD, P.H. - Spectral geometry: direct and inverse problems, Lecture Note in Math. nº 1207, Springer 1986.

[Bér 2] BÉRARD, P.H. - From vanishing theorems to estimating theorems: the Bochner technique revisited, Informes do IMPA nº 60-1986.

[Bér-Bes] BÉRARD, P.H., BESSON, G. - Integral curvature bounds for some geometric invariants by an extension of the Bochner technique, in preparation.

[C-L-Y] CHENG, S.Y., LI, P., YAU, S.T. - Heat equation on minimal submanifolds and their applications, Amer. J. Math. 106 (1984), 1033-1065.

[Re-Si] REED, M., SIMON, B. - Methods of modern mathematical physics, Vol. IV, Academic Press.

[Sa] SAKAI, T. - Comparison and finiteness theorems in Riemannian geometry, in Geometry of geodesics and related topics, Advanced Studies in Pure Math. nº 3, p. 125-181, North-Holland 1984.

[Si] SIMONS, J. - Minimal varieties in Riemannian manifolds, Annals of Math. 88 (1968), 62-105.

P.H. Bérard

Département de Mathématiques
Université de Savoie
B.P. 1104
F-73011 CHAMBÉRY Cedex
France

G. Besson

Institut Fourier, Math. Pure
Université de Grenoble I
B.P. 74
F-38402 SAINT MARTIN D'HÈRES
France

CONVERGENCE OF SOLUTIONS OF CAPILLO-VISCOELASTIC PERTURBATIONS OF THE EQUATIONS OF ELASTICITY

José Luiz Boldrini
UNICAMP - IMECC
Caixa Postal 6065
13.081 - Campinas - SP - Brazil

1. INTRODUCTION

The equations for the isothermal motion for a one-dimensional body are, in Lagrangian coordinates,

$$(1.1)\qquad \begin{cases} u_t = v_x\,, \\ v_t = T_x\,, \end{cases}$$

where $u(x,t)$ is the deformation gradient, at time t of the material point with reference position x; $v(x,t)$ represents the velocity of that point, and $T(x,t)$ represents the stress.

In Elasticity we assume that the stress depends only on the deformation gradient, and we have $T(x,t) = \sigma_o(u(x,t))$ where $\sigma_o(\cdot)$ is a material function. In this case the equations (1.1) reduce to:

$$(1.2)\qquad \begin{cases} u_t = v_x\,, \\ v_t = \sigma_o(u)_x\,, \end{cases}$$

and it is well known that the solutions of (1.2) develop discontinuities that correspond to shock waves propagating in the body.

In this paper we will study materials that incorporate in their constitutive relations viscosity and capillarity as well as a small perturbation in the elastic part of the stress, that is, $T = \sigma_o(u) + \delta\sigma_1(u) + \varepsilon v_x - \frac{1}{4}\,\varepsilon^2 u_{xx}$. Thus, equations (1.1) read

$$(1.3)\qquad \begin{cases} u_t = v_x\,, \\ v_t = \sigma_o(u)_x + \delta\sigma_1(u)_x + \varepsilon v_{xx} - \frac{1}{4}\,\varepsilon^2 u_{xxx}\,, \end{cases}$$

where $u(x,t)$, $v(x,t)$, $\sigma_o(\cdot)$ are as before; ε is a positive parameter associated to the relative strength of the viscosity and capillarity; $\delta\sigma_1(\cdot)$ is a perturbation of the elastic part of the stress, and δ is a positive parameter associated to the relative strength of the elastic perturbation. The factor $\frac{1}{4}$ is not important because it can always be achieved by scaling the system.

We will compare solutions of (1.2) and (1.3), corresponding to the same initial conditions, for small ε and δ. We assume that such initial conditions approach the equilibrium state at infinity, and, by a trivial change of variables, we can take that state at infinity as zero. We assume that this has been already done for (1.2) and (1.3), and, for simplicity, we do not change the notation, and we will keep denoting these new variables $u(x,t)$ and $v(x,t)$.

We will assume that the initial conditions $u(x,0)$, $v(x,0)$ satisfy:

$$(1.4)\qquad \begin{cases} u(\cdot,0) = u_o(\cdot) \in W^{1,\infty}(\mathbb{R}) \cap H^3(\mathbb{R}), \\ \\ v(\cdot,0) = v_o(\cdot) \in L^\infty(\mathbb{R}) \cap H^2(\mathbb{R}), \end{cases}$$

and that the material is "rubber-like", that is,

$$(1.5)\qquad \begin{cases} \sigma_o(\cdot) \text{ is } C^2\text{-smooth and } \sigma_o(0) = 0, \\ \\ \sigma_o'(\cdot) > 0, \\ \\ \sigma_o''(u) \operatorname{sign} u > 0 \quad \text{for} \quad u \neq 0, \end{cases}$$

and

$$(1.6)\qquad \begin{cases} \sigma_1(\cdot) \text{ is } C^2\text{-smooth and } \sigma_1(0) = 0 \\ \\ \sigma_o'(\cdot) + \delta\sigma_1'(\cdot) > 0 \quad \text{for small enough } \delta \\ \\ (\sigma_o''(\cdot) + \delta\sigma_1''(\cdot)) \operatorname{sign} u > 0 \quad \text{for small enough } \delta. \end{cases}$$

We observe that $\sigma_o'(\cdot) > 0$ implies that the system (1.2) is hyperbolic.

By using a result derived from the Theory of Compensated Compactness (Tartar [7], [8], Murat [3], [4]) by Di Perna [2], we will prove the following result:

THEOREM: Under the hypothesis (1.4), (1.5), (1.6), for sufficiently small δ, there are solutions $(u_\varepsilon^\delta, v_\varepsilon^\delta)$ of (1.3) such that $u_\varepsilon^\delta(\cdot,t) \in H^3(\mathbb{R}) \cap L^\infty(\mathbb{R})$ and $v_\varepsilon^\delta(\cdot,t) \in H^2(\mathbb{R})$. Moreover, as ε and δ approach zero such that $\delta = o(\varepsilon^{1/2})$, we can extract a subsequence such that along this subsequence $u_\varepsilon^\delta \to u$ in $\mathcal{D}'$ and almost everywhere (ae) and $v_\varepsilon^\delta \to v$ in $\mathcal{D}'$, where (u,v) is solution of (1.2).

The proof will be given in the following sections.

2. EXISTENCE OF SOLUTIONS

To prove that there are solutions of the Cauchy Problem (1.3), (1.4), we use a change of variables to reduce the system to another one for which we can apply the Theory of Invariant Regions (Chuch, Conley and Smoller [1]).

We introduce a new dependent variable w by

$$w = v - \frac{\varepsilon}{2} u_x . \tag{2.1}$$

Thus, we can rewrite the Cauchy Problem (1.3), (1.4) in terms of u and w as:

$$\begin{cases} u_t = \frac{\varepsilon}{2} u_{xx} + w_x \\ w_t = \frac{\varepsilon}{2} w_{xx} + \sigma_o'(u)u_x + \delta\sigma_1'(u)u_x \\ u(\cdot,0) = u_o(\cdot); \quad w(\cdot,0) = v_o(\cdot) - \frac{\varepsilon}{2} u_{ox}(\cdot) . \end{cases} \tag{2.2}$$

We observe that (1.4) implies that the initial conditions in (2.2) are uniformly L^∞-bounded. Also, (2.2) is a parabolic system with the same rate of dissipation in both variables.

The eigenvalues of the matrix

$$\begin{bmatrix} 0 & 1 \\ \sigma_o'(u) + \delta\sigma_1'(u) & 0 \end{bmatrix}$$

are $\lambda_\pm^\delta = \pm(\sigma_o'(u) + \delta\sigma_1'(u))^{1/2}$, that are real for sufficiently small δ (by (1.6)), with respective eigenvector $[1, -\lambda_\mp^\delta]$

We introduce the Riemann Invariants

$$G_+^\delta(u,w) = w + \int_o^u \lambda_-^\delta(s)ds,$$

$$G_-^\delta(u,w) = w + \int_o^u \lambda_+^\delta(s)ds.$$

Now, if we choose constantes K_1 and K_2 such that the level curves $G_+^\delta(u,w) = K_1$, $G_+^\delta(u,w) = K_2$, $G_-^\delta(u,w) = K_1$, $G_-^\delta(u,w) = K_2$ determine a bounded convex region in the (u,w)-plane that contains the points $(u(x,0),w(x,0))$ for $-\infty < x < +\infty$ (see Figure 2.1) then the Theory of Invariant Regions for parabolic systems [1] applied to our case allow us to conclude that the bounded convex region contains the points $(u(\cdot,t), w(\cdot,t))$ for $t \geq 0$ for any solution $(u(\cdot,t), w(\cdot,t)$ of (2.2)

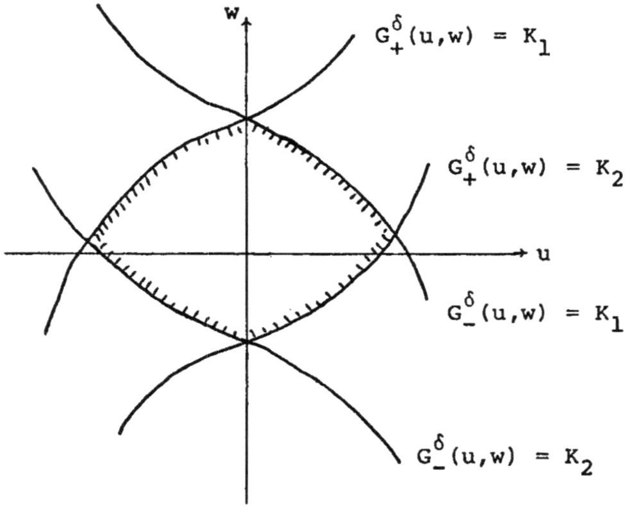

Figure 2.1.

This give us a priori L^∞-estimates for $u_\varepsilon^\delta(\cdot,t)$ and $v_\varepsilon^\delta(\cdot,t)$ for $t \geq 0$. Moreover these L^∞-estimates can be taken independent of ε and δ because, for small δ, all the bounded convex invariant region are contained in a larger bounded one that is independent of ε and δ.

Using these L^∞-estimates, standard arguments for existence of solutions of parabolic systems show us that there are global solutions $u_\varepsilon^\delta(\cdot,t) \in H^3(\mathbb{R}) \cap L^\infty(\mathbb{R})$ and $w_\varepsilon^\delta(\cdot,t) \in H^2(\mathbb{R}) \cap L^\infty(\mathbb{R})$ of (2.2) and that the L^∞-bound is independent of ε and δ. By using (2.1), we conclude that there are solutions of (1.3) as described in the statement of the theorem.

We remark that the transformation (2.1) was also found earlier by Slemrod in a independent work [5].

3. SOME ESTIMATES

Before we prove the theorem we need more estimates. Some of them can be obtained from the balance of mechanical energy. For this we multiply (1.3)(ii) by $v_\varepsilon^\delta(x,t)$ and integrate over $\mathbb{R} \times [0,T_1]$, $0 \leq T_1 \leq T < \infty$. To simplyfy the notation in what follows we will drop the indices ε and δ in u_ε^δ and v_ε^δ; an integral signal without indication of the interval of integration will mean that the integral is taken from $-\infty$ to $+\infty$, and $\mathbb{C}$ will denote a generic constant depending at most on the initial conditions (and therefore independent of ε and δ).

After some integrations by parts, using the fact that $u_t = v_x$ ((1.3)(i)), we obtain:

$$\frac{1}{2}\int v^2(x,T_1)\,dx + \int W_\delta(u,T_1))\,dx$$

$$+ \varepsilon \int_0^{T_1}\int v_x^2(x,t)\,dx\,dt + \frac{\varepsilon^2}{8}\int u_x^2(x,T_1)\,dx \leq \mathbb{C}$$

where $W_\delta(u) = \int_0^u [\sigma_0(s) + \delta\sigma_1(s)]\,ds$ is the Stored Energy.

Since $\varepsilon > 0$, $\delta > 0$ and $W_\delta(u) \geq 0$ (from (1.6) for Small enough δ) the above inequality implies that for $0 \leq T_1 \leq T$

$$(3.1)\qquad \begin{cases} \|v_\varepsilon^\delta(\cdot,T_1)\|_{L^2(\mathbb{R})} \le \mathbb{C}. \\ \|\varepsilon^{1/2} v_{\varepsilon x}^\delta(\cdot,\cdot)\|_{L^2(\mathbb{R}\times[0,T_1])} \le \mathbb{C} \\ \|\varepsilon u_{\varepsilon x}^\delta(\cdot,T_1)\|_{L^2(\mathbb{R})} \le \mathbb{C} \end{cases}$$

To obtain other estimates, we multiply (1.3)(ii) by $u_x(x,t)$ and integrate over $\mathbb{R} \times [0, T_1]$. After one integration by parts in the last term, we obtain:

$$\int_0^{T_1}\int u_x v_t dxdt = \int_0^{T_1}\int \sigma_0'(u)u_x^2 dxdt + \delta \int_0^{T_1}\int \sigma_1'(u)u_x^2 dxdt$$

$$+ \varepsilon \int_0^{T_1}\int v_{xx}u_x dxdt + \frac{\varepsilon^2}{4}\int_0^{T_1}\int u_{xx}^2 dxdt.$$

But

$$\int_0^{T_1}\int v_{xx}u_x dxdt = \int_0^{T_1}\int u_{tx}u_x dxdt = \int_0^{T_1}\int \frac{\partial}{\partial t}\left(\frac{u_x^2}{2}\right)dxdt$$

$$= \int \frac{u_x^2}{2}(x,T_1)dx - \int \frac{(u_{ox})^2}{2}(x)dx,$$

and also

$$\int_0^{T_1}\int u_x v_t dxdt = \int dx\, (u_x v\,]_{t=0}^{t=T_1} - \int_0^{T_1} u_{xt} v dt)$$

$$= \mathbb{C} + \int u_x(x,T_1)v(x,T_1)dx - \int_0^{T_1}\int u_{xt} v dxdt$$

$$= \mathbb{C} + \int u_x(x,T_1)v(x,T_1)dx - \int_0^{T_1}\int v_{xx} v dxdt$$

$$\le \mathbb{C} + [\int u_x(x,T_1)dx]^{1/2}[\int v^2(x,T_1)dx]^{1/2}$$

$$+ \int_0^{T_1}\int v_x^2 dxdt.$$

Therefore, we conclude

$$\int_0^{T_1}\int \sigma_0'(u)u_x^2dxdt + \frac{\varepsilon^2}{4}\int_0^{T_1}\int u_{xx}^2dxdt + \frac{\varepsilon}{2}\int u_x^2dx$$

$$\leq \mathbb{C} + [\int u_x^2dx]^{1/2}[\int v^2dx] + \int_0^{T_1}\int v_x^2dxdt$$

$$- \delta\int_0^{T_1}\int \sigma_1'(u)u_x^2dxdt.$$

But we can estimate

$$[\int u_x^2dx]^{1/2}[\int v^2dx]^{1/2} \leq \frac{\varepsilon}{4}\int u_x^2dx + \frac{1}{\varepsilon}\int v^2dx,$$

and obtain from the above

$$\int_0\int \sigma_0'(u)u_x^2dxdt + \frac{\varepsilon^2}{4}\int_0^{T_1}\int u_{xx}^2dxdt + \frac{\varepsilon}{4}\int u_x^2dx$$

$$\leq \mathbb{C} + \frac{1}{\varepsilon}\int v^2\,dx + \int_0^{T_1}\int v_x^2dxdt - \delta\int_0^{T_1}\int \sigma_1'(u)u_x^2dxdt.$$

Upon multiplication the above expression by $\varepsilon > 0$, and using the energy estimates (3.1)(i), (ii), we obtain

$$\varepsilon\sigma_0'\int_0^{T_1}\int u_x^2dxdt + \frac{\varepsilon^3}{4}\int_0^{T_1}\int u_{xx}^2dxdt + \frac{\varepsilon^2}{4}\int u_x^2dx$$

$$\leq \mathbb{C} - \varepsilon\delta\sigma_1'\int_0^{T_1}\int u_x^2dx\,dt,$$

where

$$0 < \sigma_0' = \min_{M_1\leq u\leq M_2}\{\sigma_0'(u)\},$$

$$0 < \sigma_1' = \max_{M_1\leq u\leq M_2}\{|\sigma_1'(u)|\},$$

and M_1 and M_2 are bounds given by the uniform L^∞-estimate obtained in Section 2.

Thus, for $\delta < \frac{\sigma_0'}{2\sigma_1'}$, the last inequality implies that we have the following uniform estimates

(3.2)
$$\begin{cases} \| \varepsilon^{1/2} u_{\varepsilon x}^{\delta} \|_{L^2(\mathbb{R} \times [0,T])} \leq \mathbb{C}, \\ \\ \| \varepsilon^{3/2} u_{\varepsilon xx}^{\delta} \|_{L^2(\mathbb{R} \times [0,T])} \leq \mathbb{C}. \end{cases}$$

4. THE PROOF OF THE THEOREM

Let's first consider the solutions $(u_\varepsilon^\delta, w_\varepsilon^\delta)$ of (2.2) as ε and δ approach zero.

Consider any entropy pair $(\eta(u,v), q(u,v))$ associated to (1.2), that is, $\eta(\cdot,\cdot)$ and $q(\cdot,\cdot)$ are C^2-smooth functions such that

(4.1)
$$\begin{cases} q_v + \eta_u = 0, \\ \\ q_u + \eta_v \sigma_0'(u) = 0. \end{cases}$$

Since we have uniform L^∞-estimates in ε and δ for u_ε^δ and w_ε^δ (Section 2), according to Di Perna's results [2], it is enough to prove that

$$\{\eta(u_\varepsilon^\delta, w_\varepsilon^\delta)_t + q(u_\varepsilon^\delta, w_\varepsilon^\delta)_x, \quad \delta, \varepsilon \downarrow 0\}$$

is contained in a compact set of H_{loc}^{-1} in order to conclude that there is $(u,x,t), v(x,t))$ and a subsequence such that along such subsequence $u_\varepsilon^\delta \to u$ ae (and L^∞-weak $*$) and $w_\varepsilon^\delta \to v$ ae (and L^∞-weak $*$) as $\varepsilon, \delta \downarrow 0$.

But, by using (4.1) and (2.1)

$$\eta(u_\varepsilon^\delta,w_\varepsilon^\delta)_t + q(u_\varepsilon^\delta,w_\varepsilon^\delta)_x$$

$$= \frac{\varepsilon}{2}(\eta_u(u,w)u_x)_x - \frac{\varepsilon}{2}\eta_{uu}(u,w)u_x^2 - \frac{\varepsilon}{2}\eta_{uv}(u,w)u_xv_x$$

$$+ \frac{\varepsilon^2}{4}\eta_{uv}(u,w)u_xu_{xx} + \delta\sigma_1'(u)\eta_v(u,w)u_x$$

$$+ \varepsilon(\eta_v(u,w)v_x)_x - \frac{\varepsilon^2}{2}(\eta_v(u,w)u_{xx})_x - \varepsilon\eta_{vu}(u,w)u_xv_x$$

$$+ \frac{\varepsilon^2}{2}\eta_{vu}(u,w)u_xu_{xx} - \varepsilon\eta_{vv}(u,w)v_x^2 + \varepsilon^2\eta_{vv}(u,w)v_xu_{xx}$$

$$- \frac{\varepsilon^3}{4}\eta_{vv}(u,w)u_{xx}^2 ,$$

where we have dropped the indices ε and δ to easy the notation.

Since we have uniform L^∞-estimates for u_ε^δ and w_ε^δ and $\eta(\cdot,\cdot)$ is C^2-smooth, we have uniform L^∞-estimates for $\eta(u_\varepsilon^\delta,w_\varepsilon^\delta)$, $\eta_u(u_\varepsilon^\delta,w_\varepsilon^\delta)$, $\eta_v(u_\varepsilon^\delta,w_\varepsilon^\delta)$, $\eta_{uu}(u_\varepsilon^\delta,w_\varepsilon^\delta)$, $\eta_{vv}(u_\varepsilon^\delta,w_\varepsilon^\delta)$, $\eta_{uv}(u_\varepsilon^\delta,w_\varepsilon^\delta)$ and, for similar reason, for $\sigma_1'(u_\varepsilon^\delta)$ and $q(u_\varepsilon^\delta,w_\varepsilon^\delta)$.

Hence, by using the above and (4.1)(i), we have that

$$\frac{\varepsilon}{2}\eta_u(u_\varepsilon^\delta,w_\varepsilon^\delta)u_{\varepsilon x}^\delta = \frac{\varepsilon^{1/2}}{2}\eta_u(u_\varepsilon^\delta,w_\varepsilon^\delta)(\varepsilon^{1/2}u_{\varepsilon x}^\delta) \xrightarrow{L^2} 0$$

as $\varepsilon, \delta \downarrow 0$, and, therefore

$$\frac{\varepsilon}{2}(\eta_u(u_\varepsilon^\delta,w_\varepsilon^\delta)u_{\varepsilon x}^\delta)_x \xrightarrow{H^{-1}} 0.$$

Also, by using (3.1)(i), as $\varepsilon, \delta \downarrow 0$

$$\varepsilon\eta_v(u_\varepsilon^\delta,w_\varepsilon^\delta)v_{\varepsilon x}^\delta = \varepsilon^{1/2}\eta_v(u_\varepsilon^\delta,w_\varepsilon^\delta)(\varepsilon^{1/2}v_{\varepsilon x}^\delta) \xrightarrow{L^2} 0,$$

and we conclude that

$$\varepsilon(\eta_v(u_\varepsilon^\delta,w_\varepsilon^\delta)v_x)_x \xrightarrow{H^{-1}} 0.$$

In a simmilar way, by using (3.2)(ii), we conclude that

$$\frac{\varepsilon^2}{2}(\eta_v(u_\varepsilon^\delta,w_\varepsilon^\delta)u_{xx})_x \xrightarrow{H^{-1}} 0 \quad \text{as} \quad \varepsilon,\delta \downarrow 0,$$

and, by using (3.2)(ii) and $\delta = o(\varepsilon^{1/2})$

$$\delta\sigma_1'(u_\varepsilon^\delta)\eta_v(u_\varepsilon^\delta,w_\varepsilon^\delta)u_{\varepsilon x}^\delta \xrightarrow{L^2} 0,$$

and, a forteriori, in H^{-1}.

Now, (3.1)(ii), (3.2)(ii) and Schwarz inequality imply that

$$\left\|\frac{\varepsilon}{2}\eta_{uv}(u_\varepsilon^\delta,w_\varepsilon^\delta)u_{\varepsilon x}^\delta v_{\varepsilon x}^\delta\right\|_{L^1} \leq \mathbb{C}\,\|\varepsilon^{1/2}u_{\varepsilon x}^\delta\|_{L^2}\,\|\varepsilon^{1/2}v_{\varepsilon x}^\delta\|_{L^2} \leq \mathbb{C}.$$

In a simmilar way we prove that $\varepsilon\eta_{uu}(u_\varepsilon^\delta, w_\varepsilon^\delta)(u_{\varepsilon x}^\delta)^2$, $\varepsilon^2\eta_{uv}(u_\varepsilon^\delta, w_\varepsilon^\delta)u_{\varepsilon x}^\delta u_{\varepsilon xx}^\delta$, $\varepsilon\eta_{vu}(u_\varepsilon^\delta,w_\varepsilon^\delta)u_{\varepsilon x}^\delta v_{\varepsilon x}^\delta$, $\varepsilon\eta_{vv}(u_\varepsilon^\delta, w_\varepsilon^\delta)(v_{\varepsilon x}^\delta)^2$, $\varepsilon^3\eta_{vv}(u_\varepsilon^\delta,w_\varepsilon^\delta)(u_{\varepsilon xx}^\delta)^2$ are $\varepsilon^2\eta_{vv}(u_\varepsilon^\delta, w_\varepsilon^\delta)v_{\varepsilon x}^\delta u_{\varepsilon xx}^\delta$ are uniformly bounded in L^1.

Therefore,

$$\{\eta(u_\varepsilon^\delta,w_\varepsilon^\delta)_t + q(u_\varepsilon^\delta,w_\varepsilon^\delta)_x, \varepsilon,\delta \downarrow 0\} \subset (S_1 + S_2) \cap W^{-1,\infty}$$

where S_1 is a compact set of H^{-1} and S_2 is a bounded set of L^1. By using Murat's Lemma [4] we conclude that

$$\{\eta(u_\varepsilon^\delta,w_\varepsilon^\delta)_t + q(u_\varepsilon^\delta,w_\varepsilon^\delta)_x,\ \varepsilon,\delta \downarrow 0\}$$

is contained in a compact set of H^{-1}_{loc} and, by Di Perna's results [2], there are $(u(x,t),v(x,t))$ and a subsequence along which

$$u_\varepsilon^\delta \to u \quad \text{ae} \quad (\text{and } L^\infty\text{-weak } *)$$

$$w_\varepsilon^\delta \to v \quad \text{ae} \quad (\text{and } L^\infty\text{-weak } *).$$

Now $\sigma_o(u_\varepsilon^\delta) \to \sigma_o(u)$ ae and, since $\sigma_o(\cdot)$ is continuous, $\sigma_o(u_\varepsilon^\delta)$ is uniformly bounded in L^∞ and, therefore in L^p_{loc} for any $p \geq 1$. These results imply $\sigma_o(u_\varepsilon^\delta) \xrightarrow{\mathcal{D}'} \sigma_o(u)$.

By using (3.1) and (3.2) we have that

$$\frac{\varepsilon}{2} u^\delta_{\varepsilon xx} \xrightarrow{H^{-1}} 0$$

and

$$\delta\sigma_1'(u_\varepsilon^\delta)u^\delta_{\varepsilon x} + \frac{\varepsilon}{2} w^\delta_{\varepsilon xx} \xrightarrow{H^{-1}} 0,$$

and we can conclude that (u,v) is solution of (1.2).

Now, since $v_\varepsilon^\delta = w_\varepsilon^\delta + \frac{\varepsilon}{2} u^\delta_{\varepsilon xx}$, the previous results imply that $v_\varepsilon^\delta \xrightarrow{\mathcal{D}'} v$, and the theorem is proved.

REFERENCES

[1] K. N. CHUEH, C. C. CONLEY and J. A. SMOLLER, Positively invariant regions for systems of nonlinear diffusion equations, Indiana University Mathematics Journal, vol. 26, Nº 2, (1977), 372 - 392.

[2] R. J. DI PERNA, Convergence of approximate solutions to conservation laws, Arch. Rational Mech. Anal. vol. 82, (1983), 27 - 70.

[3] F. MURAT, Compacité par compensation, Ann. Scuola Norm. Sup. Pisa, Sic. Fis, Math., 5 (1978), 489 - 507.

[4] F. MURAT, L'injection du cône positif de H^{-1} dans $W^{-1,q}$ est compacte pour tout $q < 2$, J. Math. Pures et appl., 60, (1981), 309 - 322.

[5] M. SLEMROD, Lax-Friedrichs and the viscosity - capillarity criterion, Proc. of U.W.Va. Conference on Physical Partial Differential Equations, July 1983, to appear.

[6] J. A. SMOLLER, Shock Waves and Reaction - Diffusion Equations, Springer Verlag, 1982.

[7] L. TARTAR, Compensated compactness and applications to partial differential equations, in Nonlinear Analysis and Mechanics: Heriot-Watt Symposion, vol. IV, Research Notes in Mathematics, 39, R. J. Knops, Ed., Pitman Publ. Co., 1979.

[8] L. TARTAR, Une nonvelle methode de resolution d'equations aux derivees partielles nonlineares, Lectures Notes in Math., vol. 665, Springer Verlag, 1, (1977), 228 - 241.

MATHEMATICAL ASPECTS OF THE MINIMUM CRITICAL MASS PROBLEM

EDUARDO BRIETZKE
Instituto de Matemática da UFRGS
Av. Bento Gonçalves, 9500
91.500 Porto Alegre, RS
Brazil

and

PEDRO NOWOSAD
Instituto de Matemática Pura e Aplicada
Estrada Dona Castorina, 110
22.460 Rio de Janeiro, RJ
Brazil

The purpose of these notes is to present a mathematically rigorous analysis of the problem of minimizing the mass of fuel in a nuclear reactor. The model considered is the multi-group approximation with extrapolated boundary under the cascade assumption. This means that the neutrons are divided into $m+1$ levels of energy with flux densities $\varphi_o, \varphi_1, \ldots, \varphi_m$ from the slow up to the fast neutrons respectively. If the domain of the reactor is $\mathcal{D}$ it is assumed that the fluxes become zero at an extrapolated boundary $\partial\Omega$ of a slightly larger domain $\Omega \supset \bar{\mathcal{D}}$ and that each flux satisfies a diffusion equation in Ω

$$\text{(1)} \qquad \frac{\partial \varphi_i}{\partial t} + L_i \varphi_i = S_i , \quad i = 0,1,\ldots,m,$$

where L_i is a second-order elliptic differential operator and S_i the external source. The cascade assumption means that when the neutrons of a given group φ_i are slowed down by the moderator, they go into the next slowest group φ_{i-1}. Therefore for $i = 2,\ldots,m$,

$S_{i-1} = c_i \varphi_i$, where $0 < c_i < 1$ are constants. For $i = 1$ the external source is composed of a part $c_o \varphi_1$ due to the slowing-down process and a negative source $-\varphi_o \sigma$ due to the absorption of slow neutrons by the nuclear fuel, where σ denotes the fuel density. So $S_o = c_o \varphi_1 - \varphi_o \sigma$.

For each neutron absorbed in the fuel an average of η ($\eta \cong 2$) fast neutrons are produced by fission. Hence $S_m = \eta \varphi_o \sigma$.

We are interested in the stationary state of this feed-back process: $\frac{\partial \varphi_i}{\partial t} = 0$, $i = 0,1,\ldots,m$. In this case if K_i denotes the Green function operator for L_i under Dirichlet boundary conditions our equations read (incorporating the constants c_i's into the operators K_i for simplicity)

$$\begin{cases} \varphi_m = K_m S_m = \eta K_m \sigma \varphi_o \\ \varphi_i = K_i S_i = \eta K_i \ldots K_m \sigma \varphi_o , \qquad i = 1,\ldots,m \\ \varphi_o = K_o(\varphi_1 - \sigma \varphi_o) = (\eta K_o \ldots K_m - K_o) \sigma \varphi_o \end{cases}$$

Denoting

(2) $$K = \eta K_o \ldots K_m - K_o ,$$

the equilibrium condition reads

(3) $$\varphi = K \sigma \varphi,$$

writing simply φ for φ_o.

Introducing the total mass $\mu = \int d\sigma$ and the specific density $\rho = \frac{\sigma}{\mu}$, (3) becomes

(4) $$\varphi = \mu K \rho \varphi .$$

By assumption ρ is a probability measure and $\varphi \geq 0$.

The optimal problem is then: Find among all non-trivial solutions $\rho, \varphi, \mu > 0$ for which (4) makes sense the one (ones) for which μ is an absolute minimum, if any.

From now on we will write μ_o, ρ_o and φ_o when the minimum is attained.

This problem has long ago been considered by physicists, mainly by Goertzel [4]. Goertzel by an informal analysis, concluded that when μ is minimized the corresponding φ_o satisfies the so called flat-flux condition:

(5) $0 \le \varphi_o \le 1, \quad \varphi_o \equiv 1$ on supp ρ (after normalization).

Goertzel did not prove the existence of an optimal density. The flat-flux condition only says that μ becomes stationary. Goertzel also proposed analytical methods to find explicit solutions when the problem has plane, cilindrical or spherical symmetry. But for that he made the following a priori suppositions: ρ is absolutely continuous with respect to the Lebesgue measure, ρ has the same type of symmetry as the data and its support has no holes.

The necessity of a mathematically rigorous analysis of this problem is evident when we consider a similar problem whose solution does not have the properties assumed to hold by Goertzel. Suppose we want to distribute a fixed amount of mass in a vibrating string with fixed end-points in such a way as to minimize the string's fundamental tone μ. The problem is again to minimize $\mu > 0$ in

$$\varphi = \mu K \rho \varphi , \quad \varphi \ge 0 , \quad \varphi \ne 0 ,$$

where K is the Green operator of $-\frac{d^2}{dx^2}$. The minimum μ in this case is attained when ρ is the Dirac measure at the mid-point of the string.

The clue to understand why in one problem the solution is

absolutely continuous while in the other it is singular is given in [3] where it is shown that, for ρ_o optimal, $L^2(\text{supp } \rho_o)$ is the closed linear spann of the constant function 1 and the non-positive eigenspace of the symmetric operator K. In the case of the vibrating string $K \geq 0$, so that $L^2(\text{supp } \rho_o)$ is one-dimensional and hence $\text{supp } \rho_o$ consists of a single point. In the case of the reactor, K being an integral operator whose kernel is negatively unbounded near the diagonal, has infinite-dimensional negative eigenspace. Hence absolutely continuous ρ's might be optimal.

Statement of the Problem. Let Ω be a bounded domain in R^N of Schauder class $C^{(1,\lambda)}$. Let $L_i = -\sum_{k,j=1}^{N} \frac{\partial}{\partial x_j} a^i_{kj} \frac{\partial}{\partial x_k} + c^i$, $i = 0,1,\ldots,m$ be uniformely elliptic, $c^i \geq 0$. Then $K_i = L_i^{-1}: H^{-1}(\Omega) \to H^1_0(\Omega)$ are continuous and there exist Green functions $g_i(x,y) \geq 0$ such that

$$(6) \qquad \begin{cases} g_i(x,y) \leq \dfrac{M}{|x-y|^{N-2}} & (N \geq 3) \\ g_i(x,y) \leq M(1 + |\ln|x-y||) & (N = 2) \end{cases}$$

and

$$K_i f(x) = \int_\Omega g_i(x,y) f(y) dy \qquad \forall f \in C^{(0,\lambda)}(\bar{\Omega}).$$

Let

$$K = K_o K_1 \ldots K_m - K_o.$$

The operator $K_o K_1 \ldots K_m$, acting on $C(\bar{\Omega})$, is an integral operator which, when $m+1 > \frac{N}{2}$, has a bounded kernel, as the result of the composition of $m+1 > \frac{N}{2}$ weakly singular kernels satisfying (6).

Denote by

$$\pi = \{\rho \in H^{-1}(\Omega) \mid \rho \geq 0, \ \int d\rho = 1\}$$

$$Lip_c = \{f \in C_c(\Omega) \mid f \ \text{is Lipschitz-continuous}\}$$

$$T_\rho: Lip_c \subset H_0^1(\Omega) \to H^{-1}(\Omega), \qquad T_\rho\varphi = \rho\varphi .$$

T_ρ is closable. Let $\overline{T}_\rho$ be its closure. Write $\rho\varphi$ for $\overline{T}_\rho\varphi$. By $\rho\varphi \in H^{-1}(\Omega)$ we mean that $\varphi \in \text{Dom}\,\overline{T}_\rho$.

Denote by

$$\pi^+ = \{\rho \in \pi \mid \exists\mu > 0, \ \exists\varphi \in H_0^1(\Omega), \ \varphi \geq 0, \ \varphi \not\equiv 0, \ \text{s.t.} \ \rho\varphi \in H^{-1} \ \text{and} \ \mu K\rho\varphi = \varphi\}.$$

Our original problem is the following

<u>Optimization Problem</u>. <u>Find</u> $\rho_o \in \pi^+$ <u>such that</u>

$$\mu_o = \mu(\rho_o) = \inf_{\rho\in\pi^+} \mu .$$

Notice that this is a very difficult problem because there is no simple characterization of the set π^-.

When K is symmetric, the optimization problem can be shown to be equivalent to the much simpler

<u>Variational Problem</u>. <u>Find</u> $\hat{\rho}_o \in \pi$ <u>such that</u>

$$Q(\hat{\rho}_o) = \sup_{\rho\in\pi} Q(\rho) ,$$

where $Q(\rho) = \langle\rho, K\rho\rangle$. The brackets here stand for the duality between $H^{-1}(\Omega)$ and $H_0^1(\Omega)$.

<u>Theorem</u>. <u>Under the assumptions above, if</u>

(7) K <u>is symmetric</u>, i.e. $\langle K\sigma, \rho\rangle = \langle\rho, K\sigma\rangle, \quad \forall \rho, \ \sigma \in H^{-1}(\Omega)$,

(8) $m+1 > \frac{3N+2}{4}$,

(9) $a_{kj}^{i}, \quad c^{i} \in C^{N+1}(\bar{\Omega})$,

(10) $\exists \rho \in \pi$ <u>such that</u> $Q(\rho) > 0$,

<u>then</u>

(11) <u>the optimization and the variational problems are equivalent and have a solution</u>,

(12) ρ_o <u>is absolutely continuous with respect to the Lebesgue measure and</u> $\rho_o \in L^\infty$.

(13) supp ρ_o <u>is bounded away from</u> $\partial\Omega$,

(14) $\varphi_o \in C^{(1,\lambda)}(\bar{\Omega}), \quad \forall \lambda \in (0,1)$,

(15) $\varphi_o \equiv 1$ <u>on</u> supp ρ_o, $0 \leq \varphi_o \leq 1$ <u>in</u> Ω (<u>after normalization</u>).

In this theorem, condition (8) meaning that the number of energy levels in the model is sufficiently high, in principle corresponds to a better approximation to the real problem.

The conclusion that supp ρ_o is bounded away from $\partial\Omega$ is important because the extrapolated boundary approximation is only valid away from the fuel.

The proof of (11), (13) and (15) appeared in [1].

We now indicate how to prove the rest of the theorem.

Assume by contradiction, that the optimal solution ρ_o is not absolutely continuous with respect to the Lebesgue measure. Then there is a compact subset S_o of S = supp ρ_o with zero Lebesgue measure but with $\rho_o(S_o) = \delta > 0$.

Let

(16) $$L = -\sum_{i,j} \frac{\partial}{\partial x_j} \left(a_{ij}^{o} \frac{\partial}{\partial x_i}\right).$$

Consider equation

(17) $$L\,\varphi_o = \mu_o K_1 \ldots K_m \rho_o - c^o\,\varphi_o - \mu_o\,\rho_o \equiv h - \mu_o\,\rho_o\,.$$

Since $h \in L^\infty(\Omega)$, there is an open neighbourhood N of S_o such that

(18) $$\int_N |h| < \frac{\mu_o\,\delta}{4} \quad \text{and}$$

(19) $$\int_N d(\rho_o)_a < \frac{\delta}{4}\,,$$

where $\rho_o = (\rho_o)_a + (\rho_o)_s$ is the decomposition of ρ_o into its absolutely continuous and singular Lebesgue components.

We can take N with smooth boundary and such that $\overline{N} \subset \Omega$, in view of (13).

Consider now any smaller open neighbourhood V of S_o again with smooth boundary ∂V and such that $\overline{V} \subset N$. Let $N_o = N \setminus \overline{V}$. Consider the capacitary potential $\tilde{u} \in H_o^1(N)$ of $\overline{V}$ with respect to the operator L in N. Recall that $\tilde{u}$ is the element of $H_0^1(N)$ which minimizes the integral $\int_N \sum_{i,j} a_{ij}^o\, \varphi_{x_i}\, \varphi_{x_j}$ in the collection of elements $\varphi \in H_o^1(N)$ satisfying $\varphi \geq 1$ in $\overline{V}$ in the $H_o^1(N)$-sense (i.e. such that there is a sequence $\theta_n \in C_c^\infty(N)$, $\theta_n \geq 1$ in $\overline{V}$ with $\theta_n \to \varphi$ in $H_o^1(N)$.). Moreover there exists a measure $\sigma \geq 0$, $\sigma \in H^{-1}(N)$, with supp $\sigma \subset \partial V$ such that

$$L\,\tilde{u} = \sigma\,,$$

$$0 \leq \tilde{u} \leq 1 \quad \text{a.e. in } N,$$

$\tilde{u} = 1$ on V in the $H_o^1(N)$-sense, i.e. $\tilde{u} \geq 1$ and $\tilde{u} \leq 1$ in V in the $H_o^1(N)$-sense.

Putting $u = \tilde{u}|_{N_o}$, we have

$$u \in C^1(\overline{N}_o)$$

$$Lu = 0 \quad \text{in} \quad N_o$$

$$u|_{\partial N} = 0$$

$$u|_{\partial V} = 1$$

The capacity of $\overline{V}$ with respect to L in N is the number $\mathrm{cap}_N \overline{V} =$ $= \int_N \sum_{i,j} a^o_{ij} \tilde{u}_{x_i} \tilde{u}_{x_j} = \int_{\partial V} \sum_{i,j} a^o_{ij} u_{x_i} n_j$, where n stands for the unit outward normal.

We extend $\tilde{u}$ to a function $\hat{u}$ on all of $\overline{\Omega}$ by setting $\hat{u} = 0$ on $\Omega \setminus N$. Thus $u \in C(\overline{\Omega}) \cap H^1_o(\Omega)$.

Since $L\,\varphi_o$ is a bounded Radon measure in Ω and $\hat{u} \in C(\overline{\Omega})$ so is $\hat{u}\, L\, \varphi_o$ a bounded Radon measure in Ω and thus, because $\hat{u} = 0$ outside N:

(20) $$\langle L\,\varphi_o, \hat{u}\rangle = \int_\Omega \hat{u}\, L\,\varphi_o = \int_N \tilde{u}\, L\,\varphi_o =$$

$$= \int_N \tilde{u}[h - \mu_o d(\rho_o)_a] - \int_N \tilde{u}\,\mu_o\, d(\rho_o)_s \le \frac{\mu_o\delta}{2} - \mu_o\delta = -\frac{\mu_o\delta}{2}.$$

In [2] it is shown how to obtain a sequence $\varphi_n \in C^N_c(\Omega)$ satisfying

(i) $0 \le \varphi_n \le 1 + \frac{1}{n}$ (ii) $\varphi_n(x) > \varphi_{n+1}(x) \quad \forall\, x \in \Omega$

(iii) $\varphi_n \to \varphi_o$ a.e. (iv) $\varphi_n \to \varphi_o$ in $H^1_o(\Omega)$.

A smooth version of the mollifiers introduced in [6] is used, which are specially adapted to the operator L and the domain Ω.

Since $\varphi_n > 1$ in S we can choose open neighbourhoods V_n of S_o such that

$$\overline{V}_n \subset N,$$

$$\varphi_n(x) \geq 1 \quad \forall\, x \in V_n ,$$

$$V_{n+1} \subset V_n ,$$

$$\partial V_n \quad \text{smooth.}$$

Consider then the capacitary potentials $\tilde{u}_n$ of $\overline{V}_n$ with respect to the operator L in N and their extensions $\hat{u}_n$ as described above.

We have

$$\|\hat{u}_n\|^2_{H^1_o(\Omega)} = \|\tilde{u}_n\|^2_{H^1_o(N)} \leq q\cdot \operatorname{cap}_N \overline{V}_n \leq q\cdot \operatorname{cap}_N \overline{V}_1 ,$$

with q a constant.

Therefore there exists a subsequence, still denoted by $\hat{u}_n$, and an element $\hat{u}_o \in H^1_o(\Omega)$ such that $\hat{u}_n \rightharpoonup \hat{u}_o$ in $H^1_o(\Omega)$.

Now since $\varphi_n \to \varphi_o$ in $H^1_o(\Omega)$ and $\hat{u}_n \rightharpoonup \hat{u}_o$ in $H^1_o(\Omega)$ the following limit exists:

$$\text{(21)} \qquad \lim_{n\to\infty} \langle L\, \varphi_n, \hat{u}_n\rangle = \langle L\, \varphi_o, \hat{u}_o\rangle = \lim_{n\to\infty} \langle L\, \varphi_o, \hat{u}_n\rangle \leq -\frac{\mu_o \delta}{2}$$

because (20) applies to $\hat{u}_n$.

On the other hand

$$\text{(22)} \qquad \langle L\, \varphi_n, \hat{u}_n\rangle = \int_\Omega \hat{u}_n\, L\, \varphi_n = \int_\Omega \sum_{i,j} a^o_{ij} (\hat{u}_n)_{x_i} (\varphi_n)_{x_j} =$$

$$= \int_{N\setminus\overline{V}_n} \sum_{i,j} a^o_{ij} (u_n)_{x_i} (\varphi_n)_{x_j}$$

because $(u_n)_{x_i} = 0$ a.e. on $(\Omega\setminus N) \cup \overline{V}_n$.

However since $u_n \in C^1(\overline{N}_n)$, where $N_n = N\setminus\overline{V}_n$, and since $\varphi_n \in C^1(\Omega)$ and ∂N and ∂V_n are smooth, we can integrate the last term in (22) by parts to get

$$\langle L\,\varphi_n, \hat{u}_n \rangle = \int_{\partial N \cup \partial V_n} \sum_{i,j} \varphi_n\, a^o_{ij} (u_n)_{x_i}\, n_j \ .$$

Since $L\,u_n = 0$ on N_n, $u|_{\partial N} = 0$, $u|_{\partial V_n} = 1$ and $u_n \in C^1(\overline{N}_n)$, Hopf's maximum principle applies so that

$$\sum_{i,j} a^o_{ij} (u_n)_{x_i}\, n_j > 0 \quad \text{on} \quad \partial V_n \ ,$$

$$\sum_{i,j} a^o_{ij}\, (u_n)_{x_i}\, n_j < 0 \quad \text{on} \quad \partial N \ ,$$

and since $1 \le \varphi_n \le 1 + \frac{1}{n}$ in V_n, we finally get

$$\langle L\,\varphi_n, \hat{u}_n \rangle \ge \left\{ \int_{\partial V_n} + \left(1 + \frac{1}{n}\right) \int_{\partial N} \right\} \sum_{i,j} a^o_{ij} (u_n)_{x_i}\, n_j =$$

$$= \int_{\partial N_n} \sum_{i,j} a^o_{ij} (u_n)_{x_i}\, n_j + \frac{1}{n} \int_{\partial N} \sum_{i,j} a^o_{ij} (u_n)_{x_i}\, n_j =$$

$$= -\int_{N_n} L\,u_n - \frac{1}{n} \operatorname{cap}_N \overline{V}_n = -\frac{1}{n} \operatorname{cap}_N \overline{V}_n \ge -\frac{1}{n} \operatorname{cap}_N \overline{V}_1 \ .$$

We thus obtain $\lim\limits_{n\to\infty} \langle L\,\varphi_n, \hat{u}_n \rangle \ge 0$, which contradicts (21), proving the absolute continuity of ρ_o.

In [2] we show that there is a representative $\hat{\varphi}_o$ in the H^1_o-class of φ_o such that $\hat{\varphi}_o \in C_o(\overline{\Omega})$ and $\hat{\varphi}_o = 1$ on $S = \operatorname{supp} \rho_o$.

Let $C = \{x \in \Omega \mid \hat{\varphi}_o(x) = 1\} \supset S$. Clearly C is compact. By assumptions (8) and (9) $\varphi_o \in C^N(\Omega \setminus S)$. It follows from the Morse-Sard Theorem that there is a sequence $1 > \epsilon_n \downarrow 0$ such that

$$S_n = \{x \in \Omega \mid \hat{\varphi}_o(x) \ge 1 - \epsilon_n\}$$

is a decreasing sequence of closed sets whose boundaries ∂S_n are compact orientable differentiable manifolds of dimension $N-1$ lying in C^c.

Define

$$\hat{\varphi}_n = \min\{\varphi_o, 1-\epsilon_n\} \cdot$$

It is shown in [2] that

$$(23) \qquad L\,\hat{\varphi}_n - \chi_{S_n^c}\, L\,\hat{\varphi}_o = \delta(\chi_{\partial S_n} - 1) \cdot \sum_{j,k} \frac{a^o_{jk} \frac{\partial \hat{\varphi}_o}{\partial x_j} \frac{\partial \hat{\varphi}_o}{\partial x_k}}{\|\operatorname{grad} \hat{\varphi}_o\|} \equiv T_n \,.$$

By the coercivity of L, $T_n \geq 0$.

It is clear that as distributions as $n \to \infty$

$$L\,\hat{\varphi}_n \longrightarrow L\,\hat{\varphi}_o$$

$$\chi_{S_n^c}\, L\,\hat{\varphi}_o \longrightarrow \chi_{C^c}\, L\,\hat{\varphi}_o \,,$$

this last one because $S_n \downarrow C$ as $n \to \infty$ and because $L\,\hat{\varphi}_o \in L^1(\Omega)$ in view of (17) and of the absolute continuity of ρ_o.

Thus from (23) we conclude that, as distributions

$$(24) \qquad \begin{aligned} &\chi_C\, L\,\hat{\varphi}_o = \lim_{n\to\infty} T_n \geq 0, \quad \text{or} \\ &\chi_C(\mu_o\, K_1 \ldots K_m\, \rho_o - c^o\, \hat{\varphi}_o) - \mu_o\, \rho_o \geq 0 \end{aligned}$$

Since $K_1 \ldots K_m$ is an integral operator with a bounded kernel, (24) implies that $\rho_o \in L^\infty(\Omega)$.

In particular

$$L\,\hat{\varphi}_o = \mu_o\, K_1 \ldots K_m\, \rho_o - \mu_o\, \rho_o \in L^p(\Omega) \qquad \forall\, p \geq 1.$$

It thus follows from the ellipticity of L_o, that [5, p.170]

$$\hat{\varphi}_o \in C^{1+\lambda}(\Omega) \cap H^2(\Omega).$$

Problems with symmetry.

Assume the domain Ω and the operators K_i are invariant under a smooth group G of rigid motions. For any optimal solution let $C_o = \{x \in \Omega \mid \varphi_o(x) = 1\}$. In this case one can prove that C_o is also invariant under G, [2]. Integrating the basic equation $\mu K\rho_o(g) = \varphi(g)$ with respect to the Haar measure on G we prove the existence of a G-invariant optimal solution $\bar{\rho}_o = \int \rho_o(g)dg$. This fact was also implicitly assumed by Goertzel and allowed him to reduce the problem to a one-dimensional problem in the particular cases of plane, cilindrical and spherical symmetry.

It is remarkable that if the group G of symmetries is discrete then C_o need not be symmetric.

REFERENCES

[1] BRIETZKE, E. and NOWOSAD, P., Existence of Minimum Critical Mass Solutions for Diffusion Equations of Reactor Theory, Nonlinear Analysis, 9, 849-860 (1985).

[2] BRIETZKE, E. and NOWOSAD, P., Mathematical Aspects of Minimum Critical Mass Problem, 23º Seminário Brasileiro de Análise, Campinas, (1986).

[3] FRIEDLAND, S. and NOWOSAD, P., Extremal Eigenvalue Problems with Indefinite Kernels, Adv. Math. 40, 128-154 (1981).

[4] GOERTZEL, G., Minimum Critical Mass and Flat-Flux, J.Nucl. Energy 2, 193-201 (1956).

[5] LEWY, H. and STAMPACCHIA, G., On the Regularity of the Solution of a Variational Inequality, Comm.Pure Appl.Math. 22, 153-188 (1969).

[6] LITTMAN, W., STAMPACCHIA, G. and WEINBERGER, H.F., Regular Points for Elliptic Equations with Discontinuous Coefficients, Annali Scu Norm.Sup. Pisa 17, 43-77 (1963).

ASYMPTOTIC TIME EVOLUTIONS FOR STRICTLY OUTGOING MULTIPARTICLE QUANTUM SYSTEMS WITH LONG-RANGE POTENTIALS

Volker ENSS

Institut für Mathematik I, Freie Universität,
Arnimallee 2-6, D-1000 Berlin 33, West-Germany.

Abstract.

Strictly outgoing multiparticle scattering states are characterized by the phase space localization of bounded subsystems relative to each other. We prove that for late times the free relative motion of the clusters is a good approximation of the interacting evolution for late times. If long-range potentials are present we analyze different modified free evolutions and we show where they approximate the true motion. Stronger correlations between position and velocity are derived with their help.

Contents.

I. Introduction.

The aim of scattering theory is to give a hopefully complete classification for the possible types of temporal asymptotic behavior for an evolving system. Here we analyze unitary linear evolutions as they arise e.g. in the quantum mechanical description of colliding atoms and molecules: the *multiparticle Schrödinger equation* in an (infinite dimensional) Hilbert space. The solutions show a very rich structure and the problem is particularly attractive because intuitive reasoning on the basis of physical experience goes hand in hand with the rigorous mathematical analysis. Even most of the intermediate mathematical steps can be understood physically, at least qualitatively. Moreover, the system is simpler and better tractable than its limiting case, the (finite dimensional) nonlinear classical N-body problem as it arises e.g. in celestial mechanics.

The problem of asymptotic completeness for systems of four or more particles is not yet settled in our own approach, see however Sigal and Soffer [12] for recent work in the short-range case which is closely related in spirit but different in the detailed techniques. The major part of our lectures at the meeting is contained in [4] (or see [5], [7] for the complete proofs) and in [3]. Therefore we use these notes to give some details of material which could only be mentioned briefly in the lectures and which is not yet available in print. We describe the asymptotic time evolution of states with specific bound state content and localization in classical phase space. A result obtained by using asymptotic observables [4] shows that a scattering state at suitable late times τ can be decomposed into a finite number of components with special properties (this and the statements to follow should be understood as "up to an arbitrarily small error"). For each of these components the particles are grouped into clusters (subsystems) and the internal state of every subsystem is a fixed bound state (eigenvector of the subsystem Hamiltonian). The clusters are far separated in configuration space. Moreover the relative velocity of the centers of mass of the clusters at time τ is restricted to a small ball in velocity space which does not contain the origin. The relative position of the centers of mass of the clusters is very well concentrated in that region where the quotient of the position and the time τ (roughly the average velocity between zero and τ) belongs to the same ball. In other words, the clusters are approximately localized such as if they had started at time zero at the origin and moved freely relative to each other with given velocities ("small bang picture"). The correlation between localization in velocity- (momentum-) space and configuration space amounts to a localization in classical phase space for the relative motion of the bounded clusters.

With this result as a starting point one has to distinguish two cases which are different mathematically and physically. The typical case is that none of the relative velocities between a pair of clusters vanishes if the total relative velocity vector for the cluster lies in the small ball. We will see below - the main content of the paper - that for such states the separation of the clusters continues to increase and that the clusters essentially move freely relative to each other. Therefore we call such states "*strictly outgoing*". The remaining degenerate case is that one or more pairs of clusters (but not all) may have zero relative velocities. For two cluster states this cannot happen and for three-particle systems one can show with different means that by a suitable choice of the parameters the degenerate components are smaller than any $\varepsilon > 0$. The same applies for systems where the only decomposition into three or more bounded clusters is the trivial one: all particles are separated. For general systems of $N \geq 4$ particles a direct proof is still missing that the parameters can be chosen in such a way. Given the results mentioned above this property is necessary and sufficient for asymptotic completeness (in its strong form).

For precise statements we need some notation which is given below. The third section gives the definition of strictly outgoing multi- particle states. Then we turn to the construction of intermediate and asymptotic time evolutions and we show that they well approximate the true evolution on these states. The potentials may be of short or of long range. Actually the treatment of long-range potentials is the main topic of this paper.

II. Notation - the Model.

We study a system of N quantum mechanical particles of masses m_i with positions $x_i \in I\!R^\nu$ which interact by pair potentials V^α. If the free center of mass motion is separated off then the state space is the Hilbert space $\mathcal{H} = L^2(I\!R^{\nu(N-1)})$. For the general setup see e.g. [11]. The time evolution is generated by the self-adjoint Hamiltonian operator

$$H = H_0 + V = H_0 + \sum_\alpha V^\alpha. \tag{2.1}$$

The free Hamiltonian H_0 is obtained from $\tilde{H}_0 = \sum_i (2m_i)^{-1}\Delta_i$ by separating off the center of mass motion, it is a Laplacian-type operator in (clustered) Jacobi coordinates. The pair potentials may contain parts of short and of long range and we assume (for simplicity of presentation only) that they are bounded multiplication operators in configuration space. The inclusion of more general potentials which are operator- or form-bounded relative to the pair-Laplacian is a well known simple exercise. We require for any pair $\alpha = (i,j)$, $x^\alpha = x^i - x^j$, the decay properties

$$V^\alpha \equiv V^\alpha(x^\alpha) = V_s^\alpha(x^\alpha) + V_\ell^\alpha(x^\alpha), \tag{2.2}$$

$$|V_s^\alpha(x^\alpha)| \le const\ (1 + |x^\alpha|)^{-(1+\varepsilon)}, \quad \varepsilon > 0, \tag{2.3}$$

$$V_\ell^\alpha \in C^1(I\!R^\nu), \quad V_\ell^\alpha(x^\alpha) \to 0 \text{ as } |x^\alpha| \to \infty, \tag{2.4}$$

$$|(\nabla V_\ell^\alpha)(x^\alpha)| \le const\ (1 + |x^\alpha|)^{-(\delta+3/2)}, \quad 0 < \delta \le 1/2. \tag{2.5}$$

Without loss of generality [9] the splitting can be made such that V_ℓ^α is smooth and in addition

$$|(D^m V_\ell^\alpha)(x^\alpha)| \le const(m) \cdot (1 + |x^\alpha|)^{-(\delta+1+|m|/2)} \tag{2.6}$$

for any multiindex m with $|m| \geq 1$ and $\delta > 0$. We will use that choice in the sequel. A further implicit condition on the potentials is (2.12) below.

We will study the components of the state corresponding to an arbitrary but fixed *decomposition* of the N particles into $1 < k \leq N$ non-empty clusters (subsystems). According to the decomposition we choose clustered Jacobi coordinates, i.e. Jacobi coordinates inside each of the clusters c_j containing $|c_j|$ particles (this gives $X(c_j) \in \mathbb{R}^{\nu(|c_j|-1)}$) and Jacobi coordinates $Y \in \mathbb{R}^{\nu(k-1)}$ for the relative position of the centers of mass of the clusters. The state space $\mathcal{H}$ can be written as a tensor product

$$\mathcal{H} = \bigotimes_{j=1}^{k} \mathcal{H}(c_j) \quad \bigotimes \mathcal{H}^{rel}$$

of the corresponding L^2-spaces, $\mathcal{H}(c_j) = \mathbb{C}$ if $|c_j| = 1$.

The internal Hamiltonian $H(c_j)$ of a cluster c_j contains a free part and all potentials between particles inside the cluster. Its explicit form does not matter here; $H(c_j) = 0$ if $|c_j| = 1$. We will always use the same symbol for an operator acting on a factor space (like $\mathcal{H}(c_j)$ above) and on the whole space $\mathcal{H}$. In each cluster we choose and fix an arbitrary eigenvector of $H(c_j)$ in $\mathcal{H}(c_j)$, i.e. each subsystem is in a particular bound state. The projection operator $\mathbf{P}$ specifies the product state in $\otimes\mathcal{H}(c_j)$, as an operator on $\mathcal{H}$ its range is isomorphic to $\mathcal{H}^{rel}$ which describes the degrees of freedom corresponding to the relative motion of the given bounded subsystems. E is the sum of the eigenvalues:

$$\sum_j H(c_j)\,\mathbf{P} = E\,\mathbf{P}. \tag{2.7}$$

The meaning of the ν-dimensional parts of $Y = (Y_{12},\ Y_3,\ \ldots,\ Y_k)$ depends on the order in which the clusters are arranged when forming Jacobi coordinates. We denote by $Y_{ij} \in \mathbb{R}^{\nu}$ the position of the i'th cluster relative to the j'th. Y_3 is the position of the third cluster relative to the center of mass of the first two, etc. Similarly we have reduced masses μ_{ij}, μ_3, etc. We will use the scalar product in $\mathbb{R}^{\nu(k-1)}$

$$< Y, Y' > := \mu_{12}\, Y_{12} \cdot Y'_{12} + \mu_3\, Y_3 \cdot Y'_3 + \cdots + \mu_k\, Y_k \cdot Y'_k \tag{2.8}$$

and denote by $\|\cdot\|$ the corresponding norm in $\mathbb{R}^{\nu(k-1)}$. They are invariant under reordering of the clusters. Analogously for $X(c_j) \in \mathbb{R}^{\nu(|c_j|-1)}$ inside the clusters. (Also the same $\|\cdot\|$ denotes norms in Hilbert spaces.) The dot means the Euclidean inner product in $\mathbb{R}^{\nu}$ with norm $|\cdot|$.

The corresponding velocity operators are $Q := (Q_{12}, Q_3, \ldots, Q_k)$ where $Q_{ij} = -i(\mu_{ij})^{-1} grad_{Y_{ij}}$, $Q_3 = -i(\mu_3)^{-1} grad_{Y_3}$, etc.

The kinetic energy of the clusters is

$$\mathbf{T} := \frac{1}{2} < Q, Q > . \tag{2.9}$$

Its explicit form as a Laplacian type operator depends on the order of the clusters. It satisfies $i[\mathbf{T}, Y] = Q$. The decomposition Hamiltonian is defined as

$$H(d) := \sum_{i=1}^{k} H(c_i) + \mathbf{T}. \tag{2.10}$$

It simplifies on the range of $\mathbf{P}$ to $H(d)\,\mathbf{P} = (E + \mathbf{T})\,\mathbf{P}$ and

$$e^{-iH(d)\,t}\,\mathbf{P} = e^{-iEt}\,e^{-i\mathbf{T}t}\,\mathbf{P}. \tag{2.11}$$

Finally the totally interacting Hamiltonian H can be expressed as

$$H = H(d) + \sum_{\beta}{}'' \; V^{\beta}$$

$$= (E + \mathbf{T}) + \sum_{\beta}{}'' \; V^{\beta} \quad \text{on} \quad \mathrm{ran}\mathbf{P}.$$

Here $\sum_{\beta}''$ means the summation only over those pairs β where the particles lie in different clusters. We will show in this paper that $\sum_{\beta}'' V^{\beta}$ has a minor effect on the asymptotic time evolution of strictly outgoing scattering states. The latter will be defined in the next section.

If the eigenvalue is not a threshold value then the corresponding eigenfunction is known to have exponential decay [8]. However, at thresholds the decay may be slow and we have to impose a technical condition like (2.12) to ensure that it is not too slow.

$$\sum_{j} \Big\| \, \|X(c_j)\|^{(2+\varepsilon)} \, \mathbf{P} \Big\| < \infty\,, \quad \varepsilon > 0. \tag{2.12}$$

Much weaker decay properties than the simple (2.12) are sufficient. We do not discuss them here because we hope to eliminate this condition eventually.

Let the particles in the pair β belong to different clusters i, j. Then we denote as

$$y^{\beta} := \pm Y_{ij} \tag{2.13}$$

the corresponding inter-cluster coordinate: $x^{\beta} = \pm Y_{ij}$ + [a bounded linear function of $X(c_i)$ and $X(c_j)$]. It follows easily from (2.12) and (2.3) - (2.5) that

$$\|V_s^{\beta}(x^{\beta})\,\mathbf{P}\,F(|y^{\beta}| > R)\| < const \;\; (1 + R)^{-(1+\varepsilon)}, \tag{2.14}$$

$$\Big\| \Big(V_{\ell}^{\beta}(x^{\beta}) - V_{\ell}^{\beta}(y^{\beta})\Big)\,\mathbf{P}\,F(|y^{\beta}| > R) \Big\| < const \;\; (1 + R)^{-(1+\varepsilon)}. \tag{2.15}$$

(As usual $F(\cdot)$ denotes a spectral projection as indicated in the parentheses, here multiplication with a characteristic function in configuration space.)

Later we will use the properties (2.14) and (2.15). They say that on the range of **P** the interactions can be replaced by a sum of effective long-range *intercluster potentials*

$$\sum_{\beta}{}'' \; V_\ell^\beta(y^\beta) \tag{2.16}$$

(which act only on $\mathcal{H}^{rel}$) and a remainder of terms each of which is of short range with respect to some intercluster separation Y_{ij}.

III. Phase Space Localization and Free Propagation.

We consider smooth cutoff functions $f \in C_0^\infty(\mathbb{R}^{\nu(k-1)})$, $|f| \le 1$, with support contained in a small ball:

$$\operatorname{supp} f \subset \left\{ Q \in \mathbb{R}^{\nu(k-1)} \middle| \, \|Q - Q^0\|^2 < [\min_{i,j} \mu_{ij}](v_0)^2 \right\}. \tag{3.1}$$

The center Q^0 is a point in (velocity) space such that for *ALL* pairs of clusters their relative velocity has positive Euclidean length

$$|Q^0_{ij}| \ge 4\, v_0 > 0 \quad \text{for all } \; i,j. \tag{3.2}$$

For the Q's arising in the support of f (3.1) we then have a minimal speed of $3v_0$ between each pair of clusters.

STRICTLY OUTGOING are the states in the range of

$$f(Q)\, f(Y/\tau)\, \mathbf{P} \tag{3.3}$$

where we are interested in large τ only. Here Q and Y are the velocity and position operators, respectively. Clearly **P** (as defined in (2.7)) commutes with the factors $f(\cdot)$. The method of asymptotic observables shows that Q and Y/τ are close for large times τ. Up to a small error any scattering state can then be decomposed into *finitely many* components of the form (3.3) and a degenerate remainder which hopefully is small for sufficiently small $v_0 > 0$ and arbitrarily large τ.

We will study good approximations for the future time evolution $(t \ge \tau)$ if the starting time τ is large, i.e. for

$$e^{-iH(t-\tau)}\, f(Q)\, f(Y/\tau)\, \mathbf{P}. \tag{3.4}$$

If the inter-cluster potentials are omitted the subsystems move freely relative to each other and their propagation properties in space and time are easy to control by the method of stationary phase.

Proposition 3.1 *Let M, $M' \subset \mathbb{R}^\nu$ be measurable sets and the Euclidean $dist(M + Q^0_{ij}\,t,\ M') - v_0 t - r > 0$. Then for any $l \in \mathbb{N}$, $t, r \geq 0$*

$$\left\| F(Y_{ij} \in M')\, e^{-i\mathbf{T}t}\, f(Q)\, F(Y_{ij} \in M) \right\| \leq C_l\, (1 + t + r)^{-l}. \tag{3.5}$$

For $M, M' \subset \mathbb{R}^{\nu(k-1)}$ and for the invariant distance (2.8) *$dist(M + Q^0 t,\ M') - [\min_{i,j} \mu_{ij}]^{1/2}\, v_0\, t - r > 0$*

$$\left\| F(Y \in M')\, e^{-i\mathbf{T}t}\, f(Q)\, F(Y \in M) \right\| \leq C_l\, (1 + t + r)^{-l}. \tag{3.6}$$

The same holds for the time evolution $\exp\{-i\, H(d)\, t\}$ *if the projection* $\mathbf{P}$ *is used as an additional factor on the right.*

The proof is essentially the same as for the two-body case (see e.g. [2]) and we omit it here. Only the notation is different.

IV. The Intermediate Evolutions.

We construct approximate time evolutions to treat the persistent effects of long-range potentials. They coincide with the free evolution if only short-range forces are present or if the effective inter-cluster potentials are of short range, i. e. if (2.16) vanishes. The latter is the case e.g. for Coulomb systems (atoms, molecules, ions) if at most one subsystem is charged and the others are neutral. A tutorial presentation of these evolutions for two-body systems with a different emphasis is given in [6], the general case for two- and three-body systems is contained in [3], here we present the straightforward generalization to N particles. In this section we show that the intermediate evolutions approximate the true one uniformly in the future. We call them "intermediate" because for even later times the Dollard evolution is a simpler approximation. The proof below will use the intermediate evolution only for an arbitrarily long but finite time interval.

The parameter ϱ has to be chosen according to the decay rate (2.5) of the long-range potential

$$1 < \varrho < 1/(1 - 2\delta). \tag{4.1}$$

Depending on an arbitrary late starting time $\tau \equiv \tau_m = m^{2\varrho} \in \mathbb{R}$ we select a sequence of times

$$\tau_k := k^{2\varrho}, \quad t_k := \tau_{k+1} - \tau_k \sim k^{2\varrho - 1}, \quad k - m \in \mathbb{N} \ \text{ or } \ \mathbb{Z}. \tag{4.2}$$

(The running label $k \in m + \mathbb{Z}$ should not be confused with the number k of clusters used above and occasionally below.)

Since the clusters will be far separated we use truncated effective intercluster potentials of long range (see (2.16))

$$V_k^\beta \equiv V_k^\beta(y^\beta) := V_\ell^\beta(y^\beta)\left[1 - \varphi(y^\beta/v_0\tau_k)\right] \tag{4.3}$$

where $\varphi \in C_0^\infty(I\!R^\nu)$, $|\varphi| \le 1$,

$$\varphi(z) = 1 \text{ if } |z| \le 1/2, \quad \varphi(z) = 0 \text{ for } |z| \ge 1. \tag{4.4}$$

The cutoff potentials have the following simple properties which we will need below:

$$\sup_z |(\nabla V_k^\beta)(z)| \le const\ (\tau_k)^{-(\delta+3/2)}, \tag{4.5}$$

$$\sup_z |(\Delta V_k^\beta)(z)| \le const\ (\tau_k)^{-(\delta+2)}. \tag{4.6}$$

For derivatives of exponential functions we conclude from (4.5)

$$\left\| \left[Q,\ \exp\{-i \sum_\beta{}'' V_k^\beta t\}\right] \right\| \le const \cdot t\ (\tau_k)^{-(\delta+3/2)} \tag{4.7}$$

and the same holds for nice functions of Q like f.

We are prepared to define some *"intermediate time evolutions"*. For our purposes it is sufficient to give them for beginning and final times from the special sequence τ_k (4.2), the generalization to arbitrary times is straightforward.

Definition 4.1 *(a) With* (4.2) *and* (4.3) *let*

$$U(\tau_n, \tau_m) := \exp\{-i \sum_\beta{}'' V_{n-1}^\beta t_{n-1}\}\ e^{-iH(d)t_{n-1}} \times \cdots$$

$$\times \exp\{-i \sum_\beta{}'' V_k^\beta t_k\}\ e^{-iH(d)t_k} \cdots \exp\{-i \sum_\beta{}'' V_m^\beta t_m\}\ e^{-iH(d)t_m}$$

$$=: \prod_{k=m}^{n-1}{}' \exp\{-i \sum_\beta{}'' V_k^\beta t_k\}\ e^{-iH(d)t_k} \equiv \prod_{k=m}^{n-1}{}' U(\tau_{k+1}, \tau_k). \tag{4.8}$$

The prime at the product sign denotes the "time ordering" of the factors as above with increasing times from right to left.
(b) With $f,\ \bar f \in C_0^\infty(I\!R^{\nu(k-1)})$, $\bar f f = f$

$$U_f(\tau_n, \tau_m) := \bar f(Q)\ \bar f(Y/\tau_n)\ U(\tau_n, \tau_m). \tag{4.9}$$

Here f *is the same as in* (3.3) *and the support of* $\bar f$ *is slightly larger, see below.*

At first glance the evolution $U(\tau_n, \tau_m)$ resembles an approximation like the Trotter product formula for $\exp\{-iHt\}$ where potentials of short range have been neglected. There is, however, an additional difference: the times in the factors become *larger*. This is possible on strictly outgoing states because the suppressed potentials and commutator terms become small due to the separation of particles which belong to different clusters. The errors can be controlled if the time intervals are not too long. On the other hand for the propagation properties under the free evolution the longer intervals allow better estimates. This is balanced in the choice (4.2).

For the estimates we need a sequence of auxiliary smooth cutoff functions which we construct now. For given Q^0 as in (3.1), (3.2) we denote as χ_k the characteristic function of the ball around Q^0

$$\left\{ Q \;\Big|\; \|Q - Q^0\|^2 \leq [\min_{i,j} \mu_{ij}]\,(c_k \cdot v_0)^2 \right\},$$

$$c_k := 1 + \frac{3}{c'} \sum^{k} r^{-\varrho} \quad < \; 2. \tag{4.10}$$

The summation over r extends over positive reals $1 \leq r \leq k$ with $r - m \in \mathbb{Z}$ (compare (4.2)). The constant c' is large enough such that the bound 2 is valid for all k. To smooth the projection operators we convolute the characteristic functions:

$$f_k := \chi_k * \psi_k$$

where $\psi \in C_0^\infty(\mathbb{R}^{\nu(k-1)})$, $\psi \geq 0$, $\int \psi = 1$, and $\psi(z) = 0$ for $\|z\| \geq 1$. Then ψ_k is the rescaled function with $\int \psi_k = 1$ and support in a ball of squared invariant radius

$$[\min_{i,j} \mu_{ij}] \cdot \left(\frac{v_0}{c'\,k^{\varrho}} \right)^2 . \tag{4.11}$$

For $\bar{f}$ we replace c_k in (4.10) by $5/2$ and use $(v_0/3)^2$ in (4.11) instead of $(v_0/c'k^{\varrho})^2$.

The relevant properties of these explicit functions are the following:

$$f_k\, f = f, \quad \bar{f}\, f_k = f_k \quad \text{for all } k \geq 1, \tag{4.12}$$

$$f_{k'}\, f_k = f_k \quad \text{for } k' > k. \tag{4.13}$$

The supports are chosen in such a way that with Proposition 3.1 we have suitable propagation properties in space and time. With (3.5) we obtain for all $l \in \mathbb{N}$

$$\left\| F\Big(|Y_{ij}| < 2v_0(\tau_k + t)\Big)\, e^{-i\mathbf{T}t}\, f(Q)\, f_k(Y/\tau_k) \right\| \leq C_l\,(1 + \tau_k + t)^{-l}, \tag{4.14}$$

and

$$\left\| F\Big(|Y_{ij}| < (4/3)v_0(\tau_n + t)\Big)\, e^{-i\mathbf{T}t}\, f(Q)\, \bar{f}(Y/\tau_n)\right\| \le C_l\,(1+\tau_n+t)^{-l}. \tag{4.15}$$

We use (3.6) and (4.12), (4.13) to derive

$$\begin{aligned} &\|[\mathbf{1} - f_{k+1}(Y/\tau_{k+1})]\, e^{-i\mathbf{T}t_k}\, f(Q)\, f_k(Y/\tau_k)\| \\ &\qquad \le C_l\,(1+k)^{-l} \quad \text{for} \quad k \ge 1. \end{aligned} \tag{4.16}$$

Although the gradient of f_k increases like k^{ϱ} as functions of Y the $f_k(Y/\tau_k)$ have decreasing slopes and we get for any $f \in C_0^\infty(\mathbb{R}^{\nu(k-1)})$ the bound of the commutator

$$\|\,[f_k(Y/\tau_k),\, f(Q)]\,\| \le const \cdot k^{-\varrho}. \tag{4.17}$$

With the set of functions f_k one can define another auxiliary time evolution which will mainly be useful in intermediate steps. It is the most complicated one, but it has the advantage to display explicitly the space-time propagation properties.

Definition 4.2 *With f, f_k as given above, τ_k as in* (4.2)

$$\begin{aligned} \tilde{U}(\tau_n,\, \tau_m) &:= f(Q) \prod_{k=m}^{n-1}{}'\, f_{k+1}(Y/\tau_{k+1})\, \exp\{-i \sum_\beta{}''\, V_k^\beta\, t_k\}\, e^{-iH(d)\,t_k} \\ &\equiv f(Q) \prod_{k=m}^{n-1}{}'\, f_{k+1}(Y/\tau_{k+1})\, U(\tau_{k+1},\, \tau_k). \end{aligned} \tag{4.18}$$

Now we are ready to show that U and U_f are good approximate time evolutions on strictly outgoing states uniformly for all future times.

Proposition 4.3 *With the notation as introduced in* (4.8),(4.9),(4.18), *and* (3.3)

$$\sup_{\tau_n > \tau_m} \left\| \left[U(\tau_n, \tau_m) - e^{-iH(\tau_n-\tau_m)}\right] f(Q)\, f(Y/\tau_m)\, \mathbf{P}\right\| \to 0 \;\text{ as }\; \tau_m \to \infty \tag{4.19}$$

and the same for $U_f(\tau_n, \tau_m)$. Moreover

$$\sup_{\tau_n > \tau_m} \left\| \left[\tilde{U}(\tau_n,\, \tau_m) - e^{-iH(\tau_n-\tau_m)}\, f(Q)\right] f(Y/\tau_m)\, \mathbf{P}\right\| \to 0 \;\text{ as }\; \tau_m \to \infty. \tag{4.20}$$

Proof. Our first step is to show

$$\begin{aligned} &\left\| \left[U(\tau_{k+1}, \tau_k) - e^{-iH\,t_k}\right] f(Q)\, f_k(Y/\tau_k)\, \mathbf{P}\right\| \\ &\qquad \le const \cdot t_k\,(\tau_k)^{-(1+\varepsilon)},\; \varepsilon > 0. \end{aligned} \tag{4.21}$$

With the standard Cook argument and (2.11) the norm is bounded by

$$\left\| \int_0^{t_k} dt \, \frac{d}{dt} \left\{ e^{iHt} \exp\{-i \sum_\beta{}'' V_k^\beta \, t\} e^{-iH(d)\,t} \right\} f(Q) \, f_k(Y/\tau_k) \, \mathbf{P} \right\|$$

$$\leq \int_0^{t_k} dt \left\| \left\{ \sum_\beta{}'' \left(V^\beta - V_k^\beta \right) + \left[\mathbf{T}, \, \exp\{-i \sum_\beta{}'' V_k^\beta \, t\} \right] \right\} \mathbf{P} \, e^{-i\mathbf{T}t} \, f \, f_k \right\|. \quad (4.22)$$

The difference of the potentials in (4.22) is a sum $\sum_\beta{}''$ over

$$V_s^\beta(x^\beta) + \left(V_\ell^\beta(x^\beta) - V_\ell^\beta(y^\beta) \right) + \left(V_\ell^\beta(y^\beta) - V_k^\beta(y^\beta) \right)$$

when multiplied by $\mathbf{P}$ the first two summands are of short range in y^β by (2.14) and (2.15). The last one vanishes for $|y^\beta| \geq v_0 \, \tau_k$ by (4.3). The estimate of the commutator term uses (2.9) and (4.5) - (4.7) giving for $0 \leq t \leq t_k$

$$\left\| \left[\mathbf{T}, \, \exp\{-i \sum_\beta{}'' V_k^\beta \, t\} \right] \bar{f}(Q) \right\| \leq const \;\; t_k \cdot (\tau_k)^{-(\delta+3/2)}.$$

Since $f = \bar{f} f$ the factor $\bar{f}(Q)$ can be pulled out of $f(Q)$. Thus the integrand of (4.22) is bounded by

$$\sum_\beta{}'' \left\{ \left\| \left[V_s^\beta(x^\beta) + \left(V_\ell^\beta(x^\beta) - V_\ell^\beta(y^\beta) \right) \right] \mathbf{P} \; F(|y^\beta| > v_0 \, \tau_k) \right\| \right.$$

$$\left. + \, const \left[t_k \, (\tau_k)^{-(\delta+3/2)} + \left\| F(|y^\beta| < v_0 \, \tau_k) \, e^{-i\mathbf{T}t} \, f(Q) \, f_k(Y/\tau_k) \right\| \right] \right\}.$$

Applying also (4.14) we get by integration the claimed estimate for (4.22) and thus for (4.21) by the choice (4.1). The point is that the bound is summable in k. As a next step we show

$$\left\| \left[U(\tau_{k+1}, \tau_k) \, f(Q) - f(Q) \, f_{k+1}(Y/\tau_{k+1}) \, U(\tau_{k+1}, \tau_k) \right] f_k(Y/\tau_k) \right\|$$

$$\leq const \cdot k^{-\varrho}. \quad (4.23)$$

The difference is bounded by

$$\left\| \left[\mathbf{1} - f_{k+1}(Y/\tau_{k+1}) \right] \exp\{-i \sum_\beta{}'' V_k^\beta \, t_k\} \, e^{-i\mathbf{T}t_k} \, f(Q) \, f_k(Y/\tau_k) \right\|$$

$$+ \left\| \left[f(Q), \, \exp\{-i \sum_\beta{}'' V_k^\beta \, t_k\} \right] \right\| + \left\| \left[f(Q), \, f_{k+1}(Y/\tau_{k+1}) \right] \right\|.$$

Estimates for the last and first norms are given in (4.17) and (4.16), respectively, because a function of Y commutes with the unitary exponential multiplication operator. Decay of the second term as $t_k \cdot (\tau_k)^{-(\delta+3/2)}$ follows from (4.7). Thus (4.23) is verified.

We have to estimate the difference of products of (non-commuting) operators whose norm is bounded by one. We use (empty products are $\mathbf{1}$)

$$\Big\|\Big(\prod_{k=m}^{n-1}{}' A_k\Big) f - f \prod_{k=m}^{n-1}{}' B_k\Big\| = \Big\|\sum_{k=1}^{n-1}\Big(\prod_{i=k+1}^{n-1}{}' A_i\Big)\Big(A_k\, f - f\, B_k\Big)\Big(\prod_{j=m}^{k-1}{}' B_j\Big)\Big\|$$

$$\leq \sum_{k=1}^{n-1}\Big\|\Big(A_k\, f - f\, B_k\Big)\Big(\prod_{j=m}^{k-1}{}' B_j\Big)\Big\|. \tag{4.24}$$

For B_k we insert the factors in $\tilde{U}$. Setting for A_k the factors in U we get

$$\|U(\tau_n, \tau_m)\, f(Q)\, f_m(Y/\tau_m) - \tilde{U}(\tau_n, \tau_m)\, f_m(Y/\tau_m)\|$$

$$\leq \sum_{k=m}^{n-1}\Big\|\Big[U(\tau_{k+1}, \tau_k)\, f(Q) - f(Q)\, f_{k+1}(Y/\tau_{k+1})\, U(\tau_{k+1}, \tau_k)\Big]$$

$$\times \prod_{j=m}^{k-1} f_{j+1}(Y/\tau_{j+1})\, U(\tau_{j+1}, \tau_j)\, f_m(Y/\tau_m)\Big\|$$

$$\leq \sum_{k=m}^{n-1}\Big\{\Big\|\Big[\mathbf{1} - f_{k+1}(Y/\tau_{k+1})\Big]\, U(\tau_{k+1}, \tau_k)\, f(Q)\, f_k(Y/\tau_k)\Big\|$$

$$+ \Big\|\Big[\exp\{-i \sum_\beta{}'' V_k^\beta\, t_k\},\ f(Q)\Big]\Big\| + \Big\|\Big[f_{k+1}(Y/\tau_{k+1}),\ f(Q)\Big]\Big\|\Big\}$$

$$\leq const \sum_{k=m}^{n-1} k^{-\varrho}.$$

In the last step we have used (4.16), (4.7), and (4.17). This bound decays as $m \to \infty$ uniformly in $n > m$. Moreover for $A_k = \exp\{-iH\, t_k\}$ we get

$$\Big\|e^{-iH(\tau_n-\tau_m)}\, f(Q)\, f_m(Y/\tau_m) - \tilde{U}(\tau_n, \tau_m)\, f_m(Y/\tau_m)\Big\|$$

$$\leq \sum_{k=m}^{n-1}\Big\{\Big\|\Big[e^{-iHt_k} - U(\tau_{k+1}, \tau_k)\Big]\, f(Q)\, f_k(Y/\tau_k)\Big\| +$$

$$+ \left\| \left[U(\tau_{k+1}, \tau_k)\, f(Q) - f(Q)\, f_{k+1}(Y/\tau_{k+1})\, U(\tau_{k+1}, \tau_k) \right] f_k(Y/\tau_k) \right\| \Big\}$$

$$\leq const \sum_{k=m}^{n-1} k^{-\varrho}.$$

Here we applied in addition (4.21). Since $f = f_m\, f$ we have verified (4.19) and (4.20). To show the corresponding statement for U_f observe that

$$\left\| \left[\mathbf{1} - \bar{f}(Q)\, \bar{f}(Y/\tau_n) \right] \tilde{U}(\tau_n, \tau_m)\, f_m(Y/\tau_m) \right\|$$

$$\leq \| \, [f(Q), \bar{f}(Y/\tau_n)] \, \| \leq const \, (\tau_n)^{-1}.$$

We can insert a factor $\bar{f}(Q)\, \bar{f}(Y/\tau_n)$ on the left of any of the time evolutions, U_f is only a special case needed below. This completes the proof of Proposition 4.3.

V. Improved Correlations in Phase Space.

The method of asymptotic observables gives decay (in the strong resolvent sense) of $(Q - Y/\tau)$ for the component in the range of $\mathbf{P}$ for any scattering state. Moreover the potentials may decay very slowly, e.g. $\delta > -1/2$ in (2.5) is sufficient. We can use the additional structure to show stronger correlations between the position and velocity of the clusters, namely $(Q - Y/\tau) \sim \tau^{(-\delta-1/2)}$ and logarithmic behavior for Coulomb potentials. The analysis of classical trajectories suggests that up to constants the estimates are optimal. They will be applied below when Dollard's modified free time evolution is studied.

We treat first the unbounded operators themselves, for bounded functions see the corollaries below.

Proposition 5.1 *(a) For* $0 < \delta < 1/2$, $n \geq m \geq 1$,

$$\|(Y - Q\, \tau_n)\, U(\tau_n, \tau_m)\, f(Q)\, f(Y/\tau_m)\| \leq C(m) \cdot (\tau_n)^{(-\delta+1/2)}. \tag{5.1}$$

(b) If $\delta = 1/2$

$$\|(Y - Q\, \tau_n)\, U(\tau_n, \tau_m)\, f(Q)\, f(Y/\tau_m)\| \leq C(m)\, \log(\tau_n/\tau_m). \tag{5.2}$$

The same bounds hold with $U(\tau_n, \tau_m)$ *replaced by* $U_f(\tau_n, \tau_m)$.

Proof. The commutation relations imply

$$(Y - Q\, \tau_{k+1})\, e^{-iH(d)\, t_k} = e^{-iH(d)\, t_k}\, (Y - Q\, \tau_k).$$

Consequently we have

$$\|(Y - Q\,\tau_n)\, U(\tau_n, \tau_m)\, f(Q)\, f(Y/\tau_m)\|$$

$$\leq \|(Y - Q\,\tau_m)\, f(Q)\, f(Y/\tau_m)\| + \sum_{k=m}^{n-1} \tau_{k+1} \left\| \left[Q,\, \exp\{-i \sum_\beta{}'' V_k^\beta\, t_k\}\right] \right\|$$

$$\leq C'(m) + const \sum_{k=m}^{n-1} \tau_{k+1}\, t_k\, (\tau_k)^{-(\delta+3/2)}.$$

The summands behave like $k^{\varrho(1-2\delta)-1}$. For $\delta < 1/2$ the sum is bounded by $const \cdot n^{\varrho(1-2\delta)} = \tau_n^{(-\delta+1/2)}$ and the logarithmic behavior follows for $\delta = 1/2$. Since

$$\| [\bar{f}(Q)\, \bar{f}(Y/\tau_n),\, (Y - Q\,\tau_n)] \| \leq const$$

the same bounds follow for $U_f(\tau_n, \tau_m)$. □

The results of Proposition 5.1 imply

$$\left\| \left[Q \cdot (\tau_n)^\lambda - Y/(\tau_n)^{1-\lambda}\right] U(\tau_n, \tau_m)\, f(Q)\, f(Y/\tau_m) \right\| \to 0 \quad \text{as } \tau_n \to \infty \tag{5.3}$$

for all $\lambda < \delta + 1/2$. The simple possibility to use the unbounded operators Y and Q themselves is special for U and U_f. For the other time evolutions one should use nice bounded functions like $g \in C_0^\infty$ or far more general ones.

Corollary 5.2 *(a) If e.g. $g \in C_0^\infty(\mathbb{R}^{\nu(k-1)})$ then for every $\lambda < \delta + 1/2$, $\varepsilon > 0$ there is a τ_m and a $T(\varepsilon, \tau_m)$ such that*

$$\left\| \left[g\left(Q \cdot (\tau_n)^\lambda - Y/(\tau_n)^{1-\lambda}\right) - g\left(0\right)\right] e^{-iH(\tau_n - \tau_m)}\, f(Q)\, f(Y/\tau_m) \right\| < \varepsilon \tag{5.4}$$

for all $\tau_n \geq T(\varepsilon, \tau_m)$. The same holds with $\tilde{U}(\tau_n, \tau_m)$. Moreover for any τ_m

$$\lim_{n\to\infty} \left\| \left[g\left(Q \cdot (\tau_n)^\lambda - Y/(\tau_n)^{1-\lambda}\right) - g\left(0\right)\right] U(\tau_n, \tau_m)\, f(Q)\, f(Y/\tau_m) \right\| = 0. \tag{5.5}$$

The same holds for $U_f(\tau_n, \tau_m)$.
(b) If $\lambda < 1/2$ then also

$$\left\| \left[g\left(Q \cdot (\tau_n)^\lambda\right) - g\left(Y/(\tau_n)^{1-\lambda}\right) \right] e^{-iH(\tau_n - \tau_m)}\, f(Q)\, f(Y/\tau_m) \right\| < \varepsilon, \tag{5.6}$$

and analogously for the other time evolutions.

Proof. (a) For late enough τ_m the time evolutions may be replaced by $U(\tau_n, \tau_m)$ up to an error of $\varepsilon/2$. The remaining estimate to derive (5.4) from (5.3) is a standard argument in strong resolvent convergence. (b) With the Baker-Campbell-Hausdorff formula the decay of (5.4) or (5.5) is equivalent to that of (5.6) for $\lambda < 1/2$. □

VI. Dollard's Asymptotic Evolution.

A modified free time evolution which is a function of the velocities alone, was introduced by Dollard [1] to show existence of modified wave operators for Coulomb potentials. We adjust it to the present situation.

We set $q^\beta := \pm Q_{ij}$ for the relative velocity of the centers of mass of the clusters c_i and c_j to which the particles of the pair β belong, analogously to (2.13). Moreover we assume a choice for the splitting of the potentials such that for the long-range parts (2.6) is satisfied at least for $1 \le |m| \le 2$.

Definition 6.1 *For $t \ge T \ge 0$ we set*

$$U^D(t,T) := \exp\Big\{-i\int_T^t ds \sum_\beta{}'' V_\ell^\beta(s\, q^\beta)\Big\}\, e^{-iH(d)\,(t-T)}. \tag{6.1}$$

The main approximation result is

Proposition 6.2 *With the notation and assumptions given above for any vector $\Psi \in \mathcal{H}$*

$$\sup_{t\ge T} \Big\| \Big\{e^{-iH(t-T)} - U^D(t,T)\Big\}\, \bar f(Q)\, \bar f(Y/T)\, \mathbf{P}\, \Psi\Big\|$$

$$\le const\, \Big\{T^{-\varepsilon} + T^{-(\delta+1/2)} \Big\| \Big[Y - T\, Q\Big]\, \bar f(Q)\, \bar f(Y/T)\, \mathbf{P}\, \Psi\, \Big\|\Big\}, \quad \varepsilon > 0. \tag{6.2}$$

Proof. Closely related arguments are well known and it is sufficient to indicate them briefly. The Cook estimate gives uniformly in $s \ge T \ge 0$

$$\Big\| \Big\{e^{-iH(s-T)} - U^D(s,T)\Big\}\, \bar f(Q)\, \bar f(Y/T)\, \mathbf{P}\, \Psi\Big\|$$

$$\le \int_T^\infty dt \sum_\beta{}'' \Big\| \Big(V^\beta(x^\beta) - V_\ell^\beta(t\, q^\beta)\Big)\, U^D(t,T)\, \bar f\, \bar f\, \mathbf{P}\, \Psi\, \Big\|$$

$$\le \int_T^\infty dt \sum_\beta{}'' \Big\{ \Big\| \Big[V_s^\beta(x^\beta) + \Big(V_\ell^\beta(x^\beta) - V_\ell^\beta(y^\beta)\Big)\Big]\, \mathbf{P}\, F(|y^\beta| > v_0\, t) \Big\|$$

$$+ \Big\| \Big[V_{(t)}^\beta(y^\beta) - V_{(t)}^\beta(t\, q^\beta)\Big]\, U^D\, \bar f\, \bar f\, \mathbf{P}\, \Psi \Big\|$$

$$+ const\, \Big\| F(|y^\beta| < v_0\, t)\, U_D(t,T)\, \bar f(Q)\, \bar f(Y/T)\, \mathbf{P} \Big\| \Big\}. \tag{6.3}$$

Here we have used for the cutoff potential the shorthand

$$V^\beta_{(t)}(z) := V^\beta_\ell(z)\,\Big[1-\varphi(z/v_0\,t)\Big]$$

with φ as in (4.4). $\bar f(Q)$ commutes with U^D and on its support $|q^\beta| > v_0$ for all arising β. Thus $[V^\beta_\ell(t\,q^\beta) - V^\beta_{(t)}(t\,q^\beta)]\ \bar f(Q) = 0$. The correction term arising from the substitution of $V^\beta_\ell(y^\beta)$ has support in $|y^\beta| < v_0\,t$. Therefore the third summand of (6.3) takes care of it. Moreover, on the range of $\bar f(Q)$ we can replace $V^\beta_\ell(s\,q^\beta)$ in the definition (6.1) of U^D by $V^\beta_{(s)}(s\,q^\beta)$. We will assume that it has been done in the sequel. Then the correction term in U^D is a convolution operator in configuration space with rapidly decaying kernel. It is easy to verify that

$$\Big\|F\Big(|y^\beta| \le v_0\,t\Big)\ \exp\Big\{-i\int_T^t ds \sum_\beta{}'' \,V^\beta_{(s)}(s\,q^\beta)\Big\}\ F\Big(|y^\beta| \ge (4/3)v_0\,t\Big)\Big\|$$

$$\le C_l\cdot t^{-l}, \tag{6.4}$$

see e.g. Corollaries 2.5 and 2.12 in [2]. Combining (6.4) with (4.15) and (2.11) we obtain

$$\|F(|y^\beta| < v_0\,t)\ U^D(t,T)\ \bar f(Q)\ \bar f(Y/T)\,\mathbf{P}\| \le C_l\cdot t^{-l}.$$

The first summand in (6.3) also vanishes as $T\to\infty$, the integral decays at least like $T^{-\varepsilon}$ by (2.14) and (2.15).

In the remaining second term we use the canonical commutation relations and the Baker-Campbell-Hausdorff formula to rewrite the difference of the potentials.

$$V^\beta_{(t)}(y^\beta) - V^\beta_{(t)}(t\,q^\beta)$$

$$= \int_0^1 ds\Big\{\Big(\nabla V^\beta_{(t)}\Big)(s[y^\beta - t\,q^\beta] + t\,q^\beta)\cdot\Big[y^\beta - t\,q^\beta\Big]$$

$$+i\Big(\Delta V^\beta_{(t)}\Big)(s[y^\beta - t\,q^\beta] + t\,q^\beta)\ \frac{t}{2\mu_{ij}}\ \Big\}.$$

Analogous to (4.6) the supremum of the Laplacian of the potential is bounded by $t^{-(2+\delta)}$. Thus the corresponding contribution to (6.3) decays like $\int_T^\infty dt\ t^{-(1+\delta)} \sim T^{-\delta}$, $\delta > 0$. Similarly with (4.5) the gradient term gives a contribution

$$const\int_T^\infty dt\ t^{-(\delta+3/2)}\,\Big\|\Big[y^\beta - t\,q^\beta\Big]\ U^D(t,T)\ \bar f(Q)\ \bar f(Y/T)\,\mathbf{P}\,\Psi\Big\|. \tag{6.5}$$

With the identity

$$y^\beta\ e^{-i\mathbf{T}(t-T)} = e^{-i\mathbf{T}(t-T)}\Big[y^\beta + (t-T)\,q^\beta\Big]$$

the norm in (6.5) is bounded by

$$\left\| \left[y^\beta - T\,q^\beta\right] \bar{f}(Q)\, \bar{f}(Y/T)\, \mathbf{P}\, \Psi \right\|$$

$$+\left\| \left[y^\beta,\ \exp\left\{-i \int_T^t ds \sum_\beta{}'' V^\beta_{(s)}(s\,q^\beta)\right\}\right] \right\|. \tag{6.6}$$

After summation over β the integral of the first term is retained in the bound (6.2). From the second we get

$$const \int_T^\infty dt\; t^{-(\delta+3/2)} \int_T^t ds\; s\; s^{-(\delta+3/2)} \quad \leq \quad const\; T^{-\delta}.$$

For sufficiently small $\varepsilon > 0$ we have verified (6.2). □

VII. The Main Results.

The construction and analysis of the "intermediate time evolutions" as given in Sections IV and V is of independent interest. With the appropriate adjustment they turn out to be useful tools also in the study of long-range magnetic fields [10]. Simpler expressions for Coulomb- and other special long-range potentials are discussed e.g. in [6]. The restriction to a special sequence of times was only for notational convenience.

These evolutions are in a sense a "semiclassical approximation": certain operators are treated as if they were classical ones which commute. The neglected commutator terms add up to a small correction thanks to the decay assumption (2.5) and the choice of times (4.1), (4.2). One could admit slower decay like $\delta > -1/2$ in (2.5) and suitable conditions on higher derivatives if a more complicated approximation with additional correction terms were used. However, the simplicity of the evolution would then be lost and the technically demanding use of Fourier integral operators would then be the appropriate tool. Instead of going into details we concentrate here on the asymptotics of (3.4) alone.

Theorem 7.1 *For the multiparticle Schrödinger operator* (2.1) *let the potentials be such that* (2.2) - (2.5), (2.12) *are satisfied. With* **P** *and* f *as characterized in* (2.7) *and* (3.1), (3.2), *respectively, and* U^D *as given in Definition 6.1 we can find for any* $\varepsilon > 0$ *a late time* τ, *and also a* $T = T(\varepsilon, \tau) \geq \tau$ *such that*

$$\sup_{t \geq T} \left\| \left\{ e^{-iH(t-T)} - U^D(t,T) \right\} e^{-iH(T-\tau)}\, f(Q)\, f(Y/\tau)\, \mathbf{P} \right\| \leq \varepsilon. \tag{7.1}$$

Proof. By Proposition 4.3 we can choose $\tau = \tau_m$ large enough such that for all $\tau_n \geq \tau_m$ from the sequence (4.2)

$$\left\| \left\{ e^{-iH(\tau_n-\tau)} - U_f(\tau_n,\tau) \right\} f(Q)\, f(Y/\tau)\, \mathbf{P} \right\| < \varepsilon/4.$$

Up to an error of $\varepsilon/2$ we can thus study instead of (7.1) the behavior for large $T = \tau_n \geq \tau$ of

$$\|\{e^{-iH(t-T)} - U^D(t,T)\}\, U_f(T,\tau)\, f(Q)\, f(Y/\tau)\, \mathbf{P}\|.$$

By Proposition 6.2 this is bounded by

$$const \left\{ T^{-\varepsilon} + T^{-(\delta+1/2)}\, \|(Y - T\,Q)\, U_f(T,\tau)\, f\, f\, \mathbf{P}\| \right\}.$$

With Proposition 5.1 it vanishes as $T = \tau_n \to \infty$. □

Actually we have shown in this proof the useful estimate:

Proposition 7.2 *Under the above assumptions for any* $\tau_m \geq 1$

$$\sup_{t \geq \tau_n} \left\| \left\{ e^{-iH(t-\tau_n)} - U^D(t,\tau_n) \right\} U(\tau_n,\tau_m)\, f(Q)\, f(Y/\tau_m)\, \mathbf{P} \right\|$$

$$\to 0 \quad \text{as } \tau_n \to \infty.$$

The same holds if U *is replaced by* U_f. *For* $\tilde{U}$ *the result is as in Theorem 7.1.*

The calculations in the previous sections can also be used to show existence of the *modified channel wave operators* like the outgoing

$$\Omega^D := s - \lim_{t\to\infty} e^{iHt}\, U^D(t,0)\, \mathbf{P}.$$

Then (7.1) obviously says that the strictly outgoing states lie approximately in the range of the wave operator. Let P_Ω denote the range projection of the partial isometry Ω^D.

Corollary 7.3 *For any* f *as in* (3.1), (3.2) *and* $\varepsilon > 0$ *there is a* $\tau = \tau(\varepsilon)$ *such that*

$$\|\{\mathbf{1} - P_\Omega\}\, f(Q)\, f(Y/\tau)\, \mathbf{P}\| < \varepsilon.$$

If the effective intercluster potentials vanish

$$\sum_\beta{}'' V_\ell^\beta(y^\beta) = 0 \tag{7.2}$$

then U and U^D coincide with the free decomposition evolution $\exp\{-iH(d)\,t\}$. Moreover we have an ordinary channel wave operator Ω instead of the modified one. By specialization we get

Corollary 7.4 *With the additional condition* (7.2)

$$\sup_{t\geq\tau} \left\| \left\{ e^{-iH(t-\tau)} - e^{-iH(d)(t-\tau)} \right\} f(Q)\, f(Y/\tau)\, \mathbf{P} \right\| \to 0 \quad \text{as } \tau \to \infty,$$

and

$$\lim_{\tau\to\infty} \| \{ \mathbf{1} - P_\Omega \}\, f(Q)\, f(Y/\tau)\, \mathbf{P} \| = 0.$$

Clearly in this case a much shorter direct proof of the result can be given.

If any given scattering state can be decomposed up to a small error into finitely many *strictly* outgoing components for arbitrary late times, then it lies approximately in the direct sum of the ranges of the channel wave operators and asymptotic completeness (including absence of singular continuous spectrum) holds. Suitable control of the degenerate components (small relative velocities of some clusters) is still missing in the general case. Moreover, we would like to eliminate completely any assumption like (2.12) about the decay of subsystem eigenfunctions.

References

[1] J. DOLLARD: (a) Asymptotic convergence and the Coulomb interaction, J. Math. Phys. **5**, 729 - 738 (1964); (b) Quantum mechanical scattering theory for short-range and Coulomb interactions, Rocky Mt. J. Math. **1**, 5 - 88 (1971).

[2] V. ENSS: Propagation properties of quantum scattering states, J. Func. Anal. **52**, 219 - 251 (1983).

[3] V. ENSS: Quantum scattering theory for two- and three-body systems with potentials of short and long range, in: *Schrödinger Operators*, S. Graffi ed., Springer LN Math 1159, Berlin 1985, pp. 39 - 176 [Proceedings C.I.M.E. Como 1984].

[4] V. ENSS: Introduction to asymptotic observables for multiparticle quantum scattering, in: *Schrödinger Operators, Aarhus 1985*, E. Balslev ed., Springer LN Math. 1218, Berlin, 1986.

[5] V. ENSS: Separation of subsystems and clustered operators for multiparticle quantum systems, preprint Serie A, Nr. 213, Mathematik, Freie Universität Berlin, 1986 (submitted for publication).

[6] V. ENSS: Quantum mechanical time evolution for Coulomb potentials, preprint Serie A, Nr. 231, Mathematik, Freie Universität Berlin 1986, to appear in Chech. Journ. of Physics [Proceedings Bechyně 1986].

[7] V. ENSS: Observables and asymptotic phase space localization of N-body quantum scattering states, preprint Serie A, Nr.226, Mathematik, Freie Universität Berlin, to appear 1986.

[8] R. FROESE and I. HERBST: Exponential bounds and absence of positive eigenvalues for N-body Schrödinger operators, Commun. Math. Phys. **87**, 429 - 447 (1982).

[9] L. HÖRMANDER: The existence of wave operators in scattering theory, Math. Z. **146**, 69 - 91 (1976).

[10] M. LOSS and B. THALLER: (a) Scattering of particles by long-range magnetic fields, preprint Serie A, Nr. 225, Mathematik, Freie Universität Berlin, 1986; (b) Short-range scattering in long-range magnetic fields: the relativistic case, preprint Mathematik, Freie Universität Berlin, 1986.

[11] M. REED and B. SIMON: *Methods of Modern Mathematical Physics, III. Scattering Theory*, Academic Press, New York, 1979.

[12] I.M. SIGAL and A. SOFFER: (a) Asymptotic completeness of short-range many-body systems, Bull. Amer. Math. Soc. **14**, 107 - 110 (1986); (b) N-particle scattering problem: asymptotic completeness for short-range systems, preprint in preparation.

A "BIRKHOFF - LEWIS" TYPE RESULT FOR NON AUTONOMOUS DIFFERENTIAL EQUATIONS

Vieri BENCI
Istituto di Matematiche Applicate
Università - PISA (Italy)

Donato FORTUNATO
Dipartimento di Matematica
Università - BARI (Italy)

1. Statement of the results

Let $V = V(x,t)(x \in \mathbb{R}^N, t \in \mathbb{R})$ be a C^2-real function and consider the equation

$$(1.1) \qquad \ddot{x} = -V'(x,t),$$

where $x \in \mathbb{R}^N$, $\ddot{x} = \frac{d^2x}{dt^2}$ and V' denotes the gradient of V with respect to $x = (x_1,\dots,x_N)$. We suppose that

V_1) V *is* $\tau = \frac{2\pi}{\Omega}$ *periodic in* t *and* $V(0,t) = 0$, $V'(0,t) = 0$ *for any* $t \in \mathbb{R}$.

In this paper we are concerned with the existence of subharmonic solutions for (1.1) i.e. we search for $k\tau$-periodic solutions $(k \in \mathbb{N})$ x_k of (1.1). The research of subharmonic solutions of Hamiltonian systems is an old problem (cp. [8] and its references). Neverthless very little is known about the existence of subharmonic solutions of (1.1) with prescribed minimal period $k\tau$. In this paper we obtain some results in this direction in the case in which V grows more than quadratically at infinity (cp. Theorem 1.1); the "subquadratic" case has been studied in [7].
Moreover, by strengthening the assumptions on V, it is possible to establish the existence (cp. Theorem 1.6) of subharmonic solutions near the origin having arbitrarily long minimal periods.
In order to state the results we need to introduce some notations.
Let

$$A = V''(0,t)$$

denote the hessian matrix of V at $x = 0$ and set

$$\hat{V}(x,t) = V(x,t) - \frac{1}{2}(Ax|x)$$

where $(\cdot|\cdot)$ denotes the inner product in $\mathbb{R}^N$.
We assume that

$$V_2)\quad A = \begin{bmatrix} \omega_1^2 & & & & & \bigcirc \\ & \ddots & & & & \\ & & \omega_\ell^2 & & & \\ & & & -\omega_{\ell+1}^2 & & \\ & & & & \ddots & \\ \bigcirc & & & & & -\omega_N^2 \end{bmatrix}$$

<u>where</u> $\ell \in \mathbb{N}$, $1 \leq \ell \leq N$, $\omega_1,\dots,\omega_N \in \mathbb{R}$

<u>with</u> $0 \leq \omega_1 \leq \dots \leq \omega_\ell$.

$V_3)$ <u>There exist</u> $p > 2$ <u>and</u> $M > 0$ <u>s.t.</u>

$\frac{1}{p}(\hat{V}'(x,t)|x) \geq \hat{V}(x,t) > 0$ <u>for</u> $|x| > M$ <u>and</u> $t \in \mathbb{R}$

$V_4)$ $\hat{V}(x,t) > 0$ <u>for any</u> $x \in \mathbb{R}^N$ <u>and</u> $t \in \mathbb{R}$

$V_5)$ $\hat{V}'$ <u>is odd</u> (i.e. $\hat{V}'(-x,-t) = -\hat{V}'(x,t)$)

$V_6)$ $(\hat{V}''(x,t)x|x) > (\hat{V}'(x,t)|x)$
<u>for</u> $x \neq 0$ <u>and any</u> $t \in \mathbb{R}$

$V_7)$ <u>If</u> $x = x(t)$ <u>is a periodic function with minimal period</u> $q\tau$, q <u>rational, and</u> $V'(x(t),t)$ <u>is periodic with minimal periodic</u> $q\tau$, <u>then</u> q <u>is an integer</u>.

Observe that $V_3)$ prescribes a superquadratic behaviour of $\hat{V}$ at infinity;in fact $V_3)$ implies that

$$\hat{V}(x,t) \geq \text{const } |x|^p \quad \text{for} \quad |x| > M.$$

Assumption $V_7)$ is satisfied, for example, if

$$\hat{V}(x,t) = g(t)\, U(x)$$

and g is a periodic function with minimal period τ.

The following theorem holds

<u>Theorem 1.1.</u> <u>Let the assumptions</u> $V_1),\dots,V_7)$ <u>be satisfied</u>.

<u>Let</u> $k \in \mathbb{N}$ <u>such that</u>[1]

$$(1.2)\qquad k_o > \sum_{j=1}^{\ell} [k\,\omega_j/\Omega] + 1$$

---

[1] If $\alpha > 0$, $[\alpha]$ denotes the greatest integer $\leq \alpha$.

<u>where</u> k_o <u>is the least prime factor of</u> k.

<u>Then</u> (1.1) <u>possesses a periodic solution with minimal period</u> $k\tau$.

In order to prove theorem 1.1 we need to find $k\tau$-periodic solutions of (1.1) having suitable bounds on their Morse indices (cf.section 2).

We recall that Morse critical point theory has already been used in the study of Hamiltonian systems in [3,4,5,9].

The following corollaries are easily deduced from Theorem 1.1.

<u>Corollary 1.2.</u> <u>Let the assumptions</u> $V_1),\ldots,V_7)$ <u>be satisfied and suppose that</u>

$$\omega_1 = \omega_2 \ldots = \omega_\ell = 0.$$

<u>Then for any</u> $k \in \mathbb{N}$, $k \geq 2$, (1.1) <u>possesses a periodic solution with minimal period</u> $k\tau$.

<u>Corollary 1.3.</u> <u>Let the assumptions</u> $V_1),\ldots,V_7)$ <u>be satisfied.</u>

<u>Let</u> $k \in \mathbb{N}$ <u>s.t.</u>

$$(1.3) \qquad k < \frac{2\pi}{\omega_\ell}, \quad \omega_\ell > 0.$$

<u>Then</u> (1.1) <u>possesses a periodic solution with minimal period</u> $k\tau$.

<u>Corollary 1.4.</u> <u>Let the assumptions</u> $V_1),\ldots,V_7)$ <u>be satisfied and suppose that</u>

$$(1.4) \qquad \delta = \sum_{j=1}^{\ell} \omega_j/\Omega < 1.$$

<u>then for any</u> k <u>prime with</u>

$$k > \frac{1}{1-\delta}$$

(1.1) <u>possesses a periodic solution with minimal period</u> $k\tau$.

<u>Remark 1.5.</u> In [2] theorem 1.1 and its corollaries have been established under the assumption that all the eigenvalues of A are non negative.

Let us now substitute the assumptions $V_3)$ and $V_4)$ by the more restrictive ones

$V_3)'$ <u>The exists</u> $p > 2$ <u>s.t.</u>

$$\frac{1}{p}\,(\hat{V}(x,t)\,|\,x) \geq \hat{V}(x,t) \qquad \forall\, x \in \mathbb{R}^N, \quad t \in \mathbb{R}$$

$V_4)'$ There exists $c_o > 0$ s.t.

$$\hat{V}(x,t) \geq c_o|x|^p \qquad \forall\, x \in \mathbb{R}^N, \quad t \in \mathbb{R}.$$

By using these more restrictive assumptions it is possible to "localize" the subharmonic solutions with large minimal periods. More precisely the following theorem holds:

Theorem 1.6. Let V satisfy $V_1)$, $V_2)$, $V_3)'$, $V_4)'$, $V_5)$ $V_6)$, $V_7)$. Moreover suppose that (1.4) is satisfied.
Then for any k prime sufficiently large there exists a subharmonic solution x_k of (1.1) with minimal period $k\tau$. Moreover x_k converges to 0 as $k \to \infty$ uniformly in $C^1(\mathbb{R}, \mathbb{R}^N)$.

2. Some preliminaries.

In this section we shall state two results which will be useful in the proof of Theorems 1.1 and 1.6.
Let us first recall some definitions.
Let f be a C^2-functional on a real Hilbert space E; we denote by $f''(x)$ the bilinear form on $E\times E$ which represents the second differential of f at x. Suppose that

(2.1) for any $x \in E$ $f''(x)$ has discrete spectrum.

If z is a critical point of f(i.e. $f'(z) = 0$) we denote by $m(z)$ its Morse-index, i.e. we set

$m(z)$ = number of the negative eigenvalues (counted with their multiplicity) of $f''(z)$.

Assume that f satisfies the Palais-Smale condition, i.e.

(2.2) Every sequence $\{x_n\} \subset E$ s.t. $\{f(x_n)\}$ is bounded and $f'(x_n) \to 0$ contains a convergent subsequence.

The following lemma permits to give, in some cases, an upper bound to the Morse index.

Lemma 2.1. Let f be a C^2-functional on a real Hilbert space E. Suppose that f satisfies (2.1), (2.2). Suppose moreover that there exists ρ, R_1, R_2, $\alpha > 0$ with $R_1 > \rho$ and a finite-dimensional subspace V of E and $\phi \in V^{\perp}$, $\|\phi\| = 1$ such that

$$(2.3) \qquad \begin{cases} f(u) \geq \alpha & \forall\, u \in V^{\perp},\ \|u\| = \rho \\ f(u) \leq 0 & \forall\, u \in \partial Q \end{cases}$$

where ∂Q is the boundary of the Hilbert manifold

$$Q = \{v + t\,\phi : v \in V, \quad \|v\| \leq R_2, \quad t \in [0, R_1]\}.$$

Under the above assumptions f possesses a critical point $\bar{u}$ with Morse index

(2.4) $\quad m(\bar{u}) \leq \dim V + 1.$

Moreover

(2.5) $\quad \alpha \leq f(\bar{u}) \leq \sup f(Q) \equiv \beta.$

The proof of lemma 2.1 (cp. [1,2]) uses a generalized version of the Morse-Conley critical point theory.
We recall that a result related to lemma 2.1 has been proved by Lazer and Solimini [6].
The following lemma is useful to get, in some situations, a lower bound to the Morse-index.

Lemma 2.3. Suppose that the potential function $V = V(x,t)$ satisfies assumption V_6) (cp. Theorem 1.1) and let $z = z(t)$ be a nontrivial solution of the boundary value problem

$$(2.6) \qquad \begin{cases} \ddot{x} + V'(x,t) = 0 & \text{in }]0,T[\ , \quad T > 0 \\ x(0) = x(T) = 0 \end{cases}$$

Suppose moreover that there exist points

$$T_o = 0 < T_1 < \ldots < T_{q-1} < T_q = T$$

s.t. $\quad z(T_i) = 0 \qquad i = 0,\ldots,q.$

Then

$$q \leq m(z)$$

where $m(z)$ denotes the Morse-index of z.

Proof. Set

$$\tau_i = [T_{i-1}\ ,\ T_i] \qquad i = 1,\ldots,q$$

and

$$\alpha_i(t) = \begin{cases} z(t) & \text{if } t \in \tau_i \\ 0 & \text{if } t \notin \tau_i \end{cases} \qquad i = 1,\ldots,q$$

Obviously α_i $(i = 1,\ldots,q)$ are linearly independent.
Let V_q denote the vector space spanned by $\{\alpha_i\}$ and consider the

functional whose critical points are the solutions of (2.6)

$$f(x) = \int_0^T \left(\frac{1}{2}\,|\dot{x}(t)|^2 - V(x(t))\right) dt \qquad x \in H_0^1\,(]0,T[).$$

We want to prove that $f''(z)$ is negative-definite on V_q. Let

$$v \in V_q \setminus \{0\} \qquad v = \sum_{i=1}^{q} c_i\,\alpha_i \qquad c_i \in \mathbb{R}$$

then:

$$\langle f''(z)v,v\rangle = \int_0^T (|\dot{v}|^2 - (V''(z,t)v|v))\,dt =$$

$$= \sum_{i=1}^{q} \int_{\tau_i} (|\dot{v}|^2 - (V''(z,t)v|v))\,dt =$$

$$= \sum_{i=1}^{q} c_i^2 \int_{\tau_i} (|\dot{z}|^2 - (V''(z,t)z|z))dt < (\text{by } V_6)) <$$

$$< \sum_{i=1}^{q} c_i^2 \int_{\tau_i} (|\dot{z}|^2 - (V'(z,t)|z))\,dt = 0.$$

The last equality follows from the fact $z = z(t)$ solves (1.1) in τ_i and $z(T_{i-1}) = z(t_i) = 0$ $(i = 1,\dots,q)$.

3. Proof of Theorem 1.1

We are now ready to prove Theorem 1.1.
Consider a positive integer k which satisfies (1.2). Set

$$2T = k\tau = k\,\frac{2\pi}{\Omega}.$$

Consider the boundary value problem

$$(3.1) \qquad \begin{cases} \ddot{x} + V'(x,t) = 0 & \text{in} \quad]0,T[\\ x(0) = x(T) = 0. \end{cases}$$

The solutions of (3.1) are the critical points of the functional

$$f(x) = \int_0^T \left(\frac{1}{2}\,(|x|^2 - V(x,t)\right) dt \qquad x \in H_0^1(]0,T[).$$

Obviously

$$f(x) = \frac{1}{2}\,(L_k\, x|x) - \int_0^T \hat{V}(x,t)\,dt$$

where L_k is the self-adjoint realization in $L^2(]0,T[)$ of the operator $x \longrightarrow -\ddot{x} - Ax$ with boundary condition $x(0) = x(T) = 0$ and

$(\cdot|\cdot)$ denotes the inner product in $L^2(]0,T[)$.
We shall use lemma 2.1.
By assumption V_3), standard calculations show that f satisfies the Palais-Smale condition (2.2).
The eigenfunctions of L_k are

$$\phi^k_{jn} = e_j \sin n \frac{\Omega}{k} t \qquad j = 1,\dots,N \qquad n \in \mathbb{N} \setminus \{0\}$$

where $\{e_j\}$ is the canonical base in $\mathbb{R}^N$.
The corresponding eigenvalues are

$$(3.2) \qquad \lambda^k_{jn} = \begin{cases} (\frac{n\Omega}{k})^2 - \omega_j^2 & \text{if} \quad j = 1,\dots,\ell \\ (\frac{n\Omega}{k})^2 + \omega_j^2 & \text{if} \quad j = \ell+1,\dots,N \end{cases}$$

where $\omega_j^2, \dots, \omega_\ell^2, -\omega_{\ell+1}^2, \dots, -\omega_N^2$ are the eigenvalues of A.
Set

$$H_+ = \overline{\text{span}\{\phi^k_{jn} : \lambda^k_{jn} > 0\}}, \qquad H_- = H_+$$

where the closure is taken in $H^1_o(]0,T_k[)$.
Since

$$\hat{V}(x,t) = 0\,(|x|^2) \quad \text{at} \quad x = 0$$

standard calculations show that there exist $\rho, \alpha > 0$ s.t.

$$(3.3) \qquad f(x) \geq \alpha \quad \text{for} \quad x \in H_+ \quad \|x\| = \rho.$$

Let R_1, R_2 be two positive numbers with $R_1 > \rho$ and consider $\phi \in H_+$ with $\|\phi\| = 1$.
Set

$$Q \equiv \{v + t\,\phi \mid v \in H_-, \quad \|v\| \leq R_2, \quad t \in [0,R_1]\}.$$

By V_4) we immediatly deduce that

$$(3.4) \qquad f(x) \leq 0 \qquad \forall\, x \in H_-.$$

Moreover, by the superquadratic growth of $\mathcal{V}$ at ∞ (cf. assumption V_3), standard calculations show that

$$(3.5) \qquad \lim_{\|x\| \to +\infty} f(x) = -\infty \qquad \text{for} \quad x \in H_- \oplus \{t\phi\} \quad t \in \mathbb{R}$$

Then, if R_1 and R_2 are sufficiently large, from (3.4) and (3.5) we easily deduce that

$$(3.6) \qquad f(x) \leq 0 \qquad \forall\, x \in \partial Q.$$

By virtue of (3.6) and (3.3), also assumption (2.3) of lemma 2.1 is satisfied. Then, by lemma 2.1, (3.1) possesses a non-trivial solution $x = x(t)$ with Morse index

$$(3.7) \qquad m(x) \le \dim H_- + 1.$$

Now, by using (3.2), we deduce that

$$(3.8) \qquad \dim H_- = \# \left\{(n,j): \frac{n\Omega}{k} \le \omega_j \, ; \; j = 1,\dots,\ell \quad n \in N \setminus \{0\}\right\} =$$

$$= \sum_{j=1}^{\ell} \left[\frac{k\omega_j}{\Omega}\right].$$

Consider now the function $\bar{x}(t)$ obtained by extending $x(t)$ by oddness to a $k\tau$-periodic function (i.e. $\bar{x}$ is the $k\tau$-periodic function s.t. $\bar{x}(t) = x(t)$ if $t \in [0,T]$ and $\bar{x}(t + T) = -x(T-t)$ if $t \notin [0,T]$).

Since V' is odd, $\bar{x}$ is a $k\tau$-periodic solution of (1.1).

Now we prove that $\bar{x}$ has minimal period $k\tau$.

Arguing by contradiction we suppose that $\bar{x}$ has not minimal period $k\tau$. Then, by V_7), $\bar{x}$ has minimal period $\le k\tau/k_o$, where k_o is the least prime factor of k. Consequently there exist points $T_o \equiv 0 < T_1 < \dots T_{k_o-1} < T \equiv T_{k_o}$ such that $x(T_i) = 0 \quad (i = 0,\dots,k_o)$.

Then, by using lemma 2.3, we get

$$(3.9) \qquad k_o \le m(x)$$

By (3.7), (3.8), (3.9) we deduce

$$k_o \le \sum_{j=1}^{\ell} \left[\frac{k\omega_j}{\Omega}\right] + 1$$

which contradicts (1.2).

4. Proof of Theorem 1.6.

We shall preserve the notations introduced in §3.

Fix $\frac{p-2}{2} < \alpha < 1$ and let us initially prove that for any $k > \Omega$ there exists a positive eigenvalue λ_k of L_k such that [(2)]

$$(4.1) \qquad \lambda_k < \frac{C_1}{k^\alpha}.$$

In fact, if $k > \Omega$, we can choose $n \in \mathbb{N}$ s.t.

(2) In the sequel $C_1, C_2, \dots$ will denote positive numbers independent on k.

$$\omega_1 \frac{k}{\Omega} < n \leq \omega_1 \frac{k}{\Omega} + (\frac{k}{\Omega})^{1-\alpha}$$

then

(4.2) $$0 < \frac{n\Omega}{k} - \omega_1 \leq (\frac{\Omega}{k})^{\alpha} .$$

From which we deduce that

(4.3) $$\frac{n\Omega}{k} + \omega_1 < (\frac{\Omega}{k})^{\alpha} + 2\omega_1 \leq 1 + 2\omega_1 .$$

Therefore, if we set

$$\lambda_k \equiv \lambda^k_{1n} = (\frac{n\Omega}{k} + \omega_1)(\frac{n\Omega}{k} - \omega_1)$$

we deduce from (4.2) and (4.3) that

$$0 < \lambda_k \leq (1 + 2\omega_1)(\frac{\Omega}{k})^{\alpha}.$$

Then we conclude that (4.1) holds with $C_1 = (1 + 2\omega_1)\Omega^{\alpha}$.
Now let ϕ a normalized eigenfunction of L_k with respect to the eigenvalue λ_k and consider the manifold Q as defined in § 3.
Let x_k be a solution of (3.1) found as in § 3. Then, by lemma 2.1, we get

(4.4) $$0 < f(x_k) \leq \sup_{x\in Q} f(x).$$

Let us now prove that

(4.5) $$f(x_k) \to 0 \quad \text{as} \quad k \to \infty .$$

Let

$$x \in Q \qquad x = x_1 + s\phi \qquad x_1 \in H_- \qquad s \in [0,R_1].$$

Then, by using (4.1) and assumption $V_4)'$, we get[3]

(4.6) $$f(x) \leq \frac{1}{2} \lambda_k s^2 - \int_0^{k\tau/2} V(x,t)dt \leq C_2 \frac{s}{k^{\alpha}} - C_0(|x_1+s\phi|^p_p) \leq$$

$$\leq \frac{1}{2} \lambda_k s^2 - C_0 (\frac{k\tau}{2})^{\frac{2-p}{2}} |x_1 + s\phi|^p_2 \leq$$

$$\leq \frac{1}{2} \lambda_k s^2 - C_3(k\tau)^{(2-p)/2} s^p \equiv g(s)$$

Easy calculations show that

(4.7) $$\max g(s) = C_4 (\lambda_k)^{p/(p-2)} k.$$

---

(3) $|\cdot|_q$ denotes the norm in L^q $(]0,k\tau/2[)$.

Then by (4.1), (4.4), (4.6) and (4.7) it follows that

$$(4.8) \qquad f(x_k) \le \frac{C_5}{k^{\mu}}, \qquad \mu = \frac{\alpha p}{p-2} - 1 > 0 .$$

Then (4.5) holds.

Now we will prove that

$$(4.9) \qquad \|x_k\| \to 0 \quad \text{as} \quad k \to \infty$$

where $\|\cdot\|$ denotes the norm in $H^1_o(]0, kT/2[)$.

Obviously

$$(4.9) \qquad \frac{1}{2}(L_k x_k | x_k) - \int_0^{kT/2} \hat{V}(x_k, t) = f(x_k)$$

$$(4.10) \qquad -\frac{1}{2}(L_k x_k | x_k) + \frac{1}{2}\int_0^{kT/2} (\hat{V}^1(x_k, t) | x_k) = 0 .$$

Then by using $V_3)'$, $V_4)'$, we get

$$(4.11) \qquad f(x_k) = \frac{1}{2}\int_0^{kT/2} (\hat{V}'(x_k,t)|x_k)dt - \int_0^{kT/2} \hat{V}(x_k,t)dt \ge$$

$$\ge \left(\frac{1}{2} - \frac{1}{p}\right)\int_0^{kT/2} (\hat{V}'(x_k, t)|x_k)dt \ge C_6 |x_k|_p^p .$$

Therefore from (4.5) and (4.11) we deduce that

$$(4.12) \qquad \int_0^{kT/2} (\hat{V}'(x_k, t)|x_k)\, dt \to 0 \qquad \text{as} \quad k \to +\infty$$

and

$$(4.13) \qquad |x_k|_p \to 0 .$$

From (4.10) and (4.12) we obtain

$$(L_k x_k | x_k) = |\dot{x}_k|_2^2 - (A x_k | x_k) \to 0 \qquad \text{as} \quad k \to \infty.$$

Then, by using (4.13), we have

$$(4.14) \qquad |\dot{x}_k|_2 \to 0.$$

From (4.13) and (4.14) we deduce that

$$(4.15) \qquad \|x_k\| \to 0 \quad \text{as} \quad k \to \infty.$$

Consider now the function $\bar{x}(t)$ obtained by extending $x_k(t)$ by oddness to a $k\tau$-periodic function.
By arguing as in § 3 we deduce that $\bar{x}_k$ is a $k\tau$-periodic solution of (1.1); moreover, since (1.4) holds, any k-prime, sufficiently large, satisfies (1.2). Then we deduce, as in § 3, that $\bar{x}_k$ has minimal period $k\tau$ (if k is prime and sufficiently large). Finally, by using (4.15), we easily get that $\bar{x}_k$ converges to 0 as $k \to \infty$ uniformly in C_1 $(\mathbb{R}, \mathbb{R}^N)$.

5. Final remarks.

Remark 5.1. Let us now observe that assumption $V_2)$ implies that $\sup \sigma(A) \geq 0$.[4]
Now assume that $V_1)$, $V_3)$, ..., $V_7)$ hold and suppose moreover that $\sup \sigma(A) \leq 0$. Then, by the arguments used in section 3, it easily follows that (1.1) possesses a periodic solution with minimal period $k\tau$ for $k \in \mathbb{N}$, $k \geq 2$.

Remark 5.2. Since theorem 1.6 has a local character it is possible to replace assumptions $V_3)'$ and $V_4)'$ by the following one:

$V^*)$ there exist $r,c > 0$ and $p > 2$

s.t.

$$\frac{1}{p} (\hat{V}' (x,t)|x) \geq \hat{V} (x,t) \geq c|x|^p$$

$$\forall\, x \in \mathbb{R}^N \quad |x| \leq r, \qquad \forall\, t \in \mathbb{R}$$

In fact, assume that V satisfies $V_1)$, $V_2)$, $V^*)$, $V_5)$, $V_6)$, $V_7)$ and carry out (see [8]) the following technical modification of the problem. Let $K \in \,]0,r[$, $\phi(s) = 1$ for $s \leq K$, $\phi(s) = 0$ for $s \geq K+1$ and $\phi'(s) < 0$ for $s \in \,]K, K+1[$ with $\phi \in C^\infty(\mathbb{R}, \mathbb{R})$. Set

$$V_k = \frac{1}{2} (Ax|x) + \phi (|x|) \hat{V}(t,x) + (1 - \phi (|x|)) R |x|^p .$$

For R sufficiently large V_k satisfies $V_1)$, $V_2)$, $V_3)'$, $V_4)'$, $V_5)$; then, as in section 3, a sequence $\{x_k\}$ of $k\tau$ periodic solutions $(k \in \mathbb{N})$ of $\ddot{x} + V_k'(x,t) = 0$ can be found. Moreover, since x_k converges to 0 in L^∞, x_k solves (1.1) for k large. Moreover, arguing as in section 3, it can be deduced that, if (1.4) holds and k is prime, x_k has minimal period $k\tau$.

(4) $\sigma(A)$ denotes the spectrum of A.

REFERENCES

[1] V. BENCI, Some Applications of the generalized Morse-Conley index to appear on "Conf.Sem.Mat.Univ.di Bari".

[2] V. BENCI - D. FORTUNATO, Subharmonic solutions of prescribed minimal period for non autonomous differential equations, preprint (1986).

[3] I. EKELAND, Une theorie de Morse pour les systèmes Hamiltoniens convexes, Ann. IHP, Analyse non lineaire, 1, 19-78 (1984).

[4] I. EKELAND - H. HOFER, Periodic solutions with prescribed period for convex autonomous Hamiltonian systems, preprint CEREMADE n. 8421, Paris (1984).

[5] M. GIRARDI - M. MATZEU, Periodic solutions of convex autonomous Hamiltonian systems with a quadratic growth at the origin and superquadratic at infinity, to appear on Ann.Mat.Pura Appl. .

[6] A. LAZER - S. SOLIMINI, Nontrivial solutions of operator equations and Morse indices of critical points of min-max type, preprint (1985).

[7] R. MICHALEK - G. TARANTELLO, Subharmonic solutions with prescribed minimal period for non autonomous Hamiltonian systems, preprint (1986).

[8] P.H. RABINOWITZ, On subharmonic solutions of Hamiltonian systems, Comm. Pure Appl.Math., 33, 609-633 (1980).

[9] C. VITERBO, Une thèorie de Morse pour les systèmes Hamiltoniens étoilés, these de III cycle, Univ. Paris Dauphine (1985).

NONRESONANCE NEAR THE FIRST EIGENVALUE OF A SECOND ORDER ELLIPTIC PROBLEM

Jean-Pierre GOSSEZ

Let Ω be a bounded open subset of $\mathbb{R}^N$ with a smooth boundary $\partial\Omega$ and let f be a continuous function from $\mathbb{R}$ to $\mathbb{R}$. We consider the problem

$$(1) \qquad \begin{cases} -\Delta u = f(u) + h(x) \text{ in } \Omega, \\ u = 0 \quad \text{on } \partial\Omega \end{cases}$$

where f is assumed to satisfy

$$(2) \qquad \liminf_{s \to \pm\infty} \frac{f(s)}{s} \geq \lambda_1 \text{ and } \limsup_{s \to \pm\infty} \frac{f(s)}{s} < \lambda_2 \ .$$

Here λ_1 and λ_2 denote the first two eigenvalues of $-\Delta$ on $H_0^1(\Omega)$. Thus our nonlinearity interferes with at most λ_1, without crossing it. Writing

$$f(u) = \lambda_1 u + g(u),$$

the conditions (2) become

$$(3) \qquad \liminf_{s \to \pm\infty} \frac{g(s)}{s} \geq 0 \quad \text{and} \quad \limsup_{s \to \pm\infty} \frac{g(s)}{s} < \lambda_2 - \lambda_1.$$

If we multiply (1) by an eigenfunction corresponding to λ_1 and integrate, we immediately see that a necessary condition for (1) to be solvable for any forcing term h is that the function $g:\mathbb{R} \to \mathbb{R}$ be unbounded from above and from below. We are interested here in the study of the converse implication.

It was proved by Dolph [3] (see also [2],[5],[7]) that if

(4) $\liminf_{s \to \pm\infty} \frac{g(s)}{s} > 0$,

then (1) is solvable for any $h \in L^2(\Omega)$. This was extended recently in at least two directions as for as nonresonance for (1) is concerned. From a result of De Figueiredo [4](see also [1]), it follows in particular that if the first condition in (3) is strengthened into $g(s)s \geq -a|s|-a'$ for some constants a,a' and all $s \in \mathbb{R}$ and if

(5) $\liminf_{s \to +\infty} g(s)=+\infty$ and $\limsup_{s \to -\infty} g(s)=-\infty$,

then (1) is solvable for any $h \in L^2(\Omega)$. Also, as a consequence of a result of De Figueiredo-Gossez [6], we have that if, for some $\eta > 0$,

(6) $$\begin{cases} \liminf_{n \to \infty} \mu\{s \in]0,n[\; ; g(s) \geq \eta s\}/n > 0 \text{ and} \\ \liminf_{n \to \infty} \mu\{s \in]-n,0[\; ; g(s) \leq \eta s\}/n > 0 , \end{cases}$$

then (1) is solvable for any $h \in L^p(\Omega)$, $p > N$; here μ denotes the Lebesgue measure on $\mathbb{R}$.

Condition (4) means that g has positive linear growth at infinity, which is not implied of course by condition (5). Condition (6) involves comparing g with a function having positive linear growth at infinity (the function $s \to \eta s$) but allows some sorts of oscillations. It is our purpose in this note to go a little bit further in the latter direction, by introducing a more general class of comparison functions. It turns out that the assumptions of the theorem below are also satisfied when condition (5) holds (see remark 3).

A function $\varphi:\mathbb{R} \to \mathbb{R}$ will belong to this class of comparison functions if it is non-decreasing with $\varphi(-\infty)=-\infty$, $\varphi(+\infty)=+\infty$ and satisfies $\varphi(s)s > 0$ for all $s\neq 0$ and

(7) $$|\varphi(2s)| \leq b|\varphi(s)|$$

for some constant b and all s with $|s|$ sufficiently large. Observe that condition

(7) is the Δ_2 condition from the theory of Orlicz spaces. It implies that for any $\alpha > \beta > 0$,

(8) $$|\varphi(\alpha s)| \leq b' |\varphi(\beta s)|$$

for some constant b' and all s with $|s|$ sufficiently large.

THEOREM. *Suppose that* (3) *holds and that one can find a comparison function* φ *as above such that*

(9) $$\liminf_{s \to \pm\infty} \frac{g(s)}{\varphi(s)} \geq 0$$

and

(10) $$\begin{cases} \liminf_{n \to \infty} \mu\{s \in]0,n[\ ; \ g(s) \geq \varphi(s)\}/n > 0 \quad \text{and} \\ \liminf_{n \to \infty} \mu\{s \in]-n,0[\ ; \ g(s) \leq \varphi(s)\}/n > 0. \end{cases}$$

Then, for any $h \in L^p(\Omega)$, $p > N$, (1) *has a solution* $u \in H_o^1(\Omega) \cap W^{2,p}(\Omega)$.

Here is an example of a nonlinearity to which this theorem applies but for which the conditions (5) and (6) do not hold :

$$g(s) = \begin{cases} \sqrt{s}\ (\sin s)^+ & \text{for } s \geq 0 \\ -(\log|s|)^+(\sin s)^+ & \text{for } s < 0; \end{cases}$$

here $(\ )^+$ denotes positive part.

The main idea in the proof is similar to that introduced in [6].

PROOF OF THE THEOREM. We consider the approximate problem

$$(11)\quad \begin{cases} -\Delta u_n - \frac{1}{n} u_n = f(u_n) + h(x) & \text{in } \Omega, \\ u_n = 0 & \text{on } \partial\Omega . \end{cases}$$

By (2) it satisfies a condition similar to (4) for each n sufficiently large and so is then solvable. Let u_n be a solution of (11). If $\|u_n\|$ remains bounded ($\| \ \|$ denotes the $H_o^1(\Omega)$ norm), then we easily obtain a solution for (1) which, by the L^p theory, will lie in $W^{2,p}(\Omega)$. So let us assume by contradiction that for a subsequence, $\|u_n\| \to \infty$. Let $v_n = u_n/\|u_n\|$. By the linear growth of f, the L^p theory and a bootstrap argument, we get, by dividing (11) by $\|u_n\|$, that v_n remains bounded in $W^{2,p}(\Omega)$. Thus v_n can be assumed to converge to v in $C^1(\bar{\Omega})$. In particular $v \not\equiv 0$ since $\|v_n\|=1$.

Using (3), we get that, for a subsequence, $g(u_n)/\|u_n\|$ converges weakly in $L^2(\Omega)$ to a function which can be written as $p(x)v(x)$, where the function p satisfies

$$0 \leq p(x) < \lambda_2 - \lambda_1 \quad \text{a.e.in } \Omega.$$

Consequently, dividing (11) by $\|u_n\|$ and going to the limit, we obtain that v satisfies

$$\begin{cases} -\Delta v - \lambda_1 v = p(x)v & \text{in } \Omega, \\ v = 0 & \text{on } \partial\Omega . \end{cases}$$

We now apply a lemma from [7] (see also [6]) to deduce that $p(x) \not\equiv 0$ is impossible. Thus $p(x) \equiv 0$ in Ω, i.e. v is an eigenfunction of $-\Delta$ on $H_o^1(\Omega)$ corresponding to λ_1. Multiplying (11) by $v/\|u_n\|$ and integrating,we then obtain

$$(12)\qquad \int_\Omega g(u_n)v < -\int_\Omega hv$$

for n sufficiently large.

From now on we assume $v > 0$ on Ω. Similar arguments hold when $v < 0$ on Ω.

Since the exterior normal derivative of v on $\partial\Omega$ is < 0 and $v_n \to v$ in $C^1(\bar{\Omega})$, we have that v_n is > 0 in Ω for n sufficiently large. This will be used later.

We first claim that there exists $d > 0$ such that

$$(13) \qquad \mu_N\{x \in \Omega;\ u_n(x)\neq 0 \text{ and } \frac{g(u_n(x))}{\varphi(u_n(x))} \geq 1\} \geq d$$

for n sufficiently large; here μ_N denotes the Lebesgue measure in $\mathbb{R}^N$. Indeed, for n sufficiently large,

$$\max u_n = \max(v_n\|u_n\|) \geq \frac{1}{2}\max v\ \|u_n\|$$

$$\geq \text{integral part of } (\frac{1}{2}\max v\ \|u_n\|) \equiv k(n) \geq \frac{1}{4}\max v\ \|u_n\|,$$

which implies range $(u_n) \supset]0,k(n)[$. On the other hand, for n sufficiently large,

$$\|\nabla u_n\|_\infty = \|\nabla v_n\|_\infty \|u_n\| \leq \frac{3}{2}\|\nabla v\|_\infty \|u_n\| \leq \frac{3}{2}\|\nabla v\|_\infty \frac{4k(n)}{\max v} .$$

We can now apply lemma 2 of [6] to the function u_n and the set

$$B_n = \{s \in]0,k(n)[\ ;\ \frac{g(s)}{\varphi(s)} \geq 1\} ,$$

using the assumption (10) that $\mu(B_n)$ tends to $+\infty$ with speed $k(n)$. This yields (13).

We now divide (12) by $\varphi(\alpha\|u_n\|)$ where $\alpha > 0$ is so chosen that $\|v_n\|_\infty \leq \alpha$ for all n. This gives

$$\limsup \int_\Omega \frac{g(u_n)}{\varphi(\alpha\|u_n\|)}\, v \leq 0 .$$

Denote by χ_n the characteristic function of $\{x \in \Omega; |u_n(x)| > s_o\}$ where $s_o > 0$ is taken such that $g(s) > -\gamma\varphi(s)$ for some $\gamma > 0$ and all $s \geq s_o$, which is clearly possible by (9). We then have

$$\limsup \int_\Omega \frac{g(u_n)}{\varphi(\alpha\|u_n\|)}\, v\, \chi_n \leq 0$$

and consequently

$$\liminf \int_{\Omega} \frac{g(u_n)}{\varphi(u_n)} \frac{\varphi(v_n\|u_n\|)}{\varphi(\alpha\|u_n\|)} v\, \chi_n\, \nu_n$$

$$+ \liminf \int_{\Omega} \frac{g(u_n)}{\varphi(u_n)} \frac{\varphi(v_n\|u_n\|)}{\varphi(\alpha\|u_n\|)} v\chi_n(1-\nu_n) \leq 0$$

where ν_n denotes the characteristic function of the set in (13). Since v_n has the same sign as v for n sufficiently large, the integrend in the first integral above is

$$\geq \frac{\varphi(v_n\|u_n\|)}{\varphi(\alpha\|u_n\|)} v\chi_n\nu_n \quad .$$

Moreover, by the choice of s_o and α , Fatou's lemma can be applied to the second integral. Using (9), we obtain

$$\liminf \int_{\Omega} \frac{\varphi(v_n\|u_n\|)}{\varphi(\alpha\|u_n\|)} v\, \chi_n\nu_n \leq 0$$

and, applying Lebesgue's theorem ,

(14) $$\liminf \int_{\Omega} \frac{\varphi(v_n\|u_n\|)}{\varphi(\alpha\|u_n\|)} v\nu_n \leq 0 \, .$$

We now introduce $\Omega' \subset \Omega$ such that $\mu_N(\Omega \setminus \Omega') < d/2$ and $v \geq \varepsilon$, $v_n \geq \varepsilon$ on Ω', for some $\varepsilon > 0$ and all n, which is possible since $v > 0$ in Ω and $v_n \to v$ uniformly; here d is the positive number found in (13). We then deduce from (14) that

$$\varepsilon \liminf \int_{\Omega'} \frac{\varphi(\varepsilon\|u_n\|)}{\varphi(\alpha\|u_n\|)} \nu_n \leq 0$$

and consequently, by (8),

$$c\, \varepsilon \liminf \int_{\Omega'} \nu_n \leq 0$$

for some constant $c > 0$. But this inequality is impossible since, by (13),

$$\mu_N[\{x \in \Omega;\ \nu_n(x)=1\} \cap \Omega'] \geq d/2 \, .$$

This completes the proof.

REMARK 1. A similar result holds when the nonlinearity f lies on the left side of λ_1 and satisfies a linear growth condition. One assume here

$$\limsup_{s \to \pm\infty} \frac{g(s)}{s} \leq 0 ,$$

$|g(s)| \leq A|s|+B$ for some constants A,B and all s,

as well as conditions (9) and (10) where g has been replaced by -g. In the approximate problem (11), $-u_n/n$ must then be changed into $+u_n/n$.

REMARK 2. A similar result holds when the operator $-\Delta$ is replaced in (1) by a second order uniformly elliptic operator of the form

$$-\sum_{i=1}^{N} \frac{\partial}{\partial x_i} (a_{ij}(x) \frac{\partial u}{\partial x_j})+a_o(x)u$$

with real-valued coefficients satisfying $a_{ij} \in C^{1+\alpha}(\bar{\Omega})$, $a_o \in C^{\alpha}(\bar{\Omega})$ and $a_{ij}=a_{ji}$. Observe that no assumption involving the unique continuation property is required.

REMARK 3. If g satisfies (5), then one can construct a comparison function φ such that the theorem applies. This follows from the lemma below and the observation that (7) holds if φ is concave for s large positive and convex for s large negative.

LEMMA. Let $g:\mathbb{R}^+ \to \mathbb{R}$ satisfy $g(s) \to +\infty$ as $s \to +\infty$. Then there exists $\varphi:\mathbb{R}^+ \to \mathbb{R}^+$ strictly increasing, concave, with $\varphi(s) \to +\infty$ as $s \to +\infty$, such that $g(s) \geq \varphi(s)$ for s sufficiently large.

PROOF. For every k=1,2,..., there exists n_k such that $g(s) \geq k$ for $s \geq n_k$. We can always assume $n_1 < n_2 <\ldots$ Put $\varphi(0)=0$, $\varphi(n_2)=1$ and extend φ linearly between 0 and n_2. Consider the line through $(0,\varphi(0))$ and $(n_2,\varphi(n_2))$ and let n_{k_o} be the greatest n_k with k=2,3,... such that this line meets the vertical above n_ℓ at a height $\leq \ell-1$ for all ℓ with $\ell=2,3,\ldots,k$. If such a n_{k_o} does not exist, then φ defined for $s \geq n_2$ by taking this line as its graph satisfies the requirements of the lemma. If such a n_{k_0} exists, then define φ between n_2 and n_{k_0} by taking this line as its graph. Put $\varphi(n_{k_0+1})=k_0$ and extend φ linearly between n_{k_0} and n_{k_0+1}. Iteration of the above construction to the right of n_{k_0+1} eventually yields a function φ with the desire requirements.

BIBLIOGRAPHY.

[1] H.BREZIS and L.NIRENBERG, Characterizations of the ranges of some nonlinear operators and applications to boundary value problems, Annali Sc.Norm.Sup.Pisa, 5 (1978), 225-326.

[2] E.DANCER, On the Dirichlet problem for weakly nonlinear elliptic partial differential equations, Proc.Royal Soc.Edinburgh, 76 (1977), 283-300.

[3] C.DOLPH, Nonlinear integral equations of the Hammerstein type, Trans.Amer.Math. Soc., 66 (1949), 289-307.

[4] D.DE FIGUEIREDO, Semilinear elliptic equations at resonance; higher eigenvalues and unbounded nonlinearities, in "Recent Advances in Differentiel Equations", Ac.Press 1981, 88-99.

[5] D.DE FIGUEIREDO and H.BERESTYCKI, Double resonance in semilinear elliptic problems, Comm.Part.Diff.Equat., 6 (1981), 91-120.

[6] D.DE FIGUEIREDO et J.-P.GOSSEZ, Conditions de non-résonance pour certains problèmes elliptiques semi-linéaires, C.R.Acad.Sc.Paris, 302 (1986), 543-545.

[7] J.MAWHIN and J.WARD, Non resonance and existence for nonlinear ellpitic boundary value problems, Nonli.Anal.,6 (1981), 677-684.

Département de Mathématique
Université Libre de Bruxelles
Campus de la Plaine C.P.214
Bd du Triomphe
1050 Bruxelles - Belgique

DIFFERENTIAL EQUATIONS IN THE SPECTRAL PARAMETER AND MULTIPHASE SIMILARITY SOLUTIONS

F. Alberto Grünbaum

Department of Mathematics
University of California
Berkeley, California 94720

CONTENTS

ABSTRACT

Multiphase similarity solutions were introduced by Flaschka and Newell [1] by imposing a specific type of differential equation in the spectral parameter for solutions of linear systems. The most general Schroedinger equation such that its solutions satisfy a very general kind of differential equation in the spectral parameter has been found in [2].

Here we draw attention to some relations between the results in these papers by writing the equations in [2] in the form

$$\Phi_x = P\Phi, \ \Phi_k = Q\Phi, \ \Phi_t = R\Phi \ \text{ with } \ P,Q,R$$

rational and by showing that all solutions in [2] have the multiphase structure introduced in [1].

We also produce families of rational multiphase similarity solutions of the m KdV equation.

Partially supported by NSF Grant #DMS84-03232
and ONR Contract #N00014-84-C-0159

1. Introduction

In [2] we raise and answer the question:

What are the potentials $V(x)$ such that one can exhibit a one parameter family of solutions $\phi(x,k)$ to the equation

$$L\phi = (-\partial_x^2 + V(x))\phi(x,k) = k^2\phi(x,k)$$

depending smoothly on the eigenvalue parameter k, and such that one also has an equation

$$B\phi \equiv \left(\sum_{i=0}^{m} b_i(k)\partial_k^i\right)\phi(x,k) = \theta(x)\phi(x,k)$$

with an eigenvalue θ a function of x?

The answer to the question posed above is rather simple: a differential operator B-depending only on k- and a function θ depending only on x such that the system

$$L\phi = k^2\phi \qquad (1)$$
$$B\phi = \theta\phi$$

is compatible, exist if and only if

$$\mathrm{ad}\, L^{m+1}(\theta) = 0 \qquad (2)$$

with m = order of B.

This converts the problem into one of finding all solutions to (2).

It turns out that the solutions to (2) are of two types. Either the system has a one dimensional vector bundle of solutions with sections

$$\phi(x,k) \quad x \in \mathbb{C} - \{\text{finite number of points}\},$$

$k \in \mathbb{C} - \{0\}$ or a 2 dimensional bundle over such base space.

In the first case it turns out that the potentials $V(x)$ coincide with the class of rational solutions of the Korteweg de Vries equation. These potentials can be described quite explicitely.

An equally explicit description is given in [2] for the potentials corresponding to the "rank two bundles". These potentials are always even functions of $x \in C$, and are referred to as the "even case" both in [2] as in our discussion below.

We remark here that Flaschka [3] has expressed the Painleve' equations in the form

$$\text{ad } L(B) \equiv [L,B] = L$$

which of course give

$$\text{ad } L^2(B) = 0$$

While the equation

$$\text{ad } L^{m+1}(\theta) = 0$$

implies

$$\text{ad } L^m(\theta) = \Sigma C_i((-L)^{i/2})_+$$

we have found that one actually has only one term in this sum. For example in the first KdV example we've

$$\text{ad } L^4(\theta) = cL^2$$

I feel that there is a deeper connection between the question raised in [2] and the Painleve' equations.

In the Kd-V case the potentials $V(x)$ are given by

$$V(x) = -2 \frac{\partial^2}{\partial x^2} \log \theta$$

where the first θ' are shown below

$$\theta_0 = 1, \quad \theta_1(x) = 0 \qquad \theta_2(x) = x^3 - p_3$$

$$\theta_3(x) = x^6 - 5p_3x^3 + 9p_5x - 5p_3^2$$

A look at these θ' shows that if one puts

$$z_i \equiv \frac{p_i^{1/i}}{x} \qquad i = 3,5,7, \ldots.$$

then the θ' are really functions of the form

$$\theta_j = x^{\frac{j(j+1)}{2}} P(z_3, z_5, \ldots)$$

This remark is useful in seeing that V are "multiphase similarity solutions" of the KdV hierarchy. See section 3.

2. The Equations $\Phi_x = P\Phi$, $\Phi_k = Q\Phi$, $\Phi_t = R\Phi$

In handling the "even case" in [2] one uses, in fact, some tools borrowed from Flaschka and Newell [4] paper on Isomonodromic Deformations.

For example if ϕ^+, ϕ^- denote the solutions of $L\phi = k^2\phi$ which look at infinity like e^{ikx} and e^{-kx} respectively, one forms the matrix

$$\Phi(x,k) \equiv \begin{pmatrix} \phi^+ & \phi^- \\ \phi^+_x & \phi^-_x \end{pmatrix}$$

satisfying

$$\partial_x \Phi = P\Phi \qquad P = \begin{pmatrix} 0 & 1 \\ V-k^2 & 0 \end{pmatrix}$$

and then shows that the matrix Q given by

$$\partial_k \Phi = Q\Phi \qquad Q = \begin{pmatrix} a(x,k) & b(x,k) \\ c(x,k) & d(x,k) \end{pmatrix}$$

has entries that are rational functions of x.

From the compatibility condition

$$\partial_k P + PQ = \partial_x Q + QP$$

one gets

$$\partial_x a = c - (V - k^2)b$$
$$\partial_x b = d - a$$
$$\partial_x c = -2k + (V - k^2)(a - d)$$
$$\partial_x d = (V - k^2)b - c$$

and eventually every entry in Q is given in terms of $b(x,k)$.

For the simplest example of the "even case" with

$$\theta = x^2 - t^2$$
$$V(x,t) = \frac{15x^4 + 18t^2x^2 - t^4}{4x^2(x^2 - t^2)^2}$$

we get

$$b(x,k) = \frac{x}{k} + \frac{t}{k^3(x-t)^2} - \frac{t}{k^3(x+t)^2} .$$

From here a, c, d are obtained from the general formulas

$$a = -\frac{1}{2}\partial_x b + \frac{1}{2k} \qquad a + d = \frac{1}{k}$$

$$C = (V - k^2)b - \frac{1}{2}\partial x^2 b$$

The existence of a pair of equations

$$\partial_x \Phi = P\Phi \qquad \partial_k \Phi = Q\Phi$$

is not necessarily linked to the "even case", although it was only used explicitly in this case in [2].

We illustrate this point with the first example of the KdV family, corresponding to

$$\theta = x^3 - p_3$$

We have

$$V(x) = -2(\log\theta)'' = \frac{6x(x^3 + 2p_3)}{(x^3 - p_3)^2}$$

and

$$\phi(x,k) = e^{ixk}\left(1 + \frac{3ix^2}{k(x^2 - p_3)} - \frac{3x}{k^2(x^2 - p_3)}\right)$$

If we set

$$\Phi = \begin{pmatrix} \phi(x,k) & \phi(x,-k) \\ \phi_x(x,k) & \phi_x(x,-k) \end{pmatrix}$$

we get

$$\partial_x \Phi = \begin{pmatrix} 0 & 1 \\ V-k^2 & 0 \end{pmatrix}\Phi$$

and

$$\partial_k \Phi = Q\Phi \equiv \begin{pmatrix} a & b \\ c & d \end{pmatrix}\Phi$$

with

$$b(x,k) = \frac{x}{k} + \frac{9p_3 x^2}{k^3(x^3 - p_3)^2}$$

while a, c, d are given in terms of b by the same relations that were given above as an expression of the compatibility condition

$$P_k - Q_x = [Q,P]$$

The equation in the x-variables is sometimes written in a different form, namely

$$\Phi_x = \begin{pmatrix} -ik & -1 \\ -V & ik \end{pmatrix} \Phi \qquad (3)$$

To achieve this we must build up the matrix Φ adroitly. The first row is still given by $\phi(x,k)$ and $\phi(x,-k)$ and this determines the second row of Φ if (3) is to be true.

The Φ constructed this second time, and the original one, satisfy

$$\det \Phi \equiv -2ik$$

If one wants to stay in the context of $SL(2,\mathbb{C})$ one can define

$$\Psi(x,k) = (-2ik)^{1/2}\Psi(x,k)$$

and obtain

$$\Psi_x = \begin{pmatrix} -ik & -1 \\ -V & ik \end{pmatrix} \Psi$$

and

$$\Psi_k = Q\Psi \equiv \begin{pmatrix} a & b \\ c & d \end{pmatrix}$$

For $b(x,k)$ we get the same expression as above, i.e.

$$b(x,k) = \frac{x}{k} + \frac{9p_3 x^2}{k^3(x^3 - p_3)^2}$$

while the expressions for a, c, d are simple modifications of the old ones in terms of b, i.e.

$$a + d = 0 \qquad (\text{i.e. } \operatorname{tr} Q = 0)$$

$$a = -\frac{1}{2}\partial_x b$$

$$c = (V - k^2)b - \frac{1}{2}\partial_x^2 b$$

In closing notice that we've

$$\Phi_x = P\Phi$$

$$\Phi_k = Q\Phi$$

and also

$$\Phi_{p_3} = R\Phi$$

with R given by the matrix

$$\begin{bmatrix} \dfrac{3x(2x^3 + p_3)}{k^2(x^3 - p_3)^3} & \dfrac{3x^2}{k^2(x^3 - p_3)^2} \\ -\dfrac{3k^2k^2 + 12}{k^2(x^3 - p_3)^2} - \dfrac{36p_3}{k^2(x^3 - p_3)^3} - \dfrac{27\,p_3^2}{k^2(x^3 - p_3)^4} & -\dfrac{3x(2x^3 + p_3)}{k^2(x^3 - p_3)^3} \end{bmatrix}$$

Notice too that all the matrices P, Q, R have entries that are rational in all its variables x, k, p_3.

This is a very strong restriction.

3. Similarity Solutions

Here we observe that the solutions $V(x)$ to our original problem are organized naturally into smooth manifolds and that as functions of these "coordinates" they are "multiphase similarity solutions" in the sense of [1]. Notice however that we do not have a natural equation floating around except in the KdV case. To stress this point we consider only the "even case" although the same arguments hold in all cases.

Put

$$V_2(x,t_2) \equiv -\frac{1}{4x^2} - 2\partial_x^2 \log(x^2 - t_2^2)$$

$$= \frac{15x^4 + 18t_2^2x^2 - t_2^4}{4x^2(x^2 - t_2^2)^2}$$

and observe that

$$V_2(x,t_2) = \frac{1}{x^2} V_2\left(1, \frac{t_2}{x}\right)$$

Similarly

$$V_3(x,t_4) \equiv -\frac{1}{4x^2} - 2\partial_x^2 \log(\frac{3}{4} x^{9/2} + t_4^4 x^{1/2})$$

$$= \frac{315x^8 - 1080t_4^4x^4 + 48t_4^8}{36x^{10} + 96t_4^4x^6 + 64t_4^8x^2}$$

and we've

$$V_3(x,t_4) = \frac{1}{x^2} V_3\left(1, \frac{t_4}{x}\right)$$

The two families of functions $V(x,\cdot)$ come from the "even family" in [2].

The next two examples illustrate the "multiphase" aspect mentioned above.

Set

$$V_4(x,t_4,t_6) = -\frac{1}{4x^2} - 2\partial_x^2 \log\left(\frac{15}{32} x^8 + \frac{15}{4} t_4^4x^4 + t_6^6x^2 - \frac{5}{2} t_4^8\right)$$

Then we have

$$V_4(x,t_4,t_6) = \frac{1}{x^2} V_4\left(1, \frac{t_4}{x}, \frac{t_6}{x}\right)$$

Finally if

$$V_5(x,t_6,t_8) = -\frac{1}{4x^2} - 2\partial_x^2 \log\left(\left(\frac{525}{2848} x^{12} + \frac{35}{8} t_8^6x^6 + \frac{3}{4} t_8^8x^4 - \frac{7}{3} t_6^{12}\right)x^{1/2}\right)$$

we get, once again,

$$V_5(x,t_6,t_8) = \frac{1}{x^2} V_5\left(1, \frac{t_6}{x}, \frac{t_8}{x}\right)$$

We could go through the list of potentials associated with the KdV equation with the same result. The fact that the θ' are homogeneous functions in the variables $x \equiv p_1, p_3^3, p_5^5, \ldots$ gives the same result as above.

For example, from

$$V_3(x,t,p_5) = -2\partial_x^2 \log(x^6 + 5t^3x^3 + 9p_5^5x - 5t^6)$$

we get

$$V_3(x,t,p_5) = \frac{1}{x^2} V_3\left(1, \frac{t}{x}, \frac{p_5}{x}\right)$$

4. Rational Solutions of m KdV

In [5] Airault produces rational solutions for the m KdV equation by observing that if

$$v(x,t) = t^{-1/3} y(z) \qquad z = xt^{-1/3}$$

is a solution of

$$6v_t = 3v^2v_x - \frac{1}{2} v_{xxx} \tag{4}$$

then $y(z)$ solves

$$y'' = 2y^3 + 4zy + \delta$$

By a change of variables this equation is reduced to the second Painlevé equation

$$y'' = 2y^3 + zy + \delta \tag{5}$$

Now Airault proves that (5) has a rational solution exactly when $\delta = n =$ integer ($\neq 0$).

The first few solutions of (5) are given below

$$y_1(z) = -\frac{1}{z}$$

$$y_2(z) = -\frac{2(z^3 - 2)}{z(z^3 + 4)}$$

$$y_3(z) = -\frac{3z^2(z^6 + 8z^3 + 160)}{(z^3 + 4)(z^6 + 20z^3 - 80)}$$

Actually one can produce all the solutions in [5] by setting - as in [5].

$$y_n(z) = -\frac{u'_n}{u_n} + \frac{u'_{n-1}}{u_{n-1}}$$

and making the observation that

$$u_n(z) = \theta_n(z, -4, 0,0, \ldots, 0) \qquad n \geq 0$$

where θ_n are the characters of appropriate representations of $GL(N,\mathbb{C})$, see [2, section 7].

Notice that in this fashion we produce a sequence of rational similarity solutions of the m KdV equation. However these are not "families of solutions" as in the KdV case. More on this below.

The first solution obtained from $y_1(z)$ is given by

$$v(x,t) = \pm \frac{1}{x}$$

The second one, obtained by putting

$$v(x,t) = t^{-1/3} \quad b \quad y(axt^{-1/3})$$

with

$$a = 4^{1/3} \qquad \text{and} \qquad b = \pm 4^{1/3}$$

is

$$v(x,t) = \pm \frac{2x^3 - t}{x(x^3 + t)}$$

$$= \pm\left[-\frac{1}{x} + \frac{1}{(x + t^{1/3})} + \frac{1}{(x + \omega t^{1/3})} + \frac{1}{(x + \omega^2 t^{1/3})}\right]$$

$$\text{with } \omega^3 = 1$$

The third one, obtained from $y_3(z)$ by setting

$$v(x,t) = t^{-1/3} \quad b \quad y_3(axt^{-1/3})$$

with

$$a = 4^{1/3} \qquad b = \pm 4^{1/3}$$

is given by

$$v(x,t) = \pm \frac{3x^2(x^6 + 2tx^3 + 10t^2)}{(x^3 + t)(x^6 + 5tx^3 - 5t^2)}$$

$$= \pm\left(-\frac{\theta_3'(x, -t, 0)}{\theta_3(x, -t, 0)} + \frac{\theta_2'(x, -t)}{\theta_2(x, -t)}\right)$$

with $\theta_2(x,t) = x^3 + t$, $\theta_3(x,t,p_5) = x^6 + 5tx^3 - 5t^2 + 9p_5x$ as one has in [2, section 7].

This expression suggests that one can keep $p_5 \neq 0$ and obtain "families of solutions" $v(x,t,p_5)$ just as in the KdV case.

Put

$$v(x,t,p_5) = \pm\left(-\frac{\theta_3'(x,-t,p_5)}{\theta_3(x,-t,p_5)} + \frac{\theta_2(x,-t)}{\theta_2(x,-t)}\right)$$

$$= \pm\frac{1}{x}\left\{\frac{3+6\left(\frac{t}{x^3}\right)+30\left(\frac{t}{x^3}\right)^2-18\left(\frac{p_5}{x^5}\right)+g\frac{p_5}{x^5}\frac{t}{x^3}}{1+6\left(\frac{t}{x^3}\right)+9\frac{p_5}{x^5}+9\frac{p_5}{x^5}\frac{t}{x^3}-5\left(\frac{t}{x^3}\right)^3}\right\}$$

$$= \pm\frac{1}{x}\,p\left(\frac{x}{t^{1/3}}, \frac{x}{p_5^{1/5}}\right)$$

i.e. we get a multiphase similarity solution of the type discussed in [1].

We close this section by proving that the expression

$$v(x,t,p_5,p_7,\ldots,p_{2n-1}) = \pm\left\{-\frac{\theta_n'(x,-t,p_3,p_5,\ldots,p_{2n-1})}{\theta_n(x,-t,p_3,p_5,\ldots,p_{2n-1})} + \frac{\theta_{n-1}'(x,-t,p_3,\ldots,p_{2n-3})}{\theta_{n-1}(x,-t,p_3,\ldots,p_{2n-3})}\right\}$$

gives a rational multiphase similarity solution of the m KdV equation (4), where θ_n are defined as in [2, section 7].

From

$$v = -\frac{\theta_n'}{\theta_n} + \frac{\theta_{n-1}}{\theta_{n-1}}$$

obtain

$$v_x + v^2 = -2\left(\frac{\theta_n'}{\theta_n}\right)^2 - \frac{\theta_n''}{\theta_n} + \frac{\theta_{n-1}''}{\theta_{n-1}} - 2\frac{\theta_n'}{\theta_n}\frac{\theta_{n-1}'}{\theta_{n-1}}$$

Now we claim that this is the same as

$$U = -2\partial x^2 \log \theta_n = -2\left(\frac{\theta_n'}{\theta_n}\right)'$$

as a consequence of the identity

$$\frac{\theta_n''}{\theta_n} + \frac{\theta_{n-1}''}{\theta_{n-1}} - 2\frac{\theta_n'}{\theta_n}\frac{\theta_{n-1}'}{\theta_{n-1}} = 0$$

which can be proved from

$$\theta_{k+1}'\theta_{k-1} - \theta_{k+1}\theta_{k-1}' = (2k+1)\theta_k^2,\ \theta_0 = 1,\ \theta_1(x) = x$$

or as in lemma 1 of [6].

Since U is a rational solution of the KdV equation and the map $U = v_x + v^2$ is the Miura transformation relating solutions of KdV and m KdV we are finished. This explicit computation is nothing but the fact that one solves the equation

$$U = v_x + v^2$$

by using the Riccati substitution

$$v = \phi_x/\phi$$

with ϕ any solution of

$$-\phi_{xx} + U\phi = 0$$

Since, see [2, section 7.7] we have

$$\phi(x,0) = \frac{\theta_{n-1}}{\theta_n}$$

the result above just illustrates the fact that the m KdV equation describes the evolution of the "Darboux factors" of the Schroedinger operator $-\partial x^2 + U$ when U evolves according to KdV. One actually gets

$$v = \partial_x \log \frac{\theta_{n-1}}{\theta_n}$$

References

[1] H. Flaschka, A. Newell, Multiphase similarity solutions of integrable evolution equations. Physica 3D (1981) 1 &. 2, 203-221.

[2] J.J. Duistermaat, F.A. Grünbaum, Differential equations in the spectral parameter. Comm. Math. Physics, 103, 177-240 (1986).

[3] H. Flaschka, A commutator representation of Painlevé equation. J. Math. Physics 21(5), 1016-1018, 1980.

[4] H. Flaschka, A. Newell, Monodromy- and spectrum-preserving deformations. Comm. Math. Physics 76, 65-116, (1980).

[5] H. Airault, Rational solutions of Painlevé equations Studies in App. Math. 61, 31-53 (1979).

[6] M. Adler, J. Moser, On a class of polynomials connected with the KdV equation. Comm. in Math. Physics 61, 1-30 (1978)

RECENT RESULTS ON SEMI-LINEAR HYPERBOLIC PROBLEMS IN BOUNDED DOMAINS

A. HARAUX(*)

1. GENERALITIES

Let Ω be a bounded open domain in $\mathbb{R}^n$, $n \geq 1$ and f,g two <u>odd</u> functions: $\mathbb{R} \to \mathbb{R}$ such that,

(1.1) $f \in C^1(\mathbb{R}),\ f' \geq 0$

(1.2) g is continuous and non-decreasing.

The partial differential equation

$$(1.3) \quad \begin{cases} \dfrac{\partial^2 u}{\partial t^2} - \Delta u + f(u) + g(\dfrac{\partial u}{\partial t}) = h(t,x), & (t,x) \in \mathbb{R}^+ x \Omega \\ \quad u(t,x) = 0 & (t,x) \in \mathbb{R}^+ x \partial\Omega \end{cases}$$

represents the small oscillations of an n-dimensional membrane with fixed edge $\{0\} x \partial\Omega$ under the action of 4 forces

- The elastic (global, distributed) force represented by Δu.
- A non linear <u>restoring force</u> $-f(u)$, of local character.
- A <u>damping force</u> $-g(\frac{\partial u}{\partial t})$, possibly non linear, also of local character.
- An exterior time-dependent force density $h(t,x)$.

The physical interpretation of (1.3) suggests that the initial value problem must be well set in the function space $C(\mathbb{R}^+, H_0^1(\Omega)) \cap C^1(\mathbb{R}^+, L^2(\Omega))$ and that the <u>total energy</u> given by

$$(1.4) \quad E(t) = \int_\Omega \left\{ \frac{1}{2} (\frac{\partial u}{\partial t})^2 + \frac{1}{2} |\Delta u|^2 + \int_0^{u(t,x)} f(s)ds \right\} dx$$

must play an important role in the definition of solutions to (1.3).

Equation (1.3) has been studied by a number of authors since 1953. More precisely, the initial-value problem has been studied successively

(*) Analyse Numerique T. 55-65, 5^{e} etage, Univ. P. et M. Curie, 4, Place Jussieu 75252 PARIS CEDEX 05 (FRANCE)

by Ficken-Fleishman [14], Amerio - Prouse [2], Lions-Strauss [27] and Brezis [7].

The existence of periodic solutions has been investigated by Ficken-Fleishman [14] for a special one-dimensional problem and then by Prodi [33], [34]. All these works except [14] consider the case $f = 0$. In this framework the existence of <u>bounded</u> (in the energy norm) and <u>almost periodic</u> solutions has been studied by G. Prouse [35], Biroli [5] and later by Biroli-Haraux [6] and Haraux ([17] - [21]). Stability questions have also been considered by several of the above quoted authors, and by Nakao [31], [32], Yamada [38], Marcati [28], [29]. Except for the stability of the zero solution, very few authors considered the case of a non linear term $f(u)$. After the pioneering work of P. Rabinowitz [36] (cf. also [37] and Brezis Coron-Nirenberg [8]) attempts were made to understand the nature of oscillations when both g and $h = 0$ (the so-called conservative case). In this direction c.f. e-g. [11], [13] and [22]. The global behavior of solutions in the general case where f,g,h are non trivial is unknown. When h does not depend on t, a theory of <u>attractors</u> have been developped by Babin-Vishik [4], Hale [16] and Ghidaglia-Temam [15]. (cf. also [20]) for more general terms $f(u)$. In this survey paper, we limit ourselves to the description of a few typical recent results obtained by the author and some of his co-workers. The proofs of most of these results can be <u>found</u> either in [24] or in [25]. For the others we refer to the corresponding research articles.

2. THE INITIAL VALUE PROBLEM

The following result, containing most of the previous existence-uniqueness theorems on (1.3), is established in [23].

<u>Theorem 2.1</u>. Let $u_o \in H^1_o(\Omega)$, $v_o \in L^2(\Omega)$ and $h \in L^1_{loc}(\mathbb{R}^+, L^2(\Omega))$. Assume that f satisfies the additional condition

$$\exists\, s \geq 0,\ (n-2)s \leq 2,\ \exists\, C \geq 0,$$

$$f'(v) \leq C(1+|v|^s),\quad \forall\, v \in \mathbb{R} \tag{2.1}$$

Then there exists a unique function u in $C(\mathbb{R}^+, H^1_o(\Omega)) \cap C^1(\mathbb{R}^+, L^2(\Omega))$ such that

$$u(0) = u_o \quad \text{and} \quad \frac{\partial u}{\partial t}(0) = v_o \tag{2.2}$$

$$g\left(\frac{\partial u}{\partial t}\right)\frac{\partial u}{\partial t} \in L^1\,(]0,T[\times\Omega) \quad \text{for all} \quad T > 0 \tag{2.3}$$

(2.4) $$\frac{\partial^2 u}{\partial t^2} - \Delta u + f(u) + g(\frac{\partial u}{\partial t}) = h \quad \text{in} \quad \mathcal{D}'(\mathbb{R}^+ x\Omega)$$

In addition, the energy $E(t)$ defined by (1.4) is absolutely continuous on each compact interval of $\mathbb{R}^+$ and we have

(2.5) $$\frac{d}{dt}(E(t)) = \int_\Omega \left[h \frac{\partial u}{\partial t} - g(\frac{\partial u}{\partial t}) \frac{\partial u}{\partial t}\right](t,x)dx$$

a.e on $\mathbb{R}^+$

The unique function u so defined will be called <u>the</u> solution of (1.3) with initial data $u(0) = u_o$ and $\frac{\partial u}{\partial t}(0) = v_o$.

3. STABILITY OF THE ZERO SOLUTION

In this section we consider the case $h \equiv 0$. Then (1.3) reduces to

(3.1) $$\begin{cases} \frac{\partial^2 u}{\partial t^2} - \Delta u + f(u) + g(\frac{\partial u}{\partial t}) = 0, & (t,x) \in \mathbb{R}^+ x\Omega \\ u(t,x) = 0 & (t,x) \in \mathbb{R}^+ x\partial\Omega \end{cases}$$

In the case $f = 0$, we have

<u>Proposition 3.1.</u> Assume that $f = 0$ and g satisfies

(3.2) $$\forall\, v \in \mathbb{R}, \quad v \neq 0 \Longrightarrow g(v)v > 0$$

Then under the hypotheses of Theorem 2.1, any solution of (3.1) satisfies

(3.3) $$\lim_{t\to+\infty} \left\{ \|u(t)\|_{H_o^1(\Omega)} + \left|\frac{\partial u}{\partial t}(t)\right|_{L^2(\Omega)} \right\} = 0 .$$

The proof is a straightforward application of La Salle's invariance principle (cf. also [24]). When f is non zero, the problem is generally open to decide whether or not (3.3) is still satisfied. In general we obtain the weaker conclusion

$$\lim_{t\to+\infty} \left\{ |u(t)|_{L^2(\Omega)} + \left|\frac{\partial u}{\partial t}(t)\right|_{L^1(\Omega)} \right\} = 0 .$$

Under suitable coerciveness and growth assumptions on g, it is possible to check that (3.3) is fulfilled. (cf. [24] for the details).

4. NON RESONANCE AND STABILITY IN THE TIME-DEPENDENT CASE

When h depends on t, the problem of asymptotic behavior as $t \to +\infty$ is much more delicate. For instance if $f = g = 0$, $\Omega =]0,\pi[$ and $h(t,x) = 2\cos t \sin x$, we have the unbounded solution $u(t,x) = t \sin t \sin x$. (This is called a resonance phenomenon). Under some coerciveness and growth conditions on the damping term $g(\frac{\partial u}{\partial t})$ we shall prove that the resonance phenomenon dissappears. The necessity of a growth assumption is not clear from the point of view of mechanics, and mathematically the problem of removing it is open since 1966. The best general result known at the present time for any forcing term $h \in L^\infty(\mathbb{R}^+, L^2(\Omega))$ can be stated as follows.

Theorem 4.1. Assume that f fulfills (1.1) and (2.1), g satisfies (1.2) and

(4.1) $\quad \exists\, \alpha \geq 0,\ \exists\, C \geq 0,\ \forall\, v \in \mathbb{R},\quad g(v)v \geq \alpha|v|^2 - C$

(4.2) $\quad \exists\, M \geq 0,\ \exists\, s \geq 0,\ (n-2)s \leq n+2$ such that $\forall\, v \in \mathbb{R},\ |g(v)| \leq C(1+|v|^s)$.

Then for any $h \in L^\infty(\mathbb{R}^+, L^2(\Omega))$ the solutions u of (1.3) are such that

(4.3) $\quad u \in L^\infty(\mathbb{R}^+, H_0^1(\Omega)),\ \frac{\partial u}{\partial t} \in L^\infty(\mathbb{R}^+, L^2(\Omega)).$

Sketch of the proof. Let $F \in C^2(\mathbb{R})$ be defined by

$$F(y) = \int_0^y f(s)ds, \qquad \forall\, s \in \mathbb{R}.$$

We have

$$E(t) = \int_\Omega \left\{\frac{1}{2}\left[\left(\frac{\partial u}{\partial t}\right)^2 + |\nabla u|^2\right] + F(u)\right\} dx$$

and as a consequence of (2.5)

$$\frac{d}{dt}(E(t)) = \int_\Omega h \frac{\partial u}{\partial t} dx - \int_\Omega g\left(\frac{\partial u}{\partial t}\right) \frac{\partial u}{\partial t} dx .$$

Therefore

$$(4.4)\quad \begin{aligned} \frac{d}{dt}\left(\frac{2}{3} E^{3/2}(t)\right) &\leq -E^{1/2}(t) \int_\Omega g\left(\frac{\partial u}{\partial t}\right) \frac{\partial u}{\partial t} dx + E^{1/2}(t)\left|\frac{\partial u}{\partial t}\right|_2 |h|_2 \leq \\ &- \frac{1}{2} E^{1/2} \int_\Omega g\left(\frac{\partial u}{\partial t}\right) \frac{\partial u}{\partial t} dx - \frac{\alpha}{2} E^{1/2} \left|\frac{\partial u}{\partial t}\right|_2^2 + E^{1/2}\left(\frac{C}{2} + \left|\frac{\partial u}{\partial t}\right|_2 |h|_2\right) \end{aligned}$$

As a consequence of (4.2) we have

$$\exists\, C_1 \geq 0, \quad \forall\, v \in L^2(\Omega),$$

(4.5) $$\|g(v)\|_{H^{-1}(\Omega)} \leq C_1\left(1 + \int_\Omega g(v)v\,dx\right).$$

We compute

$$\frac{d}{dt}\int_\Omega u\,\frac{\partial u}{\partial t}\,dx = -\int_\Omega |\nabla u|^2 dx - \int_\Omega f(u)u\,dx -$$

$$-\int_\Omega g\left(\frac{\partial u}{\partial t}\right)u\,dx + \int_\Omega hu\,dx + \int_\Omega \left|\frac{\partial u}{\partial t}\right|^2 dx.$$

Hence as a consequence of (1.1) and (4.5)

$$\frac{d}{dt}\int_\Omega u\,\frac{\partial u}{\partial t}\,dx \leq -\frac{1}{2}\int_\Omega |\nabla u|^2 dx - \int_\Omega F(u)dx + \int_\Omega \left|\frac{\partial u}{\partial t}\right|^2 dx +$$

(4.6) $$+ C_2\int_\Omega |h|^2 dx + C_1\, E^{1/2}\left(1 + \int_\Omega g\left(\frac{\partial u}{\partial t}\right)\frac{\partial u}{\partial t}\,dx\right).$$

By combining (4.4) and (4.6) we deduce that for $\varepsilon > 0$ small enough

$$\frac{d}{dt}\left(\frac{2}{3}E^{3/2} + \varepsilon\int_\Omega u\,\frac{\partial u}{\partial t}\,dx\right) \leq -\varepsilon E + C_3(|h|_{L^\infty(\mathbb{R}^+,L^2(\Omega))}).$$

By introducing

$$V(t) = \frac{2}{3}E^{3/2}(t) + \varepsilon\int_\Omega u\,\frac{\partial u}{\partial t}(t,x)dx + K,$$

for K large enough we have $V > 0$ everywhere on $[0, +\infty[$ and

$$\frac{dV}{dt} \leq \delta\, V^{2/3} + C_4(h), \text{ a.e on } \mathbb{R}^+$$

for some $\delta > 0$.

It follows at once that

$$\forall\, t \geq 0, \quad V(t) \leq \mathrm{Max}[V(0), \left(\frac{C_4(h)}{\delta}\right)^{3/2}]$$

and by the choice of $V(t)$, this clearly implies (4.3).

5. THE CONSERVATIVE CASE

When $g = h = 0$, (2.5) yields the <u>energy conservation</u>

$$\int_{\Omega} \left\{ \frac{1}{2} \left(\frac{\partial u}{\partial t}\right)^2 + \frac{1}{2} |\nabla u|^2 + F(u) \right\}(t,x)dx = \text{constant.}$$

If in addition $f = 0$, it is known (cf. [3]) that $u(t,\cdot): \mathbb{R} \to H_0^1(\Omega)$ is almost periodic. It is not difficult to check (cf. e.g [11]. Proposition 2.1) that for any u_0, v_0 in $\mathcal{D}(\Omega)$ and $x_0 \in \Omega$ the function $t \to u(t,x_0)$ is either identically zero, or oscillatory in the sense that it changes sign on any unbounded interval. In the special case $n = 1$ the function $u(t,x_0)$ is periodic with mean-value 0 and a period $2|\Omega|$. Setting $\Omega =]0, \ell[$ we conclude that either $u(t,x_0) \equiv 0$ on $\mathbb{R}$, or for any $a \in \mathbb{R}$, there exists t_1, t_2 in $]a, a+2\ell[$ with $u(t_1,x_0)\, u(t_2,x_0) < 0$. This remarkable property of the linear wave equation in one dimension has been extended in Cazenave-Haraux [9] where it is shown that the same alternative remains true for $u(t,x_0)$ when $0 < x_0 < \ell$ and u is a solution of

$$\frac{\partial^2 u}{\partial t^2} - \frac{\partial^2 u}{\partial x^2} + f(u) = 0, \quad (t,x) \in \mathbb{R} \times]0, \ell[$$

$$u(t,0) = u(t,\ell) = 0 \qquad t \in \mathbb{R}$$

as soon as f is odd and non-decreasing.

In the opposite direction, when Ω is a "generic" rectangle in $\mathbb{R}^2$, it is shown in [26] that for some points $(x_0,y_0) \in \Omega$, and for $T > 0$ arbitrarily large, we can find a solution u of

$$\frac{\partial^2 u}{\partial t^2} - \frac{\partial^2 u}{\partial x^2} - \frac{\partial^2 u}{\partial y^2} = 0 \qquad \text{on} \quad \mathbb{R} \times \Omega$$

$$u(t,x,y) = 0 \qquad \text{on} \quad \mathbb{R} \times \partial\Omega$$

$$u(t,x_0,y_0) \geq 1 \qquad \text{on} \quad [0,T].$$

In the case $n \geq 2$, other investigation of oscillatory properties of solutions is now being pursued (cf. [11], [12]).

BIBLIOGRAPHY

[1] R.R. ADAMS, Sobolev spaces, Academic press, New York (1975).

[2] L. AMERIO, G. PROUSE, Uniqueness and almost periodicity theorems for a non-linear wave equation, Atti Accad.Naz. Lincei Rend. Cl.Sci.Fis.Mat.Natur. 46 (1969), 1-8.

[3] L. AMERIO, G. PROUSE, Abstract almost periodic functions and functional equations, Van Nostrand, New-York (1971).

[4] A.V. BABIN, M.I. VISHIK, Regular attractors of semi-groups and evolution equations, J. Math. Pures et Appl. 62 (1983), 44-491.

[5] M. BIROLI, Bounded or almost-periodic solutions of the non linear vibrating membrane equation, Ricerche Mat. 22 (1973). 190-202.

[6] M. BIROLI, A. HARAUX, Asymptotic behavior for an almost periodic, strongly dissipative wave equation, J.Diff.Eq. 38, 3 (1980), 422-440.

[7] H. BREZIS, Problèmes unilatéraux, J.Math. Pure et Appl. 51 (1972), 1-168.

[8] H. BREZIS, J.M. CORON, L. NIRENBERG, Free vibrations for a non linear wave equation and a theorem of P. Rabinowitz, C.P.A. M. 33 (1980), 667-689.

[9] T. CAZENAVE, A. HARAUX, Propriétés oscillatoires des solutions de certaines equations des ondes semi-linéaires, C.R.A.S. Paris, 298 (1984), 449-452.

[10] T. CAZENAVE, A. HARAUX, Oscillatory phenomena associated to semi-linear wave equations in one spatial dimension. Trans. A.M.S., in press.

[11] T. CAZENAVE, A. HARAUX, On the nature of free oscillations associated with some semilinear wave equations, in "Non-linear partial differential equations and their applications, College de France Seminar", vol. 7(H-Brezis & J.L. Lions Editors), Research Notes in Math. nº 122, Pitman (1984), 59-79.

[12] T. CAZENAVE, A. HARAUX, Some oscillatory properties of the wave equation in several space dimensions, to appear.

[13] T. CAZENAVE, A. HARAUX, L. VAZQUEZ, F.B. WEISSLER, Nonlinear effects in the wave equation with a cubic restoring force, to appear.

[14] F.A. FICKEN, B.A. FLEISMANN, Initial value problems and time-periodic solutions for a nonlinear wave equation, C.P.A.M. 10,3 (1957), 331-356.

[15] J.M. GHIDAGLIA, R. TEMAM, Attractors for damped nonlinear hyperbolic equations, to appear in J.Math. Pure et Appl. .

[16] J.K. HALE, Asymptotic behavior and dynamics in infinite dimensions, Nonlinear Differential Equations (Hale & Martines-Amores editors),Research Notes in Math. nº 132, Pitman (1985), 1-42.

[17] A. HARAUX, Nonlinear evolution equations: Global behavior of solutions, Lecture Notes in Math. nº 841. Springer (1981).

[18] A. HARAUX, Almost periodic forcing for a wave equation with a nonlinear, local damping term. Proc. Roy. Soc. Edinburgh, 94 A(1983), 195-212.

[19] A. HARAUX, Dissipativity in the sense of Levison for a class of second order nonlinear evolution equations. Nonlinear Analysis, T.M.A. 6,11 (1982), 1207-1220.

[20] A. HARAUX, Two remarks on dissipative hyperbolic problems, in "Nonlinear partial differential equations and their applications, College de France Seminar", vol. 7 (H. Brezis & J.L. Lions editors), Research Notes in Math. nº 122, Pitman (1984), 161-179.

[21] A. HARAUX, Non-resonance for a strongly dissipative wave equation in higher dimensions, Manuscripta Math. 53 (1985), 145-166.

[22] A. HARAUX, Propriétés d'oscillation des solutions de l'équation des ondes avec conditions de Dirichlet au bord, Séminaire Bony-Sjostrand-Meyer 1985, Exp. nº 9.

[23] A. HARAUX, A new characterization of weak solutions to the damped wave equation, Publ. Lab. d'analyse numérique nº 85039 (1985), 16p.

[24] A. HARAUX, Semi-linear hyperbolic problems in bounded domains, to appear in "Mathematical reports", J. Dieudonné Editor.

[25] A. HARAUX, Nonlinear vibrations and the wave equation, Textos e Notas (L.A. Medeiros, Editor), to appear.

[26] A. HARAUX, V. KOMORNIK, Anharmonic Fourier and the wave equation. Rev.Mat. Ibero-Americana, in press.

[27] J.L. LIONS, W.A. STRAUSS, Some non-linear evolution equations Bull.Soc.Math. France 93 (1965), 43-96.

[28] P.A. MARCATI, Decay and Stability for nonlinear hyperbolic equations, J.Diff.Eq. 55, 1 (1984), 30-58.

[29] P.A. MARCATI, Stability for second order abstract evolution equations, Nonlinear Analysis, T.M.A. 8,3 (1984), 237-252.

[30] M. NAKAO, On boundedness, periodicity and almos periodicity of solutions of some nonlinear partial differential equations, J. Diff.Eq. 19, (1975), 371-385.

[31] M. NAKAO, Asymptotic stability of the bounded or almost periodic solution of the wave equation with a nonlinear dissipative term, J.Math.Anal.Appl. 58 (1977), 336-343.

[32] M. NAKAO, A difference inequality and its applications to nonlinear evolution equations, J.Math.Soc.Japan 30, 4 (1978), 747-762.

[33] G. PRODI, Soluzioni periodiche di equazioni a derivati parziali di tipo iperbolico non lineari, Ann.Mat. Pura Appl. 42 (1956) 25-49.

[34] G. PRODI, Soluzioni periodiche della equazione delle onde contermine dissipativo non lineare, Rend.Sem.Mat.Univ. Padova 36 (1966), 3749.

[35] G. PROUSE, Soluzioni quasi-periodiche della equazione delle onde con termine dissipativo non lineare, I, II, III, IV, Rend. Accad.Naz.Lincei 38,39 (1965).

[36] P.H. RABINOWITZ, periodic solutions of nonlinear hyperbolic partial differential equations, C.P.A.M. 20 (1967), 145-205.

[37] P.H. RABINOWITZ, Free vibrations for a semi-linear wave equation, C.P.A.M. 31 (1978), 31-68.

[38] Y. YAMADA, On the decay of solutions for some nonlinear evolution equations of second order, Nagoya Math.J. 73 (1979), 69-98.

Systems of Homogeneous Partial Differential Equations with Few Solutions

Howard Jacobowitz*
Department of Mathematical Sciences
Rutgers - The State University
Camden, N.J. 08102

In 1957, at a time when many other mathematicians were trying to show that every linear partial differential equation is solvable, Hans Lewy published a deceptively simple counter-example. (See [T] for interesting historical remarks in addition to an excellent exposition of the solvability problem.) The operator Lewy used comes from several complex variables and belongs to a class already exploited by Poincaré [Po]. This is the class of the induced Cauchy-Riemann operators. Let (z,w) be coordinates for $\mathbb{C}^2$ and think of $\frac{\partial}{\partial z} = \frac{1}{2}(\frac{\partial}{\partial x} - i\frac{\partial}{\partial y})$ and $\frac{\partial}{\partial w} = \frac{1}{2}(\frac{\partial}{\partial u} - i\frac{\partial}{\partial v})$ as complex vector fields. Let $M^3 \subset \mathbb{C}^2$ be a real hypersurface defined by $r(z,\bar{z},w,\bar{w}) = 0$. The complex vector field $P = r_{\bar{w}}\frac{\partial}{\partial \bar{z}} - r_{\bar{z}}\frac{\partial}{\partial \bar{w}}$ is tangent to M (in the sense that its real and imaginary parts of tangent to M) and the only other linear combinations are of $\frac{\partial}{\partial \bar{z}}$ and $\frac{\partial}{\partial \bar{w}}$ which are tangent to M are multiples of P. Any one of these multiples is called an induced CR operator. For example, in the hyperquadric given by $\operatorname{Im} w = |z|^2$ we shall consider the induced CR operator $L = \frac{\partial}{\partial \bar{z}} - iz\frac{\partial}{\partial s}$. This is the operator in the Hans Lewy counter-example [L 1]:

<u>Theorem 1</u>: There exists a function f for which $Lh = f$ has no solution in any neighborhood of the origin.

Proofs of this theorem and the related Theorem 5 are included at the end of this paper.

Because the hyperquadric is a group, any other point can be used in place of the origin. Lewy then used a Baire category argument to obtain the following result.

<u>Theorem 2</u>: There exists a function f for which $Lh = f$ has no solution on any open set in $\mathbb{R}^3$.

In the interests of brevity, we shall often not specify technical points such as the smoothness of the function. For instance, Lewy proved that there is a C^∞ function f for which there is no solution with Hölder continuous first derivatives. It is now well known that there is also no distribution solution. See, for instance, [Hö].

---

* This work was supported in part by NSF Grant DMS 8603086.

In another fundamental paper [L2], Lewy showed that induced CR operators share a behavior patterned on the classical Cauchy-Riemann operator in $\mathbb{C}$. The classical case goes as follows: Let P be the operator $A_1 \frac{\partial}{\partial x_1} + A_2 \frac{\partial}{\partial x_2}$ and assume that P and $\overline{P}$ are linearly independent. Let u be some solution to $Pu = 0$ and assume that $du|_0 \neq 0$.

Theorem: If $Pv = 0$ in some neighborhood of the origin then there is some holomorphic function H such that $v = H(u)$ on some neighborhood of the origin.

Lewy's generalization is: Let $L = A_1 \frac{\partial}{\partial x_1} + A_2 \frac{\partial}{\partial x_2} + A_3 \frac{\partial}{\partial x_3}$ and assume L, $\overline{L}$, and $[L,\overline{L}]$ are linearly independent. Let u_1 and u_2 be solutions to $Lu = 0$ and assume that at the origin $du_1 \wedge du_2 \neq 0$. Finally, let M^3 be the submanifold of $\mathbb{C}^2$ given by $(u_1(x_1,x_2,x_3), u_2(x_1,x_2,x_3))$.

Theorem 3: If $Lv = 0$ in some neighborhood of the origin and if v is thought of as a function on M^3 then there is some function F of two complex variables holomorphic on one side of M^3 and smooth up to M^3 with $F = f$ on M^3.

At the end of this paper Lewy commented that the existence of such two functions u_1 and u_2 is not known to always hold. To explain our title, if such u_1 and u_2 exist then all other solutions can be determined and $Lu = 0$ has the right "number" of solutions. But if two such solutions do not exist then $Lu = 0$ has too few solutions.

Nirenberg was the first to find operators with few solutions.

Theorem 4: Let $L = \frac{\partial}{\partial \bar{z}} - iz\frac{\partial}{\partial s}$. There exist functions ϕ_j vanishing to infinite order at the origin such that if

$$(L + \phi_1 \frac{\partial}{\partial z} + \phi_2 \frac{\partial}{\partial \bar{z}} + \phi_3 \frac{\partial}{\partial s})h = 0$$

in a neighborhood of the origin then h is a constant on a possibly smaller neighborhood of the origin.

See [N1] for this result and [N2] for an exposition of similar results.

In 1982, F. Treves and the author [JT1] were able to simplify and extend a weak version of Nirenberg's results. The idea was to prove the Lewy nonsolvability result in a form which made it easy to also obtain results about homogeneous equations. Indeed except for [LeB] every result we discuss is based on a relation between the solvability of a system $Lu = f$ and the homogeneous solvability of an associated system $Mu = 0$.

The geometric condition of strict pseudo-convexity which occurs in the statement of the next theorem is precisely the condition that L, $\overline{L}$, and $[L,\overline{L}]$ are linearly independent.

Theorem 5: Let $L = \frac{\partial}{\partial \bar{z}} + A\frac{\partial}{\partial u}$ be a CR operator for some strictly pseudo-convex real hypersurface $M^3 \subset \mathbb{C}^2$ which contains the origin. There exist functions

ϕ and ψ which vanish to infinite order at the origin such that if

$$(L + \phi \frac{\partial}{\partial z} + \psi \frac{\partial}{\partial u})h = 0$$

in a neighborhood of the origin, then $dh = 0$ at the origin.

We outline the proof at the end of this paper.

Remark: This same technique yields an operator L', which agrees with L to infinite order at the origin, such that $L'h = 0$ implies that h is a constant to infinite order at the origin.

Sometime later it was observed [JT2] that Lewy's Baire category argument could also be used in conjunction with Theorem 5. The result is a simple proof of the following generalization of Nirenberg's result.

Theorem 6: There exists an operator $L = A\frac{\partial}{\partial \bar{z}} + B\frac{\partial}{\partial z} + C\frac{\partial}{\partial u}$ defined on all of $\mathbb{R}^3$ which has the property that if $Lh = 0$ on some open connected set then h is a constant on that set.

Remark: Not only does there exist such an operator but such operators are in fact dense.

Before summarizing related results for systems of operators, it will be useful to distinguish various forms of nonsolvability for homogeneous equations. Given a system of n first order partial differential equations on $\mathbb{R}^m$

$$L_j h = 0\,, \quad j = 1,\dots,n \tag{1}$$

it is natural to look for $m-n$ independent solutions.

Definition: $\{L_1,\dots,L_n\}$ has few solutions at the point p if whenever in a neighborhood of p one has $L_j u_k = 0$ for $j = 1,\dots,n$ and $k = 1,\dots,m-n$ one has $du_1 \wedge \dots \wedge du_{m-n} = 0$ at p.

It is useful to distinguish two extreme cases.

Definition: The system is integrable at a point p if in some neighborhood of p there are $m-n$ solutions to (1) which have linearly independent gradients, over the complex numbers, at p. The system is aberrant at p if the only functions which satisfy (1) in a neighborhood of p are the constant functions.

Note that if (1) has $m-n$ independent solutions then

$$[L_j, L_k] \in \{L_1,\dots,L_n\}\,.$$

This condition on the Lie bracket is sometimes called formal integrability and is always assumed in work on homogeneous solvability. The condition is to be interpreted as saying that the Lie bracket at some point p can be expressed as a linear combination of the vectors evaluated at p, and this is true for all p in some open set.

Our first result for systems directly relates solvability and homogeneous solvability. Various versions of this result have been discovered by several mathematicians. The proof we give seems to be new.

Theorem 7: Let $L_1,\dots,L_n$ be first order partial differential operators on $\mathbb{R}^{n+1}$ with $[L_j,L_k] \in \{L_1,\dots,L_n\}$. There exist functions $h_1,\dots,h_n$ satisfying the necessary compatibility conditions such that:

$$L_j u = h_j \quad j=1,\dots,n$$

is solvable near the origin if and only if there exists some U with

$$L_j U = 0 \quad j=1,\dots,n$$

near the origin and $dU \neq 0$ at the origin.

Proof: The necessary compatibility conditions come from the usual observation that if

$$[L_j,L_k] = A^r_{jk}L_r$$

and if

$$L_j u = h_j \quad j=1,\dots,n$$

then

$$A^r_{kj}h_r = L_k(h_j) - L_j(h_k) \; . \tag{2}$$

Now, think of $L_1,\dots,L_n$ as complex vector fields and let $\omega \in \mathbb{C} \otimes T^*\mathbb{R}^{n+1}$ be any non-zero one-form which annihilates $\{L_1,\dots,L_n\}$. Our bracket condition is equivalent to

$$d\omega = \psi \wedge \omega$$

for some one-form ψ (which is not unique). Set $h_j = \psi(L_j)$.

Lemma: h_j satisfies the compatibility conditions (2).

Proof: Let $\mathscr{L}_X$ denote the Lie derivative in the direction of the complex vector field X. Recall that

$$\mathscr{L}_X\omega = X \lrcorner\, d\omega + d(X \lrcorner\, \omega)$$

and

$$\mathscr{L}_{[X,Y]} = \mathscr{L}_X \mathscr{L}_Y - \mathscr{L}_Y \mathscr{L}_X \; .$$

Now let $\mathscr{L}_j$ denote the Lie derivative in the direction L_j. So

$$\mathscr{L}_j\omega = L_j \lrcorner\, d\omega = \psi(L_j) = h_j \; .$$

Thus

$$\mathscr{L}_k \mathscr{L}_j\omega = L_k(h_j) + h_j h_k \omega$$

and

$$\mathscr{L}_{[L_k,L_j]}\omega = L_k(h_j) - L_j(h_k) .$$

Thus

$$A^r_{kj}h_r = L_k(h_j) - L_j(h_k) .$$

Lemma: If there exists U with $L_jU = 0$, $j = 1,\dots,n$ and $dU \neq 0$ then $L_ju = h_j$ is solvable.

Proof: Note that dU is another annihilator of $\{L_1,\dots,L_n\}$ and so $dU = g\omega$ with $g(0) \neq 0$. From $d^2U = 0$, we derive

$$dg \wedge \omega + g\,d\omega = 0 .$$

Contracting with L_j yields

$$L_j(g)\omega + g\,h_j\,\omega = 0$$

and so

$$L_j(-\ln g) = h_j .$$

Lemma: If there exists u with $L_ju = h_j$, $j = 1,\dots,n$ then $LU = 0$ is solvable with $dU(0) \neq 0$.

Proof: Note that for each k, $k = 1,\dots,n$

$$L_k \lrcorner\, d(e^{-u}\omega) = e^{-u}(-L_k(n) + h_k)\omega = 0 .$$

Thus the one-form $e^{-u}\omega$ is closed and there exists a function U with $dU = e^{-u}\omega$. This implies $L_jU = 0$ for $j = 1,\dots,n$ and concludes the proof of Theorem 7.

Now we discuss the higher dimensional induced Cauchy-Riemann operators.

Definition: A set of n operators $\{L_1,\dots,L_n\}$ on $\mathbb{R}^{2n+1}$ is a CR hypersurface system if

1) $L_1,\dots,L_n,\overline{L}_1,\dots,\overline{L}_n$ are linearly independent,

and

2) $[L_j,L_k] \in \{L_1,\dots,L_n\}$.

Here is the reason for this name: Let $M^{2n+1} \subset \mathbb{C}^{n+1}$ be a real hypersurface with defining function $r(z,\overline{z})$ where $z = (z_1,\dots,z_{n+1})$ are the usual coordinates on $\mathbb{C}^{n+1}$. Then

$$V = \{X \in \mathbb{C} \otimes T\mathbb{C}^{n+1} : X = \sum_{j=1}^{n+1} a_j \frac{\partial}{\partial \overline{z}_j} \text{ and } Xr = 0 \text{ along } M\}$$

restricts to a bundle over M of complex dimension n. Also call this restriction V. Any vector $X \in V$ deserves to be called an induced CR operator. Note that if X and Y are in V then so is $[X,Y]$ and also that $V \cap \overline{V}$ contains only the zero vector. Thus any basis for V provides a CR hypersurface system. Conversely, an integrable CR hypersurface system always can be realized as the set of induced CR operators from a real hypersurface.

There is a CR version of Theorem 7 which again shows a relation between solvability and homogeneous solvability: If a CR hypersurface system is integrable then a certain (0,1) form is in the range of $\bar{\partial}_b$. See [J2] for the precise statement and definitions.

Now given any CR hypersurface system choose some real vector field U so that $\{L_1,\ldots,L_n,\bar{L}_1,\ldots,\bar{L}_n,U\}$ is a linearly independent set.

The Levi matrix of the CR hypersurface system is the hermitian matrix $C = (c_{jk})$ defined by

$$[L_j,\bar{L}_k] = i\,c_{jk}U .$$

The CR system is <u>non-degenerate</u> if C has no zero eigenvalues; it is <u>strictly pseudo-convex</u> if all the eigenvalues are of the same sign; and it has <u>signature (p,n-p)</u> if C has p eigenvalues of one sign and $n-p$ of the other sign.

The deepest result about such systems is due to Kuranishi [K]:

<u>Theorem 8</u>: A strictly pseudo-convex CR hypersurface system of dimension greater than seven is integrable.

<u>Remark</u>: Recently Akahori [A] has extended this to include dimension seven. This means that only dimension five remains unresolved since CR hypersurface systems are odd dimensional and Theorem 4 shows that the result is false in dimension three.

Negative results are easier to come by [JT2].

<u>Theorem 9</u>: The aberrant CR hypersurface systems are dense among all systems of signature (n-1,1).

<u>Remark</u>: LeBrun [LeB] has used a completely different method to obtain a beautiful set of non-integrable CR systems of signature (1,1). His work is based in part on the Penrose twistor program. The simplest example of a non-integrable CR system also derives from this program, and this example is our next topic.

Let L be a CR operator for any strictly pseudo-convex three-dimensional hypersurface and let f be some function for which $Lh = f$ does not have a solution near some point p . Penrose [Pe], in reporting on joint work with Hill and Sparling, has shown how to construct from this a five-dimensional degenerate CR structure which is not integrable. To do this introduce a new complex variable ζ and let

$$L_1 = L + f\zeta\frac{\partial}{\partial\zeta} \quad\text{and}\quad L_2 = \frac{\partial}{\partial\zeta} .$$

<u>Theorem 10</u>: The CR system defined by $\{L_1,L_2\}$ is not integrable.

One proof was outlined in [Pe] and simpler proofs given in [E] and [J1]. Here we present an even simpler proof found by Treves. For the system $L_1h = 0$, $L_2h = 0$ to have three independent solutions at the point $(p,0)$, it is necessary that this system has a solution with $h_\zeta(p,0) \neq 0$. But if we differentiate the equation $L_1h = 0$ with respect to ζ we obtain

$$Lh_\zeta + fh_\zeta + f\zeta h_{\zeta\zeta} = 0 \ .$$

Thus $H(q) = -\ln h_\zeta(q,0)$ satisfies $LH = f$ in a neighborhood of p. This contradicts the original choice of f and so there cannot be any solution with $h_\zeta(p,0) \neq 0$. It is not hard to see that conversely when $Lh = f$ does have a solution then the system $\{L_1, L_2\}$ is integrable.

There is a natural generalization to higher dimensions. Let $M^{2n+1} \subset \mathbb{C}^{n+1}$ give rise to a CR hypersurface system of signature $(n-1,1)$. It is not difficult to see that one can choose linearly independent complex vector fields $L_1,\dots,L_n$ which generate the CR system of M and which satisfy $[L_j, L_k] = 0$. And it is known that the analogue of Lewy's non-solvability holds [AH]. Putting these together yields the existence of functions g_j, $j = 1,\dots,n$ which satisfy $L_j g_k = L_k g_j$ but for which there is no function h with $L_j h = g_j$, $j = 1,\dots,n$.

<u>Theorem 11</u>: The $(2n+3)$-dimensional CR hypersurface system defined by

$$P_j = L_j - g_j\zeta \frac{\partial}{\partial\zeta}, \quad (j = 1,\dots,n) \quad \text{and}$$

$$P_{n+1} = \frac{\partial}{\partial\bar\zeta}$$

is not integrable.

<u>Remark</u>: The Levi matrix for this system has $n-1$ positive eigenvalues, one negative eiganvalue, and one zero eigenvalue.

We may easily modify this example to construct examples of arbitrary co-dimension. A CR system of co-dimension d is defined precisely as the hypersurface system but with $\mathbb{R}^{2n+1}$ replaced by $\mathbb{R}^{2n+d}$. So with a system of co-dimension d we may choose real vector fields U_j so that $\{L_1,\dots,\bar L_n, U_1,\dots,U_d\}$ are linearly independent. The Levi matrix is now the matrix-valued vector $(C_1,\dots,C_d)$ defined by

$$[L_j, \bar L_k] = i\sum_r C_{rjk} U_r \ .$$

Various definitions of non-degeneracy are reasonable. Two common ones are

a) $\sum_r a_r C_{rjk} = 0$ implies each a_r is zero,

b) The vector space map $\mathbb{R}^n \longrightarrow \mathbb{R}^n \oplus \cdots \oplus \mathbb{R}^n$ given by $v = (v_1,\dots,v_n) \longrightarrow \sum_k C_{rjk} v_k$ is injective.

Note that if $M^{2n+1} \subset \mathbb{C}^{n+1}$ is non-degenerate then $M^{2n+1} \times \mathbb{R}^r \subset \mathbb{C}^{n+r+1}$ is degenerate in sense a) but is non-degenerate in sense b).

Again consider some $M^{2n+1} \subset \mathbb{C}^{n+1}$ of signature $(n-1,1)$, choose the CR operators to satisfy $[L_j, L_k] = 0$, and find g_j, $j = 1,\dots,n$ which satisfy $L_j g_k = L_k g_j$ but for which there is no function h with $L_j h = g_j$, $j = 1,\dots,n$.

Theorem 12: The (2n+d)-dimensional CR system of co-dimension d defined by

$$P_j = L_j - g_j \left(\sum_{r=1}^{d-1} t_r \frac{\partial}{\partial t_r} \right) \quad , \quad j = 1, \ldots, n$$

is not integrable. (In fact it has only $n+1$ independent solutions.)

Remarks: 1) The Levi matrix for this system is degenerate in sense a) and non-degenerate in sense b).

2) Trivial counter-examples for co-dimension d can be produced by considering $M^{2n+1} \times \mathbb{R}^{d-1}$ where M is non-realizable. These have less than $n+1$ independent solutions.

One aim of the study of homogeneous solvability is to be able to characterize those systems which have the "correct" number of solutions. For instance, when does the operator

$$L = \frac{\partial}{\partial \bar{z}} + a(z,\bar{z},s) \frac{\partial}{\partial s} \tag{3}$$

have a solution to $Lu = 0$ with $du \wedge dz \neq 0$? The answer should depend on the subset of the characteristic variety of L that controls the solvability of $Lu = f$ and on the analytic wave front set of $a(z,\bar{z},s)$. But we must permit C^∞ changes of coordinates. The following beautiful result is due to Hanges [Ha]. Consider a strictly pseudo-convex operator given by (3) and take at the origin $[L,\bar{L}] \equiv \sigma \frac{\partial}{\partial s}$ (mod $\{L,\bar{L}\}$) with $\sigma > 0$.

Theorem 13: The equation $Lu = 0$ has a solution in a neighborhood of the origin in $\mathbb{R}^3$ with $du \wedge dz \neq 0$ if and only if there is a coordinate change $w = z$, $t = f(z,\bar{z},t)$ such that L becomes $\frac{\partial}{\partial \bar{w}} + b(w,\bar{w},t) \frac{\partial}{\partial t}$ and for each fixed w close to zero, $b(w,\bar{w},t)$ is the boundary value of a function holomorphic in $\{\zeta = t + i\tau : \tau > 0 \ , \ t^2 + \tau^2 < \varepsilon\}$.

We conclude this survey by proving Theorems 1 and 5. For simplicity we work with the hyperquadric $\operatorname{Im} w = |z|^2$ instead of with an arbitrary strictly pseudo-convex real hypersurface $M^3 \subset \mathbb{C}^2$. See [JT 1] for the slight modifications necessary for the more general case and also for the proofs of the two lemmas below. Let Q be this hyperquadric and let Γ_λ denote the intersection of Q with the plane $w = \lambda$. So

$$\Gamma_\lambda = \{(z,w) : |z|^2 = \operatorname{Im} \lambda \ , \ w = \operatorname{Re} \lambda + i|z|^2\}$$

and Γ_λ is a circle, a point, or the empty set depending on whether $\operatorname{Im} \lambda > 0$, $\operatorname{Im} \lambda = 0$, or $\operatorname{Im} \lambda < 0$.

Lemma: If $Lh = 0$ in a neighborhood of some Γ_μ then $H(\lambda) = \int_{\Gamma_\lambda} h\,dz$ is a holomorphic function of λ for λ near μ .

Let Ω be any connected domain in the λ-plane and let

$$\tilde{\Omega} = \{p = (z,w) : p \in \Gamma_\lambda \text{ for some } \lambda \in \Omega\} \quad .$$

Lemma: If Ω contains a piece of the line $\operatorname{Im}\lambda = 0$ and if $Lh = 0$ on $\tilde{\Omega}$, then $H(\lambda)$ is identically zero.

We may now construct some f for which $Lh = f$ is not solvable. Let D_n be the disc in the λ-plane with center $\left(\frac{1}{n}, \frac{1}{n}\right)$ and radius $\left(\frac{1}{n}\right)^3$. Any two of these discs are disjoint. Let $\tilde{D}$ be the closure of the set

$$\{p = (z,w) : p \in \Gamma_\lambda \text{ for some } \lambda \in \cup D_n\} \quad .$$

Now let f be a non-negative C^∞ function whose support is equal to $\tilde{D}$.

Theorem: There is no C^1 function h which satisfies $Lh = f$ in a neighborhood of the origin.

Proof: Assume such a neighborhood and function do exist. Let D_n be some disc for which

$$\tilde{D}_n = \{p \in \Gamma_\lambda \text{ for some } \lambda \in D_n\}$$

is in the neighborhood. Integration by parts yields

$$2i \iiint f\, dxdyds = \iint h\, dz\, dw = \int H(\lambda)d\lambda \tag{4}$$

Here the triple integral is over $\tilde{D}_n$, the double integral is over the boundary of $\tilde{D}_n$, and the single integral is over the boundary of D_n, with the coordinates on Q being defined by $z = x + iy$, $w = s + i(x^2 + y^2)$. Let Ω be the complement of D_n in some disc centered at the origin. The previous lemma may be applied to this Ω provided the disc is sufficiently small and n sufficiently large. Thus the left-most term in (4) is positive while the right most term is zero. This contradiction proves the theorem.

From this result it is easy to prove Theorem 5. (Here we again do so only for the hyperquadric.) Take two sequences of disjoint discs D_n and D'_n and let f be a non-negative function with support equal to $\tilde{D}$ and let g be a non-negative function with support equal to $\tilde{D}'$. Consider the equation

$$Lh = fh_z + gh_s \tag{5}$$

For this to have a solution in a neighborhood of the origin we must avoid the contradiction discussed above so both $\operatorname{Re} h_z$ and $\operatorname{Im} h_z$ must have a zero in each D_n for n large enough. But then $h_z(0) = 0$ and similarly $h_s(0) = 0$. From (5) we see that always $h_{\bar{z}}(0) = 0$. Thus $dh = 0$ at 0 and we are done.

References

[A] T. Akahori, The local embedding theorem of strongly pseudo-convex CR-structures with $\dim_{\mathbb{R}} M = 2n-1 \geq 7$, to appear in Memoirs of A.M.S.

[AH] A. Andreotti and D. Hill, E. E. Levi Convexity and the Hans Lewy Problem, Part I, Ann. Sc. Norm. Sup. Pisa 36 (1972), 325-363; Part II, same journal and volume, 747-806.

[E] M. Eastwood, The Hill-Penrose-Sparlings CR-folds, Twistor Newsletter 18 (1984), 16.

[Ha] N. Hanges, The Missing First Integral, preprint.

[Hö] L. Hörmander, Linear Partial Differential Operators, Springer-Verlag, 1969.

[J 1] H. Jacobowitz, A simple example of a non-realizable CR hypersurface, to appear in Proc. of A.M.S.

[J2] ____________, The Canonical Bundle and Realizable CR hypersurfaces, to appear in Pacific J. of Math.

[JT 1] H. Jacobowitz and F. Treves, Non-realizable CR structures, Inventiones math. 66 (1982), 231-249.

[JT2] ____________, Aberrant CR structures, Hokkaido Math. J. 12:1 (1983), 276-292.

[K] M. Kuranishi, Strongly pseudoconvex CR-structures over small balls. Part III: An embedding theorem, Annals of Math. 116 (1982), 249-330.

[L 1] H. Lewy, An example of a smooth linear partial differential equation without solution, Annals of Math. 66 (1957), 155-158.

[L2] ______, On the local character of a solution ... , Annals of Math. 64 (1956), 514-522.

[LeB] C. LeBrun, Twistor CR manifolds and three-dimensional conformal geometry, Trans. of A.M.S. 284 (1984), 601-616.

[N 1] L. Nirenberg, On a question of Hans Lewy, Russian Math. Surveys 29 (1974), 251-262.

[N2] ___________, Lectures on linear partial differential equations, Conference Board of Mathematical Sciences, Regional Conference Series in Mathematics, No. 17, Amer. Math. Soc., 1973.

[Pe] R. Penrose, Physical space-time and non-realizable CR-structures, Bull. of A.M.S. 8 (1983), 427-448.

[Po] H. Poincaré, Les functions analytique de deux variables et la représentation conforme, Rend. Circ. Mat. Palermo 23 (1907), 185-220.

[T] F. Treves, On local solvability of linear partial differential equations, Bulletin of the A.M.S., 1970 (76), 552-571.

Introduction to Multiplicity Theory For Boundary Value Problems with Asymmetric Nonlinearities

by

Alan Lazer
University of Miami
Coral Gables, Florida 33124
U.S.A.

The author was partially supported by the National Science Foundation under Grant No. DMS-9S19882

Introduction

We consider the nonlinear elliptic boundary value problem

$$\begin{cases} \Delta u+g(u)=h(x), & x\in\Omega \\ u|\partial\Omega=0 \end{cases} \tag{1.1}$$

where $\Omega\subset\mathbb{R}^n$ is a smooth and bounded domain.

Let

$$0 < \lambda_1 < \lambda_2 \leq \lambda_3 < \dots$$

denote the eigenvalues of the problem

$$\begin{cases} \Delta u+\lambda u=0 & x\in\Omega \\ u|\partial\Omega=0 \end{cases} \tag{1.2}$$

where each eigenvalue is counted as often as its multiplicity. We denote the set $\{\lambda_n | n \geq 1\}$ by σ.

In [10] it was essentially shown by Dolph that if g and h are sufficiently regular and if there exists a closed interval I such that $I\cap\sigma=\Phi$ and a number $\xi_0 > 0$ such that $g(\xi)/\xi\in I$ for $|\xi| \geq \xi_0$, then (1.1) is solvable.

In recent years there has been interest in problem (1.1) where it is assumed that there exist numbers a and b such that

$$\lim_{\xi\to-\infty} g'(\xi)=a < b=\lim_{\xi\to\infty} g'(\xi)$$

and such that $[a,b]\cap\sigma$ is nonempty.

Impetus to study this problem was given by the paper [3] which considered the case $a < \lambda_1 < b < \lambda_2$. It was subsequently noted independently in [5] and in [14] that it is natural to consider the right-hand side to have the form

$$h=t\phi_1+\psi \tag{1.3}$$

where ϕ_1 is a positive eigenfunction corresponding to λ_1 and ψ is smooth and t is a parameter.

There have been many research articles dealing with the case $a < \lambda_1 < b$. We refer the reader to the list of references in the recent articles [8] and [13]. In this exposition we consider exclusively the case where $\lambda_1 < a$.

It was noted by E.N. Dancer in [7] and by S. Fucik in [11] that an important role in the study of the above problem is played by the set of pairs of numbers (c,d) for which the homogeneous problem

$$\begin{cases} \Delta u+du^+-cu^-=0 & \text{in } \Omega \\ u|\partial\Omega=0 \end{cases}$$

has a nontrivial solution. We denote the set of all such pairs by Σ. Since $(\lambda_k,\lambda_k)\in\Sigma$ for all $\lambda_k\in\sigma$, Σ may be thought of as the <u>extended spectrum</u>.

Except for a brief discussion of the O.D.E case in the next section, we shall restrict our attention to the piecewise linear problem

$$\begin{cases} \Delta u+bu^+-au^-=t\phi_1+\psi \\ u|\partial\Omega=0 \ . \end{cases} \tag{1.4}$$

A physical situation for in which this problem, in the one-dimensional case, can serve as a model is discussed in the next section.

In [15] it was observed by P.J. McKenna and the author that, in looking for multiplicity results for (1.4), it is natural to try to relate the sum of the number of solutions of (1.4) for large positive and the number of solutions for large negative t to the cardinality of $(a,b)\cap\sigma$.

We will show that if $(a,b)\notin\Sigma$, neither a nor b is in σ, $a > \lambda_1$ and $(a,b)\cap\sigma\neq\Phi$, then there exists an integer k with $1 \leq k \leq 3$ such that (1.4) has at least k solutions for t large and positive and at least 4-k solutions for t large and negative. This result, obtained by P.J. McKenna and the author, will appear in [17]. Our proof uses only standard critical point theory. E.N. Dancer [9] has recently shown using the homotopy index of Conley and cohomology that if a and b are as above and the piecewise linear function $b\xi^+-a\xi^-$ is replaced by a general C^1 function g such that $g'(-\infty)=a$, $g'(\infty)=b$, then the above multiplicity result still holds.

We give a condition under which k=3. This improves earlier work of McKenna and the author in [16].

Finally, using an idea from [19], we give an example which seems to indicate that the multiplicity results cannot be improved. Another type of example is given by Dancer in [9].

1. <u>The</u> O.D.E. <u>Case</u>

The problem

$$\begin{cases} u''+g(u)=t\sin x+\psi(x) \\ u(0)=u(\pi)=0 \end{cases}$$

was considered in [13]. It was shown that if $g\in C^1$, ψ is continuous, $g'(\xi)\to b$ as $\xi\to\infty$, $g'(\xi)\to a$ as $\xi\to-\infty$, there exist integers p and q such that

$$1 \leq p^2 < a < (p+1)^2 \leq q^2 < b < (q+1)^2$$

and $(a,b)\notin\Sigma$, <u>then there exists an integer</u> k <u>with</u> $1 \leq k \leq 2(q-p)+1$ <u>such that</u> (D) <u>has at least</u> k <u>solutions for</u> t <u>sufficiently large and positive and at least</u> 2(q-p)+2-k <u>solutions for</u> t <u>sufficiently large and negative</u>.

This result was proved by using the so called shooting method in [13]. It can also be established by means of the global bifurcation method as used in [23]. In case q-p=1 the words "at least" can be replaced by "exactly". (See Section 4.)

In order to give some feeling for the result we discuss a corresponding theorem for the autonomous Neumann problem. In this case we easily visualize why the result is true by means of phase-plane analysis.

We assume that $0 < a < b$ and consider the second-order autonomous differential equation

$$u''+bu^{+}-au^{-}=1. \tag{A}$$

We make the following simple observation:

Lemma 1.2 Let $u_0(t)$ be a non-constant solution of (A) of least period $2\pi/k$ where $k \geq 1$ is an integer. There exist two distinct solutions of the Neumann problem

$$\begin{cases} u''+bu^{+}-au^{-}=1 \\ u'(0)=u'(\pi)=0 \end{cases} \tag{N}$$

which are the restrictions of suitable translates of u_0.

Proof: If $u_0'(t_1)=0$, then $u_0(t)$ and $u_0(2t_1-t)$ are two solutions of (A) which are equal and have equal derivatives at $t=t_1$. Hence $u_0(t)=u_0(2t_1-t)$ so the graph of u is symmetric with respect to the line $t=t_1$. It follows that if $t_0 < t_1$ are two consecutive zeros of $u'(t)$, then $2(t_1-t_0)$ is a period of $u(t)$. Therefore, there exist numbers t_0 and t_1 such that $t_1-t_0=\pi/k$, $u(t)$ assumes its minimum at t_0, and $u(t)$ assumes its maximum at t_1. Hence, $u(t-t_0)$ and $u(t-t_1)$ are distinct solutions of (N).

Using this observation we can now prove

Theorem 1.2. If there exist integers p and q with $0 \leq p < q$ such that $p^2 < a < (p+1)^2 \leq q^2 < b < (q+1)^2$ and the homogeneous problem

$$u''+bu^{+}-au^{-}=0 \tag{H}$$

has no 2π-periodic solution other than $u\equiv 0$, then the total number of solutions of the problem

$$\begin{cases} u''+bu^{+}-au^{-}=1 \\ u'(0)=u'(\pi)=0 \end{cases} \tag{N_+}$$

<u>and the problem</u>

$$\begin{cases} u''+bu^+-au^-=-1 \\ u'(0)=u'(\pi)=-0 \end{cases} \qquad (N_-)$$

<u>is exactly</u> 2(q-p)+2.

<u>Proof</u>: We will only show that total number of solutions is at least 2(q-p)+2. A proof that 2(q-p)+2 is the exact number will appear in [18].

First consider the autonomous system

$$\begin{cases} u'=v \\ v'=-bu^+ +au^- +1 \end{cases} \qquad (S)$$

since all curves

$$\tfrac{1}{2}v^2+\tfrac{1}{2}b(u^+)^2+\tfrac{1}{2}a(u^-)^2-u$$

$$= \text{constant}$$

are bounded, all solutions of the conservative system are periodic. There is a unique equilibrium point $(\frac{1}{b},0)$. If we set $u(t)=\frac{1}{b}+w(t)$, where $w(t)$ is so small that $\frac{1}{b}+w(t) > 0$, then

$$w'(t)=v(t),\ v'(t)=-b(w(t)+\tfrac{1}{b})+1$$

so $w''(t)+bw(t)=0$ and $w(t)=c\sin(\sqrt{b}t+\alpha)$. Therefore, near the equilibrium point $(1/\sqrt{b},0)$ the closed orbits of (S) correspond to solutions of constant period $2\pi/\sqrt{b}$ of (A).

Consider now a solution $u(t)$ of (A) whose amplitude $|u|_\infty$ is very large. If we set $z(t)=u(t)/|u|_\infty$, then $|z|_\infty=1$ and

$$z''+bz^+-az^-=1/|u|_\infty$$

so the period of z is approximately equal to the period of a solution of (H). If $y\not\equiv 0$ is a solution of (H), then $y''+by=0$ on the intervals on which y is positive and $y''+ay=0$ on the intervals on which y is negative. If $t_0 < t_1 < t_2$ are three consecutive zeros of $y(t)$ with

$y(t) > 0$ on (t_0,t_1) and $y(t) < 0$ on (t_1,t_2), then $t_1-t_0=\pi/\sqrt{b}$ and $t_2-t_1 = \pi/\sqrt{a}$ so the period of y is $\pi/\sqrt{a}+\pi/\sqrt{b}$. It follows that the periods of solutions of (S) with very large amplitudes are close to $\pi/\sqrt{a}+\pi/\sqrt{b}$.

Since the period of a solution of (A) varies continuously with its amplitude, it follows that if $2\pi/\sqrt{b} < T < \pi/\sqrt{a}+\pi/\sqrt{b}$, then there is a solution of (A) with least period T.

If k is an integer such that

$$d\equiv 2(1/\sqrt{a}+1/\sqrt{b})^{-1} < k < \sqrt{b}$$

then

$$2\pi/\sqrt{b} < 2\pi/k < \pi/\sqrt{a}+\pi/\sqrt{b}.$$

It follows that the number of orbits of (S) corresponding to a solution having 2π as a period is at least equal to the number of integers in the interval $(d,\sqrt{b})$ so, by Lemma 1.2, the Neumann problem (N_+) has at least twice this number of solutions.

Consider now the system

$$\begin{cases} u'=v \\ v'=-bu^+ + au^- - 1 \ . \end{cases} \qquad (S_-)$$

For this system there is the unique equilibrium point $(-1/a,0)$. If we set $u(t)=-1/a+w(t)$ where $w(t)$ is so small that $u(t) < 0$, then $w''(t)+aw(t)=0$. It follows that the orbits of the system near the equilibrium point correspond to solutions with the constant period $2\pi/\sqrt{a}$. The same analysis as before shows that the solutions of (S_-) with very large amplitudes have approximately the period $\pi/\sqrt{a}+\pi/\sqrt{b}$. Hence, for each integer k satisfying $\pi/\sqrt{a}+\pi/\sqrt{b} < 2\pi/k < \pi/\sqrt{a}$ there is a periodic solution of (S_-) of least period $2\pi/k$ which gives rise to two solutions of the Neumann problem (N_-). Therefore, the number of solutions of the problem (N_-) is at least twice the number of integers in the interval $(\sqrt{a},d)$.

Therefore, the total number of nonconstant solutions of (N_-) plus the total number of nonconstant solutions of (N_+) is at least twice the number of integers k satisfying $\sqrt{a} < k < \sqrt{b}$. Since (N_+) has the constant solution $\frac{1}{b}$ and (N_-) has the constant solution $-\frac{1}{a}$, this establishes a total of $2(q-p)+2$ solutions.

Recently we have investigated the multiplicity of solutions of the problem

$$u''+bu^+-au^-=s+h(t)$$

where $h(t)\equiv h(t+2\pi)$, h is continuous and s is a large constant, subject to the periodic boundary conditions

$$u(0)=u(2\pi),\ u'(0)=u'(2\pi).$$

Although each eigenvalue of the problem

$$u''+\lambda u=0$$

$$u(0)=u(2\pi),\ u'(0)=u'(2\pi)$$

with $\lambda > 0$ has multiplicity two, the total number of solutions of this problem is, in general, still twice the number of integers in $(\sqrt{a}, \sqrt{b})$. (See Section 4 for an example.) This result is discussed for the case of a C^1 nonlinearity in [18].

We hope that some of our work may be used to study models which are essentially asymmetric in nature. One such model is the following idealization of a suspension bridge.

Consider a vibrating beam of length L where the displacement at time t and at x, with $0 < x < L$, is denoted by $U(x,t)$ and is considered positive in the downward direction. The usual boundary value problem for U is

$$U_{tt}+kU_{xxxx}=0$$

$$U(0,t)=U(L,t)=U_{xx}(0,t)=U_{xx}(L,t)=0.$$

Suppose now that the beam is suspended above by cables. Then there should be an additional restoring force of the form $cU(x,t)^+$ where $c > 0$. This is because the cables will exert no force if compressed, but will act as springs if stretched. Let us assume that there is a force of the form $F(x,t)$ where $F(0,t)=F(L,t)=0$, $F(x,t+T)\equiv F(x,t)$. Under general conditions we have

$$F(x,t)=\sum_{m=1}^{\infty} A_m(t)\sin\frac{mx}{L}$$

where $A_m(t+T)\equiv A_m(t)$. Considering all terms $A_m(t)\sin mx/L$ small for $m \geq 2$, we simply approximate $F(x,t)$ by $A_1(t)\sin x/L$. Finally, we write $A_1(t)=s+h(t)$, where the constant s is the mean value of A_1 and h is considered to be small in comparison to s.

For this model, U must satisfy the differential equation

$$U_{tt}+kU_{xxxx}+cU^{+}=\sin\frac{\pi x}{L}(s+h(t))$$

and the boundary conditions $U(0,t)=U(L,t)=U_{xx}(0,t)=U_{xx}(L,t)$. If we try a separation of variables $U(x,t)=u(t)\sin\frac{x}{L}$ we find that $u(t)$ must satisfy

$$u''+bu^{+}-au^{-}=s+h(t)$$

where $b=c+k(\pi/L)^4$, $a=k(\pi/L)^4$. In the case where $h(t)=A\sin(wt+\alpha)$, under suitable restrictions on a,b and w, we have been able to determine the exact number of solutions which have the same period as h(t). Moreover, we have been able to determine the amplitudes of these solutions and which become stable if a small damping term is added to the equation.

2. The P.D.E. case $\lambda_1 < a$ and $[a,b]$ contains an eigenvalue.

Let Ω, ϕ_1, and ψ be as in the introduction. The goal of this section is to prove

Theorem 2.1: If $a,b\notin\sigma$, $\lambda_1 < a$, $(a,b)\notin\Sigma$, and $[a,b]\cap\sigma\neq\Phi$, *then there exists an integer* m, $1 \leq m \leq 3$, *such that*

$$\Delta u+bu^{+}-au^{-}=t\phi_1+\psi \tag{2.1}$$

$$u|\partial\Omega=0$$

has at least m *solutions for* t *large and positive and at least* 4-m *solutions for* t *large and negative*.

The proof will follow from a series of lemmas. Let $p \geq 1$ and q be integers such that $\lambda_p < a < \lambda_{p+1} \leq \lambda_q < b < \lambda_{q+1}$ and let V be the finite dimensional subspace of $L^2(\Omega)$ spanned by the eigenfunctions corresponding to λ_j for $j=p+1,\ldots,q$. Let P denote the orthogonal projection of $L^2(\Omega)$ onto V and let $W=V^{\perp}=(I-P)L^2(\Omega)$.

Given $y\in C^{\alpha}(\bar{\Omega})$ and u in $\mathring{H}_1(\Omega)$ we set

$$f(u,y)=\int_{\Omega}\frac{1}{2}[|\nabla u|^2-b(u^+)^2-a(u^-)^2+2uy]dx. \tag{2.2}$$

Standard arguments show that f has a continuous derivative D_1f with respect to its first argument and that if $z\in\mathring{H}_1(\Omega)$,

$$D_1f(u,y)(z)=\int_{\Omega}(\nabla u\cdot\nabla z-bu^+z+au^-z+zy)dx. \tag{2.3}$$

Regularity arguments show that for y fixed smooth solutions of the problem

$$\Delta u+bu^+-au^-=y, \quad u|\partial\Omega=0 \tag{2.4}$$

coincide with critical points of the mapping $\mathring{H}_1(\Omega)\to\mathbb{R}$ given by $u\to f(u,y)$.

For brevity we set $g(\xi)=b\xi^+-a\xi^-$, $\xi\in\mathbb{R}$.

Lemma 2.1 *Given* $y\in C^{\alpha}(\bar{\Omega})$ *and* $v\in V$, *there exists a unique* $w\in W\cap C^{2+\alpha}(\bar{\Omega})$ *such that*

$$\begin{aligned}&\Delta w+(I-P)g(v+w)=(I-P)y\\ &w|\partial\Omega=0.\end{aligned} \tag{2.5}$$

If w *is denoted by* $\theta(v,y)$, *then* θ *is continuous in each variable*.

Using the fact that

$$\lambda_p < a \le \frac{g(\xi_2)-g(\xi_1)}{\xi_2-\xi_1} \le b < \lambda_q \tag{2.6}$$

Lemma 2.1 can be proved using the Lyapunov-Schmidt reduction method in much that same way it is used in [20, p. 118-120] or by the saddle point reduction method [1] [6]. Details will appear in [17]. (See Section 4 for a special case.)

Since V is spanned by the eigenfunction s corresponding to $\lambda_{p+1},\ldots,\lambda_q$, if $v\in V$, then $\Delta v\in V$. It follows that if $z\in W\cap\mathring{H}_1(\Omega)$, then

$$\int_\Omega \nabla z\cdot\nabla v dx=-\int_\Omega z\Delta v dx=0.$$

From this it follows from (2.5) that if $w=\theta(v,y)$ and $z\in W\cap\mathring{H}_1(\Omega)$, then

$$\int_\Omega \nabla(v+w)\cdot\nabla z-g(v+w)\cdot z+zy)dx=0.$$

Therefore from (2.3) we have

$$D_1 f(v+\theta(v,y),y)(z)=0, \quad z\in W\cap\mathring{H}_1(\Omega) \tag{2.7}$$

and $v\in V$.

<u>Lemma</u> 2.2 <u>If for</u> $v\in V$ <u>and</u> $y\in C^\alpha(\bar\Omega)$, $F(v,y)\equiv f(v+\theta(v,y),y)$, <u>then</u> F <u>is continuous</u>, F <u>has a continuous derivative with respect to its first argument</u> and

$$D_1F(v,y)(v_1)=D_1f(v+\theta(v,y),y)(v_1) \tag{2.8}$$

<u>for</u> $v_1\in V$.

<u>Proof</u>: The continuity of F follows from the continuity of θ. To establish the existence of a continuous derivative of F with respect to its first variable we use an argument similar to one used by Castro in [6]. (This could also be done by invoking abstract results of [1].)

Let W_1 denote the subspace of $L^2(\Omega)$ spanned by the eigenvalues corresponding to λ_j with $1 \le j \le p$ and let W_2 denote the set of $u\in\mathring{H}_1(\Omega)$ such that $(u,z)=0$ for $z\in W_1\oplus V$. It follows that if $w_1\in W_1$, then

$$\int_\Omega |\nabla w_1|^2dx \le \lambda_p\int_\Omega w_1^2dx \tag{2.9}$$

and if $w_2\in W_2$, then

$$\int_\Omega |\nabla w_2|^2dx \ge \lambda_{q+1}\int_\Omega w_2{}^2dx. \tag{2.10}$$

From (2.3), (2.6) and (2.10) we see that if $w_1 \in W_1$, $v \in V$, and $w_2, w_2^* \in W_2$, then

$$(D_1 f(w_1+v+w_2^*,y) - D_1 f(w_1+v+w_2,y))(w_2^*-w_2)$$
$$= \int_\Omega (|\nabla(w_2^*-w_2)|^2 - (g(w_1+v+w_2^*)-g(w_1+v+w_2))\ (w_2^*-w_2))dx$$
$$\geq m_1 \int_\Omega |\nabla(w_2^*-w_2)|^2 dx$$

where $m_1 = 1-b/\lambda_{q+1} > 0$. Similarly, using (2.3), (2.6) and (2.9) we see that if $w_1, w_1^* \in W_1$, $v \in V$ and $w_2 \in W_2$, then

$$(D_1 f(w_1^*+v+w_2,y) - D_1 f(w_1+v+w_2,y))(w_1^*-w_1)$$
$$\leq -m_2 \int_\Omega |\nabla(w_1^*-w_1)|^2 dx$$

where $m_2 = a/\lambda_p - 1 > 0$. It follows that for fixed $v \in V$ and $w_1 \in W_1 (w_2 \in W_2)$, $f(w_1+v+w_2,y)$ is strictly convex (concave) as a function of $w_2 \in W_2 (w_1 \in W_1)$. Let $v \in V$ and $\theta(v,y) = \theta_1(v,y) + \theta_2(v,y)$ with $\theta_1(v,y) \in W_1$, $\theta_2(v,y) \in W_2$. Since $D_1 f(v+\theta(v,y),y)(z) = 0$ for all $z \in W = W_1 \oplus W_2$, it follows that $f(v+w_1+\theta_2(v,y),y) \leq f(v+\theta(v,y),y) \leq f(v+\theta_1(v,y)+w_2,y)$ for $w_1 \in W_1$ and $w_2 \in W_2$. From these inequalities and the relation $F(v,y) = f(v+\theta(v,y),y)$ we see that if $v, h \in V$ and $t > 0$, then

$$f(v+th+\theta_1(v,y)+\theta_2(v+th,y),y) - f(v+\theta_1(v,y)+\theta_2(v+th,y),y)$$
$$\leq F(v+th,y) - F(v,y) \leq$$
$$f(v+th+\theta_1(v+th,y)+\theta_2(v,y),y) - f(v+\theta_1(v+th,y)+\theta_2(v,y),y).$$

From the mean value theorem, we infer the existence of numbers τ_k, $k=1,2$ with $0 < \tau_k < 1$ such that

$$D_1 f(v+\tau_1 th+\theta_1(v,y)+\theta_2(v+th,y),y)(th)$$
$$\leq F(v+th,y) - F(v,y) \leq$$
$$D_1 f(v+\tau_2 th+\theta_1(v+th,y)+\theta_2(v,y),y)(th).$$

Therefore, by continuity of D_1f and θ, it follows that

$$\lim_{t\to 0}(F(v+th,y)-F(v,y))/t=D_1f(v+\theta(v,y),y)(y).$$

This shows that F has a Gateaux derivative with respect to its first argument. Since the mapping $v\to D_1f(v+\theta(v,y),y)$ is continuous, it follows that for $v\in V$, $y\in C^{\alpha}(\bar{\Omega})$, F has a Frechet derivative $D_1F(v,y)$ with respect to its first argument given by the formula (2.8). This proves the lemma.

Lemma 2.3. Given $y\in C^{\alpha}(\bar{\Omega})$, the solutions u of (2.4) consist of functions of the form $v+\theta(v,y)$ where $v\in V$ and $D_1F(v,y)=0$.

Proof: If $D_1F(v,y)=0$, then according to (2.8) we have

$$D_1f(v+\theta(v,y),y)(v_1)=0$$

for all $v_1\in V$. Since, according to (2.7), $Df_1(v+\theta(v,y),y)(z)=0$ for $z\in W\cap\mathring{H}_1(\Omega)$ and $\mathring{H}_1(\Omega)=V\oplus(W\cap\mathring{H}_1(\Omega))$, we have that $D_1f(v+\theta(v,y),y)=0$ so $v+\theta(v,y)$ must be a solution of (2.1).

Conversely, if u is a solution of (2.1) and we write $u=v+w$ with $v\in V$ and $w\in W\cap\mathring{H}_1(\Omega)$, then (2.5) must hold so $w=\theta(v,y)$. Since $D_1f(u,y)=0$, it follows from Lemma 2.2 that $D_1F(v,y)=0$ and the lemma is proved.

Lemma 2.4. There exists $t_1 > 0$ such that if $t > t_1$ and $s > 0$, then

$$\Delta u+bu^{+}-au^{-}=s(t\phi_1+\psi)\equiv y \qquad (2.11)$$

$$u|\partial\Omega=0$$

has a solution $u^*=v^*+\theta(v^*,y)$ such that $F(v,y)$ is of class C^2 in v in a neighborhood of v^* and $F(\ ,y)$ has a nondegenerate local maximum at v^*. Similarly, there exists $t_0 < 0$ such that if $t < t_0$ and $s > 0$, then (2.11) has a solution $u_*=v_*+\theta(v_*,y)$ such that $F(v,y)$ is of class C^2 in v for v in a neighborhood of v_* and $F(\ ,y)$ has a nondegenerate local minimum at v_*.

For fixed $t > t_1$ $(t < t_0)$ $v^*\to 0$ $(v_*\to 0)$ in V as $s\to 0+$.

Proof: Since $b \notin \sigma$ there exists a unique smooth z such that

$$\Delta z+bz=\psi \text{ in } \Omega,\ z|\partial\Omega=0.$$

Since $\phi_1(x) > 0$ for all $x\in\Omega$ and $\frac{\partial\phi_1}{\partial n}(x) < 0$ for all $x\in\partial\Omega$, where $\frac{\partial}{\partial n}$ denotes differentiation in the direction of the outer normal n to $\partial\Omega$, it follows that there exists $t_1 > 0$ such that if $t > t_1$, then

$$t\phi_1(x)/(b-\lambda_1)+z(x) > 0$$

for $x\in\Omega$ and

$$t\frac{\partial\phi_1}{\partial n}(x)/(b-\lambda_1) + \frac{\partial z}{\partial n}(x) < 0$$

For all $x\in\partial\Omega$. It follows that if $t > t_1$ and $s > 0$ and we set

$$u^*(x)=s(t\phi_1(x)/(b-\lambda_1)+z(x)) \qquad (2.12)$$

then

$$\Delta u^*+bu^*=s(t\phi_1+\psi)\equiv y \qquad (2.13)$$

and $u^*(x) > 0$ in Ω so u^* is a solution of (2.11). By Lemma 2.2 $u^*=v^*+\theta(v^*,y)$ where $v^*=Pu^*$. Since $u^* > 0$ on Ω and $\partial u^*/\partial n < 0$ on $\partial\Omega$, there exists a number $\delta > 0$ such that if $h\in V$ and $|h|_1<\delta$, where $|\,|_1$ denotes $\mathring{H}_1(\Omega)$ norm, then $u^*(x)+h(x) > 0$ for $x\in\Omega$. (Recall the elements of the finite dimensional space V are smooth.)

Let $w^*=\theta(v^*,y)$. If $h\in V$ and $|h|_1 < \delta$, it follows that

$$\Delta w^*+(I-P)[b(u^*+h)^+-a(u^*+h)^-]$$

$$= \Delta w^*+(I-P)b(u^*+h)$$

$$= (I-P)[\Delta u^*+bu^*]=(I-P)y.$$

Consequently, from the uniqueness part of Lemma 2.1, it follows that for $|h|_1 < \delta$, $\theta(v^*+h,y)=w^*=\theta(v^*,y)$.

Let $h\in V$ and $|h|_1 < \delta$. Multiplying (2.13) by $-h$ and integrating over Ω, we obtain

$$\int_\Omega (\nabla u^*\cdot\nabla h - bu^*h + yh)\,dx = 0.$$

Therefore,

$$F(v^*+h,y) = f(v^*+w^*+h,y) = f(u^*+h,y)$$

$$= \int_\Omega \tfrac{1}{2}(|\nabla(u^*+h)|^2 - b((u^*+h)^+)^2 - a((u^*+h)^-)^2 + 2(u^*+h)y)\,dx$$

$$= \int_\Omega \tfrac{1}{2}(|\nabla(u^*+h)|^2 - b(u^*+h)^2 + 2(u^*+h)y)\,dx$$

$$= \int_\Omega \tfrac{1}{2}(|\nabla u^*|^2 - bu^{*2} + 2u^*y)\,dx + \int_\Omega \tfrac{1}{2}(|\nabla h|^2 - bh^2\ )\,dx.$$

Since $(u^*)^+ = u^*$, $(u^*)^- = 0$ we have for $|h|_1 < \delta$

$$F(v^*+h,y) = F(v^*,y) + \tfrac{1}{2}\int_\Omega (|\nabla h|^2 - bh^2)\,dx.$$

This shows that $F(\ ,y)$ is class C^2 in a neighborhood of v^*, and that

$$D_1^2 F(v^*,y)(h)(h) = \int_\Omega (|\nabla h|^2 - bh^2)\,dx.$$

Since V is spanned by eigenvalues corresponding to λ, with $p+1 \le j \le q$, it follows that if $h\in V$ and $h\ne 0$

$$\int_\Omega (|\nabla h|^2 - bh^2)\,dx \le (\lambda_q - b)\int_\Omega h^2 dx < 0.$$

This shows that $F(\ ,y)$ has a nondegenerate local maximum at v^* and the first part of the lemma is proved.

The same argument as used above shows that if t is sufficiently large and negative and $s > 0$, then (2.11) has a solution u_* which is strictly negative on Ω such that $v_* = Pu_*$ and $|h|_1$ is small, then

$$F(v_*+h,y) - F(v_*,y) =$$

$$\tfrac{1}{2}\int_\Omega (|\nabla h|^2 - ah^2)\,dx \ge \tfrac{1}{2}(\lambda_p - a)\int_\Omega h^2 dx > 0$$

if $h\ne 0$. Hence $F(\ ,y)$ has a nondegenerate local minimum at v_*.

From (2.12), we see that for fixed large positive t, $v^*=Pu^*\to 0$ as $s\to 0$. Similarly for fixed large negative t, $v_*=Pu_*\to 0$ as $s\to 0$ and the lemma is proved.

Lemma 2.5. *For fixed* $y\in C^\alpha(\bar{\Omega})$ *the mapping* $V\to\mathbb{R}$ *defined by* $v\to F(v,y)$ *satisfies the Palais-Smale condition*.

The idea of the proof simple - - one shows that if $\{v_m\}_1^\infty$ is a sequence in V such that $D_1F(v_m,y)\to 0$ as $m\to\infty$ and $\{v_m\}^\infty$ is unbounded, then

$$\Delta u+bu^+-au^-=0,\ u|\partial\Omega=0$$

has a nonzero solution which contradicts the assumption $(a,b)\notin\Sigma$. Details will be given in [7].

Lemma 2.6. *Exactly one of the following holds*:

(a) $F(v,0) > 0$ for all $v\neq 0$

(b) $F(v,0) < 0$ for all $v\neq 0$

(c) *There exist* $v_1\in V$ and $v_2\in V$ *such that* $F(v_1,0) > 0$ and $F(v_2,0) < 0$.

Proof: Suppose (c) does not hold. Then either

(i) $F(v,0) \geq 0$ for all $v\in V$

or

(ii) $F(v,0) \leq 0$ for all $v\in V$.

If (i) holds and there exists $v_0\neq 0$ such that $F(v_0,0)=0$, then $F(v,0)$ attains a local minimum at $v=v_0$ so $D_1F(v_0,0)=0$. According to Lemma 2.3, $u_0=v_0+\theta(v_0,0)$ is a non zero solution of $\Delta u+bu^+-au^-=0$, $u|\partial\Omega=0$ contradicting the assumption that $(a,b)\notin\Sigma$. This proves that if (i) holds, then (a) must hold.

A similar argument shows that if (ii) holds, then (b) must hold. This proves the lemma.

Proof of Theorem 2.1: Suppose case (a) of Lemma 2.6 holds. Let (,) denote the $L^2(\Omega)$ inner product and for fixed $y\in C^\alpha(\bar{\Omega})$ and $v\in V$ let $\nabla F(v,y)\in V$ be defined by

$$(\nabla F(v,y),v_1)=D_1F(v,y)(v_1).$$

Clearly ∇F is continuous in both of its arguments.

If y=0 and v=0, then w=0 is a solution of (2.5). Hence $\theta(0,0)=0$ so, according to (2.2), $F(0,0)=f(0,0)=0$. Let

$$B=\{v\in V \mid |v| < 1\}$$

where $|\,|$ denotes the $L^2(\Omega)$ norm. Since $F(v,0) > 0$ for $v\neq 0$, $F(v,0)$ has an absolute minimum at v=0 so $\nabla F(0,0)=0$. Moreover $\nabla F(v,0)\neq 0$ for $v\neq 0$; otherwise $v+\theta(v,0)$ would be a nonzero solution of (2.4) when y=0, contradicting the assumption $(a,b)\notin\Sigma$. According to a theorem of Rothe [2], [21], [22], the topological index at v=0 of the mapping $v\to\nabla F(v,0)$ is 1. Therefore, the degree of $\nabla F(\ ,0)$ with respect to B and 0, $d(\nabla F(\ ,0),B,0) = 1$.

Let t_1 be as in Lemma 2.4 and fix $t > t_1$. We may choose $s > 0$ so small that if $y=s(t\phi_1+\psi)$, then $F(v,y) > F(0,y)$ for all $v\in V$ with $|v|=1$, $\nabla F(v,y)\neq 0$ if $|v|=1$, and $d(\nabla F(\ ,y),B,0) = 1$. We may also choose s so small that $|v^*| < 1$ where v^* is as in Lemma 2.4. By the above, $F(v,y)$ assumes its minimum on the closure of B at a point $v_0\in B$. If $v=v_0$ is not an isolated zero of $\nabla F(v,y)$, then $\nabla F(v,y)=0$ has more than three solutions. Suppose v_0 is an isolated zero. Then by Rothe's theorem, the index of the map $v\to\nabla F(v,y)$ at v_0 is +1. The index at the critical point v^* where $F(\cdot,y)$ attains a local maximum is $\pm$ 1 depending on the dimension of V. If v_0 and v^* were the only solutions of $\nabla F(v,y)=0$ satisfying $|v| < 1$, then we would have $d(\nabla F(\cdot,y),B,0)=0$ or 2 which is a contradiction.

This shows that if case (a) of Lemma 2.6 holds, $t > t_1$, $s > 0$ is sufficiently small and $y=s(t\phi_1+\psi)$, then there are at least three distinct solutions v_1,v_2 and v_3 of $DF(v,y)=0$. This gives three solutions $v_k+\theta(v_k,y)$, $1 \le k \le 3$, of (2.11) and, by homogeneity of the nonlinearity, $u_k=(v_k+\theta(v_k,y))/s$ $1 \le k \le 3$ are three solutions of (2.1).

By Lemma 2.4 if $t < t_0<0$, then $(v_*+\theta(v_*,y))/s$ is a solution of (2.1).

This shows that if case (a) of Lemma 2.5 holds, then the assertion of Theorem 2.1 holds with m=3.

In case (b) of Lemma 2.6 holds, an argument parallel to the one given above proves the assertion of Theorem 2.1 is true with m=1. In this case for $t < t_0 < 0$, $s > 0$ and small, and $y=s(t\phi_1+\psi)$, one considers the function $-F(v,y)$ and shows $-F(v,y) > -F(0,y)$ for $|v|=1$ and

$d(-\nabla F(\cdot,y),0)=1$. If v_* is as in Lemma 2.4, then for s sufficiently small, v_* is contained in B and $-F(v,y)$ attains a local maximum at $v=v_*$. Using the fact that the minimum of $-F(\cdot,y)$ on the closure of B must be assumed on B, the argument used above gives the existence of at least three critical points of $v \to F(v,y)$ on B and these three critical points give three solutions of (2.1). Since, according to Lemma 2.4, we have the solution $1/s(v^*+\Theta(v^*,y)$ for $t > t_1$, the assertion of Theorem 2.1 holds with m=1.

Suppose finally that case (c) of Lemma 2.6 holds and let v_1 and v_2 be points in V such that $F(v_1,0) > 0$ and $F(v_2,0) < 0$. Let $t > t_1$ and $y=s(t\phi_1+\psi)$ with $s > 0$ and small. As shown above $F(0,0)=0$. Therefore, since according to Lemma 2.4, $v^* \to 0$ as $s \to 0$ we may choose $s > 0$ so small that

$$F(v^*,y) < F(v_1,y).$$

Since, according to Lemma 2.4, $F(v,s)$ attains a strict local maximum at $v=v^*$, $-F(v,s)$ has a strict local minimum at $v=v^*$. Therefore, since $-F(v_1,y) < -F(v^*,y)$ and $-F(\cdot,y)$ satisfies the Palais-Smale condition, we can apply the well-known mountain-pass theorem of Ambrosetti and Rabinowitz [4] to infer the existence of a critical point v_3 of $F(\cdot,y)$ with $v_3 \neq v^*$. The two critical points v^* and v_3 give the existence of two solutions of (2.1).

Now let $t < t_0$, let $s > 0$ and $y=s(t\phi_1+\psi)$. Let v_* be as in Lemma 2.4. Since $F(v_*,y) \to F(0,0)=0$ as $s \to 0$ we may choose $s > 0$ so small that

$$F(v_2,y) < F(v_*,y).$$

Since, by Lemma 2.4, $F(v,y)$ has a local minimum at $v=v_*$, we may apply the mountain pass theorem to ascertain the existence of a critical point v_4 of $F(\cdot,y)$ with $v_4 \neq v_*$. As before, this gives us two solutions of (2.1).

Therefore if case (c) of Lemma 2.6 holds, the assertion of Theorem 2.1 holds with m=2. This proves the theorem.

3. <u>The case in which</u> [a,b] <u>contains one eigenvalue of arbitrary multiplicity</u>.

In this section we assume

$$\lambda_p < a < \lambda_{p+1}=\ldots=\lambda_q < b < \lambda_{q+1} \tag{3.1}$$

where $p \geq 1$. The following result, for the piecewise linear case, is an improvement of results in [16].

Theorem 3.1. *For fixed* $a\in(\lambda_p,\lambda_{p+1})$, *there exists* $\bar{b} \in (\lambda_{p+1},\lambda_{q+1}]$ *such that for* $\lambda_{p+1} < b < \bar{b}$, $(a,b)\notin\Sigma$ and (2.1) *has at least three solutions for* t *large and positive and at least one solution for* t *large and negative*. *If* $\bar{b} < \lambda_{q+1}$, *then* $(a,\bar{b})\in\Sigma$.

Similarly, it can be shown that *if* $b\in(\lambda_{p+1},\lambda_{q+1})$ *is fixed*, *there exists* $\bar{a}$ *with* $\lambda_p \leq \bar{a} < \lambda_{q+1}$ *such that* $\bar{a} < a < \lambda_{q+1}$ *implies* $(a,b)\notin\Sigma$ *and that* (2.1) *has at least three solutions for* t *sufficiently large and negative and at least one solution for* t *sufficiently large and positive*. *If* $\lambda_p < \bar{a}$, *then* $(\bar{a},b)\in\Sigma$.

Let Z denote the subspace of $\mathring{H}_1(\Omega)$ consisting of functions in $\mathring{H}_1(\Omega)$ which are orthogonal in the $L^2(\Omega)$-sense to eigenfunctions corresponding to the eigenvalues λ_j with $j \leq p$. If $z\in Z$, then

$$\int_\Omega |\nabla z|^2 dx \geq \lambda_q \int_\Omega z^2 dx \tag{3.2}$$

and if $z\neq 0$ equality holds if and only if z is an eigenfunction corresponding to λ_q.

Let S denote the set of $z\in Z$ such that $|Z|_0=1$, where $||_0$ is the $L^2(\Omega)$ norm, and consider the functional H on Z defined by

$$H[z]=\int_\Omega (|\nabla z|^2-\lambda_q(z^+)^2-a(z^-)^2)dx$$

where a is a fixed number in (λ_p,λ_q).

Since the imbedding of $\mathring{H}_1(\Omega)$ in $L^2(\Omega)$ is compact, a standard argument based on minimizing sequences and lower semicontinuity with respect to weak convergence of the $\mathring{H}_1(\Omega)$ norm gives the existence of $z_0\in Z$ such that

$$H[z_0]=\gamma_0\equiv\inf_{z\in S} H[Z] .$$

We have

$$H[z_0]=\int_\Omega (|\nabla z_0|^2-\lambda_q z_0^2)dx+\int_\Omega (\lambda_q-a)(z_0^-)^2 dx.$$

Both terms are nonnegative and the first is zero iff z_0 is an eigenfunction corresponding to λ_q. If z_0 is an eigenfunction corresponding to λ_q, then $z_0^-\not\equiv 0$ since $q > 1$. Therefore $\gamma_0 > 0$. By homogeneity $H[z] \geq \gamma_0 |z|_0^2$ for all $z\in Z$.

We now consider the functional $f(u,y,b)$, $u\in\mathring{H}_1$, $y\in C^\alpha(\bar{\Omega})$ and $\lambda_q < b < \lambda_{q+1}$, defined as in the previous section but now allowing for variable b. For $v\in V$, $y\in C^\alpha(\bar{\Omega})$, and $\lambda_q < b < \lambda_{q+1}$, let $F(v,y,b)$ and $\theta(v,y,b)$ have the obvious meanings corresponding to those in the previous section.

For $\lambda_q < b < \lambda_{q+1}$ and $z\in Z$ we have

$$f(z,0,b)=\frac{1}{2}\int_\Omega (|\nabla z|^2-b(z^+)^2-a(z^-)^2)dx$$

$$= \frac{1}{2}H[z]+\frac{1}{2}\int_\Omega (\lambda_q-b)(z^+)^2 dx \geq \frac{1}{2}(\gamma_0+\lambda_q-b)|z|_0^2.$$

Consequently $f(z,0,b) > 0$ for $z\in Z$, $z\neq 0$, and $b < \lambda_q < \lambda_q+\gamma_0$. Let W_1 and W_2 be defined as in the previous section and for $v\in V$ and $\lambda_q < b < \lambda_{q+1}$ write

$$\theta(v,0,b)=\theta_1(v,0,b)+\theta_2(v,0,b)$$

with $\theta_k(v,0,b)\in W_k$, $k=1,2$.
Since

$$f(v+w_1+\theta_2(v,0,b),0,b) \leq f(v+\theta(v,0,b),0,b) \leq$$

$$f(v+\theta_1(v,0,b)+w_2,0,b)$$

For $w_1\in W_1$, $w_2\in W_2$ and $v\in V$, by taking $w_1=0$ we obtain

$$F(v,0,b)=f(v+\theta(v,0,b),0,b) \geq f(v+\theta_2(v,0,b),0,b)$$

Therefore, since $v+\theta_2(v,0,b)\in Z$ and $v+\theta_2(v,0,b)\neq 0$ if $v\neq 0$ $F(v,0,b) > 0$ if $v\in V$, and $\lambda_q < b < \lambda_q+\gamma_0$.

If $\lambda_q < b_1 < b_2 < \lambda_{q+1}$, then $f(u,0,b_2) \le f(u,0,b_1)$ for all $u \in \mathring{H}_1(\Omega)$. Therefore, if $b_1 < b_2$

$$F(v,0,b_2)=f(v+\theta(v,0,b_2),0,b_2) \le$$

$$f(v+\theta_1(v,0,b_2)+\theta_2(v,0,b_1),0,b_2) \le$$

$$f(v+\theta_1(v,0,b_2)+\theta_2(v,0,b_1),0,b_1) \le$$

$$f(v+\theta_1(v,0,b_1)+\theta(v,0,b_1),0,b_1)=F(v,0,b_1).$$

From this we infer the existence of a number $\bar{b}$ with $\lambda_q < \bar{b} \le \lambda_{q+1}$ such that $F(v,0,b) > 0$ for $b \in (\lambda_q, \bar{b})$ and $v \in V$ with $v \ne 0$ and, if $\bar{b} < b < \lambda_{q+1}$, there exists $\bar{v} \in V$, with $\bar{v} \ne 0$ and $F(\bar{v},0,b) \le 0$.

To complete the proof we note that for fixed $b \in (\lambda_q, \lambda_{q+1})$, $F(v,0,b)$ is positively homogeneous of degree two in v. First, it is obvious that for $k > 0$ and $u \in \mathring{H}_1(\Omega)$, $f(ku,0,b)=k^2 f(u,0,b)$. Next, we note that if $y=0$, $v \in V$ and w is a solution of (2.5), then if $k > 0$, kw is a solution of the same equation when v is replaced by kv which means $\theta(kv,0,b)=k\theta(v,0,b)$. Hence

$$F(kv,0,b)=f(kv+\theta(kv,0,b),0,b) =$$

$$k^2 f(v+\theta(v,0,b),0,b)=k^2 F(v,0,b).$$

Suppose $\bar{b} < \lambda_{q+1}$ and let Γ be the set of $v \in V$ with $|v|_0=1$. If $v \in \Gamma - \{0\}$ and $\lambda_q < b < \bar{b}$, then $F(v,0,b) > 0$. So, by continuity, $F(v,0,\bar{b}) \ge 0$. Since V is finite dimensional, Γ is compact. Therefore, if $F(v,0,\bar{b}) > 0$ for all $v \in \Gamma$, then by continuity, $F(v,0,b) > 0$ for all $v \in \Gamma$ and b slightly greater than b. But, by homogeneity, this means that $F(v,0,b) > 0$ for all $v \in V$ with $v \ne 0$ and some $b > \bar{b}$, which is a contradiction. Therefore, there exists $v_0 \in \Gamma$ such that $F(v_0,0,\bar{b})=0$. Since $F(v,0,\bar{b}) \ge 0$ for all $v \in V$, $F(\cdot,0,\bar{b})$ assumes a local minimum at v_0 so $D_1F(v_0,0,\bar{b})=0$. By Lemma 2.3, $v_0+\theta(v_0,0,\bar{b})$ is a nonzero solution of $\Delta u+\bar{b}u^+-au^-=0$, $u|\partial\Omega=0$ which means $(a,\bar{b}) \in \Sigma$.

Suppose $b \in (\lambda_q, \bar{b})$ and that contrary to the assertion of Theorem 3.1 $(a,b) \in \Sigma$. Then there exists $u \ne 0$ such that $\Delta u+bu^+-au^-=0$, $u|\partial\Omega=0$. We have $u=v_0+\theta(v_0,0,b)$ where $D_1F(v_0,0,b)=0$. Since $w=0$ satisfies (2.5) if $v=0$ and $y=0$, $\theta(0,0,b)=0$ and, hence, $v_0 \ne 0$. But $F(v,0,b)$ is positively homogeneous of degree two in v so

$2F(v_0,0,b)=D_1F(v_0,0,b)(v_0)=0$ which is a contradiction. Therefore $(a,b)\notin\Sigma$ for $(\lambda_q,\bar{b})$.

Since $F(v,0,b) > 0$ if $v\neq 0$ and $\lambda_q < b < \bar{b}$, it follows from the proof of Theorem 2.1 that (2.1) has at least three solutions for t sufficiently large and positive and at least one solution for t sufficiently large and negative. This proves the theorem.

One can show that there exists $b^*\in(\lambda_{p+1},\lambda_{q+1})$ such that for $\lambda_{p+1} < b < b^*$ and t sufficiently large and negative, (2.1) has a unique solution. (See [16], where the result is established for a smooth nonlinearity.)

4. <u>A case in which the total number of solutions for large positive</u> t <u>and large negative</u> t <u>is exactly four</u>.

Let $m > 1$ and let λ_m be a simple eigenvalue. Gallouet and Kavian [12] have shown that there exists a function $C(a,b)$ defined and continuous for $\lambda_{m-1} < a,b < \lambda_{m-1}$ such that $C(a,b)$ is strictly decreasing in each variable and (2.1) has a nontrivial solution for a and b satisfying the above restriction if and only if $C(a,b)=0$ or $C(b,a)=0$. Moreover $C(a,a)=\lambda_m a$.

Since for $a\in(\lambda_{m-1},\lambda_m)$ fixed, each of the functions $C(a,\cdot)$ and $C(\cdot,a)$ can have at most one zero on $(\lambda_m,\lambda_{m+1})$, it follows that for such an a, there can exist at most two values of $b\in(\lambda_m,\lambda_{m+1})$ such that $(a,b)\in\Sigma$. Similarly, for fixed $b\in(\lambda_m,\lambda_{m+1})$, there can exist at most two values of $a\in(\lambda_{m-1},\lambda_m)$ such that $(a,b)\in\Sigma$.

Using the results of [12], Solimini [24] has proved

Theorem 4.1. <u>If</u> $m > 2$ <u>and</u> λ_m <u>is simple</u>, $\lambda_{m-1} < a < \lambda_m$, $\lambda_m < b < \lambda_{m+1}$, <u>and</u> $(a,b)\notin\Sigma$, <u>then there exists an integer</u> k <u>with</u> $1 \leq k \leq 3$ <u>such that</u> (2.1) <u>has exactly</u> k <u>solutions for</u> t <u>sufficiently large and positive and exactly</u> 4-k <u>solutions for</u> t <u>sufficiently large and negative</u>.

In fact, Solimini proved that k=3,2 and 1 in the three cases $C(a,b) > 0$ and $C(b,a) > 0$, $C(a,b)C(b,a) < 0$, and $C(a,b) < 0$ and $C(b,a) < 0$ respectively.

<u>Remark</u>: An examination of the proof of Theorem 4.1 given in [24] shows that the statement of Theorem 4.1 remains true if one considers Neumann boundary conditions instead of Dirichlet boundary conditions. That is, if $\lambda_{m-1} < \lambda_m < \lambda_{m+1}$ are three consecutive eigenvalues of the problem

$$\begin{cases} \Delta u+\lambda u=0 & \text{in } \Omega \\ \partial u|\partial n=0 & \text{on } \partial\Omega \end{cases} \tag{4.1}$$

with λ_m simple, if $\lambda_{m-1} < a < \lambda_m < b < \lambda_{m'}$, and the problem

$$\begin{cases} \Delta u+bu^+-au^-=0 & \text{in } \Omega \\ \partial u|\partial n=0 & \text{on } \Omega \end{cases} \tag{4.2}$$

has no solution other than $u\equiv 0$, then there exists an integer k with $1 \leq k \leq 3$ such that the problem

$$\begin{matrix} \Delta u+bu^+-au^-=t+\psi & \text{in } \Omega \\ \partial u|\partial n=0 & \text{on } \partial\Omega \end{matrix} \tag{4.3}$$

has exactly k solutions for t sufficiently large and positive and exactly 4-k solutions for t sufficiently large and negative.

Using this remark and a general symmetry principle developed by P.J. McKenna in [19], we give an example of a problem of the form

$$\begin{cases} Lu+bu^+-au^-=t\theta_1+\psi \\ u|\partial\Omega=0 \end{cases} \tag{4.4}$$

where L is a smooth second-order elliptic operator in a smooth bounded domain Ω, θ_1 is a positive eigenfunction corresponding to the lowest eigenvalue of the problem

$$\begin{cases} Lu+\lambda u=0 & \text{in } \Omega \\ u|\partial\Omega=0, \end{cases} \tag{4.5}$$

the problem

$$\begin{cases} Lu+bu^+-au^-=0 \\ u|\partial\Omega=0 \end{cases} \tag{4.6}$$

has no solution, other than $u\equiv 0$, neither a nor b is an eigenvalue of the problem (4.5), a is larger than the first eigenvalue of (4.5), the interval (a,b) contains an eigenvalue of <u>multiplicity two</u>, and there exists k with $1 \leq k \leq 3$ such that (4.4) has exactly k solutions for t large and positive and exactly 4-k solutions for t large and negative. Here ψ is a smooth function defined on Ω.

We express L and Ω in terms of polar coordinates in the plane $\mathbb{R}^2$. Let r denote the radial variable and s the angular variable. Let Ω denote the annulus

$$c < r < d, \qquad 0 \leq s \leq 2\pi$$

where c and d are positive numbers on which we shall place a further restriction. If u is a C^2 function on $\bar{\Omega}$ which vanishes on $\partial\Omega$ we set

$$Lu=u_{rr}+u_{ss}$$

Our conditions on u(s,t) are

$$u(c,s)=u(d,s)=0, \ u(r,s+2\pi)\equiv u(r,s). \tag{4.7}$$

We choose $c > 0$ and $d > c$ so that if $h=d-c$, then

$$4 < 3\pi^2/h^2. \tag{4.8}$$

The eigenvalues of L subject to 4.7 are the numbers

$$\lambda_{m,n}=m^2\pi^2/h^2+n^2$$

where $m=1,2,\ldots$, $n=0,1,2,\ldots$. By varying h slightly if necessary we may assume that $\lambda_{m,n}\neq\lambda_{p,q}$ if $(m,n)\neq(p,q)$. Each eigenvalue of the form $\lambda_{m,0}$ with $m \geq 1$ is simple with a corresponding eigenfunction given by $\sin(m\pi/h)(r-c)$ and those eigenvalues of the form $\lambda_{m,n}$ with $m \geq 1$ and $n \geq 1$ are double with two independent eigenfunctions given by

$$(\sin(m\pi/h)(r-c))\sin ns, \ (\sin(m\pi/h)(r-c))\cos ns.$$

Because of condition (4.8), we have

$$\lambda_{1,0} < \lambda_{1,1} < \lambda_{1,2} < \lambda_{2,0} \tag{4.9}$$

We choose numbers a and b so that

$$\lambda_{1,0} < a < \lambda_{1,1} < b < \lambda_{1,2}. \tag{4.10}$$

It follows that if $a_1=a-\pi^2/h^2, b_1=b-\pi^2/h^2$, then

$$0 < a_1 < 1 < b_1 < 4. \tag{4.11}$$

By varying a slightly, if necessary, we may assume that

$$1/\sqrt{a_1}+1/\sqrt{b_1} \neq 2/k \tag{4.12}$$
$$k=1,2,\ldots$$

In the following we set

$$\theta_1(r)=\sin(\pi/h)(r-c),\quad \psi(r,s)=\theta_1(r)\cos s$$

and consider the boundary value problem given by (4.4).

The space spanned by the eigenfunctions corresponding to $\lambda_{1,1}$ is $V=\text{span}\{\theta_1(r)\cos s,\ \theta_1(r)\sin s\}$. Let $P:L^2(\Omega)\to V$ denote orthogonal projection and let $W=(I-P)L^2(\Omega)$.

<u>Lemma</u> 4.1 <u>Given</u> $v\in V$, <u>there exists a unique</u> $w^*=w^*(v)\in W$ <u>such that</u>

$$Lw^*+(I-P)[b(v+w^*)^+-a(v+w^*)^-]=t\theta_1 \tag{4.13}$$
$$w^*|\partial\Omega=0$$

<u>Proof</u>: Let $\gamma=(a+b)/2$ and $G(\xi)=(b-\gamma)\xi^++(\gamma-a)\xi^-$. Since for smooth $w\in W$ and $v\in V$, $Lw+(I-P)(b(v+w)^+-a(v+w)^-] =$ $(L+\gamma)w+(I-P)[b(v+w)^+-a(v+w)^--\gamma(v+w)]$, w^* satisfies the assertion of the lemma if and only if w^* is a fixed point of

$$F:(I-P)L^2(\Omega)\to(I-P)L^2(\Omega)$$

defined by

$$F(w)=(-L-\gamma)^{-1}(I-P)[G\circ(w+v)-t\theta_1]$$

where for $g\in L^2$, $(-L-\gamma)^{-1}(I-P)g$ is the weak $\mathring{H}_2(\Omega)$ solution of

$$(-L-\gamma)u=(I-P)g,\quad u|\partial\Omega=0.$$

(By standard boot strapping, fixed points of F are smooth.) Since the spectrum of the compact self adjoint linear map $(-L-\gamma)^{-1}(I-P):L^2(\Omega)\to L^2(\Omega)$ consists of the numbers $(\lambda_{m,n}-\gamma)^{-1}$ with $(m,n)\neq(1,1)$, it follows from (4.9) that

$$\|(-L-\gamma)^{-1}(I-P)\|=\max\{(\lambda_{1,0}-\gamma)^{-1},(\lambda_{1,2}-\gamma)^{-1}\}.$$

Since G is Lipshitzian with Lipshitz constant $(b-\gamma)$, we see from (4.10) that F is Lipshitzian with Lipshitz constant $\|(-L-\gamma)^{-1}(I-P)\|(b-\gamma) < 1$. Hence, there exists a unique fixed point of F and the lemma is proved.

Let S denote the closed subspace of $L^2(\Omega)$ consisting of all functions in $L^2(\Omega)$ of the form $u(r,s)=\theta_1(r)z(s)$, where z is 2π-periodic. Since S can equivalently be characterized as the closure in $L^2(\Omega)$ of the span of the eigenvalues of L corresponding to the eigenvalues $\{\lambda_{1,n}|n=0,1,2...\}$ it follows that $(-L-\gamma I)^{-1}(I-P)$ maps the subspace $(I-P)L^2(\Omega)\cap S$ into itself. If $u\in S$, $u(r,s)=\theta_1(r)z(s)$, then $G\circ u(r,s)=\theta_1(r)G(z(s))$, so $G\circ u\in S$. We note that $V\subseteq S$. Therefore, if $v\in V$ is fixed, F is defined as above and $w\in(I-P)L^2(\Omega)\cap S$, then since $v+w\in S$, $F(w)\in(I-P)L^2(\Omega)\cap S$. Therefore, since F is a contraction and $W\cap S$ is a closed subspace of W, we obtain

<u>Lemma</u> 4.2. <u>For all</u> $v\in V$, $w^*(v)\in S$.

From this, we obtain

<u>Lemma</u> 4.3. <u>If</u> u <u>is a solution of</u> (4.4), <u>then</u> $u\in S$. <u>Moreover</u> $u(r,s)=\theta_1(r)z(s)$ <u>where</u> z <u>is a solution of the problem</u>

$$\begin{cases} z''+b_1z^+-a_1z^-=t+\cos s \\ z(s+2\pi)\equiv z(s) \end{cases} \tag{4.14}$$

<u>Proof</u>: If u is a solution of (4.4), then $u=v+w$ where $v\in V$ and $w\in W$. Since $\Delta v\in V$, applying (I-P), we obtain

$$\Delta w+(I-P)(b(v+w)^+-a(v+w)^-)=t\theta_1$$

and therefore, by Lemma 4.1, $w=w^*(v)\in S$. Since $v\subset S$, we have $u=v+w^*(v)\in S$. The second statement in the Lemma follows by substituting u into the P.D.E. and cancelling a factor of θ_1. This proves the lemma.

Let $L^2_{2\pi}$ denote the set of real-valued functions which are defined and 2π-periodic on $(-\infty,\infty)$ such that if $z \in L^2_{2\pi}$ the restriction of z to $[0,2\pi]$ belongs to $L^2[0,2\pi]$. Let X denote the two-dimensional subspace of $L^2_{2\pi}$ spanned by the functions sin s and cos s and let $Q: L^2_{2\pi} \to X$ denote orthogonal projection. Let $Y=(I-Q)L^2_{2\pi}$. Given $u \in L^2_{2\pi}$ such that $u'' \in L^2_{2\pi}$, let $Au=-u''$. Since the spectrum of A consists of the numbers n^2 $n=0,1,2,\dots$, since X is the span of the eigenfunctions of A corresponding to the eigenvalue 1, and since the inequalities (4.11) hold we have the following result whose proof is completely analogous to that of Lemma 4.1.

<u>Lemma</u> 4.4 <u>Given</u> $x \in X$, <u>there exists a unique</u> $y^*=y^*(x) \in Y$ <u>such that</u>

$$y^{*''}+[I-Q][b_1(x+y^*)^+-a_1(x+y^*)^-]=t. \tag{4.15}$$

<u>The mapping</u> $F_1: Y \to Y$ <u>given by</u>

$$F_1(y)=(A-\partial_1)^{-1}[I-Q][(b_1+\gamma_1)(x+y)^+ + (\gamma_1-a)(x+y)^- - t]$$

<u>where</u> $\gamma_1=(a_1+b_1)/2$ <u>is a contraction and</u> $y^*(x)$ <u>is its unique fixed point</u>.

Let E denote the subspace of $L^2_{2\pi}$ consisting of functions $z \in E$ such that $z(-s)=z(s)$. Equivalently, E is the closure in $L^2_{2\pi}$ of the span of the functions 1, cos s, cos 2s, ... Clearly $(A-\gamma_1)^{-1}(I-Q)E \subseteq E$. It follows that if $x \in V \cap E$, then $F_1(Y \cap E) \subseteq Y \cap E$.

Therefore, we have

<u>Lemma</u> 4.5: <u>If</u> $x \in X \cap E$ <u>is fixed and</u> F_1 <u>is defined as above, then</u> $y^*(x) \in Y \cap E$.

We next prove

<u>Lemma</u> 4.6. <u>If</u> z <u>is a solution of</u> (4.14), <u>then</u> $u \in E$ <u>and consequently</u> z <u>is a solution of the Neumann problem</u>

$$\begin{cases} z''+b_1 z^+ - a_1 z^- = t+\cos s \\ z'(0)=z'(\pi)=0 \end{cases} \tag{4.15}$$

<u>Proof</u>: If $z=x+y$ with $x \in X$ and $y \in Y$, then, by applying $(I-Q)$ to (4.14), we obtain

$$y''+(I-Q)[b_1(x+y)^+-a_1(x+y)^-]=t$$

and therefore, $y=y^*(x)$. To prove the lemma, it is sufficient, by Lemma 4.5, to show that $x\in X\cap E$. Since X is the span of cos s and sin s and Y is orthogonal to X, this is equivalent to showing that

$$\int_0^{2\pi} z(s)\sin s\, ds=0 \ . \tag{4.16}$$

To this end, we note from (4.14) that

$$\frac{d}{ds}[(z'(s)^2+b_1(z(s)^+)^2+a_1(z(s)^-)^2)/2-tz(s)]$$

$$= z'(s)[z''(s)+b_1z(s)^+-a_1z(s)^--t]$$

$$= z'(s) \cos s.$$

Therefore, since $z(s)=z(s+2\pi)$, we have

$$\int_0^{2\pi} z'(s)\cos s\, ds=0$$

and integration by parts gives (4.16). This proves the lemma.

The condition (4.12) ensures that the boundary value problem

$$\begin{cases} z''+b_1z^+-a_1z^-=0 \\ z'(0)=z'(\pi)=0. \end{cases} \tag{4.17}$$

has no solution other than $z\equiv 0$.

It follows from the remark after the statement of Theorem 4.1 that there exists an integer k with $1 \le k \le 3$ such that (4.15) has exactly k solutions for t large and positive and exactly 4-k solutions for t large and negative. Thus, by what has been shown, (4.4) has exactly k solutions for t large and positive and exactly 4-k solutions for t large and negative, even though (a,b) contains an eigenvalue of multiplicity two.

REFERENCES

1. H. Amann, Saddle points and multiple solutions of differential equations, Math. Z. (1979), 127-166.

2. H. Amann, A note on the degree theory for gradient mappings, Proc. AMS 84 (1982), 591-595.

3. A. Ambrosetti and G. Prodi, On the inversion of some differentiable mappings with singularities between Banach spaces, Ann. Math. Pura. Appl. 93 (1973), 231-247.

4. A. Ambrosetti and P.H. Rabinowitz, Dual variational methods in critical point theory, J. Func. Analysis 14 (1973), 343-381.

5. M.S. Berger and E. Podolak, On the solutions of nonlinear Dirichlet problem, Indiana Univ. Math. J. 24 (1975), 837-846.

6. A. Castro, Hammerstein integral equations with indefinite kernel, International J. Math. and Math. Sci., 1 (1978), 207-211.

7. E.N. Dancer, "On the Dirichlet problem for weakly nonlinear elliptic partial differential equations, Proc. Royal Soc. Edinburgh 76A (1977), 283-300.

8. E.N. Dancer, Degenerate critical points, homotopy indices and Morse inequalities, J. Reine Angew. Math. 350 (1984), 1-22.

9. E.N. Dancer, Multiple solutions of asymptotically homogeneous problems, Preprint.

10. C.L. Dolph, Nonlinear integral equations of the Hammerstein type, Trans. Amer. Math. Soc. 60 (1949), 289-307.

11. S. Fucik, Boundary value problems with jumping nonlinearities, Casopis Pest. Mat. 101 (1975), 69-87.

12. Th. Gallouet and O. Kavian, Resultats d Existence et de Non Existence pour certain problem Demi-lineares a l'infini, Ann. Fac. Sc. de Toulouse, (1981).

13. D.C. Hart, A.C. Lazer, and P.J. McKenna, Multiple solutions of two point boundary value problems with jumping nonlinearities, J. Diff. Equa. 59 (1985), 266-282.

14. J.L. Kazdan and F.W. Warner, Remarks on some quasilinear elliptic equations, Comm. Pure Appl. Math. 28 (1975), 567-597.

15. A.C. Lazer and P.J. McKenna, Multiplicity results for a semilinear boundary value problem with the nonlinearity crossing higher eigenvalues, Nonlinear Analysis TMA 9 (1985), 335-349.

16. A.C. Lazer and P.J. McKenna, Critical point theory and boundary value problems with nonlinearities crossing multiple eigenvalues, Comm. in P.D.E. 10 (2) (1985), 107-150.

17. A.C. Lazer and P.J. McKenna, Critical point theory and boundary value problems with nonlinearities crossing multiple eigenvalues II, Comm. in P.D.E., in press.

18. A.C. Lazer and P.J. McKenna, Large scale oscillatory behavior in loaded asymmetric systems, Analyse non lineaire, IPH, to appear.

19. A.C. Lazer and P.J. McKenna, A symmetry theorem and applications to nonlinear partial differential equations, to appear.

20. L. Nirenberg, Topics in Nonlinear Functional Analysis, Courant Institute of Mathematical Sciences, 1974.

21. P.H. Rabinowitz, A note on the topological degree for potential operators, J. Math. Anal. Appl. 51 (1975), 483-492.

22. E.H. Rothe, A relation between the type numbers of a critical point and the index of the corresponding field of gradient vectors, Math. Nach. 4 (1950-51), 12-27.

23. B. Ruf, Remarks and generalizations related to a recent multiplicity result of A. Lazer and P. McKenna, Nonlinear Analysis, TMA 8 (1985).

24. S. Solimini, Some remarks on the number of solutions of some nonlinear elliptic equations, Analyse non lineare, IHP, 2 (1985), 143-156.

Regularity of Solutions of Cauchy Problems with Smooth Cauchy Data

Otto Liess

1 Introduction

1.This paper is addressed to questions of regularity for solutions of Cauchy problems of form

(1) $p(x,t,D_x,D_t)u = 0, t > 0,$

(2) $D_t^j u_{|t=0} = f_j, j = 0,...,m-1.$

Here

(3) $p(x,t,D_x,D_t) = D_t^m + \sum a_{\alpha j}(x,t)D_x^\alpha D_t^j,$

where the sum is for all α, j , such that $|\alpha| + j \leq m$, $j < m$. The coefficients $a_{\alpha j}$ are defined in a neighborhood of 0 in $R_x^n \times R_t = R^{n+1}$ and are assumed to be real-analytic there.Furthermore,the equations (1) and (2) are understood in the sense of germs , in neighborhoods of 0 in R^{n+1} and in R^n $(= R_x^n)$ respectively. Finally,even though (1) makes sense for distributions u which are defined only for $t > 0$, we shall (almost) always assume,in order to have a natural definition (in distributions,) for (2) that u is extendible across $t = 0$.
2.The main result of this paper is an extension (under a natural additional assumption) of the following result of G.Lebeau [1] :

Theorem 1.1 *Let u be an extendible solution of (1),(2) and assume that S is an analytic hypersurface in $t \geq 0$ which is tangent at 0 to $t = 0$. Assume further that $(0,\xi^o) \notin WF_A f_j$ for j=0,...,m-1.Then it follows that $(0,\xi^o) \notin WF_A D_t^k u_{|S}$ for any k. Here $WF_A f$ is the analytic wave front set of f.*

(From the notation it is clear that we assume that S is parametrized by x, so $(0,\xi^o) \in WF_A D_t^k u_{|S}$ has a natural meaning .) Before we state however results where the regularity of u is estimated in terms of the regularity of the Cauchy data f_j it is appropriate to ask what regularity u will have anyway,irrespective of any assumption on the regularity of the f_j .Since we have not made any assumption on the type of the operator in (3) ,it is clear that we may only expect regularity in (co-)normal directions. Let us then recall at first that any extendible solution u of (1) is automatically a C^∞ function in t for $t \geq 0$ small with distributional values in x. For this to hold,it is in fact not even necessary to assume that the coefficients $a_{\alpha j}$ are real-analytic near 0 : C^∞ smoothness of the coefficients would have been enough.Here however we assume that the coefficients are real-analytic , so one will expect to be able to obtain better regularity in the t-variable. This is indeed possible, in that any hyperfunction solution of (1) is , in a certain sense, a real-analytic function in t for $t \geq 0$ small with values which are hyperfunctions in x. A quantitatively precise version of what we mean by this,is the following result of K.Kataoka [1] :

Theorem 1.2 *Any hyperfunction solution u of (1) is mild from the positive side of $t = 0$.*

(A hyperfunction f defined in a neighborhood of 0 in R^{n+1} is called mild from the positive side of $t = 0$, if there are $c > 0$, k, $\xi^1, \ldots, \xi^k$,all in R^n, $|\xi^j| = 1$ and holomorphic functions $f_j, j = 1, ..., k$, defined on

$$D_{j,c} = \{(x,t) \in C^{n+1}; |(x,t)| < c \ , < Imx, \xi^j > > c|Imx| + (1/c)[|Imt| + max(0, -Ret)]\},$$

and such that $f(x,t) = \sum_j b(f_j)$, for $t > 0$ and near 0 in R^{n+1}. Here $b(f_j)$ is the hyperfunctional,i.e. cohomological,boundary value of f_j.)

3.The result of Kataoka is not directly useful for our present needs, since it does not contain any estimates on what happens when t tends to zero.When u is an extendible solution of (1) one can on the other hand obtain quantitative control on the two-microlocal regularity of u in co-normal directions:

Theorem 1.3 *(Cf.Liess [4]) Let u be an extendible solution of (1) . Then there are $d > 0, c' > 0$, such that for every b_1 we can find c and b_2 with the property that $|u(g)| < c$ for any $g \in C_o^\infty(R^{n+1})$ which satisfies*

$$(4) \quad |F(g)(\lambda)| < exp(d|Re\zeta| + d|Im\zeta| + dIm\tau_+ + b_2 ln(1 + |\zeta|) + b_1 ln(1 + |\tau|)), \quad if \ |\tau| > c'|\zeta|,$$

$$(5) \quad |F(g)(\lambda)| < exp(d|Im\zeta| + dIm\tau_+ + b_2 ln(1 + |\zeta|) + b_1 ln(1 + |\tau|)), \text{otherwise.}$$

Here $F(g)$ is the Fourier-Borel transform of g, and $\lambda = (\zeta, \tau), \zeta \in C^n$, $\tau \in C$ are the Fourier-dual variables of x and t respectively. Moreover, $a_+ = \max(0, a)$,if $a \in R$. (In Liess [4] there is an ample discussion of the relation of this result to the theorem of Kataoka and to the Sato-Hörmander regularity theorem.)

4. Having in mind that the main result which we want to obtain is a result on Gevrey regularity,we shall now also state a Gevrey version of theorem 1.3. To do so ,let $M = (M_1, \ldots, M_n)$, $M_j \geq 1$ be given and assume hat the following condition is satisfied:

(6) the sum in (3) is (only) for $< \alpha, M > +j \leq m, j < m$.

Then we can prove (cf. once more Liess [4]):

Theorem 1.3' Let u be an extendible solution of (1). Then there are $d > 0, c' > 0$ such that for any b_1 we can find b_2 and c with the property that $|u(g)| < c$ for any $g \in C^\infty(R^{n+1})$ such that, with the notation $\varphi(\xi) = \sum |\xi_j|^{1/M_j}$,

$$|F(g)(\lambda)| < exp(d\varphi(Re\zeta) + d|Im\zeta| + dIm\tau_+ + b_2 ln(1 + |\zeta|) + b_1 ln(1 + |\tau|)), \quad \text{if } |\tau| > c'(\varphi(Re\zeta) + |Im\zeta|),$$

$|F(g)(\lambda)| < exp(d|Im\zeta| + dIm\tau_+ + b_2 ln(1 + |\zeta|) + b_1 ln(1 + |\tau|))$, otherwise.

5.We now turn our attention to results in which we obtain additional regularity for u, assuming that we have already some information on additional regularity of the Cauchy data.We shall be interested in results in Gevrey regularity,so we recall that if $M = (M_1, \dots, M_n)$, $M_j \geq 1$, is given, then a germ f of a C^∞ function defined near 0 in R^n is called of (anisotropic Gevrey) class G^M at 0 if we can find c, c', such that $|D^\alpha f(x)| < c^{|\alpha|+1} (\alpha_1!)^{M_1} \cdots (\alpha_n!)^{M_n}$ for $|x| < c'$.
Of course,when $M = (1, ..., 1)$, then G^M is just the class of real- analytic functions. Note here that if the f_j are all real-analytic near 0 , then so is (by the Cauchy-Kowalewski and by Holmgren's uniqueness theorem) also the solution u of (1) and (2).The corresponding result is however not true for C^∞ -smoothness or general Gevrey-regularity. Thus for example if $p = (\partial/\partial t)^2 + (\partial/\partial x)^2 - (\partial/\partial y)^2$, $x, y \in R$, (which is the wave operator for n=2, written in "strange" notations for coordinates,) then it is not difficult to see that if $M_1 > 1$ and $M_2 > 1$, then we can find f_o and f_1 in G^M for which (1) and (2) is solvable in distributions, but for which u is not of class $G^{M'}$, with $M' = (M_1, M_2, 1)$. This example is closely related to the remark following theorem 9.6.9 in Hörmander [1] and a general approach to study the question of C^∞ smoothness of solutions of Cauchy problems with smooth Cauchy data can be obtained from §11.3 in Hörmander [1]. When one wants to study the case of Gevrey regularity,the approach from Hörmander,loc. cit.,has at least to be changed at a number of points (starting from the definition of "localizations" and up to the additional care which is needed when using "condensation-of-singularities" arguments),so we mention the following result, which one can prove using the methods from Liess [2].

Theorem 1.4 *Assume that*
a) p is constant coefficient, of form $p = p_m + r$,where the order of r is $\leq m - 2$.
b) $(\xi^o, \tau^o) \in R^{n+1}$ is given such that $p_m(\xi^o, \tau^o) = 0$,
c) a two-dimensional conic real-analytic manifold S in R^n $(= R^n_\xi)$ is given which is homogeneous and contains ξ^o.
d) p_m splits for $\xi \in S$ in the form $p_m(\zeta, \tau) = (\tau^2 + a_1(\zeta)\tau + a_2(\zeta))q(\xi, \tau)$, where $q(\xi^o, \tau^o) \neq 0$ and where the a_j are real-valued,real-analytic and positively homogeneous of degree j on S,
e) $D(\xi) = a_1^2(\xi) - 4a_2(\xi)$, vanishes at ξ^o, but the differential of D at ξ^o (computed on S) does not vanish.
Then,if $M" > 1$ is fixed and $M = (M", ..., M")$, we can find $f_j \in G^M$ for which (1) and (2) is solvable,but for which the solution is not in $G^{M'}$.

(One can prove a microlocal version of this result,but we do not state it.)
6. We now want to state the Gevrey-version of theorem 1.1. To do so,we have to recall at first the main concepts on microlocalization in anisotropic Gevrey classes.Let then,once more, $M = (M_1, ..., M_n)$, $M_j \geq 1$ be given. A natural action of R_+ on $R^n \backslash 0 = \dot{R}^n$ is then defined by $t^M\xi = (t^{M_1}\xi_1, \dots, t^{M_n}\xi_n)$. Moreover,we shall say that Γ in R^n is a quasicone if $\xi \in \Gamma$ imlies $t^M\xi \in \Gamma$ for all $t > 0$.The following definition is from Liess-Rodino [1] .

Definition 1.5 *Consider a germ f of a distribution defined in a neighborhood of 0 in R^n and fix $\xi^o \in R^n$. We shall then say that $(0, \xi^o) \notin WF_M f$ if we can find c, c', an open*

quasicone Γ which contains ξ^o and a bounded sequence f_j of distributions with compact support such that:
a) $f = f_j$ for $|x| < c$,
b) $|F(f_j)(\xi)| \le c'(c'j/\varphi(\xi))^j$ for $\xi \in \Gamma$.

A notion closely related to that of WF_M is that of the boundary wave front set "WF_M^b". Here we shall introduce such a notion following Liess [3].Before doing so we mention that,as is customary when introducing notions of tangential bounday regularity,we shall define WF_M^b only for distributions which are already known to be smooth in the t-variable. (We should mention that ,even when we restrict our notion to the analytic case $M = (1,..,1)$, our notion of boundary regularity is more explicit than other notions of boundary wave front sets considered in the literature.)

Definition 1.6 *Assume that u is a C^∞ funtion in t for $t \ge 0$ small with distributional values in x. Consider further $\xi^o \in R^n$. Then we shall say that $(0,\xi^o)$ is not in $WF_M^b u$ if we can find $d > 0, c$, an open quasicone Γ which contains ξ^o and for every b_1 in R some b_2 and c' with the property that $|u(g)| < c'$ for any g in $C_o^\infty(R^{n+1})$ which satisfies*

$$|F(g)(\lambda)| \le \exp\left(d\varphi(-Re\,\zeta) + d|Im\,\zeta| + dIm\,\tau_+ + b_1 ln(1+|\lambda|) + b_2(1+|\zeta|)\right), \quad \text{if } Re\,\zeta \in -\Gamma,$$

$$|F(g)(\lambda)| \le \exp\left(d|Im\,\zeta| + dIm\,\tau_+ + b_1 ln(1+|\lambda|) + b_2(1+|\zeta|)\right), \text{for all other } \lambda.$$

Here we recall that,at least for the case that u is C^∞ for $t \ge 0$ in (x,t), in Liess [3] there is given a charaterization of WF_M^b which is very close to the spirit of definition 1.5.
7. The main result from this paper is the following

Theorem 1.7 *Assume that p satisfies the condition (6) and let u be an extendible distribution which satisfies (1) and (2). Assume further that S is a germ of a real-analyic surface in $t \ge 0$ which is tangent at 0 to $t = 0$ and that $(0,\xi^o) \notin WF_M D_t^k u_{|t=0}$ for $k = 0,...,m-1$. Then it follows that $(0,\xi^o) \notin WF_M D_t^k u_{|S}$ for any k.*

8. The main ingredients in the proof of theorem 1.7 are theorem 1.3' from the above (which was a result on co-normal regularity) and the following complementary result on tangential boundary regularity,stated in terms of boundary wave front sets:

Theorem 1.8 *Assume that p satisfies (6) and let u be an extendible solution of (1),(2) such that $(0,\xi^o) \notin WF_M f_j$, $j = 0,...,m-1$ for some $\xi^o \in \dot{R}^n$. Then it follows that $(0,\xi^o)$ is not in $WF_M^b u$.*

(Theorem 1.8 is proved in Liess [3].)

Remark 1.9 *Theorem 1.8 gives,heuristically,good control on high order x- derivatives for u in $t \ge 0$ microlocally on the domain $\{0\} \times \{\lambda \in R^{n+1}; \xi \in \Gamma\}$. Similarily,theorem 1.3 gives good control for high-order x-derivatives of u for $t \ge 0$ microlocally near $\{0\} \times \{\lambda \in R^{n+1}; |\tau| > c\varphi(\xi)\}$. Thus,all in all,and once more heuristically,we have here good control of high-order x- derivatives of u for $t \ge 0$ microlocally near $\{0\} \times \{\lambda \in R^{n+1}; \xi \in \Gamma \,\text{or}\, |\tau| > c\varphi(\xi)\}$.*

For solutions of $pu = 0$,we can now also use the equation (1) to obtain good control on high-order t-derivatives of u on the same kind of domains. For this to become useful,we can use the following characterization of WF_M (a result related to this one appears in Rodino [1]):

Proposition 1.10 *Consider some germ f of a distribution defined near 0 in R^n and choose $\xi^o \in R^n$. Then there are equivalent :*
(i) $(0,\xi^o) \notin WF_M f$.
(ii) There is c_1 with the following property : if some constants c_i, $i = 2,..,7$ are given and if for some fixed σ with $|\sigma/|\sigma| - \xi^o/|\xi^o|| \leq c_1$ some C^∞ functions g_σ and h_σ on R^n are given such that
a) $g_\sigma(\xi) = 1$, for $|\xi - \sigma| < c_2\varphi(\sigma)$,
b) $g_\sigma(\xi) = 0$, for $|\xi - \sigma| > c_3\varphi(\sigma)$,
c) $0 < g_\sigma(\xi) \leq 1$ for all $\xi \in R^n$,
d) $h_\sigma(x) = 1$, for $|x| < c_4$,
e) $h_\sigma(x) = 0$, for $|x| > c_5$,
f) $|D^\alpha h(x)| < c_6(c_6\varphi(\sigma))^{|\alpha|}$, for $|\alpha| \leq c\varphi(\sigma) + c_7$, for some c for which $c\varphi(\sigma) > 1$, then it follows that

$$|(g_\sigma(D)D^\beta(h_\sigma f(x))| \leq c_8(c_8c)^{c\varphi(\sigma)}\varphi(\sigma)^{<\beta,M>+n}.$$

for all $|x| < c_9$ and all β such that $<\beta, M> \; < c\varphi(\sigma) + |M|$. c_8 is here some constant which depends on the c_i but not on σ.

(Here of course, $g(D)v = (2\pi)^{-n} \int exp(i < x, \xi >)g(\xi)(Fv)(\xi)d\xi$.)

2 Proofs: a technical preparation

In the proof of theorem 1.7 we shall have to estimate high order derivatives for solutions of $p(x, t, D_x, D_t)u = 0$.We describe how this is done in the present paragraph.Let then p be as in §1 for $\varphi = \sum(1+|\xi_j|)^{1/M_j}$ and consider a solution u in D'_+ of $p(x, t, D_x, D_t)u = 0$. Let further $f \in O$ be real-valued for real arguments,$f(0) = 0$ and $f(x) \geqslant 0$ for $x \neq 0$ in some small neighborhood of 0 in R^n. We denote by

$$S = \{(x, f(x)); x \text{ in some neighborhood of } 0 \in R^n\}.$$

The vector fields $X_j = (\partial/\partial x_j + f_{x_j}\partial/\partial t)$ are then tangential to S.Finally consider some multiindex α. In the next paragraph we shall then have to estimate expressions which involve $X^\alpha u = X_1^{\alpha_1} \cdots X_n^{\alpha_n} u$ on S. It will then be convenient to use repeatedly that $D_t^m u = -\sum_{j<m} a_{\alpha j}(x, t)D_x^\alpha D_t^j u$. As a consequence it is clear that we can write $X^\alpha u$ in the form

$$(1) \quad X^\alpha u = \sum_{j=0}^{m-1} \sum_{<\beta,M>\leq<\alpha,M>-j} A_{\beta j}(x, t)D_x^\beta D_t^j u, \text{ if } p(x, t, D_x, D_t)u = 0.$$

The coefficients $A_{\beta j}$ $(=(A_{\beta j}^\alpha))$ are here of course unique and one sees, e.g., inductively, that they are analytic functions for (x, t) in a complex neighborhood of 0. We need the following information on their size:

Proposition 2.1 *There are constants c_1, c_2, c_3, c_4 such that*

(2) $|\sum_\beta A_{\beta j}(x,t)\xi^\beta| \le c_1^{|\alpha|+1}(<\alpha, M>!)\, exp(c_2\varphi(\xi))$ for all complex x, t, with

$|(x,t)| \le c_3$.

Moreover

(3) $|A_{\beta j}(x,t)| \le c_4^{|\alpha|+1} <\alpha, M>!/<\beta, M>!$, for $|(x,t)| \le c_3$.

Here we denote $a! = \Gamma(a)$, with Γ the Gamma function, if $a \in R_+$ is not an integer.
Proof. The construction via induction of the $A^\alpha_{\beta j}$ shows that the fact that (x,t) is real or not plays no essential rôle.It suffices therefore to prove the estimates (2) and (3) when $(x,t) = (0,0)$, provided all constants depend only on the size of the coefficients of the operator p in a neighborhood of zero. We start from the remark that we can always write

(4) $X^\alpha = Y \circ p + \sum A_{\beta j}(x,t) D_x^\beta D_t^j$,

for some suitable $Y = \sum_{|\gamma|+j \le |\alpha|-m} B_{\gamma j}(x,t) D_x^\gamma D_t^j$. In fact it is (4) which is established first and (1) is then a consequence.We are only interested in (4) at the fixed point $(x,t) = (0,0)$, so we get (denoting by ${}^tT$ the formal adjoint of T whenever T is some given operator,and by δ the Dirac distribution at zero)

(5) ${}^t(X^\alpha)\delta = {}^t p \circ {}^t(Y_o)\delta + \sum A_{\beta j}(0)(-D_x)^\beta(-D_t)^j\delta$,

where $Y_o = \sum_{|\gamma|+j \le |\alpha|-m} B_{\gamma j}(0,0) D_x^\gamma D_t^j$.
(5) is to be regarded as an equality in distributions.The interesting thing is that it gives a decomposition of type

(6) $u = {}^t p(x,t,D_x,D_t)w + \sum v_j(x) \otimes D_t^j \delta_t$,

where $w \in E'(R^{n+1}), v \in E'(R^n)$ for $u = {}^t(X^\alpha)\delta$. This is precisely the type of decomposition encountered in Liess [3],so we can use the results from that paper to get good control on ${}^tY_o\delta$. It remains then to use (5) to estimate $\sum A_{\beta j}(0)(-\xi)^\beta(-\tau)^j$. The essential steps needed to perform this argument are stated in the following four results:

Lemma 2.2 *Assume that f is analytic on $|z| < c_5$ and that it is bounded by 1 on its domain of definition.Then we can find c_6 and c_7 such that*

$$|({}^tX^\alpha\delta) f e^{-i<z,\lambda>})| \le c_6^{|\alpha|+1} <\alpha, M>!\, exp(c_7(|\tau| + \varphi(\zeta))) \text{ for } \lambda \in C^{n+1}.$$

Here $\varphi(\zeta) = \Sigma(1 + |\zeta_j|)^{1/M_j}$.

Lemma 2.3 *For suitable constants c_8, c_9 we have*

(7) $|{}^tY_o\delta(e^{-i<z,\lambda>})| \le c_8^{|\alpha|+1} <\alpha, M>!\, exp(c_9(|\tau| + \varphi(\zeta)), \forall\lambda \in C^{n+1}$.

Lemma 2.4 *Suppose that G is an entire function of form $G(\lambda) = \sum E_{\alpha j}\zeta^{\alpha}\tau^{j}$, $E_{\alpha j} \in C$, which satisfies $|G(\lambda)| \le exp(c_9(|\tau| + \varphi(\zeta))$. Then there is c_{10} such that $|E_{\alpha j}| \le c_{10}^{|\alpha|+j+1}/(<\alpha, M> +j)!$.*

Lemma 2.5 *There are c_{11}, c_{12} such that*

$$|({}^tp \circ {}^tY_o\delta)(e^{-i<z,\lambda>})| \le c_{11}^{|\alpha|+1} <\alpha, M>!\ exp(c_{12}(|\tau| + \varphi(\zeta)), \forall\lambda \in C^{n+1}.$$

2.Proof of lemma 2.2. We note at first that $X^{\alpha}(f\, exp(-i <z, \lambda>))$ is a sum of not more than $c_{13}^{|\alpha|+1}$ terms of form

$$(8) \quad A_1D_{i_1}A_2D_{i_2}\cdots A_{|\alpha|}D_{i_{|\alpha|}}(f\, e^{-i<z,\lambda>}),$$

where the A_i are functions from the set $\{1, f_{x_1}, \ldots, f_{x_n}\}$ and the D_{i_j} are from $(D_{x_1}, \ldots, D_{x_n}, D_t)$. We can rewrite (8) as a sum of terms of form

$$A_1(D^{\gamma_1}A_2)\cdots(D^{\gamma_{|\alpha|-1}}A_{|\alpha|})(D^{\gamma_{|\alpha|}}f)(D^{\gamma_{|\alpha|+1}}e^{-i<z,\lambda>})$$

where $\sum|\gamma_i| = |\alpha|$. Many of the multiindices γ will vanish here in general.If there are $C_{k_1\cdots k_{|\alpha|},\gamma_{|\alpha|+1}}$ terms with $|\gamma_1| = k_1, \ldots, |\gamma_{|\alpha|}| = k_{|\alpha|}$ in this sum (at this moment we follow Hõrmander [2] almost up to notations), then (8) can be estimated by

$$c_{14}^{|\alpha|+1}\sum C_{k_1\cdots k_{|\alpha|},\gamma_{|\alpha|+1}}k_1!\cdots k_{|\alpha|}!|\lambda^{\gamma_{|\alpha|+1}}|$$

for $z = 0$. When $\gamma_{|\alpha|+1}$ is fixed, $\sum C_{k_1\cdots k_{|\alpha|},\gamma_{|\alpha|+1}}\ k_1!\cdots k_{|\alpha|}!$ can be estimated (cf.Hõrmande loc. cit.) by $c_{15}^{|\alpha|-|\gamma_{|\alpha|+1}|}\ (2\,|\alpha| - 2\,|\gamma_{|\alpha|+1}| - 1)!!$. All is proved therefore,if we can show that

$$(2\,|\alpha| - 2|\gamma_{|\alpha|+1}| - 1)!!\ |\lambda^{\gamma_{|\alpha|+1}}| \le c_{16}^{|\alpha|+1} <\alpha, M>!\ exp(c_{17}(|\tau| + \varphi(\zeta)).$$

Here we note that for any c_{18}

$$|\lambda^{\gamma_{|\alpha|+1}}| \le c_{18}^{-|\gamma_{|\alpha|+1}|} <\gamma_{|\alpha|+1}, (M, 1)>!\,exp(c_{18}(|\tau| + \varphi(\zeta)),$$

so the proof comes to an end if we observe that

$$(2\,|\alpha| - 2\,|\gamma_{|\alpha|+1}| - 1)!! <\gamma_{|\alpha|+1}, (M, 1)>! \le 2^{|\alpha|} <\alpha, M>!.$$

3.Proof of lemma 2.3. The situation is here as in Liess [3].This makes it possible to compute an approximation for $\mathcal{F}({}^tY_o\delta)(\lambda)$ which will lead to the desired estimate.To introduce this approximation, we consider $\sum q_j \in SF_{\varphi}^{m}$ such that $\sum q_j \circ p \sim 1$ and denote by $\Lambda(\lambda) = \{\sigma \in C;\ |\sigma| = c(|\tau| + \varphi(\zeta))\}$, anticlockwise orientation.The constant c will be assumed very large. (Cf. Liess [3] for the definition of SF_{φ}^{m}.) For fixed $0 < \chi < 1$ we then introduce

$$T(\lambda) = (1/2\pi i)\oint_{\Lambda(\lambda)} X^{\alpha}[e^{i<z,\zeta>+it(\tau+\sigma)}(1/\sigma) \sum_{j\le\chi(|\tau|+\varphi(\zeta))} q_j(x, t, -\zeta, -\tau - \sigma))]_{z=0}d\sigma.$$

It has then been proved in Liess,loc.cit.,that if χ is sufficiently small,then we can find c_{19}, c_{20} such that

$$|\mathcal{F}({}^tY_o\delta)(\lambda) - T(\lambda)| \le c_{19}\, exp(-c_{20}(|\tau| + \varphi(\zeta))).$$

This estimate shows that it suffices to estimate $T(\lambda)$ by the right hand of (7).Here we use lemma 2.2 to estimate individual terms of form

$$I_\alpha = X^\alpha[e^{i<s,\zeta>+it(\tau+\sigma)}q_j(x,t,-\zeta,-\tau-\sigma)]_{s=0}.$$

In fact,

$$|q_j(x,t,-\zeta,-\tau-\sigma)| \leq c_{21}^{j+1} j!\,|\tau+\sigma|^{-m-j} \text{ if } \sigma \in \Lambda(\lambda), \text{ and } (x,t) \text{ is small},$$

so we will have

$$(9) \quad |I_\alpha| \leq c_{22}^{|\alpha|+j+1} j! < \alpha, M >!\,|\tau+\sigma|^{-m-j} exp(c_7(|\tau+\sigma|+\varphi(\zeta))).$$

Now we observe that $|\tau+\sigma| \geq (c-1)(\varphi(\zeta)+|\tau|) \geq (c-1)j/\chi$ for $\sigma \in \Lambda(\lambda)$, such that $j!/|\tau+\sigma|^j \leq (c_{22}\chi)^j$ then .We now choose χ small and integrate the obtained estimates for each such term. Finally we use the fact that the number of such terms is smaller than $\chi(|\tau|+\varphi(\zeta))$.

4.Proof of lemma 2.4. We have that $\alpha!\,j!\,|E_{\alpha j}| \leq c_{23}^{|\alpha|+j+1}\alpha!\,j! \min_{r_1} \max_{|\zeta_1|=r_1} exp(c_9|\zeta_1|^{1/M_1})/|\zeta_1|^{\alpha_1})\cdots \min_{r_n} \max_{|\zeta_n|=r_n} (exp(c_9|\zeta_n|^{1/M_n})/|\zeta_n|^{\alpha_n}) \min_r \max_{|\tau|=r} (exp(c_9|\tau|)/|\tau|^j) \leq c_{24}^{|\alpha|+j+1}(1/(M_1\alpha_1))!\cdots(1/(M_n\alpha_n))!\,(1/j!)$, etc.

5.Proof of lemma 2.5. We denote by $Y_o(\lambda) = \sum B_{\beta j}(0)\zeta^\beta\tau^j$ the symbol of Y_o. With standard notations it follows that

$$(Y_o \circ p)(e^{-i<s,\zeta>})(0) = \sum_\gamma Y_o^{(\gamma)}(\lambda)p_{(\gamma)}(0,\lambda)/\gamma!,$$

Note here (cf. lemma 2.3 ,2.4) that

$$|(B_{\beta j}(0)\zeta^\beta\tau^j)^{(\gamma)}| \leq c_{25}^{|\beta|+|\alpha|+j+1} < \alpha, M >!\,(\beta,j)!\,|\lambda^{(\beta,j)-\gamma}|/[(<\beta,M>+j)!\,((\beta,j)-\gamma)!].$$

when $\gamma \leq (\beta,j)$. The next thing to observe is that,with the notation $\tilde{M} = (M,1) \in R^{n+1}$,

$$(\beta,j)! < (\beta,j)-\gamma, \tilde{M} >! \leq (<\beta,M>+j)!\,((\beta,j)-\gamma)!$$

and that $|p_{(\gamma)}(0,\lambda)/\gamma!| \leq c_{26}^{|\gamma|+1}\varphi(\lambda)^m$.
This gives that

$$|(Y_o \circ p)(e^{-i<s,\lambda>})(0)| \leq \varphi(\lambda)^m (<\alpha,M>!) \sum c_{25}^{|\beta|+j+1} c_{26}^{|\gamma|+1} |\lambda^{(\beta,j)-\gamma}| / < (\beta,j)-\gamma, \tilde{M} >!,$$

where the sum is for all β, j, γ with $(\beta,j) \geq \gamma$. Finally we observe that $|\lambda^{(\beta,j)-\gamma}|/ < (\beta,j)-\gamma, \tilde{M} >! \leq c_{27}\, exp(c_{28}(|\tau|+\varphi(\zeta))$.

6.Proof of proposition 2.1. (End).We start from

$$(10) \quad \sum A_{\beta,j}(0)\zeta^\beta\tau^j = X^\alpha(e^{-i<s,\lambda>})(0) - (Y_o \circ p)(e^{-i<s,\lambda>})(0).$$

The desired estimate for $\sum A_{\beta 0}(0)\zeta^\beta$ is now obtained setting here $\tau = 0$ and using the lemmas 2.2 and 2.5. To obtain the corresponding estimates for $\sum A_{\beta j}(0)\zeta^\beta$ when $j > 0$,we can now derivate (10) j-times and then set $\tau = 0$.

3 Proof of theorem 1.7

1.All assumptions and notations are as in the statement of theorem 1.7. In particular,we know that $(0,\xi^o) \notin WF^b_M\, u$ and want to show that, for every fixed j,

(1) $(0,\xi^o) \notin WF_M\, G$, where $G = D_t^j\, u_{|S}$.

Recall that G is here regarded as a distribution in the variable x. To prove (1) ,we shall check that (ii) from proposition 1.10 is satisfied. Let us then fix constants c_i as in the statement of that proposition, and fix η with $|\eta/|\eta| - \xi^o/|\xi^o|| < c_1$. We must show that for suitable choices of the c_i and with the notations from proposition 1.10 we will have

$$|(g_\eta(D)D^\alpha(h_\eta G))(x)| \leq c_8(c_8c)^{c\varphi(\eta)}\varphi(\eta)^{<\alpha,M>+n} \text{ if } <\alpha,M> \leq c\varphi(\eta) + |M|.$$

2. Our first remark is here that

$$D^\alpha(h_\eta(x)G(x)) = \sum_{\gamma\leq\alpha}(\alpha!/((\gamma!)(\alpha-\gamma)!)D^{\alpha-\gamma}h_\eta(x)D^\gamma G(x).$$

It suffices therefore to prove that for all $\gamma \leq \alpha$

$$|(g_\eta(D)(D^{\alpha-\gamma}h_\eta)(D^\gamma G)(x)| \leq c_9(c_9c)^{c\varphi(\eta)}\varphi(\eta)^{<\alpha,M>+n}.$$

We next observe that $D^\gamma G(x) = (X^\gamma u)_{|S}$, where $X^\gamma = X_1^{\gamma_1}\cdots X_n^{\gamma_n}$ and $X_j = \partial/\partial x_j + f_{x_j}\partial/\partial t$. We can now use proposition 2.1 and rewrite $X^\gamma u_{|S}$, (using notations from §2) in the form

(2) $X^\gamma u_{|S} = [\sum_{\beta,j,j<m} A^\gamma_{\beta j}(x,t)D_x^\beta D_t^j u]_{|S}$.

Here we use of course that $p(x,t,D_x,D_t)u = 0$, and in the sum from (2) we have

(3) $\gamma \leq \alpha, <\beta,M> +j+ <\alpha-\gamma,M> \leq c\varphi(\eta) + |M|$.

Moreover,the $A^\gamma_{\beta j}$ are analytic and we have for (x,t) in a small complex neighborhood of zero

$$|A^\gamma_{\beta j}(x,t)| \leq c_{10}^{|\gamma|+1} <\gamma,M>!/<\beta,M>!.$$

We are then reduced to a study of $v_{x^o,\eta}(u)$ when $|x^o|$ is small and where $v_{x^o,\eta}$ is the distribution

$$v_{x^o,\eta}(H) = g_\eta(D)[D^{\alpha-\gamma}h_\eta A^\gamma_{\beta j}(x,t)D_x^\beta D_t^j H_{|S}](x^o) \text{ for } H \in C_o^\infty(R^{n+1}).$$

To estimate $v_{x^o,\eta}(u)$, we shall rely on the following result, in which the assumptions for u are satisfied in our situation in view of the theorems 1.7 and 1.3' respectively.

Proposition 3.1 *Consider $u \in D'_+$, $\xi^o \in \dot{R}^n$ and assume that*
a) $(0,\xi^o) \notin WF^b_\varphi u$,
b) there are constants d,c' such that for every b_1 we can find b_2 and c for which $|u(g)| < c$ for any $g \in C_o^\infty$ which satisfies the inequalities from the statement of theorem 1.3'.
Then we can find constants d',c_1,c_2 such that for every c_3, b' there are c_4, b'' with the following property:

consider $\eta \in \dot{R}^n$ *such that* $|\eta/|\eta| - \xi/|\xi^\circ|| < c_2$ *and assume that* $v \in E'(R^{n+1})$ *satisfies*

$$(4) \quad |\hat{v}(\lambda)| \leq exp(d'\varphi(-Re\zeta) + d'|Im\zeta| + d'Im\tau_+ + b''ln(1+|\zeta|) + b'ln(1+|\tau|)),$$
$$\text{if } |\tau| \geq c_1(\varphi(-Re\zeta) + |Im\zeta|) \text{ or } |Re\zeta + \eta| \leq c_3\varphi(-\eta),$$

respectively

$$(5) \quad |\hat{v}(\lambda)| \leq exp(d'|Im\zeta| + d'Im\tau_+ + b''ln(1+|\zeta|) + b'ln(1+|\tau|)), \text{ for all other } \lambda.$$

Then it follows that $|v(u)| \leq c_4$.

Proposition 3.1 is a consequence of the fact that any v which satisfies the inequalities (4),(5) can be splitted in a sum of form $v_1 + v_2$ where v_1 satisfies the inequalities from the definition of $(0,\xi^\circ) \not\in WF^b_\varphi u$, and v_2 satisfies the inequalities from b).Splittings of this type can be performed with $\bar{\partial}-$ arguments. Similar situations appear e.g. in Liess [2,4] and we omit further details. What we need then is (and we state this explicitly to avoid confusion in the denomination of constants) :

Proposition 3.2 *Let C_1, C_2, d', b', be given.If the c_i used prior to the statement o proposition 3.1 are suitable,then we can find C_3, c, so that*

$$(6) \quad |\hat{v}_{x^\circ,\eta}(\lambda)| \leq C_3^{|\alpha|+1}c^{c\varphi(\eta)}\varphi(\eta)^{<\alpha,M>}exp(d'|Im\zeta| + d'Im\tau_+ + b'ln(1+|\zeta|) +$$
$$mln(1+|\tau|)), \text{if } |\tau| \leq C_1(\varphi(-Re\zeta) + |Im\zeta|) \text{ and } |Re\zeta + \eta| \geq C_2\varphi(-\eta),$$

$$(7) \quad |\hat{v}_{x^\circ,\eta}(\lambda)| \leq C_3^{|\alpha|+1}c^{c\varphi(\eta)}\varphi(\eta)^{<\alpha,M>+n}exp(d'\varphi(-Re\zeta) + d'|Im\zeta| + d'Im\tau_+$$
$$+b'ln(1+|\zeta|) + mln(1+|\tau|)), \text{for all other } \lambda.$$

(Here and later on we shall sometimes write "$\varphi(-Re\zeta)$" when this is justified from the geometry of the situation. Note however that in the present paper the function φ is symmetric. Further, we observe that when C_2 is small, then $|Re\zeta + \eta| \leq C_2\varphi(\eta)$ will imply $\varphi(-Re\zeta) \sim \varphi(\eta)$.)

Proof of proposition 3.2. To compute the Fourier transform of $v_{x^\circ,\eta}$, note that

$$(8) \quad \hat{v}_{x^\circ,\eta}(\lambda) = v_{x^\circ,\eta}(e^{-i<x,\lambda>}) = (2\pi)^{-n-1}\int\int e^{i<x^\circ,\theta>-i<x,\theta+\zeta>-i\tau f(x)}g_\eta(\theta)$$
$$A^\gamma_{\beta j}(x,f(x))D^{\alpha-\gamma}h_\eta(x)\zeta^\beta\tau^j dx\, d\theta.$$

We first observe that the integration in θ is here only for $|\theta - \eta| \leq c_3\varphi(\eta)$. A first condition on c_3 will then be that $\varphi(\theta) \sim \varphi(\eta)$ for θ in the support of g_η.

Proof of (6). **From the assumptions in (6) we have that $|Re\zeta + \eta| \geq C_2\varphi(\eta)$. It follows that $|Re\zeta + \eta| \geq C_4(\varphi(\eta) + \varphi(-Re\zeta))$, so $|\theta + \zeta| \geq C_5(\varphi(\eta) + \varphi(-Re\zeta) + \varphi(\theta) + |Im\zeta|)$ if c_3 is small. Assume then in addition, that $|\tau| \leq C_1(\varphi(-Re\zeta) + |Im\zeta|)$, and denote by**

$$(9) \quad A = \nabla_x(<x,\zeta+\theta> + f(x)\tau) = \zeta + \theta + \nabla_x f(x)\tau.$$

It follows that we can find C_6 such that

(10) $|A| \geq C_6(\varphi(\eta) + \varphi(-Re\zeta) + \varphi(\theta) + |Im\zeta|)$,

if x is in a sufficiently small neighborhood of the origin. Here we use that $\nabla_x f$ will be as small as we please, if we choose this neighborhood suitably. Next we introduce the operator $L = |A|^{-2} < A, \nabla_x >$, for which $L(exp(-i < x, \theta + \zeta > -i\tau f(x)) = exp(-i < x, \theta + \zeta > -i\tau f(x))$.
If we denote by $I = \int exp[-i < x, \theta + \zeta > -i\tau f(x)] A^{\gamma}_{\beta j}(x, f(x)) D^{\alpha-\gamma} h_\eta dx$, then we will have

$$I = \int e^{-i<x,\theta+\zeta>-i\tau f(x)} ({}^tL)^k (A^{\gamma}_{\beta j}(x, f(x)) D^{\alpha-\gamma} h_\eta(x)) dx,$$

whatever k is . We apply this for $k = < \beta\ ,\ M > +C_7 + C_8$, where C_7 is such that $\int \varphi(\theta)^{-C_7}\, d\theta\ < \infty$, and C_8 is such that

$$(\varphi(Re\zeta) + |Im\zeta|)^{-C_8+m-1} \leq (1 + |\zeta|)^{|-\nu|}.$$

(We may here assume of course that $< \beta, M > +C_7 + C_8$ is an integer.)
We can now conclude as in the proof of lemma 2.2 that

$$|({}^tL)^k (A^{\gamma}_{\beta j}(x, f(x)) D^{\alpha-\gamma} h_\eta(x))| \leq C_9^{|\gamma|+|\alpha-\gamma|+|\beta|+1} (< \gamma, M >!/ < \beta, M >!)$$
$$\varphi(\eta)^{|\alpha-\gamma|+<\beta,M>+C_7+C_8} c^{|\alpha-\gamma|+<\beta,M>} (\varphi(-Re\zeta) + \varphi(\eta) + \varphi(\theta) + |Im\zeta|)^{-<\beta,M>-C_7-C_8}$$

on the support of h_η , if c_5 was small.
Shrinking c_5 still further, we may also assume that $| < x, Im\zeta > | + Im\,\tau\, f(x) \leq d'|Im\zeta| + d'Im\tau_+$ on the support of h_η. The next thing is to note that

$$|\zeta^\beta|/[\varphi(-Re\zeta) + |Im\zeta|)]^{<\beta,M>} \leq 1,$$

and that

$$< \gamma, M >!/ < \beta, M >! \leq 2^{<\gamma,M>} (< \gamma, M > - < \beta, M >)! \leq$$
$$2^{<\gamma,M>} (c\varphi(\eta))^{<\gamma-\beta,M>}.$$

This leads for $|\tau| < C_1(\varphi(-Re\zeta) + |Im\zeta|)$ to

$$|I\, \zeta^\beta \tau^j| \leq C_{10}^{|\alpha|+1} \varphi(\eta)^{|\alpha-\gamma|+<\gamma,M>} \varphi(\theta)^{-C_7} (\varphi(-Re\zeta) + |Im\zeta|)^{-C_8+j}$$
$$c^{<\gamma-\beta,M>+|\alpha-\gamma|+<\beta,M>} exp(d'|Im\zeta| + d'Im\tau_+).$$

We can now integrate this in θ and obtain (6).

Proof of (7). First we note that

$$|A^{\gamma}_{\beta j}(x, f(x)) D^{\alpha-\gamma} h_\eta(x)| \zeta^\beta \tau^j| \leq C_{11}^{|\gamma|+1} (< \gamma, M >!/ < \beta, M >!) \zeta^\beta \tau^j |c^{|\alpha-\gamma|} \varphi(\eta)^{|\alpha-\gamma|}$$
$$\leq |\tau|^j C_{11}^{|\gamma|+1} e^{(\epsilon/2)(\varphi(Re\zeta)+|Im\zeta|)} C_{12}^{|\beta|} < \gamma, M >! c^{|\alpha-\gamma|} \varphi(\eta)^{|\alpha-\gamma|}$$
$$\leq |\tau|^j C_{11}^{|\gamma|+1} e^{(\epsilon/2)(\varphi(Re\zeta)+|Im\zeta|)} C_{12}^{|\beta|} (c\varphi(\eta))^{<\gamma,M>} c^{|\alpha-\gamma|} \varphi(\eta)^{|\alpha-\gamma|}$$

This has to be integrated over a set with volume smaller than $C_{13}\varphi(\eta)^n$ .The logarithmic term from (7) involving ζ can be absorbed in $exp((d'/2)\varphi(\zeta) + (d'/2)|Im\zeta|)$, so we are done.

References.

Hörmander,L.[1] : The analysis of linear partial differential operators, I and II, Springer Verlag, Grundlehren Series, vol.'s 256, 257, 1983. [2] : Uniqueness theorems and wave front sets..., C.P.A.M., 24, 671-704, 1971.

Kataoka,K. [1] :Microlocal analysis of boundary value problems with applications to diffraction, NATO ASI Series, vol.C65, ed. by H.Garnir, 121-133.

Lebeau,G. [1] : Une propriete d'invariance pour le spectre des traces de solutions d'op. diff., C.R. Acad. Sci. Paris, Ser.I. Math., 294:22 (1982),723-725.

Liess,O. [1] :Prolema Cauchy in doua variabile, Studii si Cerc. Mat. XXV :2 (1973), 267-281, and: The Cauchy problem for operators in two variables, II, Rev. Roum., XVIII:4, (1973), 543-561. [2] :Necessary and sufficient conditions for propagation of singularities for systems of linear partial differental operators with constant coefficients, C.P.D.E., 8:2, (1983), 89-198. [3] :Microlocality of the Cauchy problem in inhomogeneous Gevrey classes, C.P.D.E., 11:13, 1379-1437. (1986) [4] :Boundary regularity for one-sided solutions of linear partial differential equations with analytic coefficients, Springer Lecture Notes in Math., Vol.1256, 1986. Edited by Cordes-Gramsch-Widom.

Liess,O.-Rodino,L. [1] : Inhomogeneous Gevrey classes and related pseudodifferential operators, Boll. U.M.I., Ser.VI, III-C, (1984), 233-323.

Rodino,L. [1] :On the Gevrey wave front set of the solutions of a quasi-elliptic degenerate equation, Rend. Sem. Mat. Univ. Pol.Torino, vol.40, 1982.

Universita di Palermo, Dipartimento di Matematica, Palermo, Italia

Necessary and sufficient condition for maximal hypoellipticity of $\bar\partial_b$

H.-M. Maire *
Section de Mathématiques
Université de Genève
Case postale 240
CH-1211 Genève 24

1. The induced complex.

Let M be a real C^∞ hypersurface in $\mathbf{C}^{n+1}$, $T'M$ and $T''M$ the vector bundles of holomorphic and anti-holomorphic vectors tangent to M, and $\Lambda^{0,q}M = \Lambda^q(T''M)^*$ the vector bundle of q-linear skew-symmetric forms on $T''M$; the sections of $\Lambda^{0,q}M$ are called $(0,q)$-forms on M. The Cauchy-Riemann complex $\bar\partial$ on $\mathbf{C}^{n+1}$ induces a complex $\bar\partial_b : \Lambda^{0,q}M \longrightarrow \Lambda^{0,q+1}M$ defined by $\bar\partial_b\omega = \bar\partial\tilde\omega | T''M$ where $\tilde\omega \in \Lambda^{0,q}\mathbf{C}^{n+1}$ is any extension of $\omega \in \Lambda^{0,q}M$. Corresponding definition is made for ∂_b. With a hermitian metric on M we form the adjoint $\bar\partial_b{}^* : \Lambda^{0,q}M \longrightarrow \Lambda^{0,q-1}M$ of $\bar\partial_b$ and set $\Box_b = \bar\partial_b{}^*\bar\partial_b + \bar\partial_b\bar\partial_b{}^*$.

The characteristic set $\Sigma = \{ (z,\zeta) \in T^*M\, ;\, \zeta | T''_zM = 0 \}$ of $\bar\partial_b$ is a real one-dimensional subbundle of T^*M; therefore $\bar\partial_b$ is not elliptic. To analyze $\bar\partial_b$ further we need the Levi form on T''_zM :

$$\mathrm{L}_{(z,\zeta)}(U(z), V(z)) = \frac{1}{2i}<\zeta, [U, \bar V](z)> \quad , \quad (z,\zeta)\in\Sigma$$

where U, V are C^∞-sections of $T''M$. When $\mathrm{L}_{(z,\zeta)}$ is non-degenerate, *i. e.* has n non-zero eigenvalues $\lambda_j(z,\zeta)$, Folland and Kohn [3] considered the condition Y(q), $0 \le q \le n$, about the sign of $\lambda_j(z,\zeta)$ which is equivalent to the following (cf. Grigis-Rothschild [4]):

$$-\sum\lambda_j^- < \lambda_{j_1} + \cdots + \lambda_{j_q} < \sum\lambda_j^+ \ , \quad 1 \le j_1, \ldots, j_q \le n \ , \qquad \mathrm{Y(q)}$$

---

* Supported in part by the Fonds P. Moriaud, Genève.

where $\lambda_j^+ = \frac{1}{2}(|\lambda_j| + \lambda_j)$ *and* $\lambda_j^- = \frac{1}{2}(|\lambda_j| - \lambda_j)$. For $q = 0$ it means that L does not have all its eigenvalues of the same sign.

THEOREM 1. (Folland-Kohn [3]). *If the Levi form is everywhere non-degenerate then* Y(q) *is a necessary and sufficient condition for* $\bar\partial_b$ *to be subelliptic with loss of* ½ *–derivative on* $(0,q)$*–forms, i. e.*

$$\| u \|_{½} \le C(\| \bar\partial_b u \| + \| \bar\partial_b^* u \| + \| u \|), \quad u \in C_0^\infty(\Lambda^{0,q} M).$$

Suppose from now on that M is of *finite type* r or, what turns out to be equivalent, $T'M$ and $T''M$ satisfy the Hörmander condition of order r, *i. e.* the brackets of length at most r of sections of $T'M$ and $T''M$ generate TM. The following assertion justifies the next definition:

if for any distribution section u of $\Lambda^{0,q}M$: $\bar\partial_b u$ *and* $\bar\partial_b^* u \in C^\infty \Rightarrow \partial_b u \in C^\infty$, *then* $\bar\partial_b$ *is hypoelliptic on* $(0,q)$*–forms.*

DEFINITION 2. *We will say that the operator* $\bar\partial_b$ *is maximally hypoelliptic at* $z_0 \in M$ *on* $(0,q)$*–forms if*

$$\| \partial_b u \| \le C(\| \bar\partial_b u \| + \| \bar\partial_b^* u \| + \| u \|), \quad (L^2 \text{ norms})$$

for all C^∞*–sections u of* $\Lambda^{0,q}M$ *with compact support in a neighborhood of* z_0.

After Bolley-Camus-Nourrigat [1] if $\bar\partial_b$ is maximally hypoelliptic then it is subelliptic with loss of $(1-1/r)$–derivatives. When the Levi form is non-degenerate, maximal hypoellipticity and hypoellipticity coincide.

In local coordinates $(z_1, \ldots, z_{n+1})$ we may suppose that M is defined by $\phi = 0$ where ϕ is a real C^∞ function such that $\frac{\partial\phi}{\partial z_{n+1}}$ does not vanish. Then

$$Z_j = \frac{\partial}{\partial z_j} - \frac{\partial\phi/\partial z_j}{\partial\phi/\partial z_{n+1}} \frac{\partial}{\partial z_{n+1}}, \quad 1 \le j \le n, \text{ form a basis of } T'M.$$

Furthermore $\Sigma = \{(z, \frac{\alpha}{i}(d'-d'')\phi(z)); \alpha \in \mathbf{R}\}$ and

$$L_{(z, \frac{1}{i}(d'-d'')\phi(z))}(\bar Z_j, \bar Z_k) = -\frac{1}{2} < (d'-d'')\phi(z), [\bar Z_j, Z_k](z) > =: a_{jk}(z).$$

The matrix $(a_{jk}(z))$ is called the Levi matrix at z; its eigenvalues will be denoted by $\lambda_j(z)$.

Parametrizing M by $x_{n+1} = f(z_1, \ldots, z_n, t)$, $y_{n+1} = t$ we get the vector fields

$$L_j = \frac{\partial}{\partial z_j} - \frac{\partial f/\partial z_j}{1+i\,\partial f/\partial t}\, i\frac{\partial}{\partial t}$$

in an open subset U of $\mathbf{C}^n \times \mathbf{R}$ whose images are Z_j . In U

$$[\bar{L}_j, L_k](z, t) = 2a_{jk}(z, t) i \frac{\partial}{\partial t}.$$

When M is *rigid* ,*i. e.* $\frac{\partial f}{\partial t} = 0$, we have: $a_{jk} = -\frac{\partial^2 f}{\partial \bar{z}_j \partial z_k}$.

2. Statement of the results.

The first successful step to generalize Y(q) when $L_{(z,\zeta)}$ degenerates was made by Derridj [2] who proposed in the weakly pseudoconvex case *i. e.* when the eigenvalues of the Levi matrix are everywhere non-negative:

$$\exists \epsilon > 0 \text{ such that } \epsilon \sum \lambda_k(z) \leq \lambda_j(z) \ , \ 1 \leq j \leq n, \ z \in M.$$

He proved then a maximal estimate for the Neumann problem for $\bar{\partial}$ in Ω , $M = \partial\Omega$.

DEFINITION 3. *We will say that M satisfies the condition* D(q) *at $z_0 \in M$ if there exists a neighborhood U of z_0 in M and $\epsilon > 0$ such that for $1 \leq j_1, \ldots, j_q \leq n$:*

$$-\sum \lambda_j^- + \epsilon \sum |\lambda_j| \leq \lambda_{j_1} + \cdots + \lambda_{j_q} \leq \sum \lambda_j^+ - \epsilon \sum |\lambda_j| \ , \quad \text{in } U. \qquad \mathrm{D(q)}$$

We will also say that $P \in C^\infty(\mathbf{C}^n; \mathbf{R})$ satisfies the condition D(q) *at $z_0 \in \mathbf{C}^n$ if the preceeding estimates hold for the eigenvalues of the Levi matrix of P .*

Notice that the condition D(q) is homogeneous, that is P satisfies D(q) if and only if αP satisfies D(q), $\forall \alpha \in \mathbf{R}$. For $q = 0$, the medium term in D(0) is equal to 0.

THEOREM 4 (Helffer-Nourrigat [5]). *If $\bar{\partial}_b$ is maximally hypoeliptic at $z_0 \in M$ on $(0, q)$–forms then M satisfies* D(q) *at z_0.*

REMARK 5. After general results of Helffer and Nourrigat [5], $\bar{\partial}_b$ is maximally hypoelliptic if and only if $\square_b$ is.

The sufficiency of D(q) for maximal hypoellipticity has been proved by Grigis and Rothschild in the pseudoconvex case when M satisfies some special conditions (cf. [4]). Helffer and Nourrigat proved it for general pseudoconvex hypersurfaces (cf. [5], p. 112). We have still a different proof in the the spirit of Section 3 and [6].

THEOREM 6. *The operator* $\bar\partial_b$ *is maximally hypoelliptic at* $z_0 \in M$ *on* (0, 0)*–forms if and only if* M *satisfies* D(0) *at* z_0.

This result has been announced in [7] for the case of tubular hypersurfaces. The question of generalising Theorem 6 to $q > 0$ is still open.

3. D(0) is sufficient.*

We use the results and some notations of Helffer and Nourrigat [5]. Let $r \in \mathbf{N}$ be such that the brackets of length at most r of sections of $T'M$ and $T''M$ generate TM. As the problem is local we may use the description at the end of Section 1 and suppose $(z_0, t_0) = 0$.

Define a set E_0 of localized polynomials (cf. Hörmander [6]) by $P \in E_0$ if and only if

> $0 \neq P$ *is a real polynomial of degree* $\leq r$ *of the variables* $z_1, \dots, z_n, \bar z_1, \dots, \bar z_n$ *vanishing at* 0 *such that there exists a sequence* $(z_\nu, t_\nu, \tau_\nu, \epsilon_\nu) \longrightarrow 0$ *in* $\mathbf{C}^n \times \mathbf{R} \times \mathbf{R} \times \mathbf{R}^*$ *with*
>
> $$\lim_{\nu \to \infty} \frac{\Pi_{\alpha,\beta}(L, \bar L) a_{jk}(z_\nu, t_\nu)\tau_\nu^{|\alpha|+|\beta|+2}}{\epsilon_\nu} = \partial^\alpha \bar\partial^\beta \frac{\partial^2 P}{\partial \bar z_j \partial z_k}(0),$$
>
> *for* $1 \leq j, k \leq n$, $|\alpha|+|\beta| \leq r-2$, *where* $\Pi_{\alpha,\beta}(L, \bar L)$ *is any product of* α_1 *factors* $L_1, \dots, \alpha_n$ *factors* L_n *and* β_1 *factors* $\bar L_1, \dots, \beta_n$ *factors* $\bar L_n$.

After Theorems 3.2.2 and 3.2.3 p. 118 of [5] we know that $\bar\partial_b$ is maximally hypoelliptic on (0,0)–forms if and only if the following condition (called (P_2) in [5]) is satisfied:

$$\forall P \in E_0, \ P \text{ does not have a local maximum at } 0.$$

Therefore Theorem 6 is a consequence of the following next two statements.

---

* The author thanks F. Treves and M. S. Baouendi for useful remarks improving the proofs in this section.

LEMMA 7. *Suppose M satisfies* D(0) *at* $0 \in M$ *and* $P \in E_0$. *Then P also satisfies* D(0) *at* 0.

PROPOSITION 8. *Let* $0 \neq P$ *be a real polynomial in* $\mathbf{C}^n$ *vanishing at* 0. *If P satisfies* D(0) *at* 0 , *then P does not have a local maximum at* 0.

Proof of Lemma 7. First remark that all expressions occuring in D(0) are symmetric in the eigenvalues; hence they are well defined C° functions of the entries a_{jk} of the Levi matrix.

Let us first prove the lemma when M is rigid. After Section 1, $a_{jk} = -\partial^2 f / \partial \bar{z}_j \partial z_k$ is independent of t. If $P \in E_0$, then there exists a sequence $(z_\nu, \tau_\nu, \epsilon_\nu) \longrightarrow 0$, in $\mathbf{C}^n \times \mathbf{R} \times \mathbf{R}^*$ such that

$$\lim_{\nu \to \infty} \partial^\alpha \bar{\partial}^\beta a_{jk}(z_\nu) \tau_\nu^{|\alpha|+|\beta|+2} / \epsilon_\nu = \partial^\alpha \bar{\partial}^\beta \frac{\partial^2 P}{\partial \bar{z}_j \partial z_k}(0)$$

and therefore

$$\lim_{\nu \to \infty} a_{jk}(z_\nu + \tau_\nu z) \tau_\nu^2 / \epsilon_\nu = \frac{\partial^2 P}{\partial \bar{z}_j \partial z_k}(z)$$

uniformly for $|z| \leq 1$. It follows that P satisfies D(0) if f does.

In the general case, if $P \in E_0$ there exists a sequence $(z_\nu, t_\nu, \tau_\nu, \epsilon_\nu) \longrightarrow 0$ in $\mathbf{C}^n \times \mathbf{R} \times \mathbf{R} \times \mathbf{R}^*$ as in the definition of E_0. Following Helffer and Nourrigat [5], pp. 134-136, we set

$$\phi_\nu(z) = f(z, \operatorname{Im} g_\nu(z)) - \operatorname{Re} g_\nu(z)$$

where g_ν is a complex polynomial of degree at most r such that $\partial^\alpha \phi_\nu(z_\nu) = \epsilon_\nu \tau_\nu^{-|\alpha|} \partial^\alpha P(0)$, for $|\alpha| \leq r$. After Lemmma 5.3.1 of [5], we have for $|\alpha| + |\beta| \leq r$:

$$\lim_{\nu \to \infty} \tau_\nu^{|\alpha|+|\beta|} \partial^\alpha \bar{\partial}^\beta \phi_\nu(z_\nu) / \epsilon_\nu = \partial^\alpha \bar{\partial}^\beta P(0)$$

and the Taylor remainder of order r of ϕ_ν / ϵ_ν at z_ν converges to 0. Let

$$\tilde{\phi}_\nu(z) = \phi_\nu(z_\nu + \tau_\nu z) / \epsilon_\nu, \quad z \in \mathbf{C}^n, \ |z| \leq 1.$$

Then $\tilde{\phi}_\nu$ uniformly converges to P over the unit ball of $\mathbf{C}^n$.

In the conclusion of Lemma 5.2.2 of [5], we may replace z_ν by $z_\nu + \tau_\nu z$, $|z| \leq 1$. Hence, after Proposition 5.1.3 of [5]:

$$\frac{\partial^2\tilde{\phi}_\nu}{\partial\bar{z}_j\partial z_k}(z) = \frac{\tau_\nu^2}{\epsilon_\nu}\frac{\partial^2\phi_\nu}{\partial\bar{z}_j\partial z_k}(z_\nu+\tau_\nu z) = \frac{\tau_\nu^2}{\epsilon_\nu}a_{jk}(z_\nu+\tau_\nu z) + o(1)\ , \quad \nu\to\infty .$$

It follows that

$$\lim_{\nu\to\infty}\sum\frac{\tau_\nu^2}{\epsilon_\nu}|\lambda_j(z_\nu+\tau_\nu z)| = \sum|\lambda_{j,\infty}(z)|\ ,$$

$$\lim_{\nu\to\infty}\sum\frac{\tau_\nu^2}{\epsilon_\nu}\lambda_j^+(z_\nu+\tau_\nu z) = \sum\lambda_{j,\infty}^+(z)$$

where λ_j [resp. $\lambda_{j,\infty}$] are the eigenvalues of (a_{jk}) [resp. $(\partial^2P/\partial\bar{z}_j\partial z_k)$]. We have proved that P satisfies D(0) at 0 if M does.

Proof of Proposition 8. Let P be a real polynomial satisfying D(0) at 0; we have to prove that $\forall\epsilon > 0$, $\sup_{|z|\le\epsilon} P(z) > 0$. After Lemma 7 the initial form In_0P of P at 0 satisfies D(0) at 0; if $\sup_{|z|=1} In_0P(z) > 0$, we are done. Therefore we may suppose P is a real homogeneous polynomial of degree m (for the action of $\mathbf{R}_+^*$). Let us proceed by induction on m, the result being trivial when $m = 1$.

If $\sup_{|z|=1} P(z) < 0$, let $z_0\in S^{2n-1}$ (unit sphere in $\mathbf{R}^{2n}$) be such that $P(z_0) = \sup_{S^{2n-1}} P$. After Lagrange's multipliers method we have $P'(z_0)|T_{z_0}S^{2n-1} = 0$ and $P''|T_{z_0}S^{2n-1} \le 0$ where the first and second derivatives are taken in the real sense. Furthermore $m(m-1)P(z_0) = \mu_1(z_0)$ is an eigenvalue of $P''(z_0)$ because

$$P'(tz_0)z_0 = mt^{m-1}P(z_0) \quad and \quad P'(tz_0)h = 0 \text{ if } \langle h, z_0\rangle = 0\ , \forall t>0$$

$$\Rightarrow\ P'(tz_0)h = mt^{m-1}P(z_0)\langle h, z_0\rangle\ ,\ \forall t>0, \forall h\in\mathbf{R}^{2n}$$

$$\Rightarrow\ P''(z_0)(z_0, h) = m(m-1)P(z_0)\langle h, z_0\rangle\ ,\ \forall h\in\mathbf{R}^{2n}.$$

The other eigenvalues $\mu_2(z_0), \ldots, \mu_{2n}(z_0)$ of $P''(z_0)$ are non-positive. If $\lambda_1(z_0), \ldots, \lambda_n(z_0)$ denote the eigenvalues of the Levi matrix of P at z_0, we have

$$\lambda_1(z_0) + \cdots + \lambda_n(z_0) = \mu_1(z_0) + \cdots + \mu_{2n}(z_0) < 0.$$

Therefore there exists j_1 with $\lambda_{j_1}(z_0) < 0$. Using D(0) we get j_2 with $\lambda_{j_2}(z_0) > 0$; this implies that P is strictly subharmonic in some complex disc and hence $h \longmapsto P(z_0+h) - P(z_0) - P'(z_0)h$ does not have a maximum at 0. This is in contradiction with $\mu_j(z_0) \le 0, \forall j$.

Let us suppose now $\sup_{S^{2n-1}} P = 0$ and take $z_0 \in S^{2n-1}$ with $P(z_0) = 0$. Because P is homogeneous, $P'(z_0) = 0$. Let $Q = In_{z_0} P$; after Lemma 7, Q satisfies D(0) at 0. But Q has a local maximum at 0; by induction hypothesis, the case *deg* $Q <$ *deg* P does not occur. Therefore P and Q have the same degree and both are homogeneous so that:

$$P(tz_0+z) = t^m P(z_0+z/t) = t^m Q(z/t) = Q(z)\ ,\ \forall z \in \mathbf{R}^{2n}, \forall t>0.$$

Whence P depends only on $2n-1$ real variables. Take a supplementaty subspace F to z_0 in $\mathbf{R}^{2n}$ and consider $z_1 \in F$, $|z_1| = 1$ such that:

$$\sup_{z\in F,\ |z|=1} P(z) = P(z_1) \leq 0.$$

If $P(z_1) < 0$ the same argument as before leads to a contradiction; therefore $P(z_1) = 0$ and hence

$$P(sz_1 + z) = P(z)\ ,\ \forall z \in \mathbf{R}^{2n}, \forall s>0.$$

It follows that P depends only on 2n-2 real variables. The same argument applies again leading to $P = 0$, which is a contradiction.

We have proved $\sup_{|z|=1} P(z) > 0$, *i. e.* P does not have a local maximum at 0.

QED

References.

[1] Bolley P., Camus J. et Nourrigat J., La condition de Hörmander-Kohn pour les opérateurs pseudo-différentiels, *Comm. in PDE* 7 (1982), p. 197- 221.

[2] Derridj M., Sur la régularité des solutions du problème de Neumann pour $\bar\partial$ dans quelques domaines faiblement pseudo-convexes, *J. of Diff. Geometry* 13 (1978), p. 559-588.

[3] Folland G.B. and Kohn J.J., *The Neumann problem for the Cauchy-Riemamcomplex*, Ann. of Math. Study 75, Princeton 1972.

[4] Grigis A. and Rothschild L.P., L^2 estimates for the boundary Laplacian on hypersurfaces, *to appear*.

[5] Helffer B. et Nourrigat J., *Hypoellipticité maximale pour des Opérateurs Polynômes de Champs de Vecteurs* , Progress in Math., Birkäuser 1985.

[6] Hörmander L., *The Analysis of linear partial differential Operators II*, Springer, Berlin 1983.

[7] Maire H.-M., Variation et convexité maximale des fonctions de plusieurs variables, *C. R. Acad. Sc. Paris* 301 (1985), p. 431-434.

Examples of non-discreteness for the interaction geometry of semilinear progressing waves in two space dimensions.

ANTÔNIO SÀ BARRETO & RICHARD B. MELROSE

Massachusetts Institute of Technology

Abstract. The interaction geometry for the conic wave solutions of a semilinear wave equation with Cauchy data conormal at the three vertices and center of an equilateral triangle is analyzed. It is shown that there is a point of accumulation in a finite time. This illustrates a general conjecture on the occurrence of such non-discreteness in the interaction set.

§1: INTRODUCTION

In this note we give an example of the accumulation of interaction points for conic progressing waves for the semilinear wave equation

$$Pu = (D_t^2 - D_x^2 - D_y^2)u = f(t, x, y, u) \ \text{ in } \Omega \subset \mathbb{R}^3, \ f \in C^\infty(\Omega \times \mathbb{R}). \tag{1.1}$$

The analysis of such waves will not be considered here, but the problem discussed in [**Me1**] will be briefly recalled.

Suppose $u \in L^\infty(\Omega)$ is a bounded solution of (1.1) in some domain $\Omega \subset \mathbb{R}^3$ for which the linear Cauchy problem

$$\begin{aligned} &Pu = 0 \quad \text{in } \Omega \\ &u_{|t=0} = u_0, \ D_t u_{|t=0} = u_1 \quad \text{in } \Omega_0 = \Omega \cap \{t = 0\} \end{aligned} \tag{1.2}$$

is well-posed, i.e. Ω is P-convex with respect to Ω_0. The singularities of u have been related to the singularities of u_0, u_1 ([**Bo1**], [**Bo2**], [**BR**]). In general, although the non-bicharacteristic singularities weaken, they can spread to the full dependence domain, for P, of the singular support of the initial data:

$$\begin{aligned} &\text{sing.supp}(u) \subset \{(t, x, y) \in \Omega; \\ &\exists(\bar{x}, \bar{y}) \in \text{sing.supp}(u_0) \cup \text{sing.supp}(u_1) \text{ with } t^2 \le (x - \bar{x})^2 + (y - \bar{y})^2\} \end{aligned} \tag{1.3}$$

(see [**Be**]). This is to be contrasted with the linear case, in which singularities are confined to the conic surface through the initial points of singularity.

If the initial data has conormal regularity then the singular support of the solution is closer to that in the linear case, but with new conic surfaces emanating from points of triple interaction. If $p = (\bar{x}, \bar{y}, \bar{t}) \in \Omega$ set

$$C_+(p) = C_+(\bar{x}, \bar{y}, \bar{t}) = \{(x, y, t) \in \Omega; \ t - \bar{t} = [(x - \bar{x})^2 + (y - \bar{y})^2]^{\frac{1}{2}}\}, \tag{1.4}$$

the forward light-cone through p. For any finite set $L \subset \mathbb{R}^3$ the derived set L' is the smallest set with the property

$$p_1, p_2, p_3 \in L,\ C_+(p_1) \cap C_+(p_2) \cap C_+(p_3) = B \text{ finite} \implies B \subset L'. \tag{1.5}$$

If $L_0 \subset \Omega$ is any finite set then the successive derived sets are defined by adding the new points to the old:

$$L_k = L_{k-1} \cup L'_{k-1} \ \forall\, k \geq 1. \tag{1.6}$$

The interaction process corresponds to the case where the Cauchy data is conormal at a finite set $L_0 \subset \Omega_0$. If $p \in \Omega_0 \subset \mathbb{R}^2$ then $v \in C^{-\infty}(\Omega_0)$ is said to be conormal at p, written $v \in I^*(\Omega_0, \{p\})$, if for some $s \in \mathbb{R}$

$$(x - \bar{x})^k (y - \bar{y})^l v(x, y) \in H^{s+k+l}_{\text{loc}}(\Omega_0) \quad \forall\, k, l \in \mathbb{N}.$$

For a general finite set $L_0 \subset \Omega_0$

$$I^*(\Omega_0, L_0) = \sum_{p \in L_0} I^*(\Omega_0, \{p\}). \tag{1.7}$$

Clearly sing.supp$(v) \subset L_0$ if $v \in I^*(\Omega_0, L_0)$ and for such conormal functions singularity and growth properties are intimately connected. The relation between this interaction geometry and the solution to (1.1) is in terms of the successive C^k-singular supports:

$$\Omega \setminus \text{sing.supp}_k(u) = \{(t, x, y) \in \Omega; \exists \text{ a neighborhood } \omega \text{ of } (t, x, y) \text{ with } u_{|\omega} \in C^k(\omega)\}.$$

(1.8) THEOREM ([**Me1**],[**Me2**]). *Suppose $f \in C^\infty(\Omega \times \mathbb{R})$, $\Omega \subset \mathbb{R}^3$ is P-bicharacteristically convex with respect to Ω_0 and $u \in L^\infty(\Omega)$ satisfies (1.1). If the initial data u_0, u_1 is conormal with respect to a finite set L_0 then for every $k \in \mathbb{N}$ $\exists$ $N \in \mathbb{N}$ such that*

$$\textit{sing.supp}_k(u) \cap \{t > 0\} \subset \bigcup_{p \in L_N} C_+(p). \tag{1.9}$$

Conormal functions have the 'simplest' type of singularity associated to a given geometry. Results essentially equivalent to Theorem 1.8 were proved by Bony ([**Bo1**], [**Bo2**]) in case there are at most two different initial points, i.e. $\#(L_0) \leq 2$. The most significant case, $\#(L_0) = 3$ was treated in [**MR1**] and by Bony in [**Bo3**]. A more refined description of the iterated regularity of the solutions in that case is required in the proof in the general case. A discussion of the difference between iterated regularity and conormality is in [**MR2**]. The proof of Theorem 1.8 is in [**Me2**].

Interest is focused here on the geometry of the sets L_k. The finite order regularity in (1.9) appears because the full interaction set:

$$L_\infty = \bigcup_{k=0}^{\infty} L_k \tag{1.10}$$

need not be discrete in $\mathbb{R}^3$. This is shown by an example in [**MR2**] with $L_0 \subset \mathbb{R}^2$ consisting of a hexagon and its center, i.e. $\#(L_0) = 7$. A simpler example, with $\#(L_0) = 5$, is provided by a square and its center, the non-discreteness is somewhat more difficult to show in that case. Numerical experiments lead one to suspect:

(1.11) CONJECTURE. *If $L_0 \subset \mathbf{R}^2 = \{t = 0\}$, and L_0 contains four points not lying on any circle or straight line then L_∞ has a point of accumulation.*

In this note it is shown, see Proposition 4.1, that the initial set consisting of an equilateral triangle and its central point gives an example where $\#(L_0) = 4$ and L_∞ is not discrete.

The authors would like to thank Jack Lee and Niles Ritter for helpful conversations. The second author wishes to express his appreciation of the hospitality of the Institut Mittag-Leffler during the preparation of this manuscript.

§2: INTERACTION GEOMETRY

The geometry of the interaction of progressing waves with point-conormal initial data for the semilinear wave equation (1.1) is particularly simple because all the waves are conic. Thus the variety carrying the singularities, according to Theorem 1.8, consists of a union of forward characteristic cones for the wave equation. In particular these cones are fixed by the set of their vertices.

Consider three dimensional Minkowski space $\mathbf{M}^3$, with signature $(+,+,-)$. We shall always suppose that $\mathbf{M}^3$ has a fixed time-orientation. That is, if $\langle\, ,\rangle$ is the Lorentzian inner product on M^3 then the cone of light-like vectors

(2.1) $$C = \{v \in \mathbf{M}^3; \langle v, v\rangle < 0\}$$

is divided into forward and backward components:

(2.2) $$C = C_+ \cup C_-.$$

For any point $p \in \mathbf{M}^3$ the forward characteristic cone through p is just

(2.3) $$C_+(p) = \{p\} + C_+.$$

The basic property of the interaction of semilinear waves, proved in [**Bo3**] and [**MR2**], is that new waves only arise from the interaction of three or more waves. The extra cases needed to confirm this on repeated interaction can be found in [**Me2**]. More precisely a new conic expanding wave can only arise at the point $m \in \mathbf{M}^3$ if there are three expanding waves passing through m, i.e. for some three points p_1, p_2, p_3 which are sources of expanding waves

(2.4) $$m \in \bigcap_{i=1}^{3} C_+(p_i).$$

In fact this condition is necessary but not sufficient. Namely if (2.4) holds but the points p_i, $i = 1, 2, 3$ are collinear then no new wave will be produced (because this 'new' wave already exists). This leads to the full condition

(2.5) $$m \in \bigcap_{i=1}^{3} C_+(p_i) \quad \text{the } p_i \text{ not collinear.}$$

The first task is to examine the possible arrangements of three points which can produce a new interaction.

Note that if any one of the points p_i were to lie in the interior of the other cones $C_+(p_k)$ then the two forward cones $C_+(p_i)$ and $C_+(p_k)$ would not meet. This gives a simple necessary condition for the existence of a point m as in (2.5):

(2.6) LEMMA. *If there exists a point m as in (2.5) then the time-like inner products,*

$$d(p_i,p_j)^2 = \langle p_i - p_j, p_i - p_j\rangle \quad 1 \le i < j \le 3 \tag{2.7}$$

are all non-negative, with at most two of them vanishing.

PROOF: That no p_i lies within the forward cone of any other implies that none lies within the backward cone of another, so all the distances are non-negative as claimed. Moreover, if two are zero then these two points lie on light rays through the third. The intersection of the cone through either of these two with the third cone is just half this light ray. In this case the third point must be the initial point of each of the half-rays, which must be distinct because of (2.5), so the third distance must be strictly positive.

A plane $P \subset \mathbf{M}^3$ is said to be space-like, light-like or time-like as the induced metric is Euclidean, degenerate or Lorentzian. If P is Lorentzian it inherits the time-orientation from $\mathbf{M}^3$, becoming a two dimensional Minkowski space.

(2.8) LEMMA. *If $L_0 \subset \mathbf{M}^3$ is a set of three points satisfying the non-negativity condition (2.7) and spanning a plane, P, then P is time-like if and only if for some choice of ordering*

$$d(p_1,p_2) > d(p_1,p_3) + d(p_3,p_2) \tag{2.9}$$

and P is light-like if and only if for some choice of ordering there is equality:

$$d(p_1,p_2) = d(p_1,p_3) + d(p_3,p_2). \tag{2.10}$$

PROOF: Certainly validity of (2.9), the strict violation of the triangle inequality, implies that the plane P is not space-like. Nor can it be light-like since such planes are the limits of space-like planes, so (2.9) implies that P is time-like. Similarly (2.10) implies that P is either light- or time-like. Now (2.10) also shows that strict inequality (2.9) cannot be valid for any ordering of the points. Thus to complete the proof it is only necessary to show that when P is time-like (2.9) holds for an appropriate choice.

Thus if P is time-like then Lorentzian coordinates can be chosen with respect to which it is

$$P = \{x = 0\}. \tag{2.11}$$

Number the points so that $d(p_1,p_2)$ is greater than the other distances and then choose Lorentzian coordinates so that p_1, p_2 lie on the line $t = 0$ within P. Making a translation we can even assume that

$$p_1 = (0,a,0),\ p_2 = (0,-a,0) \tag{2.12}$$

Now the Lorentzian form on P is just $dy^2 - dt^2$ so (2.7) becomes

$$p_3 = (0,y,t) \text{ with } t^2 \le a^2 - y^2. \tag{2.13}$$

Then the condition (2.9) is clearly fulfilled, since

$$(2.14) \qquad d(p_1,p_3)^2 = (a-y)^2 - t^2 < (a-y)^2, \; d(p_2,p_3)^2 = (a+y)^t - t^2 < (a+y)^2.$$

This completes the proof of the Lemma.

In the two cases in Lemma 2.8 where the plane is light- or time-like the triangle carries an induced orientation. If in the normalization at the end of the proof above the third point (the one between the two short sides in (2.9) or (2.10)) lies in $t > 0$ the triangle is forward-pointing, if it lies in $t < 0$ then it is backward-pointing. Alternatively this is characterized by the condition that a time-orientation vector at the third point is directed respectively out of or into the triangle. Using this notion of orientation we can now describe the different types of interaction produced by three points.

(2.15) PROPOSITION. *If $L_0 \subset \mathbf{M}^3$ is a set consisting of three distinct points then the derived set $L_0' \neq \emptyset$ if and only if*

$$(2.16) \qquad \langle p-q, p-q\rangle \geq 0 \quad \forall\, p,q \in L_0,$$

(2.17) *L_0 spans a plane, P*

(2.18) *If P is light- or time-like the triangle L_0 is forward-directed.*

Under these conditions on L_0, L_0' consists of one point, except when P is time-like in which case it consists of two points.

PROOF: The necessity of these conditions is quite clear, except perhaps for (2.18). Starting then with the sufficiency, choose two points, in the light- or time-like case forming the bottom of the triangle with respect to the time-orientation. The intersection of the forward cones through these points lies in the plane, Q, which is the Lorentz-orthogonal bisector of the connecting segment. The intersection is in fact one component of a hyperbola with pole at the midpoint.

In the spatial case the cone through the third point meets Q in a hyperbola, or cone, with pole in the spatial sector formed by the separatrices of the first hyperbola. Clearly the three cones meet at one point. In the time-like case the pole of the hyperbola, or cone, is in the time-like sector, forwards or backwards with the orientation of the triangle. In this case there are therefore two points of intersection, given (2.18). The light-like plane is an obvious intermediate case. This discussion also completes the proof of the necessity of (2.18).

§3: NON-SPATIAL EXAMPLES

The examples of points of accumulation for the interaction geometry of semilinear progressing waves given in this, and the next, section all involve a high degree of symmetry. Although it is the symmetry which allows the repeated interaction points to be easily located it is nevertheless a regrettable special feature of these examples. Symmetry in the initial configuration of points generally leads to accidental multiplicity of the interaction points. This has the effect of reducing their number, so it is to be expected that for non-symmetrical cases, close to the symmetrical ones analyzed here, points of accumulation will still arise and possibly qualitatively earlier.

Three points $\{p_1, p_2, p_3\}$ in $\mathbf{M}^3$ will be said to form an equilateral triangle if they span a space-like plane, P, and within that plane form an equilateral triangle with respect to the induced Euclidean distance. Since $\mathbf{M}^3$ is assumed to be time-oriented the space-like plane P has a unique forward-pointing Lorentzian normal, $\nu = \nu(p_1, p_2, p_3)$, of length -1:

$$(3.1) \qquad \langle \nu, v \rangle = 0, \ \forall\, v \in P, \ \nu \in \mathrm{cvx}(C_+), \ \langle \nu, \nu \rangle = -1.$$

If $\{p_1, p_2, p_3\} \subset \mathbf{M}^3$ form an equilateral triangle let $\gamma = \gamma(p_1, p_2, p_3) \in P$ be the center. The radius of the triangle is then given by

$$(3.2) \qquad r = r(p_1, p_2, p_3) = \langle \gamma - p_i, \gamma - p_i \rangle^{\frac{1}{2}}, \quad i = 1, 2, 3.$$

We shall consider the interactions produced by waves emanating from the four initial points consisting of the triangle and

$$(3.3) \qquad p_4 = \gamma + \rho\nu, \quad \rho \in \mathbf{R}.$$

(3.4) LEMMA. *Suppose $L_0 = \{p_1, p_2, p_3, p_4\} \subset \mathbf{M}^3$ consists of an equilateral triangle $\{p_1, p_2, p_3\}$ and fourth point* (3.3). *Then if*

$$(3.5) \qquad -\frac{1}{2} < \rho \le \frac{1}{2}$$

the derived set is

$$L_0' = \{q_1, q_2, q_3, q_4\} \subset \mathbf{M}^3$$

where $\{q_1, q_2, q_3\}$ is an equilateral triangle with the same normal, ν, radius

$$(3.6) \qquad r' = r(q_1, q_2, q_3) = \frac{1 - \rho^2}{1 + 2\rho} r,$$

and center

$$(3.7) \qquad \gamma' = \gamma(q_1, q_2, q_3) = \gamma + \frac{1}{4}\left[1 + 2\rho + 3(1 + 2\rho)^{-1}\right] r\nu;$$

the fourth point is:

$$(3.8) \qquad q_4 = \gamma' + \rho' r' \nu', \ \rho' = \frac{\rho}{1 + \rho}.$$

If in (3.3)

$$(3.9) \qquad \frac{1}{2} < \rho < 1$$

then

$$L_0' = \{q_1, q_2, q_3, q_4, m_1, m_2, m_3\} \subset \mathbf{M}^3$$

with the q_i as above and $\{m_1, m_2, m_3\}$ another equilateral triangle with the same normal, radius

$$r'' = r(m_1, m_2, m_3) = \frac{1-\rho^2}{2\rho - 1} r \tag{3.10}$$

and center

$$\gamma'' = \gamma(m_1, m_2, m_3) = \gamma + \frac{1}{4}\left[2\rho - 1 + 3(2\rho-1)^{-1}\right] r\nu. \tag{3.11}$$

In both cases the new triangle is rotated by $\pi/3$ with respect to the old one after projection along ν.

PROOF: This is basically a direct computation. In view of the invariance of L_0' under spatial rotation by $2\pi/3$ it suffices to locate $\{p_1, p_2, p_3\}'$ and $\{p_2, p_3, p_4\}'$. The first of these triangles is space-like, by hypothesis. The second is space-like if $-\frac{1}{2} < \rho < \frac{1}{2}$, light-like if $|\rho| = \frac{1}{2}$ and time-like if $\frac{1}{2} < |\rho| < 1$. Clearly, under either hypothesis (3.5) or (3.9) whenever not space-like this second triangle is forward-pointing. Thus,

$$\{p_1, p_2, p_3\}' = \{q_4\}, \ \{p_2, p_3, p_4\}' = \begin{cases} \{q_1\} & -\frac{1}{2} < \rho \le \frac{1}{2} \\ \{q_1, m_1\} & \frac{1}{2} < \rho < 1 \end{cases} \tag{3.12}$$

are all well-defined, using Proposition 2.15.

Lorentzian coordinates (x, y, t) can be introduced so that

$$\gamma = (r/2, 0, 0), \ p_2, p_3 = (0, \pm\sqrt{3}/2, 0), \ \nu = (0,0,1), \ p_4 = (0, r/2, \rho r).$$

The form of the intersection of the two cones:

$$C_+(p_2) \cap C_+(p_3) = \{(x, 0, t); t^2 - x^2 = \frac{3}{4}r^2\} \tag{3.13}$$

shows that $q_4 = (x, 0, t)$ where:-

$$t^2 - x^2 = \frac{3}{4}r^2, \ t + x = \rho r + \frac{r}{2}. \tag{3.14}$$

The last equation represents one of the branches of the cone $C_+(p_4)$ in $\{y = 0\}$. In these coordinates therefore:

$$q_1 = (\frac{1}{4}\left[1 + 2\rho - 3(1+2\rho)^{-1}\right], 0, \frac{1}{4}\left[1 + 2\rho + 3(1+2\rho)^{-1}\right]). \tag{3.15}$$

Since $r' = r/2 - x$ and $\gamma' = \gamma + r\nu$, (3.6) and (3.7) follow directly. To get (3.8) just observe that $q_4 = (r/2, 0, r)$.

In case $\frac{1}{2} < \rho < 1$ the same solution exists and there is a second point of interaction, m_1, with coordinates $(x, 0, t)$ satisfying in place of (3.14)

$$t^2 - x^2 = \frac{3}{4}r^2, \ t - x = \rho r - \frac{r}{2}. \tag{3.16}$$

From this the formulæ(3.10) and (3.11) follow similarly.

This lemma allows an easy inductive argument to be used to show the existence of a point of accumulation of the iteratively derived sets of an appropriately defined set of four initial points. This initial set does not lie in a space-like plane, but such an example is considered in the next section and reduced to the present case.

(3.17) PROPOSITION. *If $\{p_1, p_2, p_3, p_4\} \subset \mathbf{M}^3$ is a set as in Lemma 3.4, then the iteratively derived sets:*

$$I_0 = L_0,\ I_1 = \{q_1, q_2, q_3, q_4\},\ I_k = I'_{k-1},\ k \geq 2 \tag{3.18}$$

form a convergent sequence:

$$I_k \longrightarrow I_\infty = \{\gamma + \tau\nu\} \quad \text{as } k \longrightarrow \infty \tag{3.19}$$

where

$$\tau = \frac{1-\rho^2}{4} \sum_{p=0}^{\infty} \frac{[(p+1)\rho+1]^2 + 3[(p-1)+1]^2}{[(p\rho+1)^2 - \rho^2][(p\rho-1)^2 - \rho^2]}. \tag{3.20}$$

PROOF: Notice first that even if $\frac{1}{2} \leq \rho < 1$ initially, L'_0 contains an equilateral triangle and fourth point for which the corresponding value of $\rho = \rho_1 = \rho/(1+\rho) < 1/2$. Thus iteratively the values are decreasing:

$$\rho_k = \frac{\rho_{k-1}}{1+\rho_{k-1}}. \tag{3.21}$$

Using Lemma 3.3, all the I_k are well-defined. Applying (3.21) iteratively,

$$\rho_k = \frac{\rho}{1+k\rho}. \tag{3.22}$$

The radius, r_k, of the equilateral triangle in I_k is given by:

$$r_k = \frac{1-\rho_{k-1}^2}{1+2\rho_{k-1}} r_{k-1} = \frac{1-\rho^2}{(1+k\rho)^2 - \rho^2} r, \quad k = 0, \ldots. \tag{3.23}$$

Certainly the $r_k \to 0$ as $k \to \infty$. Moreover, from (3.7) the centre of the kth triangle is at a time

$$\delta_k = \frac{1}{4}[1 + 2\rho_k + 3(1+2\rho_k)^{-1}] r_{k-1} \tag{3.24}$$

later than the $(k-1)^{\text{st}}$. Thus, the time elapsed when the kth triangle appears is:-

$$\tau_k = \sum_{p=0}^{k-1} \frac{1}{4} \left[\frac{(p+1)\rho+1}{(p-1)\rho+1} + 3\frac{(p-1)\rho+1}{(p+1)\rho+1} \right] \frac{1-\rho^2}{(1+p\rho)(1+(p-2)\rho)}. \tag{3.25}$$

As $k \to \infty$ this converges to the limit (3.20), completing the proof of the Proposition.

It is important to note that in the circumstances of Proposition 3.17,

$$I_k \cup I_{k+1} \not\subset (I_{k-1} \cup I_k)'. \quad (k \geq 2) \tag{3.26}$$

The missing points come from the interaction of two points in I_k with one in I_{k-1}. As will be seen below such points are necessary in order to show the non-discretness for a spatial initial set.

§4: SPATIAL EXAMPLES

In this section it is finally shown that for an initial configuration of four points in a spatial plane, comprising the vertices and the center of an equilateral triangle, interaction develops a point of accumulation in a finite time.

(4.1) PROPOSITION. *If* $L_0 = \{p_1, p_2, p_3, p_4\} \subset \mathbf{M}^3$ *consists of an equilateral triangle* p_1, p_2, p_3 *and its center,* p_4*, then the set*

$$L_\infty = \bigcup_{k=0}^{\infty} L_k,\ L_k = L_{k-1} \cup L'_{k-1},\ k \geq 1, \text{ is not discrete in } \mathbf{M}^3. \tag{4.2}$$

PROOF: After a preliminary analysis of the L_k, for $k \leq 4$, a configuration is found to which Proposition 3.17 can be applied.

From Lemma 3.4 it follows that

$$L_1 = I_0 \cup I_1, I_0 = L_0, I_1 = I'_0 = \{-T + \gamma + r\nu\} \cup \{\gamma + r\nu\} = q_1, q_2, q_3, q_4. \tag{4.3}$$

Here, $T = \{p_1, p_2, p_3\} - p_4$ is the triangle in L_0 translated back to a plane through the origin in $\mathbf{M}^3$ and $\gamma = p_4$ is the center. This argument iterates to show that

$$L_k \supset \bigcup_{j=0}^{k} I_j,\ I_k = I'_{k-1} = \{(-1)^k T + \gamma + kr\nu\} \cup \{\gamma + kr\nu\},\ k \geq 1. \tag{4.4}$$

However, after the first interaction, i. e. for $k \geq 2$, there is no equality in (4.4). Thus, at the second interaction,

$$L_2 = I_0 \cup I_1 \cup I_2 \cup J_2,\ J_2 = -\frac{1}{2}T + \gamma + \frac{3}{2}r\nu. \tag{4.5}$$

Here, J consists of the point:-

$$j_1 = \{p_2, q_1, q_4\}' \tag{4.6}$$

and its cyclic analogues. Observe that

$$j_1 = \frac{1}{2}(q_1 + q_4)$$

giving the formula for J_2 in (4.5). It is straightforward, but unnecessary for what follows, to verify that there are no other points of interaction, giving the equality in (4.5).

At the next interaction a similar phenomenon occurs, with J_3 bearing the same relationship to I_1 as J_2 does to I_0:

$$J_3 = \frac{1}{2}T + \gamma + \frac{5}{2}r\nu.$$

Moreover, there is a further interaction between $J_2 = \{j_1, j_2, j_3\}$ and $I_2 = \{n_1, n_2, n_3, n_4\}$. Namely

$$K_3 = \{s_1, s_2, s_3\} = \frac{4}{5}T + \gamma + \frac{11}{5}r\nu,$$
$$\{s_3\} = \{j_1, j_2, n_3\}', \ \{s_2\} = \{j_3, j_1, n_2\}', \ \{s_1\} = \{j_2, j_3, n_1\}'.$$

Then

$$L_3 = I_0 \cup I_1 \cup I_2 \cup I_3 \cup J_2 \cup J_3 \cup K_3 \tag{4.7}$$

Consider the location of s_1. Clearly $\{j_2, j_3, n_1\}$ is a space-like triangle, so s_1 is well-defined. If $\delta > 0$ is the distance from n_1 at which the interaction occurs then

$$s_1 = (1-\delta)[n_1 - 2r\nu - \gamma] + \gamma + (2+\delta)r\nu \tag{4.8}$$

so it suffices to find δ. In suitable Lorentzian coordinates (x, y, t), in which $\gamma = \{0\}$, $n_1 = (r, 0, 2r)$, $\nu = (0, 0, 1)$,

$$C_+(j_2) \cap C_+(j_3) = \left\{(x, 0, t); (t - \frac{3}{2}r)^2 - (x - \frac{1}{4}r)^2 = \frac{3}{16}\right\}. \tag{4.9}$$

Thus, the coordinates of s_1 satisfy:-

$$(t - \frac{3}{2}) + (x - \frac{1}{4}r) = \frac{5}{4}, \ (t - \frac{3}{2}r) - (x - \frac{1}{4}r) = \frac{3}{20}r. \tag{4.10}$$

This gives $t = 11r/5$, i. e. $\delta = 1/5$, as in (4.8). Again the proof that there is equality in (4.7) is omitted.

At the fourth iteration further extra triangles appear. As before, I_2 produces

$$J_4 = -\frac{1}{2}T + \gamma + \frac{7}{2}r\nu,$$

and interaction of J_2 and I_3 gives

$$K_4 = -\frac{4}{5}T + \gamma + \frac{16}{5}r\nu$$

as for K_3 above. More importantly there is interaction between K_3 and I_2 to produce a new triangle E, consisting of

$$e_1 = \{s_2, s_3, \gamma + 2r\nu\}', \ \gamma + 2r\nu \in I_2 \tag{4.11}$$

and its cyclic permutations. In terms of Lemma 3.4 E is obtained from the equilateral triangle K_3, of radius $4r/5$, centered at $\gamma + 11r/5$, with the fourth point $\gamma + 2r\nu$, i. e. $\rho = -1/4$. Thus, the new triangle has radius

$$r(E) = \frac{1 - \frac{1}{6}}{1 - \frac{1}{2}} \frac{4r}{5} = \frac{3r}{2}, \tag{4.12}$$

in fact:

$$E = -\frac{3}{2}T + \gamma + \frac{7}{2}r\nu. \tag{4.13}$$

The final stage of the proof is just the application of Proposition 3.17, since together with $\gamma + 4r\nu \in I_4$, E satisfies the hypothesis where $\rho = \frac{1}{3}$ and the radius is $3r/2$.

REFERENCES

[Be] Beals, M., *Nonlinear wave equations with data singular at one point*, preprint.

[BR] Beals, M. and M. Reed, *Propagation of singularities for hyperbolic pseudodifferential operators with non-smooth coefficients*, Comm. Pure Appl. Math. **35** (1982), 169–184.

[Bo1] Bony, J.-M., *Calcul symbolique et propagation des singularités pour les équations aux dérivées partielles nonlinéares*, Ann. Sci. Ec. Norm. Sup. **14** (1981), 209–246.

[Bo2] Bony, J.-M., *Interaction des singularités pour les équations aux dérivées partielles non linéares*, no. 22, Sem. Goulaouic Schwartz (1979–1980); no. 2, (1981–1982).

[Bo3] Bony, J.-M., *Second microlocalization and propagation of singularities for semi-linear hyperbolic equations*, preprint.

[Me1] Melrose, R.B., *Conormal rings and semilinear progressing waves*, in: Advances in Microlocal Analysis ed:H. G. Garnir, Reidel, 1986.

[Me2] Melrose, R.B., *Interaction of nonlinear progressing waves for semilinear wave equations III*, in preparation.

[MR1] Melrose, R.B. and N. Ritter, *Interaction of nonlinear progressing waves for semilinear wave equations*, Ann. of Math. **121** (1985), 187-213.

[MR2] Melrose, R.B. and N. Ritter, *Interaction of progressing waves for semilinear wave equations II*, Math. Scand. to appear, Inst. Mittag-Leffler Report 1985, no. 7.

On the Cauchy problem for hyperbolic equations in C^∞ and Gevrey classes

Sigeru MIZOHATA
Dept. of Mathematics, Faculty of Science
Kyoto University, Kyoto 606, Japan

1. Introduction.

The Cauchy problem for general hyperbolic partial differential equations and systems was studied by I.G. Petrowsky[23]. After then, J. Leray[12], L. Gårding[5], K.O. Friedrichs[4] clarified the work of Petrowsky and extended it from their view-points.

Now we explain it more concretely. We are concerned with general higher order single equations and the Cauchy problem for them:

$$\begin{cases} P(t,x;D_t,D_x)u(t,x) = f(t,x) \quad \text{for} \quad (t,x) \in [0,\ T] \times \mathbb{R}^\ell = \Omega, \\ D_t^j u\Big|_{t=t_0} = u_j(x), \quad 0 \le j \le m-1, \quad \text{for} \quad t_0 \in [0,\ T), \end{cases} \tag{1.1}$$

where

$$P(t,x;\tau,\xi) = \tau^m + \sum_{j=1}^{m} a_j(t,x;\xi)\tau^{m-j}, \tag{1.2}$$

and order $a_j \le j$ (Kowalewskian).

We say that the Cauchy problem for (1.1) is C^∞-wellposed, if for any $f(t,x) \in C^\infty([t_0,T] \times \mathbb{R}^\ell)$ and for any initial data $u_j(x) \in C^\infty$, there exists a unique solution $u(t,x) \in C^\infty([t_0,T] \times \mathbb{R}^\ell)$. Further we say that (1.1) is uniformly C^∞-wellposed in Ω when the above problem is well posed for any $t_0 \in [0,\ T)$.

Let us explain Petrowsky's result. In a word, he proved that (1.1) is C^∞-wellposed under the assumption that P is strictly hyperbolic. The strict hyperbolicity means the following. Let

$$P_m(t,x;\tau,\xi) = \prod_{j=1}^{m} (\tau - \lambda_j(t,x;\xi)). \tag{1.3}$$

P is called strictly hyperbolic if 1) $\lambda_j(t,x;\xi)$ are all real for any $(t,x;\xi)$; 2) they are distinct. More precisely, there exists some positive δ such that

$$|\lambda_i(t,x;\xi) - \lambda_j(t,x;\xi)| \geq \delta|\xi| \quad \text{for all} \quad (t,x;\xi) \in \Omega \times \mathbb{R}^\ell . \tag{1.4}$$

In 1957, the work of P.D. Lax[11] appeared, and this work was followed by S. Mizohata[15]. They claimed that in order that (1.1) be C^∞-wellposed, it is necessary that the characteristic roots $\lambda_j(t,x;\xi)$ be real for all $(t,x;\xi)$ irrespective of their multiplicities (Lax-Mizohata theorem).

The next step was to consider non-strictly hyperbolic equations and systems as far as the Cauchy problem concerned. Observe that, when the multiplicity of characteristic root is constant, (1.3) becomes

$$P_m = \prod_{j=1}^{m} (\tau - \lambda_j(t,x;\xi))^{m_j}, \tag{1.5}$$

and when the multiplicity is not constant (variable multiplicity), the situation becomes, in general, fairly complicated. Famous examples in this case are

$$L_0 u = \frac{\partial^2 u}{\partial t^2}(t,x) - t^{2k}a(t,x)\frac{\partial^2 u}{\partial x^2} + t^\ell b(t,x)\frac{\partial u}{\partial x} = f,$$

$$L_1 u = \frac{\partial^2 u}{\partial t^2}(t,x) - x^{2k}a(t,x)\frac{\partial^2 u}{\partial x^2} + x^\ell b(t,x)\frac{\partial u}{\partial x} = f,$$

where $a(t,x) > 0$ and both equations are considered in a neighborhood of the origin.

In the following, we explain briefly the development of these researches.

2. C^∞ and Gevrey well-posedness in the case of constant multiplicity.

We explain the historical background of researches in this domain.

i) C^∞-wellposedness.

After pioneering works of A. Lax[10] and of M. Yamaguti[26], S. Mizohata-Y. Ohya [19,20] treated the case $m_j \leq 2$ in (1.5). This gave a necessary and sufficient condition for C^∞-wellposedness. That condition can be stated in the following form: Subprincipal symbol of P, denoted by P'_{m-1}, vanishes on double characteristic sets, namely

$$P'_{m-1}(t,x;\lambda_j(t,x;\xi),\xi) \equiv (P_{m-1} - \frac{1}{2i}\sum_{i=0} \frac{\partial^2}{\partial x_i \partial \xi_i} P_m)(t,x;\lambda_j(t,x;\xi),\xi) \equiv 0 \tag{2.1}$$

for all $(t,x;\xi)$, where $x_0 = t$ and $\xi_0 = \tau$. The importance of this condition was pointed out independently by J. Vaillant[25] too.

After this work, H. Flashka-G. Strang[3] gave a necessary condition for general multiplicity m_j, and J. Chazarain[1] showed that that condition is sufficient using the theory of Fourier integral operators (see also V.Ya. Ivrii-V.M. Petkov[7]).

ii) $\gamma^{(s)}$-wellposedness.

Hereafter we denote by $\gamma^{(s)}$ the space of all functions of Gevrey class s. The systematic treatment of the Cauchy problem in Gevrey class began with the work of Y. Ohya[21], and this result was immediately extended by J. Leray-Y. Ohya[13]. J.C. De Paris[2] has shown a kind of factorization by differential polynomials and defined "bien décomposable" operators which are now known to be equivalent to satisfying the Levi condition. H. Komatsu[9], analyzing the above works, showed a general factorization theorem, and defined the notion of irregularity through this factorization. There is an important work of V.Ya. Ivrii [8].

Now we want to summarize those results in a slightly different form. First recall that

$$\begin{cases} \text{The principal symbol of } P \text{ has the form} \\ \prod_{j=1}^{k} (\tau - \lambda_j(t,x;\xi))^{m_j}, \\ \text{where } \lambda_j \text{ are } \underline{\text{real and distinct}}. \end{cases} \tag{2.2}$$

In order to consider the well-posedness in the space C^∞ or Gevrey class $\gamma^{(s)}$ $(s > 1)$, we proposed to use a fine factorization of operators, because it gives us a clear image to the well-posedness. A fine factorization means the factorization in the following form:

$$P = P_k \circ P_{k-1} \circ \cdots \circ P_1 + R, \tag{2.3}$$

where the principal symbol of P_j is $(\tau - \lambda_j)^{m_j}$. More precisely

$$P_j = (D_t - \lambda_j(t,x;D))^{m_j} + a_{1,j}(t,x;D)(D_t - \lambda_j)^{m_j-1} + \cdots$$
$$+ a_{i,j}(D_t - \lambda_j)^{m_j-i} + \cdots + a_{m_j,j}, \quad \text{order } a_{i,j} \le i - 1. \tag{2.4}$$

$$R(t,x;D,D_t) = \sum_{j=0}^{m-1} r_j(t,x;D)\, D_t^j, \quad \text{order } r_j(t,x;\xi) \le -(m-1).$$

For simplicity, we assume that all the coefficients of P are C^∞ and with all their derivatives are bounded in Ω. Under the form (2.3) and (2.4), we state

(1) Levi condition (C^∞-wellposedness).

$$\text{order } a_{i,j} \le 0 \quad \text{for } 1 \le i \le m_j,\ 1 \le j \le k. \tag{L}$$

(2) Well-posed Gevrey index s.

$$\frac{1}{s} \ge \rho, \text{ where } \rho_j = \max_{1\le i\le m_j} \text{order } a_{i,j}/i, \quad \rho = \max_{1\le j\le k} \rho_j.$$

Let us notice that, in the above factorization, the number ρ_j is invariant with respect to the order of factorization. Let us explain the above facts more precisely.

Theorem 2.1.

In order that the Cauchy problem (1.1) be uniformly C^∞-wellposed in Ω, it is necessary and sufficient that P satisfies the Levi condition.

Theorem 2.2.

In order that the Cauchy problem in Gevrey class $\gamma^{(s)}$ be uniformly well-posed in Ω, it is necessary that s satisfies $1/s > \rho$.

Theorem 2.3.

If $1/s > \rho$, there exists a unique solution $u(t,x)$ for $t \in [t_0, T]$. If $1/s = \rho$, there exists only a local solution of (1.1).

These theorems are new formulations obtained hitherto. We gave already brief direct proofs of them ([16],[17]). Let us explain about the assumptions in Gevrey class. $f(x) \in \gamma^{(s)}$ means

$$\sup_{x\in\mathbb{R}^n} |\partial^\alpha f(x)| \le M\alpha!^s C^{|\alpha|}, \quad {}^\forall\alpha \ge 0 \ ({}^\exists M,\ {}^\exists C \ge 0).$$

In Theorems 2.2 and 2.3, we assume that the coefficients of P satisfy the following type condition: $t \longrightarrow a(t,x) \in \gamma_x^{(s)}$, $t \in [0, T]$, is continuous, with all its derivatives in t. Let us notice that we don't assume the Gevrey property in t. Incidentally, if we don't assume the Gevrey property in t of the coefficients of P, we cannot get the perfect factorization in Gevrey class. On the other hand, in order to obtain the standard result on the propagation of the wave front set WF_s in Gevrey class (see for instance [18]), it seems to the author, we need to require the Gevrey property of the coefficients for (t,x).

To illustrate the advantage of the above formulation, we explain the sufficient part of Theorem 2.1. For simplicity, we assume

$$P = P_2 \circ P_1 + R.$$

Putting $(D_t - \lambda_1)^j u = u_j$, $(0 \le j \le m_1 - 1)$, $P_1 u = u_{m_1}$, $(D_t - \lambda_2)^j P_1 u = u_{m_1+j}$, $(0 \le j \le m_2 - 1)$, and denoting $U = {}^t(u_0, u_1, \cdots, u_{m-1})$, we get the (equivalent) system of $Pu = f$:

$$\partial_t U = iHU + B_0(t,x;D) + F, \tag{2.5}$$

where

(1) H is a diagonal matrix whose diagonal entries are $\lambda_i(t,x;D)$.
(2) $B_0(t,x;\xi) \in S^0$, by the Levi condition.
(3) $F = {}^t(0, \cdots, 0, f)$.

This shows that the problem is reduced essentially to consider the elementary hyperbolic operator

$$\partial_t u = i\lambda(t,x;D)u + b_0(t,x;D)u + f, \quad \text{with} \quad b_0 \in S^0, \tag{2.6}$$

and we know that for every positive integer k, and for any data $u_0(x) \in H^k$ (initial data), $f(t,x) \in C^0([0,T];H^k)$, there exists a unique solution $u(t,x) \in C^0([0,T];H^k) \cap C^1([0,T];H^{k-1})$, and it holds

$$\|u(t,\cdot)\|_k \le C(T)\Big\{\|u(0,\cdot)\|_k + \int_0^t \|f(s,\cdot)\|_k ds\Big\}.$$

This enables us to solve (2.5) by successive approximations.

Next Theorem 2.3 can be proved along the same line as above. In this case, we are led to consider (2.6) with

$$\lambda(t,x;\xi) \in \gamma_x^{(s)} \cap S^1_{1,0}, \quad b_0(t,x;\xi) \in \gamma_x^{(s)} \cap S^\rho_{1,0}.$$

By using some energy inequality in $\gamma^{(s)}_{L^2}$, we can show the existence of solution of (2.6) with $u_0 = 0$, provided that $f(t,x) \in C^0([0,T];\gamma^{(s)}_{L^2})$, and $1/s \ge \rho$.

3. C^∞-wellposedness in the case of variable multiplicity.

First we cite Oleinik's work[22]. It considers

$$\partial_t^2 u(t,x) = \sum_{i,j=1}^{n} \partial_i(a_{ij}(t,x)\partial_j u) + b_0\partial_t u + \sum_j^n b_j(t,x)\partial_j u + cu = f,$$

where $\sum a_{ij}\xi_i\xi_j \geq 0$. The Cauchy problem concerns in the space $\Omega = \{(t,x) \in [0, T] \times \mathbb{R}^n\}$, with the initial data at $t = 0$. The following condition was presented as a sufficient condition for C^∞-wellposedness. If we take two positive constants α and A suitably, it holds that

$$\alpha t(\sum b_j\xi_j)^2 \leq A \sum a_{ij}\xi_i\xi_j + \sum (a_{ij})_t'\xi_i\xi_j ,$$

for all $(t,x;\xi) \in \Omega \times \mathbb{R}^n$.

Next the research on the necessity has been done in V.Ya. Ivrii-V.M. Petkov[7]. Let us explain a part of their results. To see clearly its essence (namely, Theorems 4.1 and 7.1), we specialize the situation without much loss of generality.

Let

$$P(t,x;D_t,D) = D_t^m + \sum_{j=1}^{m} (a_j(t,x;D) + a_{j,j-1}(t,x;D) + \cdots$$
$$+ a_{j,k}(t,x;D) + \cdots + a_{j,0})\, D_t^{m-j} ,$$

where $a_{j,k}(t,x;\xi)$ is homogeneous of degree k in ξ. The principal part is $P_m = D_t^m + \sum a_j D_t^{m-j}$, and the lower order terms of order $(m - s)$ is $P_{m-s} = \sum a_{j,j-s}D_t^{m-j}$.

Suppose that

$$a_j(0,0;e_\ell) = 0 \quad \text{for} \quad 1 \leq j \leq m \quad \text{and} \quad e_\ell = (0,\cdots,0,1). \tag{3.1}$$

Next we define the <u>vanishing order</u> of $a(t,x) = a(t,x',x_\ell)$ $(x' = (x_1, \cdots, x_{\ell-1}))$ <u>with weights p, q</u>, p and q being positive real numbers.

Let

$$a(t,x) \sim \sum_{(\mu,\nu)} a_{\mu,\nu}\, t^{\mu_0}\, x'^{\mu'}\, x_\ell^\nu ,$$

then we denote

$$\text{v.o.}[a(t,x)] = \min_{(\mu,\nu),\, a_{\mu,\nu}\neq 0} (p(\mu_0 + |\mu'|) + q\nu),$$

and call it vanishing order of a (with weights p, q).

Now we assume for the principal part P_m,

$$v.o.[a_j^{(\alpha')}(t,x;e_\ell)] \geq j - |\alpha'| \quad \text{for} \quad 1 \leq j \leq m \quad \text{and} \quad |\alpha'| \leq j - 1. \tag{3.2}$$

Then in order that the Cauchy problem for P be C^∞-wellposed at the origin, it is necessary that

$$\forall_{\alpha'}, \ \forall_k, \quad v.o.[a_{j,k}^{(\alpha')}(t,x;e_\ell)] \geq j - |\alpha'| - (1 + p)(j - k), \tag{3.3}$$

where we assume that

i) $p \geq q$ (Theorem 4.1) (3.4)

ii) $1 + p > q$ (Theorem 7.1) if $m = 2, 3$.

<u>Remark</u>. Let us notice that

$$\begin{aligned} a_{j,k}(t,x;\xi) &\equiv a_{j,k}(t,x;\xi',\xi_\ell) \\ &= a_{j,k}(t,x;e_\ell)\xi_\ell^k + \sum_{1\leq|\alpha'|\leq k} \alpha'!^{-1} a_{j,k}^{(\alpha')}(t,x;e_\ell)\xi^{\alpha'}\xi_\ell^{k-|\alpha'|} \end{aligned}$$

This theorem and methods used in the proof are a core part of their work. The proof is fairly complicated. Our purpose is to try to elucidate the contents of this theorem, and to show that this theorem is true under the condition

$$1 + p > q \tag{3.5}$$

irrespective of its multiplicity (= m), and this limitation of choice of p, q would be necessary. Our method relies on the micro-local energy method. Finally we should say that T. Mandai[14] proved the same result by pursuing their methods.

4. Ivrii-Petkov's condition.

We look at the conditions (3.3)-(3.4). For simplicity, we consider the case $\ell = 1$. Then the operator P becomes

$$P = D_t^m + \sum_{j=1}^m (a_j(t,x)D_x^j + a_{j,j-1}(t,x)D_x^{j-1} + \cdots + a_{j,0}(t,x))D_t^{m-j}, \tag{4.1}$$

and we consider the Cauchy problem

$$Pu = 0, \quad \partial_t^j u\Big|_{t=0} = u_j(x) \in C^\infty, \ 0 \le j \le m-1, \tag{4.2}$$

in a neighborhood of the origin. For

$$P_m = D_t^m + \sum a_j D_x^j D_t^{m-j}, \tag{4.3}$$

we assume

$$a_j(0,0) = 0 \quad (1 \le j \le m). \tag{4.4}$$

Let Taylor expansion of a_j around the origin be

$$a_j(t,x) \sim \sum_{(\mu,\nu)} a_{j,\mu\nu} t^\mu x^\nu. \tag{4.5}$$

Then for (j,μ,ν) such that $a_{j,\mu\nu} \neq 0$, we make correspond the point $P_{j,\mu\nu}$ in Newton diagram by introducing a parameter q satisfying $0 < q < 1$,

$$P_{j,\mu\nu} = \left(\frac{j+\mu}{j}, \frac{j-q\nu}{j}\right) \quad (= (1+\mu/j,\ 1-q\nu/j)). \tag{4.6}$$

In the same way, to the term $a_{j,k}(t,x)D_x^k D_t^{m-j}$ $(0 < k < j)$, appearing in lower order terms of P, developing $a_{j,k}$:

$$a_{j,k}(t,x) \sim \sum_{(\mu,\nu)} a_{j,k,\mu\nu} t^\mu x^\nu,$$

we associate the point $P_{j,k,\mu\nu}$ if $a_{j,k,\mu\nu} \neq 0$, defined by

$$P_{j,k,\mu\nu} = \left(\frac{j+\mu}{j}, \frac{k-q\nu}{j}\right). \tag{4.7}$$

Hereafter, in Newton diagram, we consider only the points $P_{j,\mu\nu}$ or $P_{j,k,\mu\nu}$ such that $j - q\nu > 0$ or $k - q\nu > 0$. Then

Fundamental Lemma. q $(0 < q < 1)$ is fixed, put

$$p = \max_{P_m} \frac{j-q\nu}{j+\mu}. \tag{4.8}$$

Suppose $p > 0$. Then, in order that the Cauchy problem (4.2) be C^∞-wellposed at the origin, it is necessary that any lower order term satisfies

$$\frac{k-q\nu}{j+\mu} \le p. \tag{4.9}$$

Remark 1. The above condition means that the slope of $\overrightarrow{OP}_{j,k,\mu\nu}$ in Newton diagram corresponding to the lower order terms must not exceed the maximum slope of the points corresponding to the principal part. It should be noted that, when q and p_0 (> 0) are fixed, there exists only a finite number of $P_{j,k,\mu\nu}$ satisfying $(k - q\nu)/(j + \mu) \geq p_0$. In fact, this condition means that $p_0\mu + q\nu \leq k - p_0 j$, hence there exists only a finite number of (μ, ν) satisfying this inequality.

Remark 2. Condition (4.9) contains the indeterminate parameter q $(0 < q < 1)$. In this sense, this condition is indirect. Recall that q is arbitrary, and the slope $(k - q\nu)/(j + \mu)$ depends continuously on q. On the other hand, for the point $P_{j,k,\mu\nu}$ with $\nu = 0$, the slope $k/(j + \mu)$ becomes independent of q. This remark is important to see that the multiplicity of regularly hyperbolic operators can not exceed 2.

We explain briefly why this condition appears. As we explain later, we micro-licalize operator P by using cut-off functions $\{\alpha_n(\xi), \beta_n(x)\}$. These are defined as follows:

1) $\alpha_n(\xi)$. Let $\alpha(\xi) \in C_0^\infty$, $\mathrm{supp}[\alpha] \subset \{\xi;\ |\xi - 1| \leq r_0\}$, and $\alpha(\xi) = 1$ for $|\xi - 1| \leq r_0/2$. Then put $\alpha_n(\xi) = \alpha(\xi/n)$.
2) $\beta_n(x)$. Let $\beta(x) \in C_0^\infty$, $\mathrm{supp}[\beta] \subset \{x;\ |x - 1| \leq r_0\}$, and $\beta(x) = 1$ for $|x - 1| \leq r_0/2$. Then put $\beta_n(x) = \beta(n^q x)$.

We call $\{\alpha_n(D), \beta_n(x)\}$ is a micro-localizer of size r_0.

In accordance with micro-localizer, we can consider micro-localized operator of P, denoted by $P_{n,loc}$. This symbol $P_{n,loc}$ is a modification of $P(t,x;D_t,D_x)$ satisfying $P_{n,loc}(t,x;D_t,\xi) = P(t,x;D_t,\xi)$ for $(x,\xi) \in \mathrm{supp}[\beta_n(x)] \times \mathrm{supp}[\alpha_n(\xi)]$, and close to $P(t,n^{-q};D_t,n)$. Precise definition will be given later.

Now we make the change of scale

$$n^p t = s \quad \text{with} \quad p > 0. \tag{4.10}$$

Then the limit symbol of $P_{m,n,loc}$ (when r_0 tends to 0), divided by n^{-mp}, becomes

$$D_s^m + \sum_j \Big(\sum_{(\mu,\nu)} a_{j,\mu\nu}\, s^\mu n^{j-q\nu-p(j+\mu)} \Big) D_s^{m-j}\,. \tag{4.11}$$

Let us notice $0 < q < 1$. We are concerned with the power of n: $j - q\nu - p(j + \mu)$, appearing here. So, we take p by (4.8). Then (4.9) could be interpreted by saying that: In order that (4.2) be C^∞-wellposed, any lower order term can not be stronger (in some sense) than (4.11), when we regard it as a differential operator (in s) containing large parameter n. Of course in this case, we replace the lower order operators by the same rule:

$$a_{j,k}(t,x)D_x^k D_t^{m-j} \longrightarrow \sum_{(\mu,\nu)} a_{j,k,\mu\nu} s^\mu n^{k-q\nu-p(j+\mu)} D_s^{m-j} .$$

Another expression of the necessary condition (4.9).

The condition (4.8) can be written $j - q\nu \le p(j+\mu)$. Evidently we have $0 < p < 1$. So that we have

$$(1-p)j \le p\mu + q\nu, \tag{4.12}$$

for principal part. Now (4.9) can be expressed as

$$(1-p)j \le p\mu + q\nu + (j-k). \tag{4.13}$$

Now putting

$$p' = \frac{p}{1-p}, \quad q' = \frac{q}{1-p}, \tag{4.14}$$

(4.12) and (4.13) become (recalling $1/(1-p) = 1+p'$)

$$p'\mu + q'\nu \ge j \quad \text{for principal part,}$$
$$p'\mu + q'\nu \ge j - (1+p')(j-k) \quad \text{for any lower order terms.}$$

In the terminology of Section 3, if we consider the vanishing order of a_j and $a_{j,k}$ with weights p', q', these relations can be expressed by

$$\text{v.o.}[a_j(t,x)] \ge j \quad \text{for } 1 \le j \le m, \tag{4.8)'$$
$$\text{v.o.}[a_{j,k}(t,x)] \ge j - (1+p')(j-k). \tag{4.9)'$$

These are conditions of Ivrii-Petkov (3.2), (3.3).

What about the condition (3.4)? Let us recall that our condition was

$$0 < p < 1, \quad 0 < q < 1. \tag{4.15}$$

Conversely let p' and q' be given positive numbers. Then, by (4.14) we define $p = p'/(1+p')$, $q = q'/(1+p')$. We see that the condition

$0 < q < 1$ is equivalent to

$$1 + p' > q'.$$

This condition is (3.5), and since $q < 1$ is almost essential condition for our argument of micro-localization, we consider this as an essential condition.

5. Fine micro-localization.

In the work of Ivrii-Petkov, they developed the idea of changing scales of the independent variables by introducing a large parameter ρ: $\rho^{s_i} x_i = y_i$, and they concentrated their effort to the construction of several phase functions. We propose our method, which will be almost equivalent to theirs, but it looks like more intuitive than theirs as we showed it in section 4. We use only energy inequalities. Let us explain it briefly.

We are mainly concerned with the consideration of differential operators whose coefficients are degenerate at a fixed point, say $\hat{x} = 0$. First let

$$P = a(x)D'^{\alpha'} D_\ell^{\alpha_\ell}. \tag{5.1}$$

$$D' \longrightarrow \xi' = (\xi_1, \cdots, \xi_{\ell-1}), \quad D_\ell \longrightarrow \xi_\ell. \tag{5.2}$$

Let

$$a(x) \sim \sum_\beta a_\beta x^\beta = \sum_\beta a_{\beta',\beta_\ell} x'^{\beta'} x_\ell^{\beta_\ell}. \tag{5.3}$$

We are concerned with some micro-localization around $\hat{x} = 0$, and in the direction $(0,0,\cdots,1)$ with some weights.

Micro-localizer in ξ in the direction $(0,0,\cdots,1)$.

We introduce a large parameter n. We define the cut-off function. Let $\hat{\xi} = (\xi'^\circ, 1)$, $\xi'^\circ \in \mathbb{R}^{\ell-1}$, $|\xi'^\circ| = 1$. Then

$$\alpha_n(\xi) = \alpha_n(\xi',\xi_\ell) = \alpha_n(\xi')\alpha_n(\xi_\ell) = \alpha_1(\xi'/_n p')\alpha_2(\xi_\ell/n),$$

where

1) $0 < p' < 1$.

2) $\alpha_1(\xi')$ has its support in $|\xi' - \xi'^\circ| \le r_0$, and $= 1$ for $|\xi' - \xi'^\circ| \le r_0/2$.

3) $\alpha_2(\xi_\ell)$ has its support in $|\xi_\ell - 1| \le r_0$, and $= 1$ for $|\xi_\ell - 1| \le r_0/2$.

Recall that

i) $\alpha_n(\xi)$ has its support in

$$\{\xi = (\xi', \xi_\ell);\ |\xi' - n^{p'}\xi'^\circ| \le n^{p'}r_0,\ |\xi_\ell - n| \le nr_0\}.$$

ii) $|D^\mu \alpha_n(\xi)| \le C_\mu n^{-p'|\mu'|-\mu_\ell}$ where $\mu = (\mu', \mu_\ell)$.

<u>Micro-localizer in x around the origin.</u>

To make reflect the degeneracy of the coefficients on the analysis, we introduce another cut-off function $\beta_n(x)$. Let $x^\circ = (x'^\circ, x_\ell^\circ)$, where $|x'^\circ| = |x^\circ| = 1$. Put

$$\beta_n(x) = \beta_n(x')\beta_n(x_\ell) = \beta_1(n^{q'}x')\beta_2(n^q x_\ell),$$

where

i) $\beta_1(x')$ has its support in $\{x';\ |x' - x'^\circ| \le r_0\}$, and $= 1$ for $|x' - x'^\circ| \le r_0/2$.

ii) $\beta_2(x_\ell)$ has its support in $\{x_\ell;\ |x_\ell - x_\ell^\circ| \le r_0\}$, and $= 1$ for $|x - x_\ell^\circ| \le r_0/2$.

We see that

$$|D^\nu \beta_n(x)| \le C_\nu n^{q'|\nu'|+q\nu_\ell}, \quad \text{where } \nu = (\nu', \nu_\ell).$$

Finally we define the micro-localizer by cut-off function

$$\alpha_n(D)\beta_n(x).$$

In view of the above properties of $\alpha_n(\xi)$ and $\beta_n(x)$, we could say that

$$\alpha_n(\xi)\beta_n(x) \in S^0_{(p',1),(q',q)}.$$

We call r_0 the size of micro-localizer. To make this operator really micro-localizer, we make the following essential restriction:

$$0 < q < 1, \quad 0 < q' < p' < 1. \tag{5.4}$$

Micro-localized symbols.

First we remark that $\beta_n(x)$ has its support in $|x' - x'^\circ| \le n^{-q'}r_0$, $|x_\ell - x_\ell^\circ| \le n^{-q}r_0$. Let $x' \longrightarrow \tilde{x}_1(x') \in C^\infty$, keeping invariant (namely $x_1(x') = x'$) when $|x' - x'^\circ| \le r_0$, and $\tilde{x}_1(x') = x'^\circ$ when $|x' - x'^\circ| \ge 2r_0$. In the same way we define $x_\ell \longrightarrow \tilde{x}_2(x_\ell)$. For $a(x) = a(x',x_\ell)$, we define

$$a_{n,loc}(x) = a(n^{-q'}x_1(n^{q'}x'), n^{-q}x_2(n^q x)) = a(x_n(x)). \tag{5.5}$$

Since $\tilde{x}_n(x) = x$, on the support of $\beta_n(x)$, we have

$$\beta_n(x)a(x) = \beta_n(x)a_{n,loc}(x). \tag{5.6}$$

Next we remark that $\alpha_n(\xi)$ has its support in $|\xi' - n^{p'}\xi'^\circ| \le n^{p'}r_0$, and $|\xi_\ell - n| \le nr_0$. Let $\xi' \longrightarrow \tilde{\xi}_1(\xi') \in C^\infty$, keeping invariant when $|\xi' - \xi'^\circ| \le r_0$ and $\tilde{\xi}_1(\xi') = \xi'^\circ$ when $|\xi' - \xi'^\circ| \ge 2r_0$. In the same way we define $\xi_2 \longrightarrow \tilde{\xi}_2(\xi_\ell)$.

For $b(\xi) = b(\xi', \xi_\ell)$, we define

$$b_{n,loc}(\xi) = b_{n,loc}(\xi',\xi_\ell) = b(n^{p'}\tilde{\xi}_1(\xi'/n^{p'}), n\tilde{\xi}_2(\xi_\ell/n)) = b(\tilde{\xi}_n(\xi)). \tag{5.7}$$

Since $\tilde{\xi}_n(\xi) = \xi$, on the support of $\alpha_n(\xi)$, we see that

$$\alpha_n(\xi)b(\xi) = \alpha_n(\xi)b_{n,loc}(\xi). \tag{5.8}$$

Finally for $P(x,\xi)$, we define the micro-localized symbol by

$$P_{n,loc}(x,\xi) = P(\tilde{x}_n(x), \tilde{\xi}_n(\xi)). \tag{5.9}$$

We could say, by abuse of language, that if $P \in S^m_{1,0}$, then

$$P_{n,loc}(x,\xi) \in S^m_{(p',1),(q',q)}. \tag{5.10}$$

The following proposition is important in applications.

Proposition 5.1. Let

$$\alpha_n(D)\beta_n(x)P(x,D)u = \alpha_n(D)\beta_n(x)P_{n,loc}(x,D)u + r_n(x,D)u. \tag{5.11}$$

Then $\| r_n(x,D) \|_{\mathcal{L}(L^2,L^2)}$ is rapidly decreasing. Namely it is estimated by any negative power of n. It is the same as it when we replace α_n, β_n by $\alpha_n^{(\mu)}$, $\beta_{n(\nu)}$.

It is useful to consider the limit symbol of $P_{n,loc}$ when $r_0 \to 0$. This symbol is nothing but multiplication operator obtained by freezing the symbol of P at the centers of micro-localization. More precisely, rewriting (5.3),

$$a(x) \sim \sum_\beta a_\beta x'^{\beta'} x_\ell^{\beta_\ell},$$

and replacing

$$(D',D,x',x_\ell) \text{ by } (n^{p'}\xi'^\circ, n\xi_\ell^\circ\ (=0), n^{-q'}x'^\circ, n^{-q}x_\ell^\circ),$$

in (5.1) and (5.2), we get

$$P_{n,loc} \sim (\sum_\beta a_\beta n^{-q'|\beta'|-q\beta_\ell}(x^\circ)^\beta) n^{p'|\alpha'|+\alpha_\ell}(\xi^\circ)^\alpha. \tag{5.12}$$

<u>Change of scale in t.</u>

Now we consider the evolution equation

$$P = D_t^m + \sum_j a_j(t,x;D_x) D_t^{m-j}.$$

When we are concerned with the Cauchy problem with $t = 0$ as initial plane, we perform the localization in t by changing the scale:

$$n^p t = s \quad (p > 0). \quad (\text{Hence } D_t = n^p D_s). \tag{5.13}$$

Of course we consider Taylor expansion of the coefficients with respect to (t,x) around the origin. We show simplest two cases.

1) $P = -D_t^2 + at^m D_x^2 + bt^k D_x$.

$$n^{-2p} P_{n,loc} \sim -D_s^2 + as^m n^{2-p(2+m)} + bs^k n^{1-p(2+k)}.$$

2) $P = -D_t^2 + a_1 x^m t^k D_x^2 + b_1 x^j D_x$.

$$n^{-2p} P_{n,loc} \sim -D_s^2 + a_1 s^k n^{2-p(2+k)-qm} + b_1 n^{1-2p-qj}.$$

Finally, we should say that a fine micro-localization has been studied by Y. Takei in detail and applied to some problems in hypoellipticity [24].

6. Proof of Fundamental Lemma.

We explain only some essential points of our proof. We prove this lemma by contradiction. Namely we suppose that (4.9) is violated, and that the Cauchy problem (4.2) is C^∞-wellposed, then from these two assumption we derive two (micro-local) energy inequalities which are not compatible.

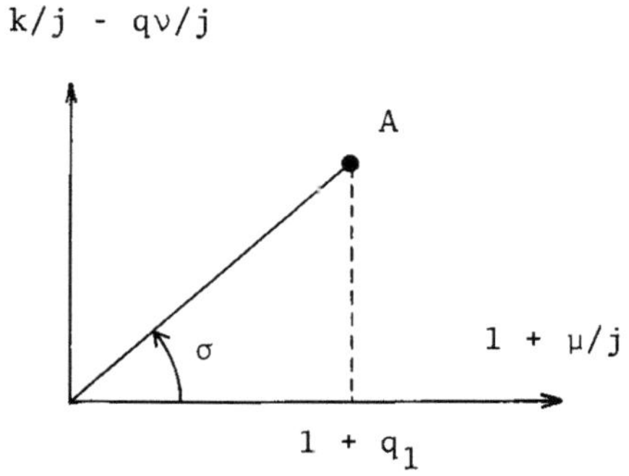

The first assumption means that the maximum slope of $\overrightarrow{OP}_{j,k,\mu\nu}$ is greater than p. Now, if necessary, by changing slightly q to $\tilde{q}$, we may assume that q is irrational. Let A be the point in the Newton diagram whose $\overrightarrow{OA}$ is the maximum slope, denoted by σ, and whose X-coordinates ($= 1 + \mu/j$) is maximum, denoted by $1 + q_1$ (See Fig.1). Now let us consider all the terms (namely $a_{j,k,\mu\nu}$) corresponding to the point A. In this case, let

$$\frac{k - q\nu}{j + \mu} = \sigma \quad (> 0), \tag{6.1}$$

Notice that

$$\mu = q_1 j . \tag{6.2}$$

Since q is irrational, in (6.1), there exists, if it exists, only one k for each j. Moreover, if necessary, by changing slightly q, we assume that there exists θ $(0 < \theta < 1)$ such that

$$k = \theta j . \tag{6.3}$$

Now, <u>the principal part of P consists only of the terms corresponding to</u> A:

$$P_A = D_t^m + \sum_j{}^A a_{j,k,\mu\nu} t^\mu x^\nu D_t^{m-j} .$$

To see more clearly the situation, changing the scale

$$n^{\sigma} t = s,$$

and considering the micro-localization in (x,ξ) explained in Section 4, we express it

$$n^{-\sigma m} P_A = D_s^m + \sum_j{}_A\, s^{q_1 j} (a_{j,k,\mu\nu} (n^q x)^{\nu} (n^{-1} D_x)^k) D_t^{m-j} \tag{6.4}$$

$$\equiv D_s^m + \sum_j{}_A\, s^{q_1 j}\, h_j(x,D)\, D_s^{m-j} .$$

Recalling $n^q x \sim 1$, and $n^{-1} D_x \sim 1$ by micro-localization, we have

$$n^{-\sigma m} P_A \sim D_s^m + \sum_j{}_A\, s^{q_1 j}\, h_j D_s^{m-j} , \tag{6.5}$$

where $h_j = a_{j,k,\mu\nu}$.

Now we call

$$\tau^m + \sum_j h_j \tau^{m-j} = 0 \tag{6.6}$$

the <u>sub-characteristic</u> <u>equation</u> <u>of</u> P.

Taking account of (6.3), possibly except for the case $\theta = 1/2$, we see that there exists at least one root, say τ_1, satisfying

$$\operatorname{Im} \tau_1 < 0. \tag{6.7}$$

In the case $\theta = 1/2$, we can assume it by changing $x \longrightarrow -x'$ (hence $\xi \longrightarrow -\xi'$).

We go back to $Pu = 0$. Putting

$$s^{q_1(m-1-j)} D_s^j u = u_j' \quad (0 \le j \le m - 1). \tag{6.8}$$

we have the following equivalent system for large s:

$$D_s U'(s,x) = s^{q_1} H_n(s,x;D) U'(s,x) + R_n(s,x;D) U' + \frac{1}{s} CU', \tag{6.9}$$

where $H_n(s,x;D)$ comes from P_A. More precisely

i) $$H_n(s,x;\xi) \sim \begin{pmatrix} 1 & & & & \\ & 1 & & & \\ & & \ddots & & \\ & & & & 1 \\ -h_m & -h_{m-1} & \cdots & & -h_1 \end{pmatrix}$$

ii) $R_n(s,x;\xi) = R_n^1(s,x;\xi) + R_n^2(s,x;\xi)$, where $R_{n,loc}^1(s,x;\xi)$ becomes small as we wish if s becomes large, and $R_{n,loc}^2(s,x;\xi)$ tends uniformly to 0 in $[s_0, s_n]$ when n tends to ∞, and $s_n = n^{\theta'}$, θ' being small positive constant.

iii) C is a constant diagonal matrix.

Then we follow the argument explained in Mizohata[17]. Let us notice that Proposition 4.1 plays an important role in our argument, and that we assign the initial data at $s = s_0$, and by using the solutions of the micro-localized equation

$$D_s\tilde{U}(s,x) = s^{q_1}H_{n,loc}(s,x;D)\tilde{U} + R_{n,loc}(s,x;D)\tilde{U} + \frac{1}{s}C\tilde{U}, \tag{6.10}$$

we define the initial data of u at $s = 0$.

7. General case.

The above argument can be carried out in the general case without significant modification. We use the same notations as in Section 3.

Our condition becomes: For any lower order terms arising from $a_{j,k}^{(\alpha')}(t,x;\, e_\ell)$,

$$\frac{(k - |\alpha'|) - q\nu}{(j - |\alpha'|) + \mu_0 + |\mu'|} \le \max_{P_m} \frac{(j - |\alpha'|) - q\nu}{(j - |\alpha'|) + \mu_0 + |\mu'|} \tag{7.1}$$

Then denoting the right-hand side by $p\ (>0)$, we obtain the relation for the principal part:

$$p(\mu_0 + |\mu'|) + q\nu \ge (1 - p)(j - |\alpha'|),$$

and for lower order terms

$$p(\mu_0 + |\mu'|) + q\nu \ge (1 - p)(j - |\alpha'|) + (j - k).$$

These are equivalent to (3.2), (3.3).

For the proof, we consider another ratio, by taking p', q', each close to p satisfying $(0 < q' < p' < 1)$,

$$\frac{(k - |\alpha'|) + p'|\alpha'| - q'|\mu'| - q\nu}{j + \mu_0},$$

and we choose q and q' in the form $q' = \gamma q$, q being irrational and γ is rational.

Finally we remark that the following fact (Corollary 3 in Ivrii-Petkov) follows immediately from the above. This corollary claims that, if P is regularly hyperbolic, then the multiplicity must be at most double. In fact, we consider the operator $P = P_m + D_\ell^{m-1}$. Then (7.1) claims that, since the left-hand side is $(m - 1)/m$, by taking $q = (m - 1)/m$, there exists at least one term in the principal part satisfying

$$j - |\alpha'| - \frac{m-1}{m}\nu \geq \frac{m-1}{m}(j - |\alpha'| + \mu_0 + |\mu'|),$$

namely

$$(m - 1)(\mu_0 + |\mu'| + \nu) \leq j - |\alpha'|.$$

On the other hand, Lax-Mizohata theorem claims that

$$\mu_0 + |\mu'| + \nu \geq j - |\alpha'|,$$

which shows that unless $m = 1$ or 2, the above inequality does not hold.

References

[1] J. Chazarain, Opérateurs hyperboliques à caractéristiques de multiplicité constante, Ann. Inst. Fourier, 24(1974), 173-202.

[2] J.-C. De Paris, Problème de Cauchy analytique à données singulières pour un opérateur différentiel bien décomposable, J. Math. pures et appl., 51(1972), 465-488.

[3] H. Flashka-G. Strang, The correctness of the Cauchy problem, Advances in Math., 6(1971), 347-379.

[4] K.O. Friedrichs, Symmetric hyperbolic system of linear differential equations, Comm. Pure Appl. Math., 7(1954), 345-392.

[5] L. Gårding, Cauchy's problem for hyperbolic equations, Lecture Notes Univ. Chicago, 1957.

[6] L. Hörmander, The Cauchy problem for differential equations with

double characteristics, J. d'Analyse Math., 32(1977), 118-196.

[7] V.Ya. Ivrii-V.M. Petkov, Necessary conditions for the Cauchy problem for non strictly hyperbolic equations to be well posed, Uspehi Mat. Nauk, 29(1974), 3-70.(Russian Math. Surveys, 29(1974),1-70).

[8] V.Ya. Ivrii, Conditions for correctness in Gevrey classes of the Cauchy problem for weekly hyperbolic equations, Siberian J. of Math., 17(1976), 422-435. (English Translation).

[9] H. Komatsu, Linear hyperbolic equations with Gevrey coefficients, J. Math. pure et appl., 59(1980), 145-185.

[10] A. Lax, On Cauchy's problem for partial differential equations with multiple characteristics, Comm. Pure Appl. Math., 9(1956),135-169.

[11] P.D. Lax, Asymptotic solutions of oscillatory initial value problems, Duke Math. J., 24(1957), 627-646.

[12] J. Leray, Hyperbolic differential equations, Institute for adv. study, Princeton, 1953, (Notes mimeo graphiees).

[13] J. Leray-Y. Ohya, Systèmes linéaires, hyperboliques non stricts, Colloque de Liège, 1964, C.B.R.M., 105-144.

[14] T. Mandai, Generalized Levi conditions for weakly hyperbolic equations — An attempt to treat the degeneracy with respect to the space variables, Publ. RIMS. Kyoto Univ., 22(1986), 1-23.

[15] S. Mizohata, Some remarks on the Cauchy problems, J. Math. Kyoto Univ., 1(1961), 109-127.

[16] S. Mizohata, Sur l'indice de Gevrey, Séminaire sur les équations aux dérivées partielles hyperboliques et holomorphes (J. Vaillant) Hermann(1984), 106-120.

[17] S. Mizohata, On the Levi condition, Séminaire sur les équations aux dérivées partielles hyperboliques et holomorphes (J. Vaillant) to appear.

[18] S. Mizohata, Propagation de régularité au sens de Gevrey pour les opérateurs différentiels à multiplicité constante, Séminaire sur les équations aux dérivées partielles hyperboliques et holomorphes (J. Vaillant), Hermann(1984), 106-133.

[19] S. Mizohata-Y. Ohya, Sur la condition de E.E. Levi concernant des équations hyperboliques, Publ.RIMS. Kyoto Univ.,4(1968/69),511-526.

[20] S. Mizohata-Y. Ohya, Sur la condition d'hyperbolicité pour les équations à caractéristiques multiples II, Jap.J.Math., 40(1971),63-104.

[21] Y. Ohya, Le problème de Cauchy pour les équations hyperboliques à caractéristique multiple, J. Math. Soc. Japan, 16(1964), 268-286.

[22] O.A. Oleinik, On the Cauchy problem for weakly hyperbolic equations Comm. Pure Appl. Math., 23(1970), 569-586.

[23] I.G. Petrowsky, Über das Cauchysche Problem für Systeme von partiellen Differentialgleichungen, Mat. Sb., Vol. 2(1937), 815-870.

[24] Y. Takei, Mizohata's micro-localization, Master Thesis, Kyoto Univ., 1986.

[25] J. Vaillant, Données de Cauchy portées par une caractéristique double, Dans le cas d'un système linéaire d'équations aux dérivées partielles, Rôle des bicaractéristiques, J. Math. pures et appl., 47(1968), 1-40.

[26] M. Yamaguti, Le problème de Cauchy et les opérateurs d'intégrale singulière, Mem. Coll. Sci. Univ. Kyoto, Ser. A, 32(1959), 121-151.

Positivity and Stability for Cauchy Problems with Delay

by

Wolfgang Kerscher and Rainer Nagel

Many partial differential equations with delay can be written as

$$\text{(ACPd)} \qquad \dot{u}(t) = Bu(t) + \Phi u_t \,, \quad u_o = f \,,$$

for a function $u(\cdot)$ with values in a Banach space X and for linear operators B and Φ. In this paper we use semigroup theory on the space $E := C([-r_o,0],X)$ in order to solve (ACPd) and to discuss qualitative properties such as positivity and stability of the solutions.

In section 1 we fix the appropriate assumptions on B and Φ in order to obtain the solutions of (ACPd) through a strongly continuous semigroup $\mathcal{T}_{B,\Phi}$ in E. Some regularity properties of this semigroup are proved in section 2. Since positivity of $\mathcal{T}_{B,\Phi}$ will play an essential role in our approach we characterize it in theorem 3.4 through properties of B and Φ.

With these tools we are well prepared to determine the stability of the solutions of (ACPd) through spectral properties: It is well known that the spectrum of the generator $A_{B,\Phi}$ of $\mathcal{T}_{B,\Phi}$ can be described by certain operators on X. (For example, this is done by the "characteristic equation" for systems of ordinary differential equations with delay.) In our setting positivity facilitates this task in three aspects:

1. In order to locate the spectral bound
   $$s(A_{B,\Phi}) = \sup \{ \operatorname{Re}\lambda : \lambda \in \sigma(A_{B,\Phi}) \}$$
   it suffices to look at a <u>single</u> operator $B + \Phi_o$ on X only.
2. It suffices to look at the <u>real</u> spectral values of $B + \Phi_o$ only.
3. Positivity permits to conclude stability of $\mathcal{T}_{B,\Phi}$ from the inequality $s(A_{B,\Phi}) < 0$.

All this is discussed in section 4 and theorem 4.1 gives a simple stability criterion for positive semigroups $T_{B,\Phi}$. A particular interesting feature is the fact that positivity also implies that the stability remains independent of the size of the delay (see e.g., theorem 4.3). Various applications are made to non-positive semigroups (e.g., cor. 4.4) and non-linear equations (section 5).

Many authors have studied problems as (ACPd) using semigroup theory (e.g., [9],[13],[18],[24],[34]). Similarly, the independence of stability of the delay has been treated in many papers (e.g., [7], [17],[19],[20],[26],[27]). But the aspect of positivity and its surprising consequences seems to have been realized first in [22] (see also [23], section 4), which contains most of the results presented here.

1. Well posedness for linear Cauchy problems with delay

It is well established that semigroup theory may be used in order to solve linear Cauchy problems. More precisely, the (abstract) Cauchy problem

(ACP) $\quad \dot{u}(t) = Bu(t), \quad u(0) = x ,$

for a closed linear operator B with dense domain $D(B)$ in some Banach space X is well-posed if and only if B generates a strongly continuous semigroup $(S(t))_{t\geq 0}$ of bounded linear operators on X (see [11], thm. II. 1.2).

In many concrete situations the derivative $\dot{u}(t)$ does not only depend on the momentary value $u(t)$ (via B) but also on the past values $u(t+s)$, $s \leq 0$ (via some functional Φ). Therefore we consider Cauchy problems with delay of the following form:

(ACPd) $\quad \dot{u}(t) = Bu(t) + \Phi u_t , \quad u_0 = f ,$

where $u_t(s) := u(t+s)$, s in some maximal delay interval $[-r_0,0]$ and Φ a continuous linear operator from $E := C([-r_0,0],X)$ into X (see section 5 for concrete examples).

Again it is well known that (ACPd) can be transformed into a Cauchy problem of type (ACP) in the larger Banach space E. In fact, on E we consider the differential operator

$$A: f \to f'$$

defined for all continuously differentiable functions f on $[-r_o,0]$ with values in X. Its restriction to the domain

$$D(A_{B,\Phi}) := \{ f \in C^1([-r_o,0],X) : f(0) \in D(B),\ f'(0) = Bf(0) + \Phi(f) \}$$

will be denoted by $A_{B,\Phi}$.

The following theorem shows under which assumptions on B and Φ the Cauchy problem defined by $A_{B,\Phi}$ in E is well-posed and in which sense it yields the solutions of the original Cauchy problem with delay (ACPd).

<u>Theorem 1.1</u> (well-posedness): Assume that B generates a strongly continuous semigroup $(S(t))_{t\geq 0}$ on X and let Φ be a bounded linear operator from $E := C([-r_o,0],X)$ into X. Then the following assertions hold:

(i) The operator $A_{B,\Phi}$ generates a strongly continuous semigroup $\mathcal{T}_{B,\Phi} = (T_{B,\Phi}(t))_{t\geq 0}$ on E.

(ii) This semigroup satisfies the "translation property", i.e.

$$T_{B,\Phi}(t)f(s) = \begin{cases} f(t+s) & \text{if } t+s \leq 0, \\ T_{B,\Phi}(t+s)f(0) & \text{if } t+s \geq 0. \end{cases}$$

(iii) The solutions of (ACPd) are obtained as

$$u(t) := \begin{cases} f(t) & \text{if } t \leq 0, \\ T_{B,\Phi}(t)f(0) & \text{if } t \geq 0, \end{cases}$$

for $f \in D(A_{B,\Phi})$.

(iv) The semigroup $\mathcal{T}_{B,\Phi}$ satisfies

$$T_{B,\Phi}(t)f(0) = S(t)f(0) + \int_0^t S(t-s)\Phi(T_{B,\Phi}(s)f)\,ds$$

for every $f \in E$.

The proof can be found in ([29], B-IV, thm. 3.1 and cor. 3.2).

We proceed by studying the qualitative behavior of the above solution semigroup. To that purpose we need information on the spectrum and the resolvent of $A_{B,\Phi}$. Therefore we introduce certain auxiliary operators.

Definition 1.2: For $\lambda \in \mathbb{C}$, $\varepsilon_\lambda(s) := e^{\lambda s}$ and $\varepsilon_\lambda \otimes x := (s \to e^{\lambda s} \cdot x)$ we define the following operators:

(i) $H_\lambda \in L(E)$ by $H_\lambda g(t) := \int_t^0 e^{\lambda(t-s)} g(s)ds$, $t \in [-r_o, 0]$,

(ii) $\Phi_\lambda \in L(X)$ by $\Phi_\lambda(x) := \Phi(\varepsilon_\lambda \otimes x)$, $x \in X$.

Since each Φ_λ is bounded on X the operators $B + \Phi_\lambda$ generate semigroups on X. The spectrum of $B + \Phi_\lambda$ can be used to characterize the spectrum of $A_{B,\Phi}$ (see [29], B-IV, prop. 3.4).

Theorem 1.3 (spectral characterization): Under the above assumptions on B and Φ the following holds:

(i) $\lambda \in \sigma(A_{B,\Phi})$ if and only if $\lambda \in \sigma(B+\Phi_\lambda)$.

(ii) If $\lambda \in \rho(A_{B,\Phi})$ then the resolvent is given as

$$R(\lambda, A_{B,\Phi})g = \varepsilon_\lambda \otimes [R(\lambda, B+\Phi_\lambda)(g(0)+\Phi H_\lambda g)] + H_\lambda g \quad \text{for} \quad g \in E.$$

2. Properties of the solution semigroup

We now always suppose that B generates a semigroup $(S(t))_{t\geq 0}$ on X and take $\Phi \in L(E,X)$. The semigroup on E generated by $A_{B,\Phi}$ will be denoted by $\mathcal{T}_{B,\Phi}$. In order to study the qualitative properties of $\mathcal{T}_{B,\Phi}$, and hence of the solutions of (ACPd), we recall that this semigroup can be obtained by a perturbation - in the sense of Greiner, see [15] - of the semigroup $\mathcal{T}_{B,0}$ defined as

$$T_{B,0}(t)f(s) := \begin{cases} f(t+s) & \text{if } t+s \leq 0, \\ S(t+s)f(0) & \text{if } t+s \geq 0, \end{cases}$$

and having generator $A_{B,0}f = f'$ with domain

$$D(A_{B,0}) = \{ f \in C^1([-r_o,0],X) : f(0) \in D(B), f'(0) = Bf(0) \}.$$

Therefore we may expect that the properties of $(S(t))_{t\geq 0}$ are somehow inherited by $(T_{B,\Phi}(t))_{t\geq 0}$. In fact, the subsequent results bear a striking analogy to the "stable under bounded perturbations" theorems of Phillips (see [31]).

<u>Proposition 2.1</u>: Assume that $(S(t))_{t\geq 0}$ is norm continuous for $t>0$. Then $(T_{B,\Phi}(t))_{t\geq 0}$ is norm continuous for $t > r_o$.

<u>Proof</u>: Define $\psi(t) \in L(E,X)$ by

$$\psi(t)f := \int_0^t S(t-s)\Phi(T_{B,\Phi}(s)f)\,ds \ , \ f \in E .$$

From

$$\|\psi(t+h) - \psi(t)\| \leq c_1 \cdot \int_0^t \|S(t+h-s) - S(t-s)\|ds + c_2 \cdot \int_t^{t+h} \|S(t+h-s)\|ds$$

and the norm continuity of $(S(t))_{t\geq 0}$ we conclude that $t \to \psi(t)$ is norm continuous on $(0,\infty)$. Take now $t > r_o$, $0 < h < t$ and $f \in E$. Then

$$\|T_{B,\Phi}(t+h)f - T_{B,\Phi}(t)f\| = \sup_{s\in[-r_o,0]} \|T_{B,\Phi}(t+h)f(s) - T_{B,\Phi}(t)f(s)\|$$

$$= \sup_{s\in[-r_o,0]} \|T_{B,\Phi}(t+h-s)f(0) - T_{B,\Phi}(t+s)f(0)\| \qquad \text{by thm. 1.1(ii)},$$

$$= \sup_{s\in[-r_o,0]} \|S(t+h+s)f(0) + \psi(t+h+s)f - S(t+s)f(0) - \psi(t+s)f\|$$

by thm. 1.1(iv),

$$\leq \sup_{s\in[-r_o,0]} \|S(t+h+s) - S(t+s)\|\,\|f\| + \sup_{s\in[-r_o,0]} \|\psi(t+h+s) - \psi(t+s)\|\,\|f\|$$

which converges to zero if $h \to 0$. □

Without proof we state the analogous result for differentiable and compact semigroups (see [22]).

<u>Proposition 2.2</u>: Assume that $(S(t))_{t\geq 0}$ is analytic (resp., compact for $t > 0$). Then $(T_{B,\Phi}(t))_{t\geq 0}$ is eventually differentiable (resp., eventually compact).

<u>Remark</u>. If $\dim X < \infty$ then (ACPd) corresponds to a finite system of ordinary differential equations with delay. Then the assumptions of proposition 2.2 are always fulfilled and it follows that the solution semigroup on E is eventually differentiable and eventually compact. All of the above properties imply that the <u>spectral bound</u>

$$s(A_{B,\Phi}) := \sup \{ \operatorname{Re}\lambda : \lambda \in \sigma(A_{B,\Phi}) \}$$

coincides with the <u>growth bound</u>

$\omega := \inf \{ w \in \mathbb{R}: \|T(t)\| \le Me^{wt}$ for some M amd all $t \ge 0 \}$
and therefore $s(A_{B,\Phi}) < 0$ implies uniform exponential stability of the solutions of (ACPd). We refer to [29], A-IV, sect. 1 (in particular p.106) for a complete discussion of this phenomenon and recall that the same also holds for all positive semigroups on certain Banach lattices (see [29], C-IV, thm. 1.1). Therefore it is of interest to investigate the order properties of the semigroup $(T_{B,\Phi}(t))_{t\ge 0}$.

3. Positivity and domination

The Cauchy problem with delay (ACPd) has positive solutions $u(t)$ for all $t \ge 0$ in some ordered Banach space X for each positive initial value f if and only if the semigroup $\mathcal{T}_{B,\Phi}$ consists of positive operators. Since this occurs frequently in concrete models we try to give a general characterization.

To that purpose we assume that X is a Banach lattice (e.g., $\mathbb{R}^n$, $C(K)$, $L^p(\mu)$). Then $E := C([-r_o,0],X)$ is also a Banach lattice with the natural pointwise order and the sup-norm. As for some basic concepts from order theory we refer to [32] or [29], B-I and C-I and recall only the following.

Definition 3.1: For operators $T_1, T_2 \in \mathcal{L}(E)$, resp., for semigroups $(T_1(t))_{t\ge 0}$, $(T_2(t))_{t\ge 0}$ on E we say that T_1 dominates T_2 , resp., $(T_1(t))_{t\ge 0}$ dominates $(T_2(t))_{t\ge 0}$ if

$$|T_2 f| \le T_1 |f|, \quad \text{resp.,} \quad |T_2(t) f| \le T_1(t)|f|$$

for every $f \in E$, $t \ge 0$.

Clearly, the dominating operator (semigroup) is always positive. It has been shown in [29], C-II, thm. 4.1 that $(T_1(t))_{t\ge 0}$ dominates $(T_2(t))_{t\ge 0}$ if and only if the resolvents $R(\lambda,A_1)$ dominate $R(\lambda,A_2)$ for large λ . From this and the explicit representation of the resolvent given in theorem 1.3(ii) one obtains immediately the following result.

Proposition 3.2: For $i = 1,2$ assume that B_i generates a strongly continuous semigroup $\mathcal{S}_i$ on X and that $\Phi_i \in \mathcal{L}(E,X)$. If $\mathcal{S}_1$ dominates $\mathcal{S}_2$ and Φ_1 dominates Φ_2 then the semigroup $\mathcal{T}_{B_1,\Phi_1}$ generated by A_{B_1,Φ_1} dominates the semigroup $\mathcal{T}_{B_2,\Phi_2}$ generated by A_{B_2,Φ_2}.

In particular, if B generates a positive semigroup and Φ is positive then $T_{B,\Phi}$ is a positive semigroup.

Clearly, the generator $A_{B,\Phi}$ does not determine B and Φ in a unique way. In order to obtain necessary and sufficient conditions for domination we need an additional concept.

Definition 3.3: The operator $\Phi \in L(E,X)$ has no mass in zero if for every $\varepsilon > 0$ there exists $\delta > 0$ such that

$$\|\Phi f\| \leq \varepsilon\|f\|$$

for every $f \in E$ having support in $[-\delta,0]$.

Theorem 3.4: Let X be a Banach lattice and take $E = C([-r_o,0],X)$. Assume that B_1 , B_2 generate semigroups S_1 , S_2 on X and that Φ_1 , $\Phi_2 \in L(E,X)$ have no mass in zero. Then the following assertions (a) and (b) are equivalent:

(a) The semigroup T_{B_1,Φ_1} dominates T_{B_2,Φ_2} on E .

(b) (i) The semigroup S_1 dominates S_2 .
  (ii) The operator Φ_1 dominates Φ_2 .

Proof: It only remains to show that (a) implies (b)(i) and (b)(ii): For every $x \in X$ and some large λ and μ we define

$$y_\mu := [R(\lambda,A_{B_1,\Phi_1})|\varepsilon_\mu \otimes x| - |R(\lambda,A_{B_2,\Phi_2})(\varepsilon_\mu \otimes x)|](0)$$

which is in X_+ by hypothesis.
From theorem 1.3(ii) and since $\lim_{\mu\to\infty} H_\lambda(\varepsilon_\mu \otimes x) = 0$ we conclude that

$$y := \lim_{\mu\to\infty} y_\mu = R(\lambda,B_1+\Phi_{1,\lambda})|x| - |R(\lambda,B_2+\Phi_{2,\lambda})x| \in X_+ ,$$

i.e., $R(\lambda,B_1+\Phi_{1,\lambda})$ dominates $R(\lambda,B_2+\Phi_{2,\lambda})$.
Since Φ_1 , Φ_2 have no mass in zero we have $\lim_{\lambda\to\infty}\Phi_{i,\lambda}x = 0$ for $i = 1,2$ and every $x \in X$. Therefore $\|\lambda R(\lambda,B_i+\Phi_{i,\lambda})\|$ is bounded for large λ and one obtains through the usual power series argument that $R(\mu,B_1+\Phi_{1,\lambda})$ dominates $R(\mu,B_2+\Phi_{2,\lambda})$ for $\lambda_o < \mu \leq \lambda$. Since $R(\mu,B_i) = \lim_{\lambda\to\infty} R(\mu,B_i+\Phi_{i,\lambda})$ the assertion (b)(i) is proved.
In order to show that $\Phi_1|f| \geq |\Phi_2 f|$ for all $f \in E$ we observe first that it suffices to prove this inequality for all $g \in E$ satisfying $g(0) = 0$ (Φ_1 and Φ_2 have no mass in zero).
Then we collect the following properties:

- $\lim_{\lambda\to\infty}\lambda H_\lambda g = g$ for every $g \in E$, $g(0) = 0$.

(*) - $\|\lambda R(\lambda,B+\Phi_\lambda)\|$ is bounded for large λ ,

- $\lim_{\lambda\to\infty}\lambda R(\lambda,B+\Phi_\lambda)f = f$ for every $f \in E$.

From theorem 1.3(ii) it follows that for $g \in E$ with $g(0) = 0$

$$R(\lambda,A_{B_1,\Phi_1})|g|(0) = R(\lambda,B_1+\Phi_{1,\lambda})(\Phi_1 H_\lambda|g| + H_\lambda|g|)$$
$$= (1/\lambda^2)[\lambda R(\lambda,B_1+\Phi_{1,\lambda})(\Phi_1\lambda H_\lambda|g|)] + H_\lambda|g| ,$$ and
$$|R(\lambda,A_{B_2,\Phi_2})g(0)| = (1/\lambda^2)|[\lambda R(\lambda,B_2+\Phi_{2,\lambda})(\Phi_2\lambda H_\lambda g)] + H_\lambda g| .$$

By assumption, $R(\lambda,A_{B_1,\Phi_1})|g|(0) - |R(\lambda,A_{B_2,\Phi_2})g(0)| \geq 0$, which implies $\Phi_1|g| - |\Phi_2 g| \geq 0$ by the convergence properties (*). □

Corollary 3.5: Let $X = \mathbb{R}^n$ and take $B = (b_{ij})_{n\times n}$ and $\Phi = (\eta_{ij})_{n\times n}$ where each $\eta_{ij} \in M[-r_o,0]$ has no mass in zero. Then the semigroup $T_{B,\Phi}$ is dominated by the semigroup $T_{\tilde{B},|\Phi|}$, where $\tilde{B} = (\tilde{b}_{ij})_{n\times n}$ with $\tilde{b}_{ii} = b_{ii}$ and $\tilde{b}_{ij} = |b_{ij}|$ for $i \neq j$ and $|\Phi| = (|\eta_{ij}|)_{n\times n}$.

Proof: Clearly, $|\Phi|$ dominates Φ . Moreover it follows from [8] that $(\exp(\tilde{B}t))_{t\geq 0}$ is the smallest positive semigroup dominating $(\exp(Bt))_{t\geq 0}$. □

Remark: In [2] it is shown that $T_{\tilde{B},|\Phi|}$ is the smallest strongly continuous semigroup, called the "modulus", dominating $T_{B,\Phi}$.

4. Spectrum and Stability

It is well known that in many cases the stability (i.e., $\lim_{t\to\infty}T(t) = 0$) of a strongly continuous semigroup $(T(t))_{t\geq 0}$ is determined by the location of the spectrum $\sigma(A)$ of its generator A. In particular,

$s(A) < 0$ for the spectral bound $s(A) := \sup\{\mathrm{Re}\lambda : \lambda\in\sigma(A)\}$

is a necessary and often sufficient condition in order to have (uniform) exponential stability (see [29], A-IV.1). Therefore considerable effort was spent to calculate or estimate $\sigma(A_{B,\Phi})$ or $s(A_{B,\Phi})$ for the generators $A_{B,\Phi}$ of the solution semigroup of (ACPd). The characterization given in theorem 1.3(i)

(4.1) $\lambda \in \sigma(A_{B,\Phi})$ if and only if $\lambda \in \sigma(B+\Phi_\lambda)$

transfers the problem from the space $E = C([-r_o,0],X)$ into the smaller space X , but even for finite dimensional X the verification of the right hand side of (4.1) can be quite difficult:

For $X = \mathbb{R}$, $B = b \in \mathbb{R}$ and $\Phi = \eta \in M[-r_o,0]$ we obtain from (4.1) that the spectral bound of $A_{b,\eta}$ is

$$s(A_{b,\eta}) = \sup\{ \operatorname{Re}\lambda : \lambda \in \mathbb{C} \text{ such that } \xi(\lambda) = 0 \} ,$$
where $\xi(\lambda) = -\lambda + b + \int_{-r_0}^{0} e^{\lambda s} d\eta(s)$.

Even for simple measures η this supremum is not easy to calculate.

The situation changes drastically when we consider positive semigroups (see also [23]). First, we recall from [29] that for the generator A of a positive semigroup $(T(t))_{t\geq 0}$ the estimate $s(A) < 0$ always implies exponential stability (l.c., C-IV, thm. 1.3) and often implies uniform exponential stability (l.c., C-IV, thm 1.1 and cor. 1.2). Even more useful is the fact that for positive semigroups the spectral bound is always contained in the spectrum, i.e., $s(A) \in \sigma(A)$ (l.c., C-III, cor. 1.4). Therefore, in order to determine $s(A)$ it suffices to look at the real spectral values of A only.

For instance, if the measure η in the above example is positive it suffices to look at $\lambda \to \xi(\lambda)$ as a real function. Clearly, this function is continuous, strictly decreasing and has a unique zero $\lambda_0 \in \mathbb{R}$. Since $A_{b,\eta}$ generates a positive semigroup by prop. 3.2 we obtain $s(A_{b,\eta}) = \lambda_0$.

Moreover, these simple properties of ξ imply the following stability criterion:

(4.2) $\quad \xi(0) = b + \|\eta\| < 0 \quad$ if and only if $\quad \lambda_0 = s(A_{b,\eta}) < 0$.

Hence already in the case $X = \mathbb{R}$ we obtain a useful stability criterion using simple results from the theory of positive semigroups. The same arguments can also be applied to the general situation and we quote the main result from [29], B-IV, thm. 3.7 or [23], section 4.

<u>Theorem 4.1</u>: Let X be a Banach lattice. Assume that B generates a positive semigroup and that Φ is a positive operator from $E = C([-r_0,0],X)$ into X . For the generator $A_{B,\Phi}$ of the positive semigroup $\mathcal{T}_{B,\Phi}$ on E and for $\lambda \in \mathbb{R}$ the following holds:

(i) If $s(B+\Phi_\lambda) < \lambda$ then $s(A_{B,\Phi}) < \lambda$.

(ii) If $s(B+\Phi_\lambda) = \lambda$ then $s(A_{B,\Phi}) = \lambda$.

(iii) If B has compact resolvent and $\sigma(B+\Phi_\mu) \neq \emptyset$ for some $\mu \in \mathbb{R}$ then

(4.3) $\quad s(B+\Phi_\lambda) \lesseqgtr \lambda \quad$ if and only if $\quad s(A_{B,\Phi}) \lesseqgtr \lambda$.

In particular,

(4.4) $\quad s(B+\Phi_0) < 0 \quad$ if and only if $\quad s(A_{B,\Phi}) < 0$.

As explained above the equivalence (4.4) gives a simple stability criterion for the solutions of (ACPd): it suffices to estimate the real spectral values of the single operator $B + \Phi_o$ on X.

But there is a second aspect of this criteron: if the delay Φ varies in such a way that $\Phi_o(x) = \Phi(1 \otimes x)$, $x \in X$, remains unchanged then we always arrive the same necessary and sufficient stability criterion.

For example, if we take B as in (iii) and $\Phi_t f := f(t)$, $t \in [-r_o, 0]$, then $s(B+Id) < 0$ if and only if $s(A_{B,\Phi_t}) < 0$ independently of the size of t.

This "stability independence of the delay" is typical for positive semigroups $T_{B,\Phi}$ and will now be treated in a general context.

Definition 4.2: A continuous map $r: [-1,0] \to \mathbb{R}_-$ satisfying $\min_{-1 \le s \le 0} r(s) = -r_o$ is called a delay function on $[-r_o, 0]$. If Φ is a bounded linear operator from $C([-1,0],X)$ into X we obtain the delayed operator $\Phi_r \in L(C([-r_o,0],X),X)$ as $\Phi_r f := \Phi(f \circ r)$, $f \in C([-r_o,0],X)$.

As before we denote by Φ_o the operator $\Phi_o(x) := \Phi(1 \otimes x)$ for $x \in X$.

Theorem 4.3: Assume that B generates a positive semigroup on a Banach lattice X and take a positive operator Φ from $C([-1,0],X)$ into X. If $s(B+\Phi_o) < 0$ then $s(A_{B,\Phi_r}) < 0$ for every delay function r.

Proof: For $x \in X$ we have

$$(\Phi_r)_o x = \Phi_r(1_{[-r_o,0]} \otimes x) = \Phi(1_{[-1,0]} \otimes x) = \Phi_o(x),$$

i.e., $(\Phi_r)_o = \Phi_o$ is independent of the delay function r. Therefore we obtain the assertion from theorem 4.1(i). □

By [29], C-IV, thm. 1.3 the above positivity assumptions on B and Φ yield that "$s(B+\Phi_o) < 0$" implies exponential stability of the solutions of (ACPd). Obviously, this criterion is independent of the delay function r.

Moreover, since uniform exponential stability (see [29], A-IV, def. 1.1) is inherited from a dominating semigroup we can extend this result to non positive semigroups as soon as spectral bound and growth bound coincide for the dominating semigroup.

Corollary 4.4: Let B be the generator of a semigroup S on the Banach lattice X and take $\Phi \in L(C([-1,0],X),X)$. Assume that there exists a semigroup $\tilde{S}$ with generator $\tilde{B}$ which dominates S and an operator $\tilde{\Phi}$ dominating Φ . Then the semigroup T_{B,Φ_r} is uniformly exponentially stable for all delay functions r if the spectral bound $s(\tilde{B}+\tilde{\Phi}_o) < 0$ and one of the following conditions is satisfied:

(i) $X = C(K)$, K compact.

(ii) $\tilde{S}$ is norm continuous for $t > 0$.

Proof: In case (i) $T_{\tilde{B},\tilde{\Phi}_r}$ is a positive semigroup on $C([-r_o,0]\times K) \cong C([-r,0],X)$, hence $s(A_{\tilde{B},\tilde{\Phi}_r}) = \omega$ by [29], B-IV, thm. 1.1. In case (ii) $T_{\tilde{B},\tilde{\Phi}_r}$ is eventually norm continuous (prop. 2.1), hence spectral and growth bound coincide by [29], A-IV, (1.7). In both cases the assertion follows from theorem 4.3. □

In case $X = \mathbb{R}^n$ an even weaker condition guarantees stability independence of the delay function.

Corollary 4.5: Let $X = \mathbb{R}^n$ and take $B = (b_{ij})_{n\times n}$ and $\Phi = (\eta_{ij})_{n\times n}$ with $\eta_{ij} \in M[-1,0]$. The semigroup T_{B,Φ_r} on $C([-r_o,0],\mathbb{R}^n)$ is uniformly exponentially stable for all delay functions r if the following conditions are satisfied:

(i) $s(B+\Phi_o) < 0$, and

(ii) $s(\tilde{B}+|\Phi|_o) \le 0$, where $\tilde{B}$, $|\Phi|$ are as in corollary 3.5 and $|\Phi|_o = (\|\eta_{ij}\|)_{n\times n}$.

Proof: The semigroup T_{B,Φ_r} is eventually norm continuous, hence $s(A_{B,\Phi_r}) < 0$ implies uniform exponential stability. If $s(\tilde{B}+|\Phi|_o) < 0$ the result follows from corollary 4.4. Therefore assume $s(\tilde{B}+|\Phi|_o) = 0$. Again from the eventual norm continuity of T_{B,Φ_r} it follows that $\sigma(A_{B,\Phi_r})$ is bounded in right semiplanes ([29], A-II, thm.1.20), hence in order to obtain $s(A_{B,\Phi_r}) < 0$ we show $\sigma(A_{B,\Phi_r}) \cap i\mathbb{R} = \emptyset$.
By (4.1) this is equivalent to $i\alpha \in \rho(B+\Phi_{r,i\alpha})$ for every $\alpha \in \mathbb{R}$. Decompose now $\tilde{B} = B_1 + |B_2|$ where $B_1 = \mathrm{diag}(b_{ii})_{n\times n}$ and $B_2 = B-B_1$ and consider $B_\# := B_1 + \|B_1\|\cdot\mathrm{Id} + |B_2| + |\Phi|_o$. Then $B_\#$ is a positive matrix with spectral radius $r(B_\#) = \|B_1\|$ since $s(\tilde{B}+|\Phi|_o) = 0$. Therefore

$$r(B_1 + \|B_1\|\cdot\mathrm{Id} + B_2 + \Phi_{r,i\alpha}) \le \|B_1\| , \qquad \text{and}$$
$$\sigma(B + \Phi_{r,i\alpha}) \subset \{ \lambda \in \mathbb{C}: |-\|B_1\| - \lambda| \le \|B_1\| \} .$$

For $0 \ne \alpha \in \mathbb{R}$ we obtain $i\alpha \in \rho(B+\Phi_{r,i\alpha})$ while $0 \in \rho(B+\Phi_o)$ by assumption (i). □

Remarks: 1. The matrix $\tilde{B}$ generates the smallest semigroup dominating $(\exp(Bt))_{t\geq 0}$. Using [8] the same result holds for regular operators B on an order complete Banach lattice X.
2. In the situation of corollary 4.5 we have to consider the matrices

$$C = (c_{ij})_{n\times n} \quad \text{with} \quad c_{ij} := \begin{cases} |b_{ij}| + \|\eta_{ij}\| & \text{if } i \neq j, \\ b_{ij} + \|\eta_{ij}\| & \text{if } i = j, \end{cases}$$

$$D = (d_{ij})_{n\times n} \quad \text{with} \quad d_{ij} := b_{ij} + \int_{-1}^{0} d\eta_{ij}(s) .$$

If $s(C) < 0$ or if $s(C) = 0$ but $s(D) < 0$ we obtain stability independence of the delay. Hence it suffices to estimate the eigenvalues of two matrices only. This should be compared with the results in [5].

Simple examples on $X = \mathbb{R}^n$ with $\Phi = 0$ show that the above conditions are not necessary. However, there have been many attempts (e.g., [5],[7],[19],[20],[26],[27]) to find necessary and sufficient conditions for "stability independence of the delay". Most of them are quite complicated. We first give a necessary condition due to [5] and then use it for $X = \mathbb{R}$ in order to extend the characterization of [27] to more general delays.

Proposition 4.6: Let B be the generator on the Banach space X and take $\Phi \in L(C([-1,0],X),X)$. If $s(A_{B,\Phi_r}) < 0$ for every delay function r then

(i) $\sigma(B+\Phi_0) \subset \{ \lambda \in \mathbb{C}: \mathrm{Re}\lambda < 0 \}$, and

(ii) $\sigma(B+\Phi_\gamma) \cap i\mathbb{R} \subset \{0\}$ for every

$\gamma \in C[-1,0]$, $|\gamma| = 1$ and $\Phi_\gamma(x) := \Phi(\gamma \otimes x)$, $x \in X$.

Proof: For every $\lambda \in \mathbb{C}$, $\mathrm{Re}\lambda \geq 0$ and every delay function r we have $\lambda \in \rho(A_{B,\Phi_r})$, hence $\lambda \in \rho(B+\Phi_{r,\lambda})$ by (4.1). Taking $r \equiv 0$ we obtain $\lambda \in \rho(B+\Phi_0)$, i.e., property (i).

In order to show (ii) we take $\alpha \in \mathbb{R}\setminus\{0\}$. Again

$$i\alpha \in \rho(A_{B,\Phi_r}) \quad \text{and therefore} \quad i\alpha \in \rho(B+\Phi_{r,i\alpha}) .$$

But $\Phi_{r,i\alpha}(x) = \Phi(e^{i\alpha r(\cdot)} \otimes x) = \Phi(\gamma \otimes x) = \Phi_\gamma(x)$ for

$$\gamma(s) := e^{i\alpha r(s)}, \quad -1 \leq s \leq 0 .$$

Since every continuous unimodular function γ can be obtained as $\gamma(\cdot) = e^{i\alpha r(\cdot)}$ for some delay function r we proved the assertion.

□

Corollary 4.7: Let $X = \mathbb{R}$ and $b \in \mathbb{R}$, $\eta \in M[-1,0]$ such that $\eta \geq 0$ or $\eta \leq 0$. The semigroup T_{b,η_r} on $C[-r_o,0])$ is uniformly exponentially stable for all delay functions r if and only if the following conditions hold:

(i) $b + \int_{-1}^{0} d\eta(s) < 0$, and

(ii) $b + \|\eta\| \leq 0$.

Proof: If η is positive then $\int_{-1}^{0} d\eta(s) = \|\eta\|$ and the assertion follows from our general result (see (4.4)). If η is negative then the sufficiency follows from corollary 4.5. Condition (i) is necessary by proposition 4.6(i). We finally assume that (i) is true while (ii) is false. Then

$$0 < b + \|\eta\| = b - \int_{-1}^{0} d\eta(s) .$$

Define the function

$$\xi(\alpha) := b + \int_{-1}^{0} e^{i\pi\alpha} d\eta(s) \qquad \text{for } \alpha \in [0,1] .$$

Then $\xi(0) < 0$ and $\xi(1) > 0$ and there exists $\alpha_o \in (0,1)$ such that $\mathrm{Re}\,\xi(\alpha_o) = 0$ and $\beta_o := \mathrm{Im}\,\xi(\alpha_o) \neq 0$.
For the constant function $\gamma(s) := e^{i\pi\alpha_o}$, $s \in [-1,0]$, we obtain

$$i\beta_o = b + \int_{-1}^{0} \gamma(s)\, d\eta(s)$$

which contradicts prop. 4.6(ii). □

5. Applications

In this final section we show how the results of the previous sections can be applied in order to obtain stability for the solutions of linear and nonlinear Cauchy problems involving a delay. Clearly, we only present some typical examples.

Example 5.1: (The Goodwin oscillator).
The system

$$\begin{aligned} \dot{x}_1(t) &= -b_1 x_1(t) + k_1 [1 + x_n^a(t)]^{-1} \\ \dot{x}_i(t) &= -b_i x_i(t) + k_i x_{i-1}(t) \qquad \text{for } i = 2,\dots,n, \end{aligned} \tag{5.1}$$

and with $b_i, k_i \in \mathbb{R}_+$, $a \in \mathbb{N}$ has been introduced by Goodwin [12] as a model for a control by repressive genes of biochemical reaction chains in living cells. Later Landahl [25] and MacDonald [28] introduced certain delays in (5.1) and studied the stability of the system.
Our results allow to treat the following equations:

$$(5.2)\qquad \begin{aligned} \dot{x}_1(t) &= -b_1x_1(t) + k_1[1 + (\textstyle\int_{-1}^{0} x_n(t+r(s))\,d\eta_1(s))^a\,]^{-1} \\ \dot{x}_i(t) &= -b_ix_i(t) + k_i\textstyle\int_{-1}^{0} x_{i-1}(t+r(s))\,d\eta_i(s) \quad \text{for } i=2,\dots,n, \end{aligned}$$

with positive probability measures $\eta_i \in M[-1,0]$ and a continuous delay function $r\colon [-1,0] \to \mathbb{R}_-$. It is known that there always exists a positive stationary solution $\bar{x} = (\bar{x}_1,\dots,\bar{x}_n) \in \mathbb{R}^n$ satisfying

$$(5.3)\qquad \begin{aligned} (b_1\cdot\ldots\cdot b_n)(k_1\cdot\ldots\cdot k_n)^{-1}\cdot\bar{x}_n &= (1 + (\bar{x}_n)^a)^{-1} \\ \bar{x}_{i-1} &= (b_i/k_i)\bar{x}_i \quad \text{for } i=2,\dots,n\,. \end{aligned}$$

<u>Proposition</u>: The stationary solution of (5.2) is exponentially asymptotically stable for all delay functions r if

$$(5.4)\qquad (b_1\cdot\ldots\cdot b_n)(k_1\cdot\ldots\cdot k_n)^{-1} > (a-1)^{1/a}(a-1)/a\,.$$

In particular, this is always true in case $a = 1$.

<u>Proof</u>: Linearization of (5.2) in $\bar{x}$ yields the linear system

$$(5.5)\qquad \dot{x}(t) = Bx(t) + \Phi_r(t),$$

where $B = \operatorname{diag}(-b_i)$ and Φ_r is obtained (see def. 4.2) from the operator Φ mapping $f = (f_1,\dots,f_n) \in C([-1,0],\mathbb{R}^n)$ into

$$\Phi f := \left(-k_1\beta\textstyle\int_{-1}^{0} f_n(s)\,d\eta_1(s)\,,\ k_2\int_{-1}^{0} f_1(s)\,d\eta_2(s)\,,\dots,\ k_n\int_{-1}^{0} f_{n-1}(s)\,d\eta_n(s)\right)$$

with $\beta := a(\bar{x}_n)^{a-1}(1 + (\bar{x}_n)^a)^{-2}$.

Clearly, B generates a positive semigroup on $\mathbb{R}^n$. Moreover, the absolute value $|\Phi|$ of Φ applied to the functions $1\otimes x$ with $x = (\xi_1,\dots,\xi_n) \in \mathbb{R}^n$ yields $|\Phi|(1\otimes x) = (\beta k_1\cdot\xi_n, k_2\cdot\xi_1,\dots,k_n\cdot\xi_{n-1})$, hence

$$|\Phi|_o = \begin{pmatrix} 0 & \cdot & \cdot & \cdot & \beta k_1 \\ k_2 & \cdot & \cdot & \cdot & \cdot \\ \cdot & k_3 & \cdot & \cdot & \cdot \\ \cdot & \cdot & \cdot & \cdot & 0 \\ 0 & \cdot & \cdot & k_n & 0 \end{pmatrix}.$$

By cor. 4.4 we have to estimate the eigenvalues of $B + |\Phi|_o$ which are the zeros of the polynomial

$$p(\lambda) := (\lambda+b_1)(\lambda+b_2)\cdot\ldots\cdot(\lambda+b_n) - \beta k_1\cdot\ldots\cdot k_n\,.$$

Therefore the spectral bound $s(B+|\Phi|_o)$ is smaller than zero if

$$(5.6)\qquad (b_1\cdot\ldots\cdot b_n)(k_1\cdot\ldots\cdot k_n)^{-1} > \beta.$$

In order to show this inequality we use (5.3) and the definition of β which yields the equivalent inequality

$$(5.7)\qquad a(\bar{x}_n)^a < 1 + (\bar{x}_n)^a .$$

For $a = 1$ this always holds. If $a \geq 2$ we transform it into

$$(5.8)\qquad \bar{x}_n < (a-1)^{-1/a} .$$

Write $c := (b_1\cdot\ldots\cdot b_n)(k_1\cdot\ldots\cdot k_n)^{-1}$ and define y as in (5.3) by

$$(5.9)\qquad cy = (1 + y^a)^{-1} .$$

Then y is strictly increasing in c and $c_o := (a-1)^{1/a}(a-1)/a$ solves (5.9) with $y_o := (a-1)^{-1/a}$. Therefore if $c > c_o$ (i.e., (5.4)) then $y < y_o$ and (5.8) holds. □

<u>Example 5.2</u> (A semilinear population equation):
The following widely used equation (see [6], [14], [33]) describes the growth of a spatially distributed population with delay in the birth process:

$$(5.10)\qquad \dot{u}(x,t) = d\cdot\Delta u(x,t) + a\cdot u(x,t)[1-b\cdot u(x,t)-\int_{-1}^{0} u(x,t+r(s))\,d\eta(s)].$$

Here we take $x \in [0,1]$, $\Delta = \frac{d^2}{dx^2}$ with Dirichlet boundary conditions, $a,b,d \in \mathbb{R}_+$, $0 \leq \eta \in M[-1,0]$ such that $b + \|\eta\| = 1$ and r a delay function on $[-1,0]$. If $a > d$ then there exists a stationary solution $h \in C^2[0,\pi]$ of (5.10) which is strictly positive on $(0,\pi)$ (see [33]), i.e., $d\cdot h'' + a(1-h)\cdot h = 0$.
In order to apply our methods we consider

$$X = \{ f \in C[0,\pi]: f(0) = f(\pi) = 0 \} \quad \text{and}$$

$$B_o := \frac{d^2}{dx^2} \text{ with maximal domain in } X .$$

Then we define

$$B := d\cdot B_o + M_{a(1-h-bh)} ,$$

where M_g denotes the bounded multiplication operator with the function $g \in X$. Next we take $\Phi \in L(C([-1,0],X),X)$ defined by

$$\Phi f := -M_{ah}\cdot\int_{-1}^{0} f(s)\,d\eta(s).$$

Then the linearization of (5.10) at the stationary solution h can be written as

(5.11) $\dot{u}(t) = Bu(t) + \Phi_r(u_t)$ for $u(t) \in X$,

or

(5.12) $\dot{v}(t) = A_{B,\Phi_r}\, v(t)$ for $v(t) \in C([-r_o,0],X)$.

Since X is a Banach lattice and B generates a positive semigroup on X our theory applies and we obtain the following stability criterion.

<u>Proposition</u>: If $a > d$ and $b > \|\eta\|$ then the solution semigroup corresponding to (5.12) is uniformly exponentially stable independently of the delay function r .

<u>Proof</u>: It is well known that the operator B_o generates a positive irreducible semigroup on X which is compact for $t > 0$. The same properties hold for the semigroup generated by $B_o + M_f$, $f \in X$ real. Since the function h is strictly positive it follows from $d\cdot h'' + a(1-h)\cdot h = 0$ and [29], B-III, thm. 3.6(e) that the spectral bound of $B_1 := d\cdot B_o + M_{a(1-h)}$ is zero. On the other hand

$$B_2 := B + |\Phi|_o = d\cdot B_o + M_{a(1-h)} - M_{a(b-\|\eta\|)h} \leq B_1 ,$$

hence $s(B_2) \leq s(B_1) = 0$.

Assume that $s(B_2) = 0$. Then there exists a strictly positive fixed function $g \in X$ for the semigroup generated by B_2 which satisfies $S_1(t)g \geq g$, $t \geq 0$, for the semigroup $(S_1(t))_{t\geq 0}$ generated by B_1 . But $(S_1(t))_{t\geq 0}$ possesses a strictly positive invariant linear form ([29], B-II, sect. 3), hence $S_1(t)g = g$ and $B_1 g = 0 = B_2 g$ which is impossible since $M_{a(b-\|\eta\|)h} \neq 0$.

Hence we conclude that $s(B_2) < 0$ and the assertion follows from corollary 4.4. □

<u>Remark</u>: It seems to be unknown in which sense a "principle of linearized stability" (see [1], thm. 15.6) holds in this situation.

Example 5.3 (A model from epidemology):
In the equations proposed by Capasso-Maddalena [4] describing the spread of an infectious disease over a region Ω we introduce a delay term due to the incubation time. Thus we obtain the following:

$$\begin{aligned} \dot{u}_1(x,t) &= d_1\cdot\Delta u_1(x,t) - a_{11}\cdot u_1(x,t) + a_{12}\cdot u_2(x,t), \\ \dot{u}_2(x,t) &= d_2\cdot\Delta u_2(x,t) - a_{22}\cdot u_2(x,t) + \int_{-1}^{0} g[u_1(x,t+r(s))]d\eta(s), \end{aligned} \tag{5.13}$$

where g is an increasing differentiable function, $0 \leq \eta \in M[-1,0]$ and r a delay function. Choosing an appropriate domain for the Laplacian Δ we arrive at the following situation:
Let X be a Banach lattice (e.g., $X = C_o(\Omega)\times C_o(\Omega)$ in (5.13)) and B the generator of a positive semigroup on X, which is uniformly continuous for $t > 0$. Assume that the map $G: X \to X$ is Fréchet differentiable with <u>positive</u> derivative $G'[x]$ for every $x \in X_+$. Consider the semilinear equation

$$\dot{u}(t) = Bu(t) + G[u(t)] \tag{5.14}$$

and assume that $\bar{u} \in X_+$ is a stationary solution of (5.14). Then $\bar{u}$ is also a stationary solution for the delayed equation

$$\dot{u}(t) = Bu(t) + \int_{-1}^{0} G[u(t+r(s))]d\eta(s) \tag{5.15}$$

for every probability measure $\eta \in M[-1,0]$ and every delay function r on $[-1,0]$. Now, the linearization of (5.15) in $\bar{u}$ is

$$\dot{u}(t) = Bu(t) + \int_{-1}^{0} G'[\bar{u}]u(t+r(s))d\eta(s) . \tag{5.16}$$

This is a Cauchy problem of type (ACPd) and using the terminology of definition 4.2 we have $(\Phi_r)_o = G'[\bar{u}]$ which by assumption is a positive operator. Therefore the results of section 4 can be applied immediately.

<u>Proposition</u>: If the semigroup generated by $B + G'[\bar{u}]$ is uniformly exponentially stable (i.e., if $s(B+G'[\bar{u}]) < 0$), then the solution semigroup of (5.16) is uniformly exponentially stable independently of the delay function r.

<u>Proof</u>: The semigroup generated by $B + G'[\bar{u}]$ is eventually norm continuous ([29], A-II, thm. 1.30), hence uniformly exponentially stable if and only if $s(B+G'[\bar{u}]) < 0$ ([29], A-IV, (1.7)). Therefore and

since the solution semigroup is norm continuous by prop. 2.1 the assertion follows from thm. 4.3.

Example 5.4 (The linear Boltzmann equation):
The linear Boltzmann equation

$$(5.17)\quad \dot{u}(x,v,t) = -v\cdot \mathrm{grad}_x u(x,v,t) - \sigma(x,v)\cdot u(x,v,t) + \int_V k(x,v,v')u(x,v',t)\,dv'$$

describes the flow of particles with position $x \in \Omega$ and speed $v \in V$ and subject to absorption and scattering. Many authors (e.g., [16], [21], [35]) have studied this equation using positive semigroups on the Banach lattice $L^1(\Omega \times V)$. Physically however (see [3], p. 463), it seems to be more realistic to introduce a delay term into the scattering term, e.g.,

$$(5.18)\quad \dot{u}(x,v,t) = \ldots + \int_{-1}^{0}\int_V k(x,v,v')\xi(s)u(x,v',t+s)\,dv'ds\ ,$$

where $0 \le \xi \in L^1[-1,0]$.
Clearly this yields a Cauchy problem of the form (ACPd) and all assumptions of theorem 4.3 are satisfied. Therefore we obtain our final result using [29], C-IV, thm. 1.1(a).

Proposition: Assume that the generator corresponding to (5.17) has spectral bound less than zero. Then for every $0 \le \xi \in L^1[-1,0]$ such that $\int_{-1}^{0}\xi(s)ds \le 1$ the spectral bound of the generator corresponding to (5.18) is less than zero and the solutions are exponentially stable.

References:

[1] Amann, H.: Gewöhnliche Differentialgleichungen, Berlin-New York, de Gruyter 1983.

[2] Becker, I., Greiner, G.: On the Modulus of One-parameter Semigroups. Semigroup Forum (to appear).

[3] Bell, G.I., Glasstone, S.: Nuclear Reactor Theory, p. 463. New York, Van Nostrand 1970.

[4] Capasso, V., Maddalena, L.: Convergence to Equilibrium States for a Reaction-Diffusion System Modelling the Spatial Spread of a Class of Bacterial and Viral Diseases. J. Math. Biol. 13 (1981), 173-184.

[5] Cooke, K.L., Ferreira, J.M.: Stability Conditions for Linear Retarded Functional Differential Equations. J. Math. Anal. Appl. 96 (1983), 480-504.

[6] Cushing, J.M.: Integrodifferential Equations and Delay Models in Population Dynamics. Lecture Notes in Biomathematics 20, Berlin, Springer-Verlag 1977.

[7] Datko, R.: A Procedure for Determination of the Exponential Stability of Certain Differential Difference Equations. Quart. Appl. Math. 36 (1978), 279-292.

[8] Derndinger, R.: Betragshalbgruppen normstetiger Operatorenhalbgruppen. Arch. Math. 42 (1984), 371-375.

[9] Di Blasio, G., Kunisch, K., Sinestrari, E.: Stability for Abstract Linear Functional Differential Equations. Israel J. Math. 50 (1985), 231-263.

[10] Dickerson, J., Gibson, J.: Stability of Linear Functional Differential Equations on Banach Spaces. J. Math. Anal. Appl. 55 (1976), 150-155.

[11] Goldstein, J.A.: Semigroups of Linear Operators and Applications. Oxford University Press 1985.

[12] Goodwin, B.C.: Temporal Organization in Cells. New York, Academic Press 1963.

[13] Grabosch, A., Moustakas, U.: A Semigroup Approach to Retarded Differential Equations in C([-1,0]). Section B-IV in [29].

[14] Green, D., Stech, H.W.: Diffusion and Hereditary Effects in a Class Population Models. In: Differential Equations and Applications, Academic Press 1981.

[15] Greiner, G.: Perturbing the Boundary Condition of a Generator. To appear in Houston J. Math..

[16] Greiner, G.: Spectral Properties and Asymptotic Behaviour of the Linear Transport Equation. Math. Z. 185 (1984), 167-177.

[17] Hadeler, H.P.: Delay Equations in Biology. In: Peitgen, H.O., Walther, H.O. (eds.): Functional Differential Equations and Approximations of Fixed Points. Lecture Notes Math. 730, Springer Verlag 1979.

[18] Hale, J.K.: Functional Differential Equations. New York-Berlin, Springer Verlag 1977

[19] Hale, J.K., Infante, E.F., Tsen, F.P.: Stability in Linear Delay Equations. J. Math. Anal. Appl. 105 (1985), 533-555.

[20] Infante, E.F.: A Note of Stability in Retarded Delay Systems for all Delays. In: Dynamical Systems II, edited by A. Bednarek and L. Cesari, 121-133, New York, Academic Press 1982.

[21] Kaper, H.G., Lekkerkerker, C.G., Hejtmanek, J.: Spectral Methods in Linear Transport Theory. Basel, Birkhäuser Verlag 1982.

[22] Kerscher W.: Majorisierung und Stabilität der Lösungen retardierter Cauchyprobleme. Dissertation, Tübingen 1986.

[23] Kerscher W., Nagel, R.: Asymptotic Behavior of One-parameter Semigroups of Positive Operators. Acta Appl. Math. 2 (1984), 297-309.

[24] Kunisch, K., Schappacher, W.: Necessary Conditions for Partial Differential Equations with Delay to generate C_o-Semigroups. J. Diff. Equations 50 (1983), 49-79.

[25] Landahl, H.D.: Some Conditions of Sustained Oscillations in Biochemical Chains with Feedback Inhibition. Bull. Math. Biophysics 31 (1969), 775-787.

[26] Lenhart, S.M., Travis, C.C.: Stability of Functional Partial Differential Equations. J. Diff. Equations 58 (1985), 212-227.

[27] Lenhart, S.M., Travis, C.C.: Global Stability of a Biological Model with Time Delay. Proc. Amer. Math. Soc. 96 (1986), 75-78.

[28] MacDonald, N.: Time Lag in a Model of a Biochemical Reaction Sequence with End Product Inhibition. J. theoret. Biol. 67 (1977), 549-556.

[29] Nagel, R. (editor): One-parameter Semigroups of Positive Operators. Lecture Notes Math. 1184, Springer-Verlag 1986.

[30] Nagel, R.: What can Positivity do for Stability? In: Functional Analysis, Surveys and Results III, 145-154, North Holland 1984.

[31] Phillips, R.S.: Perturbation Theory for Semigroups of Linear Operators. Trans. Amer. Math. Soc. 74 (1953), 199-221.

[32] Schaefer, H.H.: Banach Lattices and Positive Operators. Berlin, Springer Verlag 1974.

[33] Stech, H.W.: The Effect of Time Lags on the Stability of the Equilibrium State of Population Growth Equations. J. Math. Biology 5 (1978), 115-120.

[34] Travis, C., Webb, G.: Existence and Stability for Partial Functional Differential Equations. Trans Amer. Math. Soc. 200 (1974), 395-418.

[35] Voigt, J.: Positivity in Time Dependent Linear Transport Theory. Acta Appl. Math. 2 (1984), 311-331.

□ □ □

Wolfgang Kerscher
Rainer Nagel

Mathematisches Institut
Auf der Morgenstelle 10

D-7400 Tübingen

ON THE RESONANCES AND THE INVERSE SCATTERING PROBLEM FOR PERTURBED WAVE EQUATIONS*

Gustavo Perla Menzala

ABSTRACT

We consider finite energy solutions of the perturbed wave equation $\Box u + q(x,t)u = 0$ where $x \in \mathbb{R}^3$, $t \in \mathbb{R}$. We analyse two type of problems: First, we give suitable conditions on q and we prove that there exist infinite many "resonances" λ_j associated with q. Secondly, we study the problem of determining q from the scattering operator associated with the above equation. We describe a uniqueness result on the inverse scattering problem and state some open problems on the subject.

§ 1) INTRODUCTION

In this paper we will discuss some properties of finite energy solutions of the perturbed wave equation

$$\Box u + q(x,t)u = 0 \tag{1.1}$$

where $x \in \mathbb{R}^3$, $t \in \mathbb{R}$ and $q = q(x,t)$ is a real-valued function, non-negative, depending on position and time. In (1.1) $\Box$ denotes the d'Alembertian operator, i.e., $\Box = \partial^2/\partial t^2 - \Delta$ where Δ is the Laplacian operator.

We observe that problem (1.1) with q depending on time, it is closely related to the following perturbed problem: Consider the wave equation

$$\Box u = 0 \tag{1.2}$$

where $x \in \Omega(t)$, $t \in \mathbb{R}$. Here $\Omega(t) \subseteq \mathbb{R}^3$ is the exterior of a bounded region which is allow to change its shape and position. Some boundary conditions have to be imposed, usually Dirichlet or Neumann boundary conditions.

There are two type of results which we will describe in this paper. Both have to do with the behavior of the solutions of (1.1) as time evolves.

---

* This is an expanded version on a one-hour invited Lecture presented by the author at the Latin American School of Mathematics (ELAM) held at IMPA (July 1986), Rio de Janeiro, RJ, Brasil.

Problem I. Suppose that $q \geq 0$, q is periodic in time, $q \equiv 0$ for $|x| \geq R$ and all times. Furthermore, suppose that $q \to q(\cdot,t)$ is continuous on $\mathbb{R}$ with values in $L^p(\mathbb{R}^3)$ for some $p > 3$. Is there a sequence $\{\lambda_j\}_{j=1}^{\infty}$ of complex numbers λ_j with the property that if u is a finite energy solution of (1.1) then $u(x,t)$ has the asymptotic expansion

$$u(x,t) \sim \sum_{j=1}^{\infty} p_j(x,t) \exp(i\lambda_j t) \tag{1.3}$$

say, for fixed x and large t? If so, we call those complex numbers λ_j the resonances of q.

Problem II. Suppose that $q \geq 0$, $q \equiv 0$ for $|x| \geq R$ and all time, $q \in C^1$ and has bounded derivatives. Furthermore, let us suppose that $q_t = O(|t|^{-\alpha})$ for some $0 < \alpha \leq 1$, uniformly in x. It is known that under the above conditions the scattering operator S exists (see [8] or [9]). Does S determines q uniquely?

Our interest in problems I and II is both theoretically and for their applicability to a variety of situations. Problem I is known in the engineering literature [1] as a subject included in "target identification" and Problem II occupies a significant role in inverse acoustic and electromagnetism.

Finally we remark that in the special case when q does not depend on time, then there is a close relation between problems I and II because is well known [7] that there exists an expansion like (1.3) and λ_j are the poles of the scattering matrix $S(z)$ obtained from S and extended meromorphically to the whole complex plane.

This paper is a expanded version of an invited lecture given by the author at the Latin American School of Mathematics (ELAM) at IMPA during July 1986. I would like to thank the Organizing Committee for their kind invitation.

In the next two sections we describe some of our recent joint work on Problems I and II with my colleagues and friends W.A. Strauss, J. Cooper and J.A. Ferreira. None of the new results presented in this paper have appeared until now.

§ 2) RESONANCES FOR TIME DEPENDENT POTENTIALS

First, let us describe a known technique to obtain the asymptotic expansion (1.3) when q does not depend on time.

Suppose for simplicity that $q \geq 0$, $q \in C^2$ and $q(x) \leq C \exp(-\alpha|x|^2)$ for some $\alpha > 0$. Let us consider the Cauchy problem

$$\begin{aligned} &\square u + q(x)u = 0, \quad x \in R^3, \quad t \in \mathbb{R} \\ &u(x,0) = f(x) \\ &u_t(x,0) = g(x) \end{aligned} \tag{2.1}$$

where f and g are C^∞ functions with compact support.

Let $v(x,k) = \int_0^\infty u(x,t) \exp(i\,k\,t)dt$. Multiplying equation (2.1) by u_t and then integrating from zero to ∞ in time gives us that v satisfies the elliptic problem

$$\begin{aligned} &k^2 v + \Delta v - q(x)v = F, \quad x \in \mathbb{R}^3 \\ &|x| \left|\frac{\partial v}{\partial |x|} - i\,k\,v\right| \longrightarrow 0 \quad \text{as} \quad |x| \longrightarrow +\infty \end{aligned} \tag{2.2}$$

where $F = -g + i\,k\,f$, (here $i = \sqrt{-1}$). Formally v $(k^2+\Delta-q)^{-1} F = J*F$ where J is an integral operator whose kernel is the Green function $G(x,y,k)$ associated with (2.2). Under the above conditions on q it is well known (See [12], [7]) that G can be extended meromorphically to the complex plane which is analytic in the upper-half-plane and the poles λ_j are such that $|\text{Im}\,\lambda_j| \longrightarrow \infty$ as $j \longrightarrow +\infty$.

It follows that the solution $v(x,k)$ of (2.2) can be extended meromorphically to the complex plane and

$$u(x,t) = (2\pi)^{-1} \int_{-\infty}^{\infty} v(x,k) \exp(-i\,k\,t)dk. \tag{2.3}$$

Moving the contour of integration in (2.3) followed by the use of the theorem of residues we obtain that

$$u(x,t) = \sum_{j=1}^{N} p_j(x,t) \exp(i\,\lambda_j\,t) + O(\exp(-|\text{Im}\,\lambda_N|)).$$

As $t \longrightarrow +\infty$, where $p_j(x,t) \exp(-i\,k_j\,t) = \operatorname*{Res}_{k=k_j} v(x,k) \exp(-i\,k\,t)$.

If the poles are simple then p_j does not depend on t.

Roughly, this is the method to obtain expansion (1.3). The reader can easily see from the above discussion that the method fails in case q depends on time. Some improvement was given recently by the author and T. Schonbek to include the case where q may assume negative values (See [10]). From the mathematical point of view some of the open problems in the subject are: 1) Are the complex poles λ_j simple or not? For problem (1.2) is known that if $\Omega(t) = \Omega$ is the exterior of a ball then the poles are simple. In general the problem remains open. 2) Can we recover (at least part of) q from the complex poles of the Green function G? This problem has to be posed in an appropriate form because for the interior problem it is well known that the poles of the Green function do not determine q uniquely.

Now we want to describe briefly a partial answer to problem I in a simple situation. The result was obtained recently by the author in a joint work with W.A. Strauss and J. Cooper [4]. Consider a potential term $q_\varepsilon(x,t)$ of the form

$$q_\varepsilon(x,t) = q_o(x) + \varepsilon q_1(x,t) \tag{2.4}$$

where $\varepsilon > 0$. We assume the following conditions: $\exists\, R, r > 0$ such that

1) $q_o \in L^\infty(\mathbb{R}^3)$, $q_o \geq 0$, $q_o(x) \geq r > 0$ on some open set

2) $q_o \equiv 0$ for $|x| \geq R$

3) $q_1 \equiv 0$ for $|x| \geq R$ and all t, $q_1 \geq 0$

4) $\exists\, T > 0$, $q_1(x,t+T) = q_1(x,t)$ for all t and all x

5) $q_1 \in C(\mathbb{R}, L^p(\mathbb{R}^3))$ for some $p > 3$.

<u>Definition</u>. Let q_ε given by (2.4) where q_o and q_1 satisfy all the above conditions. A complex number λ is called a <u>resonance</u> for q_ε if there exists a function $u(x,t)$ (of locally finite energy for each t) such that

a) $\Box u + q_\varepsilon(x,t)u = 0, \quad x \in \mathbb{R}^3, \quad t \in \mathbb{R}$ (2.5)

b) $u \exp(-i\lambda t)$ has period T

c) u is outgoing, that is, if at time $t = s$ we remove the perturbation $q_\varepsilon u$, then $u \equiv 0$ in some forward propagation cone (or equivalently, $u \equiv 0$ in an increasing ball).

We observe that when q is time independent, the above definition agrees with the usual notion of outgoing for time-harmonic solutions of $\Box u + q(x)u = 0$.

<u>Theorem 1</u>. Let q_ε given by (2.4) where q_o and q_1 satisfy conditions 1)-5). Then q_ε has a discrete, non-empty set of resonances for ε sufficiently small.

Proof. We consider the space H to be the completion of pairs $(f_1,f_2) \in C_o^\infty(\mathbb{R}^3) \times C_o^\infty(\mathbb{R}^3)$ with respect to the norm

$$\|(f_1,f_2)\|_H^2 = 1/2 \int_{\mathbb{R}^3} (|\Delta f_1|^2 + f_2^2)dx.$$

Let q_ε given by (2.4) and let u be the solution of (2.5) with initial data in $C_o^\infty(\mathbb{R}^3)$. We consider the propagator $U(t,s): H \to H$ which maps $(u(\cdot,s),u_s(\cdot,s)) \to (u(\cdot,t),u_t(\cdot,t))$. It is easy to show that

$$\|U(t,s)\| \leq \exp(k_\varepsilon(t-s)), \quad t \geq s$$

where

$$k_\varepsilon = \text{const} \sup_{0\leq t\leq T} \|q_\varepsilon(\cdot,t)\|_{L^3(\mathbb{R}^3)} . \tag{2.6}$$

Now, let us denote by C the cylinder $C = \{(x,t), |x| < R\}$ and $C_1 = \{(x,t) \in C, \ 0 < t < 1\}$. Let X and Y the following spaces

$$X = \{g:C \to \mathbb{R}, \ g \in L^r(C_1), \ g(x,t+1) = g(x,t) \ \text{for all } (x,t) \in C\}.$$

and

$$Y = \{g \in X, \ \|g\|_Y = \|g\|_{H^1(C_1)} < +\infty\}.$$

The norm in X is given by $\|g\|_X = \|g\|_{L^r(C_1)}$ where $2^{-1} = p^{-1} + r^{-1}$. It is not difficult to see that the embedding $Y \to X$ is compact because $r < 6$. We define for each complex number ξ the operator $L_{\xi,\varepsilon}$ on X by

$$(L_{\xi,\varepsilon}g)(t) = -T \int_{-\infty}^{t} W_o(T(t-\tau))\exp(-i\xi T(t-\tau))q_\varepsilon(\cdot,\tau T)g(\tau)d\tau \tag{2.7}$$

where $W_o(t)h$ (for $h \in L^2(\mathbb{R}^3)$) denotes the first component of U $U_o(t)(0,h)$. Here $\{U_o(t)\}$ denotes the group of unitary operators in H associated with the free wave equation, i.e. (2.1) with $q \equiv 0$. Since $q_\varepsilon \, g$ has support on the ball $\{|x| \leq R\}$ we see that (2.7) reduces to

$$(L_{\xi,\varepsilon} g)(t) = -T \int_{t-2R/T}^{t} W_o(T(t-\tau))\exp(-i\xi T(t-\tau))q_\varepsilon(\cdot,\tau T)g(\tau)d\tau.$$

Let K denote the space of compact operators on X with operator norm. The following can be verified: 1) The map $\xi \to L_{\xi,\varepsilon}$ is entire analytic with values in K. 2) The map $(\xi,\varepsilon) \to L_{\xi,\varepsilon}$ is jointly continuous on $\mathbb{C} \times \mathbb{R}$ with values on K and 3) $\xi \in \mathbb{C}$ is a resonance for q_ε if and only if $\lambda = 1$ belongs to the spectrum of $L_{\xi,\epsilon}$. We use 1) to show that there are no resonances for q_ϵ for $\operatorname{Im} \xi \leq -K_\epsilon$ (here K_ϵ is given by (2.6)), therefore by 2), (3) and Steinberg's result on analytic families of compact operators (See [11]) the proof is complete. Technical details of the above proof is a more general context will appear in a joint paper of the author with J. Cooper and W.A. Strauss [4].

I was told by W.A. Strauss that recently he, M. Wei and G. Majda [13] obtained some numerical results for computing some resonances of (1.1).

§ 3) INVERSE SCATTERING PROBLEMS

Suppose we consider finite energy solutions of (1.1) and give suitable assumptions on $q(x,t)$. Then it is not very difficult to show that the scattering operator S associated with equation (1.1) exists (See [8] or [9]). This means, roughly, that given a finite energy solution u of (1.1) there exist two solutions u_+ and u_- of the free wave equation (i.e. $\Box u_\pm = 0$) such that $u - u_\pm \to 0$ in energy nor as $t \to \pm\infty$.

In fact, under the assumptions on q in Problem II (See § 1) we can show that there exists such scattering operator S. In what follows we will summarize recent results we obtained with J.A. Ferreira on the inverse scattering problem associated with equation (1.1), that is, to obtain information on q from the knowledge of S.

Let $R(x,t) = \delta(|x|-t)/4\pi t$ for $t > 0$ where δ denotes the three-dimensional delta function. Let $q(x,t)$ satisfy all the assumptions in Problem II (See § 1).

Let us define $P_\infty : L^\infty(\mathbb{R}^4) \to L^\infty(\mathbb{R}^4)$ given by

$$P_\infty f(x,t) = \int_{-\infty}^{\infty} \int_{\mathbb{R}^3} R(x-y,t-s)q(y,s)f(y,s)dyds. \tag{3.1}$$

Similarly, we consider the linear operators $P_\pm : L^\infty(\mathbb{R}^4) \to L^\infty(\mathbb{R}^4)$ given by

$$P_- f(x,t) = \int_{-\infty}^{t} \int_{\mathbb{R}^3} R(x-y,t-s)q(y,s)f(y,s)dyds \tag{3.2}$$

and

$$P_+ f(x,t) = \int_{t}^{+\infty} \int_{\mathbb{R}^3} R(x-y,t-s)q(y,s)f(y,s)dyds. \tag{3.3}$$

Then it is not difficult to show that P_∞ and $P_\pm$ are linear bounded operators in $L^\infty(\mathbb{R}^4)$ and $\|P_\infty\| \leq 2\|P_\pm\|$, $\|P_\pm\| \leq (R^3 + \sqrt{3})/3 \; \|q\|_{L^\infty(\mathbb{R}^4)}$

where $\| \ \|$ denotes the operator nor in $L(L^\infty, L^\infty)$.

<u>Theorem 2</u>. Let $q: \mathbb{R}^3 \times \mathbb{R} \to \mathbb{R}$ as above and $\|q\|_{L^\infty} < 3/2(R^3 + \sqrt{3})^{-1}$. Let S be the scattering operator associated with equation (1.1) as describe above. If S is the identity operator then q has to be identically zero.

<u>Proof</u>. Our assumptions on q imply that there exist $(I + P_\pm)^{-1}$. There is a simple way to obtain bounded solutions of (1.1):
Let $v = v(x,t)$ a bounded solution of $\Box v = 0$ then it follows that $u = (I + P_-)^{-1} v$ is a bounded solution of (1.1). Let $\epsilon_j \to 0$, $\epsilon_j > 0$. We choose $v = v^{\epsilon_j}$ such that

$$\Box v^{\epsilon_j} = 0, \quad x \in \mathbb{R}^3, \quad t \in \mathbb{R}$$

$$v^{\epsilon_j}(x,t_o) = 0, \quad \frac{\partial v^{\epsilon_j}}{\partial t}(x,t_o) = \phi_j(x)$$

where (y_o, t_o) is some point in $\mathbb{R}^3 \times \mathbb{R}$ and ϕ_j is given by $\phi_j(x) = 3[4\pi\epsilon_j^3]^{-1}$ if $|x - v_o| \leq \epsilon_j$ and $\phi_j(x) = 0$ otherwise. We can show that if $P_- v^{\epsilon_j} \equiv 0$ for all j then $q(y_o,t_o) = 0$. Also, since S is the identity operator we know that $P_\infty u^{\epsilon_j} = 0$ where $u^{\epsilon_j} = (I + P_-)^{-1} v^{\epsilon_j}$. By iteration we obtain that

$$P_\infty v^{\epsilon_j} = P_\infty (I + P_-)^{-1} P_- v^{\epsilon_j}$$

therefore

$$\|P_\infty v^{\epsilon_j}\|_{L^\infty} \leq 2\|P_-\|(1 - \|P_-\|)^{-1} \|P_- v^{\epsilon_j}\|_{L^\infty} .$$

There are two possibilities: 1) If $\|P_- v^{\epsilon_j}\|_{L^\infty} = 0$ for some subsequence of ϵ_j approaching to zero, then because of the above observations it follows that $q(y_o,t_o) = 0$. Since (y_o,t_o) was arbitrary, then $q \equiv 0$. 2) If there exist $\epsilon_o > 0$ and $N_o > 0$, such that $\|P_- v^{\epsilon_j}\|_{L^\infty} \neq 0$ for all $0 < \epsilon_j < \epsilon_o$ and $j > N_o$.

In this case we obtain the inequality

$$\sup_{\substack{0<\epsilon_j<\epsilon_o \\ j>N_o}} \|P_\infty v^{\epsilon_j}\|_{L^\infty} \|P_- v^{\epsilon_j}\|_{L^\infty}^{-1} \leq 2\|P_-\|(1-\|P_-\|)^{-1} \tag{3.4}$$

However, in this case we can obtain a lower bound for the left hand side of (3.4) (this is a delicate technical lemma):

$$2 \leqq \sup_{\substack{0<\epsilon_j<\epsilon_0 \\ j>N_0}} \|P_\infty v^{\epsilon_j}\|_{L^\infty} \|P_- v^{\epsilon_j}\|_{L^\infty}^{-1} .$$

Combining this inequality with (3.4) we deduce that $\|P_-\| \geqq 1/2$ which is a contradiction because our assumptions on q imply that $\|P_-\| < 1/2$.

Similarly we can show the following uniqueness type theorem:

<u>Theroem 3</u>. Let q_1 and q_2 satisfy all of the assumptions of Theorem 2. In addition we assume that 1) $q_1 \geqq q_2$ for all $x \in \mathbb{R}^3$, $t \in \mathbb{R}$, 2) $\|q_j\|_{L^\infty} \leqq [2(R^3+3)]^{-1}$, $j = 1,2$, and

$$\sup_{\substack{z\in \mathbb{R}^3 \\ t\in \mathbb{R}}} \int_{|y|\leqq R} |y-z|^{-1} q_1(y,t-|y-z|)dy < 5\pi\alpha(6R)^{-1}$$

for some α such that $2/3 < \alpha < 1$. Let $S(q_1)$ and $S(q_2)$ be the scattering operators associated with $\Box u + q_1 u = 0$ and $\Box u + q_2 u = 0$ respectively. Then $S(q_1) = S(q_2)$ implies $q_1 \equiv q_2$.

FINAL REMARKS

1) Huygens' principle and the positivity of the Riemann function for the three-dimensional free wave equation were important tools in our proof of theorems 2 and 3. We suspect that the same conclusion of uniqueness may be true without the smallness assumption we imposed.

2) Using different techniques, I and my colleague J.A. Ferreira obtained recently some partial results in the even dimensional case (See [6]).

REFERENCES

[1] C.E. BAUM - Emerging technology for transient and broad-band analysis and synthesis of antennas and scatterers, Proc. IEEE 64(1976) 1598-1616.

[2] J. COOPER and W.A. STRAUSS - The leading singularity of a wave reflected by a moving boundary, J. Diff. Equations, 54(2), (1984), 175-203.

[3] J. COOPER and W.A. STRAUSS - Abstract scattering theory for time-periodic systems with applications to electromagnetism, Ind. Univ.Math.J., 34(1) (1985), 33-84.

[4] J. COOPER; G. PERLA MENZALA & W.A. STRAUSS - On the scattering frequencies of time dependent potentials (to appear).

[5] J.A. FERREIRA and G. PERLA MENZALA - Time dependent approach to the inverse scattering problem for wave equations with time dependent coefficients (to appear).

[6] J.A. FERREIRA and G. PERLA MENZALA - Inverse scattering for wave equations with time-dependent coefficients: The even dimensional case (to appear).

[7] P. LAX and R. PHILLIPS - Decaying models for the wave equation in the exterior of an obstacle, Comm. Pure Appl.Math. 22 (1969), 737-787.

[8] G. PERLA MENZALA - Sur l'opérateur de diffusion pour l'equation des ondes avec des potentiels dependant du temps, C.R.Sc. Paris, t. 300, nº 18, (1985), 621-624.

[9] G. PERLA MENZALA - Scattering properties of wave equations with time-dependent potentials, Comp. and Math. with Appls., 12A, (1986), 457-475.

[10] G. PERLA MENZALA and T. SCHONBEK - Scattering frequencies for the wave equation with a potential term, J. Funct.Anal., 55(3), (1984), 297-322.

[11] S. STEINBERG - Meromorphic families of compact operators, Arch.Rat. Mech.Anal., 13, (1968), 372-379.

[12] D. THOE - On the exponential decay of solutions of the wave equation, J.Math.Anal.Appl., 16, (1966), 333-346.

[13] M. WEI; G. MADJA and W.A. STRAUSS - Numerical computation of the scattering frequencies for acoustic wave equations (to appear).

THE INITIAL VALUE PROBLEM FOR EULER AND NAVIER-STOKES EQUATIONS IN $L^p_s(\mathbb{R}^2)$

Gustavo Ponce

Department of Mathematics
University of Chicago
Chicago, Illinois 60637, U.S.A.

Consider the abstract Cauchy problem

$$\begin{cases} \dfrac{du}{dt} = F(u,\lambda) & t > 0 \\ u(0) = u_o \end{cases} \tag{1.1}$$

where $\lambda \in A \subseteq R$. Suppose there are two Banach spaces Y, X such that $Y \hookrightarrow X$, and $F: Y \times A \to X$ is continuous with $F(y,\cdot)$ differentiable in A. We say that the problem (1.1) is locally well-posed in Y if

I) for each $(u_o,\lambda) \in Y \times A$ there exists $T > 0$, and a unique function

$$u^\lambda \in C([0,T]: Y)$$

satisfying the problem (1.1) (existence, uniqueness, and persistence property),

II) the map $u_o \mapsto u^\lambda(t)$ is continuous from Y to $C([0,T]: Y)$ (continuous dependence on the data),

III) the map $\lambda \to u^\lambda(\cdot)$ is continuous from A to $C([0,T]: Y)$ (continuous dependence on the parameter λ).

We say that problem (1.1) is globally well-posed in Y if in I)-III) T can be taken arbitrarily large.

This notion of well-posedness is rather strong, and for many problems its proof is not available in the published literature.

In the present paper we are mainly interested in establishing global well-posedness for the Euler and Navier-Stokes equations for incompressible fluids in $\mathbb{R}^2$ in any vector Lebesgue (Sobolev) spaces ([3], [2])

$$L_s^p = (1-\Delta)^{-s/2} L^p(\mathbb{R}^2), \quad \text{with} \quad s > 1 + \frac{2}{p}, \quad \text{and} \quad 1 < p < \infty.$$

Here we will use the notation $L_s^p(\mathbb{R}^2)$ for scalar or vector valued functions.

Thus we consider the initial value problem,

$$\begin{cases} \partial_t u + (u\cdot\nabla)u = \nu\Delta u - \partial\pi & x \in \mathbb{R}^2, \quad t > 0 \\ \operatorname{div} u = 0 \\ u(x,0) = u_o(x) \end{cases} \tag{1.2}$$

where $\partial_t = \frac{\partial}{\partial t}$; $\nabla = \text{grad} = (\partial_1, \partial_2) = (\frac{\partial}{\partial x_1}, \frac{\partial}{\partial x_2})$; $u = u(x,t) = (u_1(x,t), u_2(x,t))$ is the velocity field, $\pi = \pi(x,t)$ is the pressure; $(u\cdot\nabla)u$ has j-componente $u_k\partial_k u_j$ (with summation convention), and $\nu \geq 0$ is the viscosity, with $\nu > 0$ in the Navier-Stokes problem, and $\nu = 0$ (ideal flow) for the Euler problem.

In these equations the pressure π is determined (up to a function of t) by u; indeed $\partial\pi = -(1-P)(u\cdot\nabla)u$, where P is a pseudo-differential operator with matrix symbol $(\delta_{ij} - \xi_i\xi_j/|\xi|^2)$. Thus the operator P, which projects into solenoidal vectors and annihilates gradients is a matrix of singular integral operators of Calderón-Zygmund [4] type except for some delta-function components in diagonal elements. In particular P is continuous from L_s^p into L_s^p for any $p \in (1,\infty)$.

The existence and uniqueness for the Navier-Stokes equation in $\mathbb{R}^2$ was proved by Leray [12]; here persistence holds with $Y = PL^2$. The existence, uniqueness and persistence property for the Euler equation was proved by Wolibner [17] for data u_o such that $\nabla \times u_o$ is Hölder continuous and bounded by $c(1+|x|)^{-(1+\epsilon)}$. Analogous results were proved by Golovkin [5], and McGrath [13] for the Navier-Stokes and Euler equations together with the problem as $\nu \to 0$. The function space Y used by the latter authors were more restricted than Wolibner's and persistence property was not proved.

More recently, Kato [8] has shown that the property I) of global well-posedness holds for $Y = PL^2_s(\mathbb{R}^2)$ for any s real > 2. In [14] the author proved similar result for $Y = L^2 \cap PL^p_s$ with $(s,p) \in [2,\infty)\times[2,\infty)$ except the case $s = p = 2$. Finally in a joint work [10], Kato and the author have extended these results to all $Y = PL^p_s$ with $s > 1 + \frac{2}{p}$, and $1 < p < \infty$.

The main results in [10] may be summarized in the following theorem.

<u>THEOREM A</u>. Let $s > 1 + \frac{2}{p}$, and $1 < p < \infty$. Then for any $u_o \in PL^p_s(\mathbb{R}^2)$ the I.V.P. (1.2) has a unique global solution u^ν such that $u^\nu \in C([0,\infty): PL^p_s)$. Moreover, for any $T > 0$

$$\sup_{[0,T]} \|u^\nu(t)\|_{s,p} \le K = K(T,\|u_o\|_{s,p}).$$

<u>Remarks</u>

- For the case $p > 2$ the solution may not have finite energy.
- By embedding theorem and persistence property the solution obtained are classical. For the Euler equation this result is sharp.
- Theorem A implies that the Euler equation does not possess any regularization nor deregularization effects in L^p_s (backward techniques).

<u>Proof</u>: Although we do not carry out the details of the proof we would like to point out the main ideas there.

First, the following L^p_s-energy estimate was proved for the solutions of the Navier-Stokes equations

$$\frac{d}{dt} \|u^\nu(t)\|_{s,p} \le C\|\nabla u^\nu(t)\|_\infty \cdot \|u^\nu(t)\|_{s,p} \tag{1.3}$$

for any $s \in [1,2]$, and $p \in (1,\infty)$, where the constant C does not depend on $\nu > 0$.

For $p = 2$ and s integer this type of energy estimate is well-known for many non-liner evolution equations. However for s non-integer they are hard to obtain. In fact, we do not know whether or not the inequality (1.3) holds for s non-integer larger than 2. (see Remark 4.1.a in [8], and page 494 in [14]).

Writing the equation for the vorticity $\omega^\nu(t) = \nabla \times u^\nu(t)$, i.e.

$$\begin{cases} \partial_t \omega^\nu + (u\cdot\nabla)\omega^\nu = \nu\Delta\omega \\ \omega(x,0) = \nabla\times u_o(x) = \omega_o(x) \end{cases}$$

it is easy to prove that for any $q \in (1,\infty]$

$$\|\omega^\nu(t)\|_q \le \|\nabla\times u_o\|_q . \tag{1.4}$$

Using the Biot-Savart Law, $u^\nu(t) = \nabla\times(-\Delta)^{-1}\,\omega^\nu(t)$ and the L^p-continuity of the Riesz Transform it follows that for $q \in (1,\infty)$

$$\|\partial u^\nu(t)\|_q \le c\|\partial u_o\|_q$$

in any time interval $[0,\bar{T}]$ where the existence of the classical solution is known.

However, in order to have control of the growth of the higher derivatives of $u^\nu(\cdot)$ we need to estimate $\|\partial u^\nu(t)\|_\infty$, (see (1.3)). In [10], Kato and the author proved the following inequality. If $\omega^\nu = \nabla\times u^\nu(t)$ then

$$\|\nabla u^\nu(t)\|_\infty \le c\{\|\omega^\nu(t)\|_p + \|\omega^\nu(t)\|_\infty \cdot \log^+(\|\omega^\nu(t)\|_{s-1}/\|\omega^\nu(t)\|_\infty)\} \tag{1.5}$$

with the notation $\log^+(\lambda) = \max\{1, \log(\lambda)\}$.

Here we present a different version of the above estimate which generalizes analogous results given in [1,5,8,10,15], (all these results involves a logarithmic function).

Consider operator of the form

$$(Tf)(x) = k\cdot f(x) + \lim_{\epsilon\to 0}\int_{|y|\ge\epsilon} \frac{\Omega(y)}{|y|^n} f(x-y)\,dy = k\cdot f(x) + (\tilde{T}f)(x),$$

where k is a given constant, and Ω satisfies:

a) Ω is homogeneous of degree zero; thus Ω is determined by its values on the unit sphere S^{n-1},

b) Ω has mean value zero on the unit sphere, i.e.

$$\int_{S^{n-1}} \Omega(x')dx' = 0,$$

c) $\Omega \in L^{\infty}(S^{n-1})$.

LEMMA. An operator T of the type described above satisfies the following estimate:

$$\|Tf\|_{\infty} \le \tilde{c}\{\|f\|_p + \|f\|_{\infty} \cdot \log^+(\|f\|_{\beta}/\|f\|_{\infty})\} \tag{1.6}$$

where $\beta \in (0,1)$, $p \in [1,\infty)$, the constant $\tilde{c}$ only depends on k, β, p, n, and the L^{∞}-norm of Ω, and $\|\cdot\|_{\beta}$ denotes the norm of the Hölder space $C^{\beta}(\mathbb{R}^n)$.

Remarks

- Since no smoothness conditions are assumed on Ω, the class of operator given above are more general than the classical singular integral operator, (see [16]).

- In particular, when T is a double Riesz transform by using Sobolev Embedding Theorem in (1.6) we otain (1.5).

Proof: Dividing the domain of integration in three parts we have that,

i) $$\left|\int_{|y|\ge 1} \frac{\Omega(y)}{|y|^n} f(x-y)dy\right| \le c\|f\|_p$$

for any $p \ge 1$,

ii) $$\left|\int_{\delta\le|y|\le 1} \frac{\Omega(y)}{|y|^n} f(x-y)dy\right| \le c\|f\|_{\infty}(-\log \delta)$$

for any $\delta \in (0,1]$,

iii) $$\left|\int_{\delta\le|y|\le\delta}\frac{\Omega(y)}{|y|^n}f(x-y)dy\right| = \left|\int_{\varepsilon\le|y|\le\delta}\frac{f(x-y)-f(x)}{|y|^{\beta}}dy\right|$$
$$\le c\cdot\delta^{\beta}|||f|||_{\beta}\ .$$

Thus,

$$\|\tilde{T}f\|_{\infty} \le c\{\|f\|_p + \|f\|_{\infty}(-\log\delta) + \delta^{\beta}\cdot|||f|||_{\beta}\}.$$

Minimizing the above expression respect to $\delta \in (0,1]$, and taking $\tilde{c} = c+k$, we complete the proof of the estimate (1.6).

Using the estimates (1.3), (1.5) and (1.6) we have that

$$\frac{d}{dp}\|u^{\nu}(t)\|_{s,p} \le c\{\|\omega_o\|_p + \|\omega_o\|_{\infty}\ \log^{+}(\|u^{\nu}(t)\|_{s,p}/\|\omega_o\|_{\infty})\cdot\|u^{\nu}(t)\|_{s,p}\},$$

for $s \in [1,2]$, and $p \in (1,\infty)$.

Therefore a global a priori estimate for the classical solution of the Navier-Stokes equation (uniformly in the viscosity $\nu > 0$) has been proved if an appropriate local existence theorem is available.

Thus we state,

THEOREM B. Let s, p and u_o be as in Theorem A. Then there exists $T > 0$, and a unique solution u^{ν} to the I.V.P. for the Navier-Stokes equation such that

$$u^{\nu} \in C([0,T]: PL^p_s) \cap C((0,T]: L^p_{\infty})$$

where T depends on $s,p,\nu > 0$, and $\|u_o\|_{s,p}$.

In fact Theorem B holds in any dimension n and with appropriate external forces. However the main difficulty is to overcome the dependence on $\nu > 0$ of T.

Also, the assumption on u_o can be weakened if $u(t) \in L^p_s$ is not required to be continuous up to $t = 0$.

The proof of Theorem B is based in the method of integral equations given by Kato-Fujita in [9] (see also [6], [7]).

To estimate higher derivatives we use a simple boot-strap argument.

The remaider of the proof of Theorem A is concerned with the convergence of the solutions $u^\nu(t)$ to the ideal one $u^0(t)$, (viscosity method).

Finally we would like to remark that in a coming paper [11], Kato and the author prove the conditions II) III) of the well-posedness stated above. More precisely there the continuous dependence of the solution respect to the initial data $u_o \in PL^p_s(R)$ $(s > 1 + 2/p,$ $p \in (1,\infty))$ is established. Also the convergence with vanishing viscosity of Navier-Stokes flow to ideal flow is shown in $C([0,T]:PL^p_s(R^2))$ for any $T > 0$, and for a fixed data $u_o \in PL^p_s(R^2)$.

References

[1] J.T. Beal, T. Kato, and A. Majda, Remarks on the breakdown of smooth solutions for the 3-D Euler Equations, Comm. Math. Phys. 94 (1984), 61-66.

[2] J. Bergh and J. Löfström, Interpolation spaces, Berlin, Heidelberg, New York, Springer 1970.

[3] A.P. Calderón, Lebesgue spaces of differentiable functions and distributions, Partial Differential Equations, Proc. Symp. Pure Math. Vol. IV, Amr. Math. Soc., Providence 1961, pp. 33-49.

[4] A.P. Calderón and A. Zygmund, On singular integrals, Amer. J. Math. 78 (1956), 289-309.

[5] K.K. Golovkin, Vanishing viscosity in Cauchy's problem for hydrodynamic equations, Trud. Math. Inst. Steklov. 92 (1966), 31-49; Amer. Math. Soc. Translation, 1968, pp. 33-53.

[6] T. Kato, Nonstationary flows of viscous and ideal fluids in $\mathbb{R}^3$, J. Functional Anal. 9 (1972), 296-305.

[7] T. Kato, Strong L^p-solutions of the Navier-Stokes equation in $\mathbb{R}^m$, with applications to weak solutions, Math. Z. 187 (1984), 471-480.

[8] T. Kato, Remarks on the Euler and Navier-Stokes Equations in $\mathbb{R}^2$, Proc. Symp. Pure Math. 45, part 2, 1-8, Providence, R.I. Amer. Math. Soc. 1986.

[9] T. Kato and H. Fujita, On the non-stationary Navier-Stokes system, Rend. Sem. Math. Univ. Padova 32 (1962), 243-260.

[10] T. Kato and G. Ponce, Well-posedness of the Euler and Navier-Stokes equations in Lebesgue spaces $L^p_s(\mathbb{R}^2)$, Revista Matematicas Iberoamericana, vol. 2 (1986), pp. 73-88.

[11] T. Kato and G. Ponce, in preparation.

[12] J. Leray, Étude de diverses équations intégrales non linéaires et du quelques problèmes que pose l'hydrodynamique, J.Math. Pures Appl. 12 (1933), 1-82.

[13] F.J. McGrath, Nonstationary plane flow of viscous ideal fluids, Arch. Rational Mech. Anal. 27 (1968), 329-348.

[14] G. Ponce, On two dimensional incompressible fluids, Comm. P.D.E., Comm. P.D.E. 11, 5, 483-511, 1986.

[15] G. Ponce, Global existence of solutions to a nonlinear evolution equation with nonlocal coefficients, J. Math. Anal. Appl., to appear.

[16] E. Stein, Singular integrals and differenciability properties of functions, Princeton University Press, Princeton, 1970.

[17] W. Wolibner, Un theorème sur l'existence du mouvement plan d'un fluide parfait, homogène, Incompressible, pendent un temps infiniment long, Math. Z. 37 (1933), 698-726.

PERIODIC SOLUTIONS OF PRESCRIBED ENERGY OF HAMILTONIAN SYSTEMS

Paul H. Rabinowitz
Mathematics Department and
Mathematics Research Center
University of Wisconsin
Madison, Wisconsin, U.S.A. 53706

A Hamiltonian system of ordinary differential equations models the motion of a discrete mechanical system when no frictional forces are present. Such systems have the form:

$$\dot{p} = -H_q(p,q)$$
(1)
$$\dot{q} = H_p(p,q) \quad .$$

Here $H : R^{2n} \to R$, $p,q \in R^n$, $\dot{p} = \frac{dp}{dt}$, etc. The system (1) may also be written as

(2) $$\dot{z} = JH_z(z)$$

where $z = (p,q) \in R^{2n}$ and $J = \begin{pmatrix} 0 & -id \\ id & 0 \end{pmatrix}$, id denoting the $n \times n$ identity matrix.

One of the important properties of (2) is that if $z(t)$ is a solution, then $H(z(t))$ is independent of t. Thus H is an integral for the system (2) or said another way, "energy is conserved" and solutions of (2) lie on an energy surface $H \equiv$ constant. A major question of interest for (2) is what sort of conditions does one have to impose on H so that the energy surface, say $H^{-1}(1)$, possesses a periodic solution of (2). During the past ten years, a lot of progress has been made on this question. Restricting ourselves to a fairly general setting, we will briefly survey this work and

---

This research was sponsored in part by the United States Army under Contract No. DAAG29-80-C-0041 and by the National Science Foundation under Grant No. MCS-8110556. Reproduction in whole or in part is permitted for any purpose of the United States Government.

then describe some recent research of our own in this direction. Before doing so, it is worth emphasizing now that although there have been several approaches to this question, all of the recent progress involves applications of the calculus of variations in some form or other.

Our survey begins with an older theorem due to Seifert [1] which is the first global result in a general setting that we know of for (2).

Theorem 1: Suppose $H(p,q) = \sum_{i,j=1}^{n} a_{ij}(q)p_ip_j + V(q)$ where

(V_1) $V \in C^2(\mathbb{R}^n,\mathbb{R})$ and $\mathcal{D} \equiv \{q \in \mathbb{R}^n \mid V(q) < 1\}$ is diffeomorphic to the unit ball in $\mathbb{R}^n$ and $V_q(q) \neq 0$ on $\partial\mathcal{D}$

and

(K_1) $a_{ij} \in C^2(\mathbb{R}^n,\mathbb{R})$ and the matrix $(a_{ij}(q))$ is uniformly positive definite in $\mathcal{D}$.

Then there is a $T > 0$, points $Q_1 \neq Q_2 \in \partial\mathcal{D}$, and a solution $(p(t), q(t))$ of (2) on $H^{-1}(1)$ such that $(p(0), q(0)) = (0,Q_1)$ and $(p(T), q(T)) = (0,Q_2)$.

Observing that H is even in p, if we extend $p(t)$ as an odd function about 0 and T and $q(t)$ as an even function about 0 and T, the resulting function is a $2T$ periodic solution of (2) on $H^{-1}(1)$. Thus solutions of the type found by Seifert are special periodic solutions of (2) whose projection in $\mathbb{R}^n$ bounces back and forth between two points on $\partial\mathcal{D}$. Roughly speaking Seifert obtained these "bouncing orbits" as geodesics in a Riemannean metric associated with the kinetic energy term $\sum_{i,j=1}^{n} a_{ij}(q)p_ip_j$.

Seifert's work was generalized by A. Weinstein [2] who proved

Theorem 2: Suppose $H(p,q) = K(p,q) + V(q)$ where V satisfies (V_1) and K satisfies

(K_2) $0 = K(0,q)$, K is even and strictly convex in p, and $K(\alpha p,q) \to \infty$ as $|\alpha| \to \infty$ uniformly for $p \in S^{n-1}$ and $q \in \mathcal{D}$.

Then the conclusions of Theorem 1 hold.

Weinstein called these boundary orbits "brake orbits". He used an existence proof related to that of Seifert with the Riemannean metric replaced by a Finsler metric associated with $K(p,q)$. By an ingeneous reduction to Theorem 2, Weinstein went on to prove [2]:

Theorem 3: Suppose $H \in C^2(R^{2n},R)$ and $H^{-1}(1)$ bounds a compact convex neighborhood of 0 in R^{2n}. Then (2) possesses a period solution on $H^{-1}(1)$.

A more elementary proof of Theorem 3 using ideas from convex analysis and optimization theory was given by F. Clarke [3].

We also studied (2) using a rather different variational approach than those mentioned above and proved [4]:

Theorem 4: Suppose $H \in C^1(R^{2n},R)$ and $H^{-1}(1)$ bounds a compact starshaped neighborhood of 0 (with $z \cdot H_z \neq 0$ on $H^{-1}(1)$). Then (2) possesses a periodic solution on $H^{-1}(1)$.

The proof of Theorem 4 involves minimax methods from the calculus of variations. Some of the ideas used in this approach will be described shortly. Motivated by [2], we further proved [5]:

Theorem 5: Suppose $H(p,q) = K(p,q) + V(q)$ where V satisfies (V_1) and K satisfies

(K_3) $0 = K(0,q)$, $p \cdot K_p(p,q) > 0$ for $p \neq 0$, and $K(\alpha p,q) \to \infty$ as $|\alpha| \to \infty$ uniformly for $p \in S^{n-1}$ and $q \in \mathcal{D}$.

Then (2) possesses a periodic solution on $H^{-1}(1)$.

In another direction, working towards treating more general potential energy terms, Gluck and Ziller have proved [6]:

Theorem 6: Suppose $H(p,q) = K(p,q) + V(q)$ where V satisfies

(V_2) $\mathcal{D} \equiv \{q \in R^n | V(q) < 1\}$ is compact and $V_q \neq 0$ on $\partial\mathcal{D}$

and (K_2). Then the conclusions of Theorem 1 hold.

Their proof again uses geodesic ideas from geometry. Special cases of Theorem 6 with (V_2) replaced by (V_1) have been obtained independently by Hayashi [7] and Benci [8].

An interesting open question in the direction of the above results is how far can one go in weakening hypotheses on H and still find periodic solutions of (2) on $H^{-1}(1)$. E.g. suppose $H^{-1}(1)$ is the boundary of a neighborhood Ω of 0 in R^{2n} and Ω is diffeomorphic to the unit ball in R^{2n}. Nothing is known concerning the existence or nonexistence of periodic solutions of (2) on $H^{-1}(1)$ for this setting.

Observe that Theorems 1-6 fall into two classes: (a) results based on arguments from geometry for Hamiltonians which are even in p and which establish the existence of "bouncing orbits" of (2) and (b) Theorems 4 and 5 which use minimax arguments to merely get periodic solutions of (2) but under milder conditions on H than for (a). Recently we became interested in the question of whether H is also even in p, direct minimax methods would be used to find "bouncing orbits". This turns out to be the case at least in the settings of Theorems 4 and 5. We also believe it is true for the case represented by Theorem 6. We will illustrate the new results by means of the following variant of Theorem 4:

Theorem 7: Suppose H satisfies the hypotheses of Theorem 4 and is also even in p. Then there exists a $T > 0$ and a $2T$ periodic solution (p,q) of (2) on $H^{-1}(1)$ with p odd about 0 and T and q even about 0 and T.

Before sketching the proof of Theorem 7, part of the approach used in the proof of Theorem 4 will be described and we further indicate why it is inadequate to prove Theorem 7. To begin it is convenient to make a change of time scale so that the period of the solution we seek - which is a priori unknown - becomes 2π. Then (2) transforms to

$$\dot{z} = \lambda J H_z(z) \tag{3}$$

(where $2\pi\lambda$ is the period for (2)). Working in the class of 2π periodic functions and ignoring precise topologies for now, let $z = (p,q)$ and

$$A(z) \equiv \int_0^{2\pi} p\cdot\dot{q}\,dt \quad .$$

Thus A is the so-called action integral. Let

(4) $$\Psi(z) \equiv \frac{1}{2\pi}\int_0^{2\pi} H(z(t))dt \quad .$$

A formal calculation shows that a critical point of A on $M \equiv \Psi^{-1}(1)$ is a 2π periodic solution of (3), λ occurring as a Lagrange multiplier due to the constraint that $z \in M$. Thus we can try to find the desired solution of (3) as a critical point of $A|_M$. This approach is complicated by the fact that A need not be bounded from above or below on M. E.g. if $H(z) = |z|^2$, $n = 1$, $k \in \mathbf{Z}$, and $(p_k, q_k) = (\sin kt, \cos kt)$, then

$$A(z_k) = -\frac{k}{2} \quad .$$

This indefiniteness of $A|_M$ makes existence difficult to establish but nevertheless it can be done in the setting of Theorem 4. A key ingredient in the proof is the use of an S^1 symmetry possessed by $A|_M$. Namely if $z \in M$, then for all $\theta \in [0,2\pi]$, $g_\theta(z)(t) \equiv z(t+\theta)$ also belongs to M and $A(g_\theta(z)) = A(z)$.

Turning now to Theorem 7, since solutions of the special form p odd about 0 and π and q even about 0 and π are sought, it is natural to try to prove existence directly in this class of functions. However if we attempt to duplicate the proof of Theorem 4, we see immediately that this class is not invariant under the family of mappings g_θ, $\theta \in [0,2\pi)$. Hence the earlier existence machinery breaks down and the main difficulty here becomes that of finding a new existence mechanism.

In what follows we will sketch the framework of the existence argument for Theorem 7. For more details see [9]. To simplify matters, we further assume $H \in C^2(\mathbf{R}^{2n},\mathbf{R})$. Let $W^{1/2,2}(S^1,\mathbf{R}^{2n})$ denote the (Hilbert) space of $2n$-tuples of 2π periodic functions z for which $\|z\|_{W^{1/2,2}} < \infty$. Here if

$$z = \sum_{j\in\mathbf{Z}} a_j e^{ijt} \quad , \quad a_j \in \mathbf{C}^n \text{ and } a_{-j} = \overline{a}_j \quad ,$$

then

(5) $$\|z\|^2_{W^{1/2,2}} = \sum_{j\in\mathbf{Z}} (1+|j|)|a_j|^2 \quad .$$

Let

$$X = \{z = (p,q) \in W^{1/2,2}(S^1,\mathbf{R}^{2n}) \mid p \text{ is odd about } 0 \text{ and } \pi \text{ and } q \text{ is even about } 0 \text{ and } \pi\} \quad .$$

The hypotheses of Theorem 7 allow us no control over H near ∞. Hence Ψ as given by (4) may not be defined on all of X nor be differentiable on X. Therefore to properly pose (3) as a variational problem in X, it is necessary to redefine H. Note that by the hypotheses of Theorem 4, for each $z \in R^{2n}\setminus\{0\}$, there is a unique $w(z) \in H^{-1}(1)$ and $\alpha(z) > 0$ such that $z = \alpha(z)w(z)$. Set $\overline{H}(0) = 0$ and for $z \neq 0$, $\overline{H}(z) = \alpha(z)^2$. Then $\overline{H} \in C^{1,Lip}(R^{2n},R) \cap C^2(R^{2n}\setminus\{0\},R)$, $\overline{H}^{-1}(1) = H^{-1}(1)$ and $\overline{H}$ is positively homogeneous of degree α. An elementary lemma [4] shows if z is a periodic solution of

$$\dot{z} = \lambda J\overline{H}_z(z) \quad , \tag{6}$$

a reparametrization of z is a periodic solution of (3) (and conversely). Hence to prove Theorem 7, it suffices to find a 2π periodic solution of (6).

Now for $z \in X$, define

$$\Psi(z) = \frac{1}{2\pi}\int_0^{2\pi} \overline{H}(z(t))dt \tag{7}$$

and let $M \equiv \Psi^{-1}(1)$. It is not difficult to establish that M is a $C^{1,Lip}$ manifold in X which bounds a starshaped neighborhood of 0 in X and is bounded in $L^2(S^1,R^{2n})$. (A good model case to keep in mind is $H(z) = |z|^2$.) Moreover by above remarks, any critical point of $A|_M$ leads to a solution of the desired type of (3). Thus we focus our attention on obtaining critical points of $A|_M$.

Next X can be decomposed into three complementary subspaces: $X = X^+ \oplus X^0 \oplus X^-$ where X^+, X^0, X^- are subsapces of X on which A is respectively positive definite, null, and negative definite. In fact such spaces can be written down explicitly. Let $e_1,\dots,e_{2n}$ denote the usual orthonormal bases in R^{2n}. Then

$$X^+ \equiv \text{span}\{(\sin jt)e_k - (\cos jt)e_{k+n} \mid 1 < j < \infty,\ 1 < k < n\} ,$$

$$X^0 \equiv \text{span}\{e_k \mid 1 < k < 2n\} ,$$

$$X^- \equiv \text{span}\{(\sin jt)e_k + (\cos jt)e_{k+n} \mid 1 < j < \infty,\ 1 < k < n\} .$$

Any $z \in X$ can be written as $z = z^+ + z^0 + z^- \in X^+ \oplus X^0 \oplus X^-$. In fact by choosing a new inner product for X whose norm is equivalent to that given by (5), the spaces X^0, $X^\pm$ are mutually orthogonal in X and in $L^2(S^1,\mathbb{R}^{2n})$ and $A(z)$ takes the simple form

$$A(z) = \|z^+\|^2 - \|z^-\|^2 \quad .$$

It is now a straightforward computation to verify that $A|_M$ satisfies the $(PS)^+$ condition, i.e. for any sequence $(z_m) \subset M$ such that $A(z_m) \to c$ and

$$A|_M'(z_m) = A'(z_m) - \lambda(z_m)\Psi'(z_m) \to 0$$

has a convergent subsequence. (Here A' denotes the Frechet derivative of A,

$$\lambda(z) = \frac{\langle A'(z),\Psi'(z)\rangle_{X^*}}{\|\Psi'(z)\|^2_{X^*}} \quad ,$$

etc.).

With these preliminaries in hand, we can indicate how to obtain a critical point of $A|_M$. The idea is to characterize a corresponding critical value of $A|_M$ as a minimax over an appropriate class of sets. The choice of the class of sets may seem strange but it is guided by experience with related situations. To define these sets, first two subsets of M are singled out. Let $v \in X^+$ with e.g. $\|v\| = 1$. Define $M^+ \equiv M \cap X^+$ and $M^- \equiv M \cap (\text{span}\,\{v\} \oplus X^0 \oplus X^-)$.

Set

$$\alpha \equiv \inf_{z\in M^+} A(z) \quad .$$

Since M is the boundary of a neighborhood of 0 in X, M^+ is closed in X and $A(z) = \|z\|^2$ for $z \in X^+$, it follows that $\alpha > 0$. Next set

$$\beta \equiv \sup_{z\in M^-} A(z) \quad .$$

Since $z \in M^-$ implies $z = \rho(z)v + z^0 + z^-$, where $\rho(z) \in \mathbb{R}$,

$$A(z) = \rho(z)^2 - \|z^-\|^2 \le \rho(z)^2 \quad .$$

The boundedness of M in $L^2(S^1,\mathbb{R}^{2n})$ implies there is an $r > 0$ such that

$$\|z\|^2_{L^2} = \rho(z)^2\|v\|^2_{L^2} + \|z^0\|^2_{L^2} + \|z^-\|^2_{L^2} \le r \quad .$$

Hence $\rho(z)^2 \le r\|v\|_{L^2}^{-2}$ and $\beta < \infty$.

Next let P^- denote the orthogonal projector of X onto X^- and define

$$\Gamma \equiv \{h \in C(X,X) \mid h : \mathcal{M} \to \mathcal{M},\ h(z) = z \text{ if } A(z) \notin [0,\beta+1] \text{ or if } |\Psi(z) - 1| > \tfrac{1}{2}, \text{ and } P^-h(z) = e^{\theta(z)}z^- + K(z) \text{ where } 0 \le \theta(z) \le \gamma = \gamma(h) \text{ and } K \text{ is compact}\} .$$

Note that the family of maps Γ is closed under composition.

Finally we define

$$c \equiv \inf_{h\in\Gamma} \sup_{z\in\mathcal{M}^-} A(h(z)) \quad . \tag{8}$$

We claim that c is a critical value of $A|_{\mathcal{M}}$. To prove this fact, note first that since $id \in \Gamma$,

$$c \le \sup_{z\in\mathcal{M}^-} A(z) = \beta < \infty \quad . \tag{9}$$

To continue, we need an important "intersection theorem". In fact the structure required to obtain this result dictates some of the properties of the mappings in Γ.

<u>Proposition 8</u>: If $h \in \Gamma$, $h(\mathcal{M}^-) \cap \mathcal{M}^+ \neq \emptyset$.

The proof of this proposition involves a finite dimensional approximation argument. The related finite dimensional problem is solved by a degree theoretic argument. Then the form of $\mathcal{M}^\pm$ and properties of h allow passage to a limit.

By Proposition 8, for each $h \in \Gamma$, there is a $w = w(h) \in h(\mathcal{M}^-) \cap \mathcal{M}^+$. Hence

$$\sup_{z\in\mathcal{M}^-} A(h(z)) \ge A(w) \ge \inf_{z\in\mathcal{M}^+} A(z) = \alpha \quad . \tag{10}$$

Consequently $c \ge \alpha$.

Using the fact that $A|_{\mathcal{M}}$ satisfies $(PS)^+$, it is possible to prove the following variant of standard "deformation theorems". Let $\mathcal{a}_r \equiv \{z \in \mathcal{M} \mid A(z) \le r\}$ and $K_c \equiv \{z \in \mathcal{M} \mid A(z) = c \text{ and } A|'_{\mathcal{M}}(z) = 0\}$.

<u>Proposition 9</u>: If $c, \bar{\varepsilon} > 0$, there exists an $\varepsilon \in (0,\bar{\varepsilon})$ and an $\eta \in C([0,1] \times X,X)$ such that

1° $\eta(s,\cdot)$ is a homeomorphesm of X onto X and M onto M for each $s \in [0,1]$.

2° $\eta(1,z) = z$ if $A(z) \in [c-\bar{\varepsilon}, c+\bar{\varepsilon}]$ or if $|\Psi(z) - 1| > \frac{1}{2}$.

3° $P^-\eta(1,z) = e^{-\theta(z)} z^- + K(z)$ where $\theta \in C(X,R^+)$ is bounded and K is compact.

4° If $K_c = \emptyset$, $\eta(1,A_{c+\varepsilon}) \subset A_{c-\varepsilon}$.

Observe that properties 1° - 3° of Proposition 9 imply $\eta(1,\cdot) \in \Gamma$ (where we choose e.g. $\bar{\varepsilon} = \frac{\alpha}{2}$). It now quickly follows that c is a critical value of $A|_M$ for if not, let $h \in \Gamma$ such that

$$\sup_{z \in M^-} A(h(z)) < c + \varepsilon \quad .$$

Then by 4° of Proposition 9,

$$\sup_{z \in M^-} A(\eta(1,h(z)) < c - \varepsilon \quad . \tag{11}$$

But $\eta(1,h(\cdot)) \in \Gamma$ so (11) is contrary to the definition of c (8). Hence c is a critial value of $A|_M$ and our sketch of the proof of Theorem 7 is complete.

<u>Remark</u>: An interesting open question for the setting of Theorem 7 concerns the multiplicity of bouncing orbits of (3) on $H^{-1}(1)$. If $H^{-1}(1)$ bounds a convex region and $H^{-1}(1)$ is nested between two concentric spheres of radius r and R respectively with $1 < \frac{R}{r} < \sqrt{2}$, a result of Ekeland and Lasry [10] implies there are at least n geometrically distinct periodic orbits of (3) on $H^{-1}(1)$. Related results were obtained by van Groesen [11] and Girardi [12]. These multiplicity results rely on S^1 or Z_2 symmetries of the corresponding variational problem and cannot be used in our setting.

REFERENCES

[1] Seifert, H., Periodische Bewegungen mechanischen Systeme, Math. Z., 51, (1948), 197-216.

[2] Weinstein, A., Periodic orbits for convex Hamiltonian systems, Ann. Math. 108, (1978), 507-518.

[3] Clarke, F., A classical variational principle for periodic Hamiltonian trajectories, Proc. Amer. Math. Soc. 76, (1979), 186-188.

[4] Rabinowitz, P. H., Periodic solutions of Hamiltonian systems, Comm. Pure Appl. Math., 31, (1978), 157-184.

[5] Rabinowitz, P. H., Periodic solutions of a Hamiltonian system on a prescribed energy surface, J. Diff. Eq., 33, (1979), 336-352.

[6] Gluck, H. and W. Ziller, Existence of periodic solutions of conservative systems, Seminar on Minimal Submanifolds, Princeton Univ. Press, (1983), 65-98.

[7] Hayashi, K., Periodic solution of classical Hamiltonian systems, Tokyo J. Math., 6, (1983), 473-486.

[8] Benci, V., Closed geodesics for the Jacobi metric and periodic solutions of prescribed energy of natural Hamiltonian systems, Ann. Inst. H. Poincare. Analy. Nonlineaire.

[9] Rabinowitz, P. H., On the existence of periodic solutions for a class of symmetric Hamiltonian systems, to appear: Nonlinear Analysis, T.M.A.

[10] Ekeland, I. and J. M. Lasry, On the number of periodic trajectories for a Hamiltonian flow on a convex energy surface, Ann. Math., 112, (1980), 283-319.

[11] van Groesen, E. W. C., Existence of multiple normal mode trajectories on convex energy surfaces of even classical Hamiltonian systems, to appear, J. Diff. Eq.

(12) Girardi, M., Multiple orbits for Hamiltonian systems on starshaped surfaces with symmetries, Ann. Inst. H. Poincare, Analy. Nonlineaire, 1 (1984), 285-294.

SEMI-CLASSICAL APPROXIMATIONS IN SOLID STATE PHYSICS

J.-C. Guillot, J. Ralston and E. Trubowitz

This note shows how familiar asymptotic methods can be used to justify and systematize commonly used approximations in solid state physics.

BACKGROUND

Solid state physics is an edifice built on rather firm sand. Solving Schrödinger's equation for the motion of a number of particles on the order of Avogadro's number is presently unfeasible. In response to this difficulty simplified models are used. The most standard of these is known as "the independent electron approximation" and it will be used here. One simply assumes that in a solid electrons move independently of each other in the periodic potential produced by the lattice of atoms stripped of their outermost (valence) electron shells. Thus (throughout this discussion we ignore spin) each individual electron is governed by the Schrödinger equation

$$i\hbar\frac{\partial u}{\partial t} = \frac{-\hbar^2}{2m}\Delta u + V(x)u, \quad x \in R^3,$$

where V is smooth and $V(x + d) = V(x)$ for all vectors d in the atomic lattice.

The full spectrum of the hamiltonian $(-\hbar^2/2m)\Delta + V$ as an operator on $L^2(R^3)$ is the union over k of the spectra of this operator on a fundamental domain for the lattice with the boundary conditions

$$u(x + d) = \exp(\frac{-ik\cdot d}{\hbar})u(x).$$

To avoid repetitions it is conventional to restrict k to a fundamental domain for the dual lattice. Writing $u = \exp(-ik\cdot x/\hbar)\varphi$, the eigenvalue equation becomes

$$L(k)\varphi_n \equiv \frac{1}{2m}(i\hbar\frac{\partial}{\partial x} + k)^2\varphi_n + V(x)\varphi_n = E_n(k)\varphi_n$$

where $\varphi_n(x + d,k) = \varphi_n(x,k)$. The convention here is that $E_n(k)$ is the nth -largest eigenvalue of $L(k)$ on a fundamental domain with periodic boundary conditions. The functions $E_n(k)$ are called "band functions" and the spectrum of the hamiltonian in $L^2(R^3)$,

$$\bigcup_n \bigcup_k E_n(k),$$

is known as the "Bloch spectrum" in honor of one of the pioneers of the subject.

While one ignores interactions between electrons, one retains the Pauli exclusion principle, and assumes that no two electrons occupy the same energy level. Taking the restriction of the hamiltonian to a block of N fundamental domains with suitable conditions on the boundary, so that the energy levels become discrete, one considers the situation where the lowest N $\times$ (valence) energy levels are filled. The limit of the highest filled level as the block expands to fill R^3 is known as the "Fermi energy", E_f. The Fermi energy plays the role of the ground state energy for the physical solid, and it is quite important - particularly in low temperature work. The set in k-space,

$$\sum_f = \{k: E_n(k) = E_f \text{ , for some } n\}$$

is known as the Fermi surface. Since the band functions extend to functions on R^3 periodic with respect to the dual lattice, one canconsider the Fermi surface as a set in R^3 periodic with respect to the dual lattice.

MAGNETIC FIELDS

When the solid is subjected to an external magnetic field, B, the Schrödinger equation for a single electron becomes

$$(1) \qquad i\hbar\frac{\partial u}{\partial t} = \frac{1}{2m}\left(i\hbar\frac{\partial}{\partial x} + \frac{e}{c}A\right)^2 u + V(x)u = Hu$$

where $\nabla \times A = B$, e is the electron charge, and c is the speed of light. The vector potential A is not unique, since $A + \nabla\psi$ works as well. However, this change only transforms H to

$$(2) \qquad \exp\left(\frac{ie}{\hbar c}\psi\right)\left(H\left(\exp\left(\frac{-ie}{\hbar c}\psi\right)\right)\right),$$

an obviously unitary transformation.

In 1933 R. Peierls proposed that the motion of a suitably prepared wave packet governed by (1) could be approximated by

(3)
$$\dot{x} = \frac{\partial E_n(k + A(x))}{\partial k}$$
$$\dot{k} = -\frac{\partial E_n(k + A(x))}{\partial x}$$

in other words $E_n(k + A)$ is a hamiltonian for the motion of the packet. To the best of our knowledge such packets were first constructed in 1966 by R. Chambers, [2], but by that time this approximation was standard and known as the "semi-classical model".

To understand the sense in which (3) is a valid approximation one needs first to consider the scales in the equation (1). The potential V(x) is on the order of a few electron volts (1 ev = 1.6×10^{-12} erg) and the separations of the atoms in the lattice are on the order of Ångstroms (1 Å = 10^{-8}cm). Using these units with t in seconds, (3) takes the form

(4)
$$i\mu \frac{\partial u}{\partial t} = (i\frac{\partial}{\partial x} + \varepsilon a(x))^2 u + v(x)u$$

where $\varepsilon = 1.5 \times 10^{-9} \times G$, G is the typical magnetic field strength in gauss, $\mu = 1.7 \times 10^{-16}$, a(x) and v(x) are order 1, and the periods of v(x) are order 1. To consider the motion of wave packets on the scale of ε^{-1} Å, we introduce $y = \varepsilon x$ and $s = \varepsilon\mu^{-1}t$ (note $t = 1 \iff s = 10^7 \times G^{-1}$). This transforms (4) to

$$\frac{i}{\varepsilon}\frac{\partial u}{\partial s} = (i\frac{\partial}{\partial y} + a(\frac{y}{\varepsilon}))^2 u + \frac{1}{\varepsilon^2} v(\frac{y}{\varepsilon})u$$

Finally, we assume that the field B can be treated as constant on the support of the packet, i.e. on sets of diameter order 1 in the y-scale. This assumption is always valid in experiments and it means that we can choose a(x) linear in x. This leads to the form of the scaled equation that we will use:

(5)
$$\frac{i}{\varepsilon}\frac{\partial u}{\partial s} = (i\frac{\partial}{\partial y} + \frac{1}{\varepsilon} a(y))^2 u + \frac{1}{\varepsilon^2} v(\frac{y}{\varepsilon})u.$$

The asymptotic analysis that follows treats ε as a small parameter. This is quite reasonable since the strongest magnetic fields available for laboratory experiments are around 5×10^4 gauss, which corresponds to $\varepsilon = 7.5 \times 10^{-5}$.

Once one has the form (5) for the Schrödinger equation the asymptotic analysis becomes standard. Following Chapter V of the monograph of Benssousan, Lions, and

Papanicolaou [1], one make the ansatz

$$u(y,s) = e^{-\frac{i}{\varepsilon} S(y,s)} m(\tfrac{y}{\varepsilon}, y, s, \varepsilon)$$

where $m(x,y,s,\varepsilon) = m_o(x,y,s) + \varepsilon m_1(x,y,s) + \ldots\ldots$ and m is assumed to have the periodicity of v in x. Substituting $u(y,s,\varepsilon)$ into (5) and equating the coefficient of ε^{-2} to zero, one sees that we must have

$$\text{i)} \quad \frac{\partial S}{\partial s} = E_n\left(\frac{\partial S}{\partial y} + a(y)\right)$$

(5) and

$$\text{ii)} \quad m_o(x,y,s) = \varphi_n\left(x, \frac{\partial S}{\partial y} + a(y)\right) h_o(y,s)$$

where h_o is an arbitrary function of y and s.

The coefficient of ε^{-1}, when one substitutes $u(y,s,\varepsilon)$ into (5), is a sum of an operator in x applied to m_1 and a term depending only on m_o. Since the operator in x has all functions of the form $\varphi_n(x, \partial S/\partial y + a(y))h(y,s)$ as its null space -- we assume here that $E_n(\partial S/\partial y + a(y))$ is a <u>simple</u> eigenvalue of $(i\partial_x + \partial S/\partial y + a(y))^2 + v(x)$ -- it follows that we will be able to solve for m_1, provided that the term in m_o is orthogonal to this null space (by the Fredholm alternative). After some computation, this orthogonality condition reduces to the following transport equation for $h_o(y,s)$:

$$\frac{\partial h_o}{\partial s} = \frac{\partial E_n}{\partial k} \cdot \frac{\partial h_o}{\partial y} + \frac{1}{2}\left(\frac{\partial}{\partial y} \cdot \frac{\partial E_n}{\partial k}\right) h_o + i b(y) h_o, \tag{7}$$

where b is the (real-valued) function

$$b = 2i \int_\Gamma \left(i\frac{\partial}{\partial x} + k\right)\varphi_n \cdot A \overline{\frac{\partial}{\partial k}\varphi_n}\, dx.$$

Here A is the anti-symmetric part of the jacobian of $a(y)$, Γ is a fundamental domain for the lattice over which $\varphi_n(x,k)$ is normalized,and $k = \partial S/\partial y + a(y)$ everywhere. The formulas (6) and (7) make the validity of the semi-classical approximation clear. One can, of course, continue equating coefficients of powers of ε to zero, so that u will satisfy the Schrödinger equation to order ε^N for any N. The leading term in u, $\exp(\frac{i}{\varepsilon} S)\, m_o$, is determined by (6) and (7). By Hamilton-Jacobi theory the solutions to

$$\dot{y} = \frac{\partial E_n}{\partial k} \left(\frac{\partial S}{\partial y}(y) + a(y)\right)$$

are solutions of

$$\dot{y} = \frac{\partial E_n}{\partial k} (k + a(y))$$
$$\dot{k} = -\frac{\partial E_n}{\partial y} (k + a(y))$$

and hence the support of h_o in space-time does propagate along the trajectories predicted by the semi-classical model. The hamiltonian $E_n(k + a(y))$ is not invariant whan one replaces $a(y)$ by $a(y) + \nabla \psi$. However, this is perfectly natural, since this replacement corresponds to replacing S by $S + \psi$, which in turn corresponds to the unitary transpormation in (2). The coefficient b in (7) is invariant when we add a gradient to $a(y)$, since the gradient does not contribute to the anti-symmetric part of the jacobian. We do not know the physical significance of $b(y)$, but it causes a variation in the phase of u.

THE DE HAAS-VAN ALPHEN EFFECT

For bismuth at temperatures close to absolute zero in strong (constant) magnetic fields, de Haas and van Alphen observed in the 1930's that various thermodynamic variables (they studied the "magnetic susceptibility") depend periodicallyon the reciprocal of the magnetic field strength. Since one is working at very low temperatures, this really means that, if one considers the spectral density for the hamiltonian H in (1) as a function of energy and field strength, then at Fermi energy the spectral density is a periodic function of the reciprocal of the field strength. Ln 1952, combining the semi-classical model with Bohr's quantization rule, Onsager [5] related the period of the de Haas-van Alphen oscillations to the geometry of the Fermi surface. Onsager predicted the following: the periods in the oscillations (there may be more than one) are proportional to the extremal areas of cross-sections of the Fermi surface by planes perpendicular to the magnetic field. In the 1960's this was developed into an efficient tool for measuring Fermi surfaces (of more metals than bismuth!) that has only recently been supplanted by more direct methods using positrons.

To understand Onsager's relation we need to construct approximate eigenfunctions for the hamiltonian H. Such eigenfunctions were constructed by Chambers in [2], but we hope that the construction here is more transparent. We will assume for simplicity that the atomic lattice is cubic, and consider the restriction of the hamiltonian to a block N cubes on a side so that the spectrum becomes discrete. We will also assume for simplicity that the boundary conditions on the faces of the block are just periodicity across parallel faces. Imitating Fedoriuk - Maslov [4] and Duistermaat [3], we want to build families of approximate eigenfunctions associated with invariant tori for the semi-classical flow. We assume that $B = (0,0,G)$ so that we may choose $a(y) = (0,y_1,0)$, and that the cross-section of the Fermi surface by the plane $k_3 = \xi_3$ contains a simple closed curve that we can represent as $k_1 = f_i(k_2,\xi_3)$, $\underline{k} \leq k_2 \leq \overline{k}$, $i = 1,2$. Then a famliy of invariant tori, parameterized by ξ_2 and ξ_3, is given by

$$\left\{ (y,k) : k_1 = f_i(\xi_2 + y_1),\ k_2 + \xi_2,\ k_3 = \xi_3,\ i = 1,2 \quad \underline{k} - \xi_2 \leq y_1 \leq \overline{k} - \xi_2 \right\}$$

To build eigenfunctions associatedwith these tori, when y_1 is strictly between $\underline{k} - \xi_2$ and $\overline{k} - \xi_2$ we make the ansatz

$$u(y,\varepsilon) = e^{\frac{i}{\varepsilon} S(y)} m(\tfrac{y}{\varepsilon}, y, \varepsilon)$$

where $S(y) = F_i(\xi_2 + y_1) + \xi_2 y_2 + \xi_3 y_3$, and F_i is an indefinite integral of f_i, $i = 1,2$. For y_1 near $\underline{k} - \xi_2$ or $\overline{k} - \xi_2$ we need to modify this ansatz in the usual way for a fold singularity, replacing u of the ansatz above by

$$u = \int e^{\frac{i}{\varepsilon} S(y,\alpha)} m(\tfrac{y}{\varepsilon}, y, \alpha, \varepsilon)\, d\alpha$$

To insure that these local asymptotic solutions can be patched together to give a globally defined approximate eigenfunction one must satisfy the Maslov quantization condition. In this case this condition takes the following form. The parameters ξ_2 and ξ_3 must be integer multiples of $2\pi\varepsilon N^{-1}$, and the area enclosed by the curve $k_1 = f_i(k_2,\xi_3)$, $i = 1,2$ must be an odd integer multiple of $\pi\varepsilon$. If one recalls that ε is a constant multiple of G, the field strength, one sees that the Maslov conditions will only be satisfied for a discrete set of values of the field strength. If one assumes that all eigenfunctions with energies close to Fermi energy can be asymptotically constructed by the method above, then one can conclude that as one varies ε and hence G the density of energy levels at Fermi energy will have peaks at intervals determined by the areas of the cross-sections of the Fermi surface which are extremal with respect to ξ_3.

REFERENCES

[1] A. Benssousan, J.L. Lions, and G. Papanicolaou, Asymptotic Analysis for Periodic Structures, North-Holland, Amsterdam, 1978.

[2] R. G. Chambers, Proc. Phys. Soc. 89, 695 (1966).

[3] J. Duistermaat, Oscillatory integrals, Lagrange immersions, and unfolding of singularities, CPAM 27 (1974),207 -281.

[4] V.P. Fedoriuk and M. V. Maslov, Semi-Classical Approximation in Quantum Mechanics, D. Reidel, Dordreht, 1980.

[5] L. Onsager, Phil. Mag. 43, 1006 (1952).

J.-C. Guillot
University of Paris Nord

J. Ralston
UCLA

E. Trubowitz
ETH - Zentrum - Zurich

MICROLOCAL ANALYSIS FOR

INHOMOGENEOUS GEVREY CLASSES

Otto Liess
Institut für Angewandte Mathematik
Universität Bonn
Wegelerstr. 10
D - 5300 Bonn 1
West - Deutschland

Luigi Rodino
Dipartimento di Matematica
Università di Torino
Via Carlo Alberto 10
I - 10123 Torino
Italia

Introduction

In the microlocal study of the pseudo differential operators with multiple characteristics it is not possible in general to base on the homogeneous geometry of the principal symbol and one is often led to replace the standard (homogeneous) wave front set with more general wave front sets of inhomogeneous type, modelling the total symbol of the operator.

As for the C^∞ category, inhomogeneous wave front sets were already introduced by Hörmander (1) to study constant coefficients equations with first order localizations, and subsequently used by many authors in other contexts, under different definitions and degrees of generality (see Rodino (5) and the references there).

In the study of the Gevrey singularities, the use of inhomogeneous wave front sets is still more natural. In fact, there is often a direct link among the geometric structure of the total symbol of a given operator with multiple characteristics, the corresponding inhomogeneous microlocalization and the Gevrey class on which the results find their precise formulation.

In the next § 1 we shall begin by defining the inhomogeneous Gevrey class G_φ related to a fixed "weight function" $\varphi(\xi)$. We shall then recall from Liess-Rodino (3) the notion of the φ-inhomogeneous Gevrey wave front set of a distribution and the main lines of the calculus of the pseudo differential operators with S^m_φ-symbol.

Such inhomogeneous machinery will allow us to state in § 2 a precise result of Gevrey microsolvability for the evolution model

$$(0\text{-}1) \qquad P = D_t - a_1(t,x,D_x) + a_0(t,x,D_t,D_x),$$

where $a_1(t,x,\xi)$ is an analytic family of real-valued symbols in S^1_φ and $a_0(t,x,\tau,\xi)$ is a symbol of order zero in the class determined by the weight function $|\tau| + \varphi(\xi)$.

The proof, which is detailed in Liess-Rodino (4), bases on a suitable calculus of Gevrey inhomogeneous Fourier integral operators and leads to the explicit construction of a parametrix of P.

As it will be shown in a future paper, this result is the starting point for the study of a wide class of linear partial differential operators with multiple characteristics which can be reduced to the form (0-1) through conjugation by classical analytic Fourier integral operators and factorization, once a suitable weight $\varphi(\xi)$ has been chosen describing the inhomogeneity of the total symbol.

Other applications of classes G_φ and φ-inhomogeneous wave front sets are given in Liess (2).

1 - Classes G_φ, related wave front sets and pseudo differential operators.

Our arguments are based on a given weight function $\varphi(\xi)$ which we assume uniformly Lipschitzian on R^n, i.e.:

$$(1\text{-}1) \qquad |\varphi(\xi) - \varphi(\eta)| \leqslant c\,|\xi - \eta| \quad \underline{\text{for all}} \quad \xi,\eta \in R^n$$

for some constant c; moreover we require that

$$\varphi(\xi) \geqslant c'(1 + |\xi|)^{\delta}, \tag{1-2}$$

for some positive constants c', δ.

For simplicity, we shall choose in the following the space of C^{∞} functions as our universe set (a generalization to Schwartz distributions would be not difficult, however).

The class G_{φ} is then defined locally near a point $x_0 \in R^n$ as the set of all C^{∞} functions u for which there are c, c' > 0 and a bounded sequence of distributions u_j such that $u = u_j$ for $|x - x_0| < c'$ and such that for all $\xi \in R^n$

$$|\hat{u}_j(\xi)| \leqslant c(cj/\varphi(\xi))^j, \qquad j = 1, 2, \ldots \quad . \tag{1-3}$$

Thus, when

$$\varphi(\xi) = (1 + |\xi|)^{\rho}, \qquad 0 < \rho \leqslant 1,$$

then $G_{\varphi}(U)$, $U \subset R^n$, is the standard Gevrey class $G^s(U)$, $s = 1/\rho$, of all functions $u \in C^{\infty}(U)$ satisfying in every $K \subset\subset U$ the estimates

$$\sup_K |D^{\alpha}u(x)| < C_K^{|\alpha|+1}(\alpha!)^s.$$

In particular, if $\rho = 1$, then we obtain the class of the real-analytic functions in U. More generally, if

$$\varphi(\xi) = 1 + \sum_{j=1}^{n} |\eta_j|^{1/M_j}, \tag{1-4}$$

where $M = (M_1, \ldots, M_n)$ is a n-tuple of rational numbers $\geqslant 1$, then $G_{\varphi}(U)$ is the class $G^M(U)$ of all $u \in C^{\infty}(U)$ such that for every $K \subset\subset U$

$$\sup_K |D^{\alpha} u(x)| < C_K^{|\alpha|+1}(\alpha_1!)^{M_1} \ldots (\alpha_n!)^{M_n}.$$

The definition in (1-3) can be microlocalized in a natural way. Let us in fact fix $\Gamma \subset R^n$ and set for $\varepsilon > 0$

$$\Gamma_{\varepsilon\varphi} = \{\xi \in R^n, \ \underline{\text{dist}} \ (\xi,\Gamma) < \varepsilon\varphi(\xi)\}.$$

We say that a C^∞ function u is G_φ -smooth at $\{x_0\} \times \Gamma$, and use the formal notation $WF_\varphi u \cap (\{x_0\} \times \Gamma) = \emptyset$, if the estimates (1-3) are satisfied in $\Gamma_{\varepsilon\varphi}$ for a sufficiently small $\varepsilon > 0$.

Let us then introduce the space of the φ-microfunctions at $\{x_0\} \times \Gamma$, $x_0 \in U$, $\Gamma \subset R^n$:

$$M_\varphi \ (\{x_0\} \times \Gamma) \ = \ C^\infty(U)/ \approx,$$

where $u \approx v$ means that $WF_\varphi (u - v) \cap (\{x_0\} \times \Gamma) = \emptyset$.

We now define a class of micro-pseudo differential operators on $U \times \Gamma$ by the standard formula

$$\text{(1-5)} \quad Au(x) = (2\pi)^{-n} \int e^{ix\xi} \ a(x,\xi) \ \chi(\xi) \ \hat{u}(\xi) \ d\xi,$$

where χ is some C^∞ function in R^n such that $\chi(\xi) = 0$ if $\xi \in \Gamma_{\varepsilon'\varphi}$, $\chi(\xi) = 1$ if $\xi \in \Gamma_{\varepsilon''\varphi}$; ε', ε'' are small constants, $0 < \varepsilon'' < \varepsilon'$, and we assume moreover $|D^\alpha \chi(\eta)| \leqslant c_\alpha$.

As for the symbol $a(x,\xi)$, it is an element of $S^m_\varphi \ (U,\Gamma)$, set of all the C^∞ functions defined on $U \times \Gamma_{\varepsilon\varphi}$, for some $\varepsilon > 0$, satisfying there the estimates

$$\text{(1-6)} \quad |D^\alpha_x \ D^\beta_\xi \ a(x,\xi)| \leqslant C^{|\alpha|+|\beta|+1} \ \alpha! \ \beta! \ \varphi(\xi)^{m-|\beta|}.$$

Note that (1-6) is equivalent with the fact that $a(x,\xi)$ admits an analytic extension as function of x in a complex neighborhood of U in $\mathbb{C}^n$, and $\zeta = \xi + i\vartheta$ in the region $|\vartheta| < \varepsilon\,\varphi(\xi)$, for a sufficiently small ε, with $a(x,\zeta)$ bounded there by $\varphi(\mathrm{Re}\zeta)^m$.

If $x_0 \in U$, then A in (1-5) defines by factorization a map

$$(1\text{-}7) \qquad A : \mathcal{M}_\varphi(\{x_0\} \times \Gamma) \to \mathcal{M}_\varphi(\{x_0\} \times \Gamma).$$

The same property (1-7) is valid if we consider the more general class of symbols $\tilde{S}^m_\varphi(U,\Gamma)$, of all the C^∞ functions which satisfy the estimates (1-6) for all α, β with x, ξ in the domain

$$(1\text{-}8) \quad \{x \in U,\ \xi \in \Gamma_{\varepsilon\varphi},\ |\xi| > c_1 \ \underline{\text{and}}\ \ \varphi(\xi) > c_2|\beta|\},$$

for some large constants c_1, c_2. We also introduce the corresponding class $\widetilde{SF}^m_\varphi(U,\Gamma)$ of formal sums $\sum_{j=0}^\infty a_j$, where

$$|D^\alpha_x D^\beta_\xi a_j(x,\xi)| < C^{|\alpha|+|\beta|+j+1}\ \alpha!\ \beta!\ j!\ \varphi(\xi)^{m-|\beta|-j},$$

for $x \in U$, $\xi \in \Gamma_{\varepsilon\varphi}$, $|\xi| > c_1$ and $\varphi(\xi) > c_2(|\beta|+j)$.

Obviously $\tilde{S}^m_\varphi(U,\Gamma)$ can be regarded as subset of $\widetilde{SF}^m_\varphi(U,\Gamma)$. An equivalence relation is then defined in a natural way on $\widetilde{SF}^m_\varphi(U,\Gamma)$, so that for any given $\sum_{j=0}^\infty a_j \in \widetilde{SF}^m_\varphi(U,\Gamma)$ we may always construct $a \in \tilde{S}^m_\varphi(U,\Gamma)$ with $a \sim \sum_{j=0}^\infty a_j$. Moreover, if $a \sim b$ then

$$(1\text{-}9) \quad |D^\alpha_x((a(x,\xi) - b(x,\xi))| \leqslant c_1^{|\alpha|+1}\ \alpha!\ \exp(-c_2\varphi(\xi))$$

for all $x \in U$, $\xi \in \Gamma_{\varepsilon\varphi}$, $|\xi| > c_3$, with suitable positive constants c_1, c_2, c_3. It follows that the operator with symbol $a(x,\xi) - b(x,\xi)$ is φ-regularizing, i.e. it maps any $u \in \mathcal{M}_\varphi(\{x_0\} \times \Gamma)$ into the zero microfunction.

The standard rules of the symbolic calculus are valid for pseudo differential operators with symbol in $\tilde{S}^m_\varphi(U, \Gamma)$, as detailed in (3).

Quite often we are forced to work with symbols from $S^m_\varphi(U, \Gamma)$, however; in particular, amplitudes of type (1-6) are necessary for the inhomogeneous Fourier integral calculus in (4). It is then natural to assume the following condition is satisfied:

(1-10) for any $a \in \tilde{S}^m_\varphi(U, \Gamma)$ we can find $b \in S^m_\varphi(U', \Gamma)$ such that for some positive constants c_1, c_2, c_3 the estimates (1-9) are satisfied by $a(x,\xi) - b(x,\xi)$ for all x in a neighborhood $U' \subset U$ of x_0, $\xi \in \Gamma_{\varepsilon\varphi}$, $|\xi| > c_3$.

The operator with symbol $a(x,\xi) - b(x,\xi)$ is therefore φ-regularizing; we may then transfer all results from $\tilde{S}^m_\varphi(U, \Gamma)$ symbols to symbols in $\tilde{S}^m_\varphi(U, \Gamma)$.

The property (1-10) is known to hold in the classical analytic case, i.e. for $\varphi(\xi) = 1 + |\xi|$, $\Gamma = R^n$. The property (1-10) is also valid for $\varphi(\xi)$ as in (1-4), Γ given by any M-parabolic ray, under the additional assumption that $a(x,\xi)$ be initially defined on a full M-parabolic neighborhood of Γ (the proof is given in (4), where sufficient conditions for (1-10) are discussed in detail).

2. Parametrix for an evolution model

Let now X be an open subset of R^{n+1}, with variables (t,x) and dual variables (τ,ξ). For simplicity we shall assume that $X = (-T,T) \times U$ for some $T > 0$ and for some bounded neighborhood U of the origin in R^n. We then consider in R^{n+1} a weight function of the form

$$\Psi(\tau,\xi) = |\tau| + \varphi(\xi), \tag{2-1}$$

where φ is a weight function in R^n. All our arguments in this paragraph will refer to sets of the form

(2-2) $$\{(\tau,\xi) \in R^{n+1};\ |\tau| < C\ \varphi(\xi)\},$$

for some $C > 0$. Note that on such sets automatically $\varphi(\xi) \sim \Psi(\tau,\xi)$.

Let us now also fix $\Gamma \subset R^n$ and write

(2-3) $$\Gamma' = \{(\tau,\xi) \in R^{n+1};\ \xi \in \Gamma,\ \ |\tau| < C\ \varphi(\xi)\}.$$

We want to study the pseudo differential model (0-1) of the Introduction:

(2-4) $$P = D_t - a_1(t,x,D_x) + a_0(t,x,D_t, D_x),$$

with symbol p given by

(2-5) $$p(t,x,\tau,\xi) = \tau - a_1(t,x,\xi) + a_0(t,x,\tau,\xi).$$

<u>For</u> a_1 <u>we assume that it is real-valued and that we can find a complex neighborhood</u> $X^{¢}$ <u>of</u> X <u>in</u> $¢^{n+1}$, <u>a constant</u> $\varepsilon > 0$ <u>and an analytic extension of</u> a_1 <u>to the set</u> $X^{¢} \times \Gamma^{¢}$, <u>with</u>

$$\Gamma^{¢} = \{\zeta \in ¢^n,\ \zeta = \xi + i\vartheta,\ \xi \in \Gamma_{\varepsilon\varphi},\ |\vartheta| < \varepsilon\ \varphi(\xi)\}.$$

<u>We assume for this extension the bound</u>

(2-6) $$|a_1(t,x,\zeta)| \leqslant c\ \varphi(\mathrm{Re}\,\zeta),\quad (t,x,\zeta) \in X^{¢} \times \Gamma^{¢},$$

<u>for some</u> $c > 0$.

In other words: a_1 is an analytic family of symbols in $S^1_\varphi(U,\Gamma)$, depending on the parameter $t \in (-T,T)$. We may however also regard a_1 as a symbol in $S^1_\Psi(X,\Gamma')$, once we have fixed some C in (2-2), (2-3); in fact $\Psi \sim \varphi$ on Γ' and hence also in $\Gamma'_{\varepsilon\varphi}$, if ε is small.

As for a_0, we shall suppose it is a remainder term of order zero. To express this in a more precise way, we now observe that all difficulties in the study of (2-4) should came from those points (t,x,τ,ξ) for which $\tau - a_1(t,x,\xi)$ is small, but (τ,ξ), and there - fore in fact ξ, is large. We shall also restrict attention to the origin $(t=0,\ x=0) = 0$ in R^{n+1} and set consequently

$$\Sigma = \{(\tau,\xi) \in R^{n+1};\quad \xi \in \Gamma,\ |\xi| > c_1,\ \tau = a_1(0,0,\xi)\},$$

with a large constant c_1. We have $\Sigma \subset \Gamma'$, if the constant C in (2-3) is sufficiently large.

Our assumption on a_0 in now: $a_0 \in S^o_\Psi(X,\Sigma)$ (in practice, we are allowed to shrink X and Γ in order to have this satisfied).

From § 1 we know that P is well defined as a map

$$\text{(2-7)} \qquad P : M_\Psi(\{o\} \times \Sigma) \to M_\Psi(\{o\} \times \Sigma)$$

A further hypothesis in required, on the geometry of the set

$$\Sigma^* = \{(\sigma,\eta) \in R^{n+1},\quad \sigma = 0,\quad \eta \in \Gamma,\quad |\eta| > c_1^*\}.$$

In the sequel we need to know that the symbols from $S^m_\Psi(X,\Sigma^*)$ lead to a class of operators which is as rich as the one from $\tilde{S}^m_\Psi(X,\Sigma^*)$; therefore henceforth we assume the property (1-10) is valid at the origin for Ψ and Σ^*.

Under the proceding assumptions, we have the following micro-solvability result.

Theorem 2-1. For all $v \in M_\Psi(\{o\} \times \Sigma)$ there is $u \in M_\Psi(\{o\} \times \Sigma)$ such that $Pu = v$.

Actually, we may find a right parametrix for P, i.e. a linear map $E : M_\Psi(\{o\} \times \Sigma) \to M_\Psi(\{o\} \times \Sigma)$ such that

$$\text{(2-8} \qquad PE = \text{identity on } M_\Psi(\{o\} \times \Sigma).$$

This is an obvious consequence of the following proposition, reducing the problem to the trivial study of the equation $D_t u = \upsilon$.

Proposition 2-2. There exist two linear maps A and B,

A acting from $M_\Psi(\{o\} \times \Sigma^*)$ to $M_\Psi(\{o\} \times \Sigma)$,

B acting from $M_\Psi(\{o\} \times \Sigma)$ to $M_\Psi(\{o\} \times \Sigma^*)$,

such that

(i) BA and AB are the identity map in

$M_\Psi(\{o\} \times \Sigma^*)$ and $M_\Psi(\{o\} \times \Sigma)$;

(ii) $PA = A D_t$ on $M_\Psi(\{o\} \times \Sigma^*)$.

3. Proof of proposition 2-2

Before proving proposition 2-2, we shall study the Hamiltonian flow H_λ associated to

$$p_1(t,x,\tau,\xi) = \tau - a_1(t,x,\xi). \tag{3-1}$$

Let us in fact consider $X' = (-T',T') \times U'$, with $0 < T' < T$ and U' some neighborhood of the origin in $\mathbb{R}^n$ with $U' \subset\subset U$. We also fix $\delta > 0$ so small that $(\Gamma_{\delta\varphi})_{\delta\varphi} \subset \Gamma_{\epsilon\varphi}$ anf we set $\Sigma_1 = \Sigma_{\epsilon',\varphi}$ for some ϵ' which is chosen such that $\Pi(\Sigma_1) \subset \Gamma_{\delta\varphi}$. For $(s,y) \in X'$, $(\sigma,\eta) \in \Sigma_1$, $H_\lambda(s,y,\sigma,\eta) = (t,x,\tau,\xi)$ is then defined for small λ as the solution of

$$\left\{\begin{aligned} &\frac{d\xi_j}{d\lambda} = (a_1)_{x_j}(t,x,\xi),\ j = 1, \ldots, n. \\ &\frac{dx_j}{d\lambda} = -(a_1)_{\xi_j}(t,x,\xi),\ j = 1, \ldots, n. \\ &\xi(0) = \eta,\ x(0) = y. \end{aligned}\right. \tag{3-2}$$

with $t = \lambda + s$. to which we add

(3-3) $\frac{d\tau}{d\lambda} = (a_1)_t\ (t,x,\xi), \qquad \tau(0) = \sigma.$

Here we consider (3-2), (3-3) in the complex domain $X^{\mathbb{C}} \times \Gamma^{\mathbb{C}}$ and we use the extension of a_1 satisfying (2-6) for $(t,x) \in X^{\mathbb{C}}$, $\zeta \in \Gamma^{\mathbb{C}}$ To obtain bounds for x,ξ and for the domain of existence of solutions, it is here convenient to replace temporarily ξ_j by $\tilde{\xi}_j = \xi_j / \varphi(\eta)$. With the notation $\tilde{a}_1(t,x,\tilde{\xi}) = a_1(t,x,\varphi(\eta)\tilde{\xi}) / \varphi(\eta)$, (3-2) is then replaced by

$$\frac{d\tilde{\xi}_j}{d\lambda} = (\tilde{a}_1)_{x_j}\ (t,x,\tilde{\xi}),\ j = 1, \ldots, n,$$

$$\frac{dx_j}{d\lambda} = (\tilde{a}_1)_{\xi_j}\ (t,x,\tilde{\xi}),\ j = 1, \ldots, n,$$

$$\xi(0) = \eta/\varphi(\eta),\ x(0) = y.$$

From the classical theory of ordinary differential equations in the complex domain applied first for $(x,\tilde{\xi})$, we obtain immediately

<u>Proposition 3-1 The solutions</u> $x = x(\lambda,s,y,\eta)$, $\xi = \xi(\lambda,s,y,\eta)$, <u>of (3-2) are well-defined as functions of</u> $(s,y,\eta) \in X' \times \Gamma_{\delta\varphi}$ and $\lambda \in (-\lambda_o,\lambda_o)$ <u>for a suitable</u> $\lambda_o > 0$ <u>which does not depend on</u> (s,y,η). <u>Writing</u>

(3-4)

$$x_j(\lambda,s,y,\eta) = y_j + x_{oj}(\lambda,s,y,\eta),\ j = 1,\ldots, n,$$

$$\xi_j(\lambda,s,y,\eta) = \eta_j + \xi_{oj}(\lambda,s,y,\eta),\ j = 1,\ldots, n,$$

<u>we have that the</u> x_{oj}, ξ_{oj} <u>can be extended as analytic functions of</u> λ,s,y, <u>regarded as complex variables in some open neighborhood</u> $((-\lambda_o,\lambda_o) \times X')^{\mathbb{C}}$ <u>of</u> $(-\lambda_o,\lambda_o) \times X'$ <u>in</u> $\mathbb{C}^{n+2}$, <u>and of</u> $\zeta = \eta + i\theta$ <u>in the set</u>

$$\Gamma_1^{\mathbb{C}} = \{\zeta \in \mathbb{C}^n;\ \zeta = \eta + i\theta,\ \eta \in \Gamma_{\delta\varphi} \text{ and } |\theta| < \delta\varphi(\eta)\}$$

<u>For these extensions we have the bounds</u>

(3-5) $|x_{oj}(\lambda,s,y,\eta)| \leq C\ |\lambda|,\quad j = 1, \ldots, n,$

(3-6) $|\xi_{oj}(\lambda,s,y,\eta)| \leq C\ |\lambda|\ \varphi(\eta),\ j = 1, \ldots, n,$

<u>for a suitable constant</u> C <u>which does not depend on</u> λ,s,y,η.
As for the solution of (3-3), we have

(3-7) $\tau = \tau(\lambda,s,y,\sigma,\eta) = \sigma + \tau_o(\lambda,s,y,\eta)$

where $\tau_o(\lambda,s,y,\eta)$ can also be extended analytically to $((-\lambda_o,\lambda_o) \times X')^{\mathbb{C}} \times \Gamma_1^{\mathbb{C}}$, with

(3-8) $|\tau_o(\lambda,s,y,\eta)| \leq C|\lambda|\varphi(\eta)$.

From (3-6) and from Cauchy's inequalities we conclude that ξ_{oj} is in $S^1_\Psi((-\lambda_o,\lambda_o) \times X', \Gamma')$ for all j: similarily we obtain $x_{oj} \in S^o{}_\Psi((-\lambda_o,\lambda_o) \times X', \Gamma')$, $j = 1, \ldots, n$ and $\tau_o \in S^1_\Psi((-\lambda_o,\lambda_o) \times X',\Gamma')$.
It is also clear from (3-5),(3-6),(3-8) that $H_\lambda(s,y,\sigma,\eta)$ takes actually values in $X \times \Sigma_{\epsilon\varphi}$ if λ_o is sufficiently small.
We may now prove proposition 2-2. Following a standard argument,we begin by considering the Fourier integral operator

(3-9) $$Fu(t,x) = \int\int e^{i\omega(t,x,\sigma,\eta)}\chi(\sigma,\eta)\ \hat{u}(\sigma,\eta)\ d\sigma\ d\eta,$$

where the cut - off function is defined = 1 in $(\Sigma^*)_{\epsilon'\Psi}$ and where ω is determined by

(3-10) $$\omega(t,x,\sigma,\eta) = t\sigma + \tilde{\omega}(t,x,\eta)$$

and

(3-11) $$\partial_t\tilde{\omega} = a_1(t,x,\tilde{\omega}_x)$$
$$\tilde{\omega}|_{t=0} = \langle x,\eta\rangle.$$

Thus, in fact, $\tilde{\omega}$ is the solution of the eikonal equation associated with p_1. The solution of (3-11) can therefore be expressed in terms of the Hamilton - Jacobi system (3-2) in which we take $t = \lambda$(i.e.s = 0). We denote these solutions by $x = X(\lambda,y,\eta)$, $\xi = \Xi(\lambda,y,\eta)$. Let us recall the standard construction of $\tilde{\omega}$, starting from X and Ξ. First we solve $x = X(\lambda,y,\eta)$ with respect to y, obtaining $y = G(\lambda,x,\eta)$.(This is always possible since, as a consequence of uniqueness results for solutions of ordinary differential equations, the map $y \to X(\lambda,y,\eta)$ is injective if λ and η are fixed). We can then show that G is defined for all small complex λ, for all x in a full complex neighborhood of 0 and for all η on a set of type $\Gamma^{\mathbb{C}}$. Moreover, G is analytic in all variables and is bounded.

Also note that for fixed λ,η, $G(\lambda,x,\eta)$ is precisely that initial value for x(0) in (3-2) which, when combined with η as an initial value for $\xi(0)$, gives $x(\lambda) = x = X(\lambda,y,\eta)$. The next thing to consider is the function $L(\lambda,x,\eta) = \Xi(\lambda,G,\eta)$. Thus, roughly speaking, L associates to the x - component on some integral curve of (3-2) the corresponding ξ - component.

From the Hamilton-Jacobi theory it is then clear that we must have $L(\lambda,x,\eta) = \tilde{\omega}_x(\lambda,x,\eta)$ if $\tilde{\omega}$ satisfies (3-10). If we write

$$(3\text{-}12) \quad \tilde{\omega}(t,x,\eta) = \langle x,\eta \rangle + \omega_o(t,x,\eta)$$

we therefore have

$$(3\text{-}13) \quad \omega_o(t,x,\eta) = \int_o^t a_1(\lambda,x,L(\lambda,x,\eta))\, d\lambda .$$

Conversely, it is not difficult to prove that (3-12) and (3-13) define a solution of (3-11).

Taking here into account the estimates from (2-6) and from proposition 3-1 (and using similar notations), we see easily that $\omega_o(t,x,\eta)$ is well-defined in $X' \times \Gamma_{\delta\varphi}$ and can be extended as an analytic function in $((-\lambda_o,\lambda_o) \times U')^{\mathbb{C}} \times \Gamma_1^{\mathbb{C}}$, satisfying there

$$(3\text{-}14) \quad |\omega_o(t,x,\eta)| \leq C|t|\ \varphi(\eta),$$

for a suitable constant C. It follows that $\omega_o \in S^1_\varphi \quad ((-\lambda_o,\lambda_o) \times U',\Gamma')$ and therefore the estimates (1), §1.4 of (4) are satisfied.
Now we observe that $|\det \omega_{(t,x),(\sigma,\eta)}(t,x,\sigma,\eta)| = 1 + R(t,x,\sigma,\eta)$, where $R \in S^o_\varphi\ (X',\theta)$ whenever θ is as in (2-2).
Moreover, $R(0,x,\sigma,\eta) = 0$, so we conclude that

$|\det \omega_{(t,x),(\sigma,\eta)}(t,x,\sigma,\eta)| \geq 1/2$ (say) for all $|t| < \lambda_o'$ if λ_o is small. This shows that we can solve the equations

(3-15) $\quad (s,y) = \omega_{(\sigma,\eta)}(t,x,\sigma,\eta)$

locally with respect to (t,x) as well as the equation

(3-16) $\quad (\tau,\xi) = \omega_{(t,x)}(t,x,\sigma,\eta)$

with respect to (σ,η).
If we first solve (3-15) and the introduce (t,x) as functions of (s,y,σ,η) into (3-16) we get a canonical transformation:

(3-17) $\quad (t,x,\tau,\xi) = T(s,y,\sigma,\eta)$.

It is well-known from the Hamilton-Jacobi theory that actually we have $x = X(s,y,\eta)$, $\xi = \Xi(s,y,\eta)$.
Moreover, $t = s$ and for the τ - component of T, which we call τ, we have

$$\tau = \tau(s,y,\sigma,\eta) = \sigma + \tilde{\omega}_t(s,x,\eta) =$$
$$= \sigma + a_1(s,X(s,y,\eta),\ \Xi(s,y,\eta)).$$

It follows in particular that T, which at first was defined locally is now defined globally on $X' \times R \times \Sigma^*_{C\Psi}$. It is also clear from the above that $T(0,0,0,\eta) = a_1(0,0,\eta)$ so the map $(\sigma,\eta) \to T(0,0,\sigma,\eta)$ maps $(0,0,0,\Sigma^*)$ to $0 \times \Sigma$.
Moreover , the inverse of T can be computed, at first locally, if we first solve (3-16) with respect to (σ,η) and then combine the result with (3-15). It follows that for $(s,y,\sigma,\eta) = T^{-1}(t,x,\tau,\xi)$, we have $t = s$, $y = x + y_0(t,x,\xi)$, $\eta = \xi + \eta_o(t,x,\xi)$, $\sigma = \tau + \sigma_o(t,x,\xi)$, where the components y_{oj}, $j = 1, \ldots, n$ are symbols in $S^o_\Psi\ (X',\Sigma)$ for a small

neighborhood X' of $0 \in R^{n+1}$, and the η_{oj}, $j = 1, \ldots, n$, σ_o are symbols in $S^1_\Psi(X',\Sigma)$.

Our next remark is that $\Psi(\omega_{t,x}(0,0,\sigma,\eta)) = |\omega_t| + \varphi(\omega_x) = |\sigma + a_1(0,0,\eta)| + \varphi(\eta) \sim \varphi(\eta) \sim \Psi(\sigma,\eta)$ if (σ,η) remains in a set of type θ (and $\eta \in \Gamma$). This shows that condition (6) from §1.1 in (4) is valid.

After all this discussion it is now clear that F defined in (3-9) is a Fourier integral operator of the type considered in this paper. Note then that

(3-18) $|\omega_x(t,x,\sigma,\eta) - \eta| \leq c_1|t| \; \varphi(\eta)$

(3-19) $|\omega_t(t,x,\sigma,\eta) - \sigma - a_1(0,x,\eta)| \leq c_2|t| \; \varphi(\eta)$,

(3-20) $\omega_\sigma(0,x,\sigma,\eta) = t$

(3-21) $\omega_\eta(t,x,\sigma,\eta) - x| \leq c_3|t|$.

From (3-20) and (3-21) it follows in particular that (t,x), $(s,y) = 0$ and Σ^* are ω - compatible in the sense of (4). Moreover, both, C and D, from §1.4 in (4) will hold for ω if t is small.

We may now end the proof of proposition 2-2. We can conclude in fact from the above that F and F* are welldefined as maps from $M_\Psi(\{0\} \times \Sigma^*)$ to $M_\Psi(\{0\} \times \Sigma)$ respectively $M_\Psi(\{0\} \times \Sigma)$ to $M_\Psi(\{0\} \times \Sigma^*)$. We now look at PF. It follows from §3.1, §3.2 in (4) that the principal symbol of PF, when written as a F.I.O. with phase funcion ω is $\omega_t(t,x,\sigma,\eta) - a_1(t,x,\omega_x(t,x,\sigma,\eta)) = \sigma$. Combining this with §3.7 in (4), we conclude that we must have

$$PF = FD_t + F_o$$

where for some $q \in \tilde{S}^o_\Psi(X',\Sigma^*)$,

$$F_o u(t,x) = \iint e^{i\omega(t,x,\sigma,\eta)} \chi(\sigma,\eta) q(t,x,\sigma,\eta) \hat{u}(\sigma,\eta) d\sigma \, d\eta.$$

Since (1-10) is valid, we may assume $q \in S^o_\Psi(X',\Sigma^*)$.

We now claim that there is a pseudodifferential operator K with symbol in $S^o_\Psi(X'',\Sigma^*)$, such that

(3-22) $F_o = FK$,

if $X'' \subset\subset X'$ is a small neighbourhood of 0.

In fact, F^*F is an elliptic pseudodifferential operator in $M_\Psi(\{0\} \times \Sigma^*)$, so $(F^*F)^{-1}$ exists as a pseudodifferential operator. Therefore (3-22) is valid with $K = (F^*F)^{-1}F^*F_o$.
It is easily checked that the principal symbol r_o of K is related to the principal symbol q_o of F_o by $r_o(t,\tilde{\omega}_\eta(t,x,\eta),\ \sigma,\eta) = q(t,x,\sigma,\eta)$.
Summing up, we have

$$PF = F(D_t + K).$$

We now apply the following

<u>Lemma 3-2.</u> <u>If</u> K <u>is a pseudodifferential operator with symbol in</u> $S^o_\Psi(X_1,\Sigma^*)$, <u>then there exists a pseudodifferential operator</u> $\tilde{K}$ <u>with symbol in</u> $S^o_\Psi(X_2,\Sigma^*)$ <u>such that</u>

$$(D_t + K)\ \tilde{K} = \tilde{K}\ D_t.$$

The proof of lemma 3-2 is standard.
In fact an asymptotic expansion for the symbol r of $\tilde{K}$ is obtainable by solving suitable transport equations. In particular, the principal symbol $\tilde{r}_o$ of $\tilde{K}$ is given by

$$\tilde{r}_o(s,y,\sigma,\eta) = \exp\ (-i \int_o^s r_o(\lambda,y,\sigma,\eta)\ dx).$$

This is easily verified to be a symbol in $S^o_\Psi(X_1,\Sigma^*)$, etc.
Returning to the proof of proposition 2-2, we set finally $A = F\tilde{K}$ and (ii) is satisfied. Since we have had control on principal symbols at each step, it is also immediate that A is an elliptic F.I.O.. But then B^* is just $(A^*A)^{-1}\ A$, where A^*A is inverted as a peudodifferential operator.
This completes the proof.

R E F E R E N C E S

(1) L. Hörmander, On the existence and regularity of solutions of linear pseudo differential operators, L'Enseignement Math., 17 (1971), 99-163.

(2) O. Liess, Microlocality of the Cauchy problem in inhomogeneous Gevrey classes, Comm. Partial Differential Equations, to appear.

(3) O. Liess - L. Rodino, Inhomogeneous Gevrey classes and related pseudo differential operators, Boll. Un. Mat. Ital., Ser. VI, 3 - C (1984), 233-323.

(4) O. Liess - L. Rodino, Fourier integral operators and inhomogeneous Gevrey classes, Ann. Mat. Pura Appl., to appear.

(5) L. Rodino, Microlocal analysis for spatially inhomogeneous pseudo differential operators, Ann. Scuola Norm. Sup. Pisa, Ser. IV, 9 (1982), 211-253.

Estimates on the number of resonances for semiclassical Schrödinger operators

Johannes Sjöstrand

Dept. of Mathematics
University of Lund
Box 118, S-221 00 Lund
Sweden

We are interested in the behavior of the resonances near 0 for the Schrödinger operator $P = -h^2\Delta + V(x)$ on $\mathbb{R}^n$, when $h \to 0$. A microlocal approach to these problems was developed by B. Helffer and the author in [8], and was based on the idea of complex scaling due to Aguilar-Combes [2] and Balslev-Combes [3]. So far, resonances near 0 have been analyzed rather thoroughly in a number of relatively simple cases; in [8] we studied the resonances created by a potential well in an island through tunneling. Similar results were obtained with complex scaling techniques by Combes-Duclos-Klein-Seiler [5], and by Hislop-Sigal [9]. In [11] we analyzed the resonances created by a non-degenerate critical point of the potential. The special case of a potential maximum was also treated by Briet-Combes-Duclos [4]. In [6] C. Gérard and the author studied the case when the set of classically trapped points form a closed trajectory of hyperbolic type. These results have been reviewed elsewhere ([7], [12]), so we shall not recall the detailed statements here.

In more complicated cases, there is little hope to obtain complete asymptotic expansions for the individual resonances in a certain region, and one has to look for weaker results. So far, we have studied two such questions:

1° To find resonance-free regions near the real axis (Work in progress with C. Gérard).

2° To find upper bounds for the number of resonances in certain regions of the complex plane.

We shall here report on some recent results about the second problem. They are still preliminary in the sense that we have not yet analyzed completely the various geometric situations, where they apply.

Assume that V is analytic and real-valued on $\mathbb{R}^n$ and that there are functions r, $R \in C^\infty(\mathbb{R}^n)$ such that

(1) $r \geq 1$, $r R \geq 1$ and $\partial^\alpha r = O(r R^{-|\alpha|})$, $\partial^\alpha R = O(R^{1-|\alpha|})$ uniformly on $\mathbb{R}^n$ for all $\alpha \in \mathbb{N}^n$.

(2) There exists $C > 0$ such that V has a holomorphic extension to $\{x \in \mathbb{C}^n;\ |\mathrm{Im}\, x| < C^{-1}R(\mathrm{Re}\, x)\}$, satisfying $|V(x)| \leq C\, r(\mathrm{Re}\, x)^2$.

We also assume

(3) There exists a real-valued (escape-) function $G \in C^\infty(\mathbb{R}^{2n})$ with $\partial_x^\alpha \partial_\xi^\beta G = O(\tilde{r}^{1-|\beta|} R^{1-|\alpha|})$ for $|\alpha| + |\beta| \geq 1$, such that $H_p G \geq r^2/C$ in $p^{-1}(0) \setminus K$. Here C is some constant and K is some compact, $p = \xi^2 + V(x)$ and H_p is the corresponding Hamilton field. Finally $\tilde{r}(x, \xi) = (\xi^2 + r(x)^2)^{1/2}$.

After a suitable modification of G in the region where $|\xi| >> r(x)$, we can define certain Sobolev spaces $H(\Lambda_{tG}, m)$, when $t > 0$ and $h > 0$ are small enough (See [8] for details.) Here m is a suitable weight function and $\Lambda_{tG} \subset \mathbb{C}^{2n}$ is given by: $\mathrm{Im}(x, \xi) = t\, H_G(\mathrm{Re}(x, \xi))$. In [8] we obtained the following basic result:

<u>Theorem 1</u>. For $t > 0$ sufficiently small, there exists $h_o = h_o(t) > 0$ and a neighborhood $\Omega = \Omega_t \subset \mathbb{C}$ of 0, such that for $0 < h \leq h_o$, we have:

For all $z \in \Omega$, the operator $P - z$: $H(\Lambda_{tG}, \tilde{r}^2) \to H(\Lambda_{tG}, 1)$ is Fredholm of index 0. Moreover, there is a discrete set $\Gamma(h) \subset \Omega$ such that $P - z$ is bijective for $z \in \Omega \setminus \Gamma(h)$, and when $z \in \Gamma(h)$, then $P - z$ splits in a natural way into a direct sum of one bijective operator and one nilpotent operator: $F_z \to F_z$. Here $F_z \neq 0$ is finite dimensional.

The elements of $\Gamma(h)$ are called resonances and if $z \in \Gamma(h)$, then $\dim(F_z)$ is the corresponding multiplicity. In [8], we showed that a different choice of (t, G) gives rise to the same resonances and the same spaces F_z for z in a sufficiently small neighborhood of 0. We also showed that the resonances belong to the closed lower half-plane.

Let $\epsilon_o > 0$ be so small that the conclusion of (0.3) remains valid also in $p^{-1}(\epsilon)$, for $\epsilon \in [-\epsilon_o, \epsilon_o]$. We then know that $H_p(G) \geq r^2/C$ outside a compact set in $p^{-1}([-\epsilon_o, \epsilon_o])$. We shall need some control on G, but less regularity, also inside a compact. For that purpose we introduce the following refined version of the assumption (3):

(4) There exists a real valued function G on $\mathbb{R}^{2n}$ and an open subset $W \subset\subset p^{-1}([-2\varepsilon_o, 2\varepsilon_o])$, such that:

(a) G satisfies all the assumptions of (3) outside W and $H_pG \geq r^2/C$ in $p^{-1}([-\varepsilon_o, \varepsilon_o]) \setminus W$.

(b) G and H_pG belong to $C^{1,1}$ in a neighborhood of $\overline{W}$ and $H_pG \geq C^{-1}\|\nabla G\|^2$ in $W \cap p^{-1}([-\varepsilon_o, \varepsilon_o])$.

Here we recall that $C^{1,1}$ is the space of C^1-functions with Lipschitz gradient.

For $0 \leq \delta \leq r \leq 1$, $\delta \leq C_o^{-1}$, let $M(r, \delta)$ be the number of resonances of P in the disc $\overline{D(ir, r+\delta)}$ of center i r and radius r + δ. Put

(5) $$V(r, \delta) = \mathrm{Vol}(\{\rho \in \mathbb{R}^{2n};\ p^2 + r\, H_pG < r\delta\}) .$$

Then the main result is:

<u>Theorem 2</u>. Assume (1), (2), (4) and let C_o, $C > 0$ be sufficiently large constants. Then, for $C_oh \leq \delta \leq r \leq 1$, $\delta \leq C_o^{-1}$, $r\delta \geq C_oh$, we have

(6) $$M(r, \delta) \leq C\, V(r, C\delta)h^{-n} .$$

In the proof of this theorem, we work in the Hilbert space $H(\Lambda_{tG}, 1)$, where it turns out that some classical inequalities of H. Weyl can be applied, that compare the eigen-values of P-ir with those of $\sqrt{(P-ir)^*(P-ir)}$. The work is then to

(a) estimate the lowest eigenvalue of $(P-ir)^*(P-ir)$ from below,

(b) estimate the number of eigenvalues of $(P-ir)^*(P-ir)$, which are smaller than or equal to $(r + \delta)^2$.

Using the techniques of [8], [13], this becomes rather easy.

Depending on the behavior of the classical flow, one wants of course to choose G so that V becomes as small as possible, i.e. so that H_pG is as large as possible. In cases when the set of classically trapped points is not a manifold, it seems essential that we can allow G to be only of class $C^{1,1}$ in a bounded domain (See Proposition 5 below).

Under the assumptions (1) - (3) it is always possible to find a smooth function G as in (4), (see the appendix of [6]), but we may have $H_p(G) = 0$ on a set of non-zero volume. We then get the following result, which could be viewed as a semiclassical analogue of a result of Melrose [10] in acoustical scattering.

Corollary 3. Under the general assumptions (1) - (3), there are constants $\delta_o > 0$ and $C > 0$ such that

(7) $$M(1, \delta_o) \leq C h^{-n} .$$

If we assume in addition that $dp \neq 0$ on the set $p = 0$, we get the sharper estimate,

(8) $$M(r, \delta) \leq C\sqrt{r\delta}\, h^{-n} ,$$

when $C_o h \leq \delta \leq r \leq 1$, $\delta \leq 1/C_o$, $r\delta \geq C_o h$.

Back to Theorem 2, we notice that if $\delta \geq \sqrt{C_o h}$ we get with a new constant C:

(9) $$M(\delta, \delta) \leq C\, \tilde{V}(C\delta)h^{-n} ,$$

where,

(10) $$\tilde{V}(\delta) = \mathrm{Vol}(\{\rho \in \mathbb{R}^{2n};\ |p| \leq \delta,\ H_p G \leq \delta\}) .$$

Theorem 2 remains uniformly valid, if we replace P by $P - \varepsilon$, $\varepsilon \in [-\varepsilon_o/2, \varepsilon_o/2]$. If $I \subset [-\varepsilon_o/2, \varepsilon_o/2]$, and $\sqrt{C_o h} \leq \delta \leq 1/C_o$, we can cover the rectangle $I - i[0, \delta]$ by discs $\overline{D(\varepsilon_j + 2i\delta, 4\delta)}$, with $\varepsilon_{j+1} - \varepsilon_j = 2\sqrt{3}\delta$. If $M(I, \delta)$ denotes the number of resonances in the rectangle $I - i[0, \delta]$, we obtain quite easily from Theorem 2:

Corollary 4. Assume (1), (2), (4) and let C_o, $C > 0$ be sufficiently large. Then for $C_o h \leq \delta \leq 1/C_o$ and all intervals $I \subset [-\varepsilon_o/2, \varepsilon_o/2]$, we have

(11) $$M(I, \delta) \leq C\, \mathrm{Vol}(R(I + [-C\tilde{\delta}, C\tilde{\delta}], C\delta))h^{-n} .$$

Here $R(J, \varepsilon) = \{\rho;\ p(\rho) \in J,\ H_p G \leq \varepsilon\}$, $\tilde{\delta} = \max(\delta, \sqrt{C_o h})$.

In order to discuss examples, we first recall an easy geometric discussion from [6], valid under the general assumptions (1) - (3): We work in $p^{-1}([-\varepsilon_o, \varepsilon_o])$, for $\varepsilon_o > 0$ sufficiently small. Let $]T_-(\rho), T_+(\rho)[$ be the maximal interval of definition for $t \to \Phi_t(\rho) = \exp(tH_p)(\rho)$ and put: $\Gamma_\pm = \{\rho \in p^{-1}([-\varepsilon_o, \varepsilon_o]);\ \Phi_t(\rho) \not\to \infty,\ t \to T_\mp(\rho)\}$. Then $\Gamma_\pm$ are closed and $\Gamma_+ \cap \{G \leq T\}$, $\Gamma_- \cap \{G \geq -T\}$ are compact for every $T \in \mathbb{R}$. For some $T_o > 0$ we have $\Gamma_+ \subset \{G \geq -T_o\}$, $\Gamma_- \subset \{G \leq T_o\}$. Here G is as in (3). In particular, $K = \Gamma_+ \cap \Gamma_-$ is compact. We also let $\Gamma^o_\pm$, K^o denote the intersection with $\{p = 0\}$.

Example 1. Assume that K^o is a point. After a translation, we may assume that $K^o = \{(0, 0)\}$ so that

(12) $$V(0) = 0 \quad , \quad V'(0) = 0 \; .$$

Then in suitable coordinates,

(13) $$p(x,\xi) = \sum_1^k \frac{1}{2}\lambda_j(\xi_j^2+x_j^2) + \sum_{k+1}^{n-d} \xi_j^2 + \sum_{n-d+1}^{n} \lambda_j\frac{1}{2}(\xi_j^2-x_j^2) + O(|(x,\xi)|^3),$$

where $\lambda_j > 0$. Here (k, d) is the signature of $V''(0)$. The linearization of H_p at $(0, 0)$ then has the eigen-values $\pm i\lambda_j$ for $1 \leq j \leq k$, ± 0 for $k+1 \leq j \leq n-d$ and $\pm\lambda_j$ for $n-d+1 \leq j \leq n$. Let Σ be a manifold of class C^r, $r \geq 2$ defined in a neighborhood of $(0, 0)$, passing through $(0, 0)$, such that

(14) H_p is tangent to Σ at every point,

(15) $$T_{(0,\,0)}\Sigma = \mathbb{R}^{2(n-d)}_{(x',\xi')} \quad , \quad x' = (x_1,\dots,x_{n-d}) \; .$$

(See [1], appendix C by A. Kelley.) Then it is easy to construct G as in (4) of class C^r such that

(16) $$H_pG \sim d_\Sigma^2 \quad , \quad G = O(d_\Sigma^2) \quad \text{near } \Sigma \; ,$$

where d_Σ denotes the euclidean distance to Σ. It follows rather easily that if $\varepsilon_o > 0$ is sufficiently small, then

(17) $$K \subset \Sigma \; .$$

In the non-degenerate case; $k+d = n$, the H_p-trajectories in $\Sigma \cap p^{-1}([-\varepsilon_o, \varepsilon_o])$ are trapped near $(0, 0)$, so in that case,

(18) $$K = \Sigma \cap p^{-1}([-\varepsilon_o, \varepsilon_o]) \; .$$

Choosing new symplectic coordinates; $(y, \eta) = (y', y'', \eta', \eta'')$ centered at $(0, 0)$ such that $\Sigma : y'' = \eta'' = 0$, we get

(19) $$p = p^\Sigma(y', \eta') + O((y'', \eta'')^2) \; .$$

If I is an interval in $[-\varepsilon_o/2, \varepsilon_o/2]$ and $\delta > 0$ is small, we get,

(20) $$\mathrm{Vol}(R(I + [-C\tilde\delta, C\tilde\delta], C\delta))$$
$$\leq \hat C\, \delta^d\, \mathrm{Vol}_\Sigma(\{(y', \eta');\ p^\Sigma(y', \eta') \in I + [-\tilde C\tilde\delta, \tilde C\tilde\delta]\}) \; ,$$

where $\hat C$ and $\tilde C$ are new constants. We can use this in Cor. 4 to get estimates on $M(I,\delta)$. In the degenerate case, the resulting estimate

Proposition 5. Assume (1) - (3) and that there is a neighborhood Ω of K in $p^{-1}([-\varepsilon_o, \varepsilon_o])$ such that

(25) $\quad d_K \sim d_{\Gamma_+} + d_{\Gamma_-} \quad \text{in } \Omega \quad ,$

(26) $\quad d(\exp(\pm tH_p)(\rho), \Gamma_\pm) \leq C\, e^{-t/C} d(\rho, \Gamma_\pm)$ if

$\exp(\pm sH_p)(\rho) \in \Omega$ for all $s \in [0, t]$.

Then (possibly after decreasing ε_o slightly), there exist G and W as in (4) such that $G = O(d_K^2)$, $H_p G \sim d_K^2$ in $W \cap p^{-1}([-\varepsilon_o, \varepsilon_o])$.

This with Cor. 4 gives estimates on the number of resonances depending on a suitable "dimension" of K. In general, we cannot hope for better regularity of G than $C^{1,1}$, and we suspect very strongly that the potential $V = \mathrm{Re}(x_1 + ix_2)^3 - 1$ on $\mathbb{R}^2$ gives an example of this.

References.

1. R.ABRAHAM, J.ROBBIN, Transversal mappings and flows, Benjamin, New York (1967).
2. J.AGUILAR, J.M.COMBES, Comm. Math. Phys. 22 (1971), 269-279.
3. E. BALSLEV, J.M. COMBES, Comm. Math. Phys. 22 (1971), 280-294.
4. Ph. BRIET, J.M. COMBES, P.DUCLOS, On the location of resonances. II. Preprint.
5. J.M. COMBES, P. DUCLOS, M. KLEIN, R. SEILER, The shape resonance, Preprint.
6. C. GÉRARD, J. SJÖSTRAND, Semiclassical resonances generated by a closed trajectory of hyperbolic type, Preprint.
7. B. HELFFER, J. SJÖSTRAND, in "Advances in Microlocal Analysis", Reidel, Ser. C, Vol. 168 (1986).
8. B. HELFFER, J. SJÖSTRAND, Résonances en limite semiclassique, Bull. de la SMF, to appear.
9. P.HISLOP,I.SIGAL, In the proceedings of the conference on Math. Physics, Birmingham, Alabama 1986.
10. R. MELROSE, In the proceedings of "Journées des EDP de St Jean des Monts 1984".
11. J. SJÖSTRAND, Semiclassical resonances generated by non-degenerate critical points, Preprint.
12. J. SJÖSTRAND, Semiclassical resonances in some simple cases, Proceedings of "Journées des EDP de St Jean des Monts 1986".
13. J. SJÖSTRAND, Astérisque n° 95, (1982).

will not always be optimal; the reason is that K need not be equal to Σ and that it should then be possible to find an even better escape function, by taking into account the dynamics in Σ. In the non-degenerate case we have $p^{\Sigma}(y', \eta') \sim |(y', \eta')|^2$ so the order of magnitude of the last volume in (20) can easily be determined. When $\delta \geq h$ and $\lambda \geq \tilde{\delta}$ are small, we get

(21) $$M([0, \lambda], \delta) \leq C\, \delta^{d}\lambda^{n-d}h^{-n} ,$$

(22) $$M([\lambda-\tilde{\delta}, \lambda+\tilde{\delta}], \delta) \leq C\, \delta^{d}\tilde{\delta}\lambda^{n-d-1}h^{-n} ,$$

(23) $$M([-\tilde{\delta}, \tilde{\delta}], \delta) \leq C\, \tilde{\delta}^{n-d}\delta^{d}h^{-n} .$$

In [11] we described rather completely the resonances in $[-Ch, Ch]-i[0, Ch]$ for any $C > 0$. Those results indicate that the exponents in (21) - (23) are optimal.

Example 2. Suppose that K^{o} is the image of a closed trajectory γ^{o} of minimal period $T^{o} > 0$. Then $dp \neq 0$ on $\{p = 0\}$. Let ρ^{o} be a point in the image of γ^{o} and let A be the differential of $\exp(T^{o}H_p)$ at ρ^{o}. Let 2d be the number of eigenvalues of modulus $\neq 1$ of this symplectic transformation. We have $0 \leq d \leq n-1$. Let $H = T_{\rho o}(p^{-1}(0))$ and let S be the sum of the eigenspaces corresponding to the eigenvalues of modulus 1. Then S is a symplectic $2(n-d)$-dimensional space transversal to H. Let Σ be a manifold of class C^{r}, $r \geq 2$, defined in a neighborhood of γ^{o}, invariant under the H_p-flow, containing γ^{o} and such that $T_{\rho o}(\Sigma) = S$. Again $K \subset \Sigma$ and we can find a function G as in (4), such that G is of class C^{r} and $G = O(d_{\Sigma}^{2})$, $H_pG \sim d_{\Sigma}^{2}$ near Σ. Using also the fact that $dp \neq 0$, we obtain the estimate,

(24) $$M(I, \delta) \leq C\, \delta^{d}(|I| + \tilde{\delta})h^{-n} ,$$

where $|I|$ is the length of I. In the case of a trajectory of hyperbolic type, we have $d = n-1$, and in [6] we obtained "complete results" about the resonances in $I-i[0, Ch]$, which indicate that the exponents in (24) are optimal in that case.

In cases where the Hamilton flow is a more general dynamical system of hyperbolic type (for instance with a Cantor set structure on K, Γ_+, Γ_-), the following result seems applicable:

Heat-flow methods for harmonic maps of surfaces and applications to free boundary problems

Michael Struwe
ETH - Zentrum - Zurich

Abstract

In [17] the Eells-Sampson method for constructing harmonic maps between manifolds was extended to maps from a surface to an arbitrary compact manifold. We review the results in [17] and present several applications: First a new proof of the Sacks-Uhlenbeck results is given. Then we study minimal surfaces and surfaces of constant mean curvature with free boundaries on a supporting surface in $\mathbb{R}^3$.

AMS classification code: 53 A 10, 53 C 20, 58 E 20, 58 G 11.

1. Basic definitions

Let M , N be compact Riemannian manifolds of dimensions $\dim M = 2$, $\dim N = n$, and with metrics γ , g respectively.

For differentiable maps $u = M \to N$ let

$$E(u) = \int_M e(u)\,dM$$

be the energy of u with energy density $e(u)$ in local coordinates being given by

$$e(u) = \frac{1}{2} g_{ik}(u)\gamma^{\alpha\beta}(x)\partial_\alpha u^i \partial_\beta u^k .$$

Here and in the sequel $\partial_\alpha u^i = \frac{\partial}{\partial x^\alpha} u^i$, $(\gamma^{\alpha\beta})$ denotes the inverse of the coefficient matrix $(\gamma_{\alpha\beta})$ of γ , and (g_{ik}) is a local representation of g . By convention double Greek indeces will be summed from 1 to 2 , double Latin indeces from 1 to n . Of course, locally the volume element on M

$$dM = \sqrt{\gamma}\, dx$$

where $\gamma = \det(\gamma_{\alpha\beta})$.

u is harmonic, iff E is stationary at u , i.e. iff

$$(1.1) \qquad \delta E(u) := -\Delta_M u - {}^N\Gamma(u)(\nabla u, \nabla u)_M = 0$$

with

$$(-\Delta_M u)^\ell = -\frac{1}{\sqrt{\gamma}} \partial_\alpha (\sqrt{\gamma}\gamma^{\alpha\beta}\partial_\beta u^\ell) \ , \qquad 1 \leq \ell \leq n$$

denoting the Laplace-Beltrami operator on M and

$$({}^N\Gamma(u)(\nabla u, \nabla u)_M)^\ell =$$

$$= \Gamma^\ell_{ik}(u)\gamma^{\alpha\beta}\partial_\alpha u^i \ \partial_\beta u^k \ , \qquad 1 \leq \ell \leq n \ ,$$

where

$$\Gamma^\ell_{ik} = \frac{1}{2} g^{\ell j}\left(\frac{\partial}{\partial u^i} g_{jk} + \frac{\partial}{\partial u^k} g_{ij} - \frac{\partial}{\partial u^j} g_{ik}\right)$$

are the Christoffel symbols of the metric g .

Note that formally at a smooth map $u : M \to N$ for any smooth variation vector φ

$$\frac{d}{d\varepsilon} E(u+\varepsilon\varphi)\Big|_{\varepsilon=0} =$$

(1.2)

$$= \int_M g_{k\ell}(u)(-\Delta_M u - {}^N\Gamma(u)(\nabla u, \nabla u)_M)^k \varphi^\ell \, dM \quad ,$$

whence δE may be regarded as the L^2-gradient of E with respect to the metric $g(u)$.

By the Nash embedding theorem we may assume that $N \subset \mathbb{R}^N$ isometrically. A natural space on which to consider E then is the space

$$H^{1,2}(M,N) = \{u \in H^{1,2}(M;\mathbb{R}^N)\,|\,u(M) \subset N\}$$

of L^2-functions from M into N having a distributional derivative in L^2.

By a result of Schoen and Uhlenbeck [13] and since M is 2-dimensional, $H^{1,2}(M,N)$ is the closure of the space of smooth functions from M into N in $H^{1,2}(M;\mathbb{R}^N)$. Hence we may continuously extend E to $H^{1,2}(M;N)$. Moreover, E is coercive with respect to the norm induced by $H^{1,2}(M;\mathbb{R}^N)$, and therefore $H^{1,2}(M;N)$ is precisely the space of measurable functions $u : M \to N$ a.e. with $E(u) < \infty$.

However, note that $H^{1,2}(M;N)$ is not a manifold and hence E cannot be differentiable on this space.

2. Existence of harmonic maps

Given the coerciveness of E on $H^{1,2}(M,N)$ it is natural to apply the direct methods in the calculus of variations to obtain harmonic maps from M into N as (relative) minimizers of E in this space.

Unfortunately, E trivially assumes its global minimum 0 at any constant map $u : M \to N$. Moreover (although as a consequence of the Schoen-Uhlenbeck result homotopy classes of (continuous) maps between M and N are stable in the strong $H^{1,2}(M;N)$-topology) homotopy classes in general will not be weakly closed in $H^{1,2}(M;N)$. Therefore, the existence of non-trivial harmonic maps and, in particular, the existence of harmonic representants of homotopy classes of maps between M and N does not follow from the direct methods.

Instead, in their pioneering paper [2] Eells and Sampson consider the evolution problem

$$(2.1) \qquad \partial_t u - \Delta_M u - {}^N\Gamma(u)(\nabla u, \nabla u)_M = 0$$

$$(2.2) \qquad u\big|_{t=0} = u_o$$

for harmonic maps from M into N. Their fundamental result is the following

<u>Theorem 2.1</u> (Eells-Sampson): Suppose the sectional curvature K of N is non-positive. Then for any smooth map $u_o : M \to N$ there exists a smooth solution $u : M \times [0,\infty[\to N$ of (2.1), (2.2) which as $t \to \infty$ converges to a smooth harmonic map $u_\infty : M \to N$.

Remark that by (1.2) for any solution u of (2.1), (2.2) $E(u(\cdot,t))$ is non-increasing in t, so that in particular the result of Eells and Sampson guarantees the existence of a smooth harmonic map which minimizes E (relatively) in any homotopy class of maps $u : M \to N$, if N is non-positively curved.

Theorem 2.1 was later generalized to arbitrary target manifolds assuming the range $u(M \times [0,\infty[)$ to be a-priori bounded in a geodesic ball of radius $R < \frac{\pi}{2\sqrt{K}}$ where $K > 0$ is an upper bound for the sectional curvature of N, cp. [8], [19]. For complete M, however, only constant maps can be harmonic and have range in a ball of radius $< \frac{\pi}{2\sqrt{K}}$, and the above results are useful only for harmonic maps of manifolds with boundaries.

While Theorem 2.1 is valid for all dimensions, Sacks and Uhlenbeck in 1981 obtained significantly new results for $\dim(M) = 2$, cp. [12]. In particular, they proved [1)]

<u>Theorem 2.2</u> (Sacks-Uhlenbeck): If N is compact, and $\pi_2(N) = 0$, there exists an energy-minimizing harmonic map in every homotopy class of maps from $M \to N$.

For arbitrary targets N they can assert:

<u>Theorem 2.3</u> (Sacks-Uhlenbeck): Every conjugacy class of homomorphisms from $\pi_1(M)$ into $\pi_1(N)$ is induced by an energy-minimizing harmonic map from M into N.

If $M = S^2$ moreover they obtain

<u>Theorem 2.4</u> (Sacks-Uhlenbeck): i) If the universal covering space of N is not contractible, there exists a non-trivial harmonic map of S^2 into N.

ii) There is a set Λ_i of free homotopy classes of maps from S^2 into N generating $\pi_2(N)$ acted on by $\pi_1(N)$ and represented by relatively energy-minimizing harmonic maps $u_i : S^2 \to N$.

<u>Remarks:</u> i) By results of Gulliver, Osserman, and Royden a non-trivial harmonic map $u : S^2 \to N$ is a branched minimal immersion [5].

ii) Note that in general not every homotopy class of maps from M into N has a harmonic representant, cp. [3].

---

1) Independently, related results were obtained by Luc Lemaire.

In order to obtain their results, Sacks and Uhlenbeck construct approximate solutions as relative minimizers of a family of perturbed functionals E_α , $\alpha > 1$. As $\alpha \to 1$ approximate solutions will either converge towards a smooth harmonic map or harmonic spheres will tend to separate at finitely many points of the domain M in the sense that a sequence of blown-up solutions will tend to a smooth, non-constant harmonic map $u : \overline{\mathbb{R}^2} \cong S^2 \to N$, locally uniformly in the tangent space $T_xM \cong \mathbb{R}^2$ at a point $x \in M$.

In [17] we have extended the heat flow method to maps from a compact surface into an arbitrary compact Riemannian manifold. In the next section we review these results and indicate how the theorems of Sacks and Uhlenbeck may be derived from them.

3. The evolution of harmonic maps of surfaces

Our basic result is the following [17, Section 4]:

Theorem 3.1: Let M be a smooth, compact Riemannian surface, and let N be a smooth, compact Riemannian manifold of dimension n . For any smooth $u_o : M \to N$ there exists a global (distribution) solution $u : M \times [0,\infty[\to N$ to (2.1), (2.2) which is regular on $M \times [0,\infty[$ with exception of at most finitely many points $(x_1,t_1), \ldots (x_k,t_k)$, $0 < t_j \leqq \infty$, and which is unique in this class.

At a singularity $(\bar{x},\bar{t})$ of u a smooth non-constant harmonic map $\bar{u} : S^2 \to N$ separates in the sense that for suitable $r_m \to 0$, $t_m \to \bar{t}$, $x_m \to \bar{x}$ in local coordinates

$$u(x_m+r_mx,t_m) \to \tilde{u} \quad \text{in} \quad H^{2,2}_{loc}(\mathbb{R}^2;N) .$$

$\tilde{u}$ has finite energy and may be extended to a smooth, non-constant harmonic map $\bar{u} : S^2 \cong \overline{\mathbb{R}^2} \to N$. Finally, as $t \to \infty$,

$u(\cdot,t)$ converges weakly in $H^{1,2}(M,N)$ to a smooth harmonic map $u_\infty : M \to N$. If all $t_j < \infty$, $1 \leq j \leq k$, $u(\cdot,t) \to u_\infty$ even strongly in $H^{2,2}(M,N)$.

Remarks: [5] also applies in the case of Theorem 3.1 and we conclude that the harmonic spheres separating at singular points of the flow must be C^∞ conformal branched immersions of S^2 into N.

We also remark that the time needed to create a singularity of the solution u to (2.1), (2.2) can be uniformly estimated from below on an $H^{1,2}$-neighborhood of the initial value u_o. This permits to extend the existence result stated in Theorem 3.1 to initial values of class $H^{1,2}(M,N)$, cp. [17, Theorem 4.1]. (Now, of course, we only obtain "interior" regularity on $M \times]0,\infty[\setminus\{(x_j,t_j) | 1 \leq j \leq k\}$.)

By the non-existence result of Eells and Wood [3] e.g. for initial maps $u_o : T^2 \to S^2$ of degree 1 the solution u to (2.1), (2.2) has to become singular at some point $(\bar{x},\bar{t})$. However, it is an open question if singularities of the flow (2.1), (2.2) appear in finite time.

Let us sketch for example how Theorem 2.2 may be deduced from Theorem 3.1.

Proof of Theorem 2.2: Let

$$\varepsilon_o = \{\inf E(u) | u : S^2 \to N \text{ is a smooth, non-constant harmonic map}\} .$$

It is well-known that $\varepsilon_o > 0$; this also follows from Lemma 4.2 below.

Let $[u_o]$ be a homotopy class of maps from M into N, represented by u_o. We may suppose that u_o is smooth and

$$E(u_o) \leq \inf_{u\in[u_o]} E(u) + \frac{\varepsilon_o}{4} .$$

Let $u : M \times [0,\infty[\to N$ be the solution to the evolution problem (2.1), (2.2), guaranteed by Theorem 3.1. Suppose u becomes singular at a point $(\bar{x},\bar{t})$. Then for $t_m < \bar{t}$, $t_m \to \bar{t}$, $x_m \to \bar{x}$ and some $r_m \to 0$

$$u_m(x) :\equiv u(x_m+r_m x, t_m) \to \tilde{u} \quad \text{in} \quad H^{2,2}_{loc}(\mathbb{R}^2;N)$$

where $\tilde{u}$ may be extended to a smooth non-constant harmonic map $\bar{u} : S^2 \to N$. In particular, we can find a sequence of radii $R_m \to \infty$ such that

$$R_m \int_{\partial B_{R_m}} (|\nabla u_m|^2 - |\nabla\tilde{u}|^2)\, do \to 0$$

and

$$\int_{\mathbb{R}^2\setminus B_{R_m}} |\nabla\tilde{u}|^2 dx \to 0$$

as $m \to \infty$. Inverting $\tilde{u}$ in ∂B_{R_m} we may replace the large almost spherical surface $u_m(B_{R_m})$ having energy $\geq \frac{3\varepsilon_o}{4}$ by a small disc (essentially $\tilde{u}(\mathbb{R}^2\setminus B_{R_m})$) of energy $\leq \frac{\varepsilon_o}{4}$ to obtain a new map $v_m : M \to N$. Since $\pi_2(N) = 0$ v_m lies in the same homotopy class as $u(\cdot,t_m)$, i.e. $v_m \in [u_o]$.

On the other hand, since $E(u(\cdot,t))$ is non-increasing in t, cp. Lemma 4.3 below, we have:

$$E(v_m) \leq E(u(\cdot,t_m)) - \frac{\varepsilon_o}{2}$$

$$\leq E(u_o) - \frac{\varepsilon_o}{2} \leq \inf_{u\in[u_o]} E(u) - \frac{\varepsilon_o}{4} .$$

The contradiction proves that the solution u through u_o is globally regular and as $t \to \infty$ converges to a smooth harmonic map homotopic to u_o.

q.e.d.

Similarly, Theorem 2.3 can be derived from Theorem 3.1 and the observation that the conjugacy class of homomorphisms from $\pi_1(M)$ into $\pi_1(N)$ induced by the maps $u(\cdot,t)$ envolving from a given map u_o will not change if harmonic spheres separate. Finally, also Theorem 2.4 can be proved using Theorem 3.1 by arguing as in [12, Section 5].

4. Sketch of the proof of Theorem 3.1

It may be instructive to recall the essential features of the proof of Theorem 3.1.

Let u be a smooth solution to (2.1), (2.2) in $M \times [0,T]$. We can formulate a-priori estimates for u in terms of the quantity

$$\varepsilon(R,T) = \sup_{x\in M,\, t\leq T} \int_{B_R(x)} e(u(t))\,dM$$

where $0 < R < R_o$, R_o : the injectivity radius of M, and $B_R(x)$ is the geodesic ball of radius R around $x \in M$.

Lemma 4.1: There exists $\varepsilon > 0$, $c_o \in \mathbb{R}$ such that

$$\int_0^T \int_M |\nabla^2 u|^2 dM dt \leq c_o\, E(u_o)\, (1+TR^{-2}) ,$$

whenever $\varepsilon(R,T) \leq \varepsilon$.

The proof uses the following fundamental Sobolev-type inequality, cp. [17, Lemma 3.1]:

Lemma 4.2: For any $v \in L^\infty([0,T];L^2(M,\mathbb{R}^N))$ with $|\nabla v| \in L^2(M \times [0,T])$ we have $v \in L^4(M \times [0,T];\mathbb{R}^N)$ and for any $R < R_o$ there holds

$$\int_0^T \int_M |v|^4 dMdt \leqq c^* \cdot \sup_{x \in M, t \leqq T} \left(\int_{B_R(x)} |v(t)|^2 dM \right) \cdot \left(\int_0^T \int_M |\nabla v|^2 dMdt + \frac{1}{R^2} \int_0^T \int_M |v|^2 dMdt \right)$$

with a uniform constant $c^* = c^*(M)$.

Moreover, we need:

Lemma 4.3: There holds the energy estimate

$$E(u(\cdot,T)) + \int_0^T \int |\partial_t u|^2 dMdt \leqq E(u_o) \ .$$

Proof: Simply compute, using (1.1), (2.1)

$$\frac{d}{dt} E(u(\cdot,t)) = \int_M (-\Delta_M u - {}^N\Gamma(u)(\nabla u,\nabla u)_M) \cdot \partial_t u \, dM$$

$$= - \int_M |\partial_t u|^2 dM \ .$$

q.e.d.

Proof of Lemma 4.1: Multiply (2.1) by $\Delta_M u$ and integrate to obtain for sufficiently small $\varepsilon > 0$:

$$\int_0^T \int_M \partial_t\left(\frac{|\nabla u|^2}{2}\right) + |\Delta_M u|^2 \, dMdt$$

$$\leq c(\delta) \int_0^T \int_M |\nabla u|^4 dMdt + \delta \int_0^T \int_M |\nabla^2 u|^2 dMdt$$

$$\leq (c(\delta)c^*\varepsilon(R,T) + \delta) \int_0^T \int_M |\nabla^2 u|^2 dMdt$$

$$+ c(\delta)c^*\varepsilon(R,T) \, R^{-2} \int_0^T \int_M |\nabla u|^2 dMdt$$

$$\leq 2\delta \int_0^T \int_M |\nabla^2 u|^2 dMdt + c(\delta)TR^{-2} \sup_{t\leq T} E(u(\cdot,t)) \quad ,$$

for any pre-assigned $\delta > 0$.

Note that after integrating by parts for a.e. $t \leq T$

$$\int_M |\nabla^2 u|^2 dM \leq c \int_M |\Delta_M u|^2 dM + c \int_M |\nabla u|^2 dM \quad .$$

Moreover, by Lemma 4.3

$$E(u(\cdot,t)) \leq E(u_o) \quad , \quad \forall \, t \leq T \quad .$$

The claim follows.

q.e.d.

From Lemma 4.1 full regularity may be deduced. Differentiate (2.1) in t and multiply by $\partial_t u$. Integrating one obtains:

$$\int_0^T \int_M \partial_t\left(\frac{|\partial_t u|^2}{2}\right) + |\nabla \partial_t u|^2 dMdt$$

$$\leq c(\delta) \int_0^T \int_M |\nabla u|^2 |\partial_t u|^2 dMdt + \delta \int_0^T \int_M |\nabla \partial_t u|^2 dMdt$$

$$\leq c(\delta) \left(\int_0^T \int_M |\nabla u|^4 dMdt\right)^{1/2} \left(\int_0^T \int_M |\partial_t u|^4 dMdt\right)^{1/2}$$

$$+ \delta \int_0^T \int_M |\nabla \partial_t u|^2 dMdt \quad ,$$

for any $\delta > 0$.

Let $\int_M |\partial_t u|^2 dM\big|_t$ achieve its supremum at $t = t_o$.
We may assume $t_o = T$. By Lemma 4.1 and 4.2:

$$\int_M |\partial_t u|^2 dMdt\big|_{t=T} - \int_M |\partial_t u|^2 dM\big|_{t=0} + \int_0^T \int_M |\nabla\partial_t u|^2 dMdt \leq$$

$$\leq c(\delta)\left(\int_0^T \int_M |\nabla u|^4 dMdt\right)^{1/2} \cdot \left(\sup_{t\leq T}\left(\int_M |\partial_t u|^2 dM\right) \cdot \left(\int_0^T \int_M |\nabla\partial_t u|^2 dMdt + \int_0^T \int_M |\partial_t u|^2 dMdt\right)\right)^{1/2}$$

$$+ \delta \int_0^T \int_M |\nabla\partial_t u|^2 dMdt \leq$$

$$\leq \left(c(\delta)\left(\int_0^T \int_M |\nabla u|^4 dMdt\right)^{1/2} + \delta\right)\left(\int_M |\partial_t u|^2 dM\big|_{t=T} + \int_0^T \int_M |\nabla\partial_t u|^2 dMdt + E(u_o)\right).$$

By absolute continuity of the Lebesgue integral of $|\nabla u|^4$ upon choosing δ , $T > 0$ small enough we can achieve that

$$c(\delta)\left(\int_0^T \int |\nabla u|^4 d\ dt\right)^{1/2} + \delta \leq 1/2$$

and obtain that

$$\sup_{t\leq T} \int_M |\partial_t u|^2 dM \leq c \int_M |\partial_t u|^2 dM\big|_{t=0} + c\, E(u_o) \quad .$$

Back to (2.1) and using Lemma 4.2 for $v(x,t) \equiv \nabla u(x,T)$ this implies

$$\int_M |\nabla^2 u|^2 dM\big|_{t=T} \leq c \int_M |\nabla u|^4 dM\big|_{t=T} + c \int_M |\partial_t u|^2 dM\big|_{t=T}$$

$$\leq c \int_M |\partial_t u|^2 dM\big|_{t=0} + c\, E(u_o)$$

$$+ c\, \varepsilon(R,T)\left(\int_M |\nabla^2 u|^2 dM\big|_{t=T} + R^{-2} E(u_o)\right) .$$

I.e.

$$\sup_{t\leq T} \int_M |\nabla^2 u|^2 dM \leq c \int_M |\partial_t u| dM \Big|_{t=0} + c\, E(u_o)(1+R^{-2})$$

Since $H^{2,2}(M;\mathbb{R}^N) \hookrightarrow C^\alpha(M;\mathbb{R}^N)$, $\forall\, \alpha < 1$, this estimate implies continuity of u on $M \times [0,T]$. The regularity theory of Ladyženskaya, Solonnikov and Ural'ceva [10] now is applicable, and we obtain a-priori bounds of all derivatives of u on $M \times]0,\tau]$ in terms of $E(u_o)$, R, $\int_M |\partial_t u|^2 dM\Big|_{t=0}$, where $\tau > 0$ is determined by the condition that $\int_0^\tau \int_M |\nabla u|^4 dM dt$ be sufficiently small. These estimates are translation invariant. Moreover, by Fubini's theorem and Lemma 4.3

$$\inf_{\substack{t_o-\tau\leq t\leq t_o \\ t\geq 0}} \int_M |\partial_t u|^2 dM \leq \inf\{t_o,\tau\}^{-1} \int_{\sup\{0,t_o-\tau\}}^{t_o} \int_M |\partial_t u|^2 dM dt \leq \inf\{t_o,\tau\}^{-1} E(u_o) \, .$$

Hence we obtain a-priori bounds on any interval $[t_o,T]$ in terms of $E(u_o)$, R, t_o, T and τ. Using Lemma 4.1 and a compactness argument, one finally can show that the modulus τ is completely determined by $E(u_o)$ and R. This gives

<u>Proposition 4.4:</u> If $u : M \times [0,T] \to N$ is a smooth solution to (2.1) with smooth initial value u_o and $\varepsilon(R,T) < \varepsilon$ then u and its derivatives are uniformly bounded on $M \times [0,T]$ in terms of T, R, and u_o. On any set $M \times]t_o,T]$, $t_o > 0$, bounds will depend only on T, R, $E(u_o)$, and t_o.

Control of the energy density function $\varepsilon(R,T)$ therefore is crucial for Theorem 3.1:

Let

$$E_R(v;x) = \int_{B_R(x)} e(v) dM \, .$$

Lemma 4.5: For any solution u of (2.1), (2.2) in $M \times [0,T]$ and any $R < R_o$ there holds the estimate

$$E_R(u(\cdot,T);x) \leqq E_{2R}(u_o;x) + c_1 \, TR^{-2} \, E(u_o) \; .$$

c_1 is a constant depending only on M and N .

Proof: Multiply (2.1) by $\partial_t u\varphi^2$ where $\varphi \in C_o^\infty(B_{2R}(x))$ is $\equiv 1$ on $B_R(x)$ and integrate to obtain

$$\int_0^T \int_M |\partial_t u|^2 \varphi^2 + \frac{d}{dt} \, e(u(\cdot,t))\varphi^2 dM dt \leqq c \int_0^T \int_M |\nabla u| \, |\partial_t u| \, |\nabla\varphi| \varphi dM dt$$

$$\leqq \frac{1}{2} \int_0^T \int_M |\partial_t u|^2 \varphi^2 dM dt + cTR^{-2} E(u_o) \; .$$

The lemma follows.

q.e.d.

Remark 4.6: i) In particular, if for some $R > 0$

$$\sup_x E_R(u_o;x) \leqq \frac{\varepsilon}{2} \, ,$$

for any solution u to (2.1), (2.2) we will have

$\varepsilon(R,T) \leqq \varepsilon$ for $T = \dfrac{\varepsilon R^2}{2c_1 E(u_o)}$. Thus, by Proposition 4.4 we will have uniform bounds on u in $M \times [0,T]$ in terms of u_o alone. This permits to construct local solutions to (2.1) for smooth initial data u_o . By the same token we obtain uniform a-priori bounds locally on $M \times]0,T]$ for solutions u with initial values in an $H^{1,2}$-neighborhood of u_o . Hence we may extend the flow u also to initial values u_o of class $H^{1,2}(M;N)$. By Lemma 4.3 u assumes its initial values continuously in $H^{1,2}$.

ii) By Proposition 4.4 and Lemma 4.5 a local solution to (2.1) as above may be continued to an interval $[0,T_1[$, where T_1 is characterized by the condition that

$$\liminf_{t \nearrow T_1} (\sup_{x \in M} E_R(u(\cdot,t);x)) \geq \varepsilon$$

for all $R > 0$. If $\{x_1,\ldots,x_k\}$ are points in M where

$$\limsup_{t \nearrow T_1} (E_R(u(\cdot,t),x_j) \geq \varepsilon \ , \ 1 \leq j \leq k \ ,$$

let $R < \frac{1}{2} \inf_{i \neq j} \operatorname{dist}(x_i,x_j)$. Then for all t close to T_1

$$E_{2R}(u(\cdot,t);x_j) \geq E_R(u(\cdot,t_j);x_j) - c_1 \frac{t_j-t}{R^2} E(u_o)$$

and by suitable choice of t_j the right hand side can be made $\geq \varepsilon/2$ for all $j = 1,\ldots,k$.

But by choice of R

$$k \cdot \frac{\varepsilon}{2} \leq \sum_{j=1}^{k} E_{2R}(u(\cdot,t);x_j) \leq E(u(\cdot,t)) \leq E(u_o)$$

and $k \leq \frac{2E(u_o)}{\varepsilon}$ is uniformly a-priori bounded.

Moreover, $u(\cdot,t) \rightharpoonup: u_1 \in H^{1,2}(M;N)$ weakly, and we may simply continue u to some larger interval $[0,T_2[$ by letting u solve (2.1) on $[T_1,T_2[$ with initial value u_1 , etc.

Since $u(\cdot,t) \to u_1$ in $L^2(M,N)$ as $t \to T_1$, the extended function u will also be a distribution solution to (2.1) on $M \times [0,T_2]$.

Note that $E(u_1) \leq \lim_{R \to 0} \int_{M \setminus \bigcup_j B_{2R}(x_j)} e(u_1)dM$

$$\leq \lim_{R \to 0} \lim_{t \to t_1} \int_{M \setminus \bigcup_j B_{2R}(x_j)} e(u(\cdot,t))dM \leq E(u_o) - \frac{k\varepsilon}{2} \ .$$

In particular, by iteration of our above argument there can be at most finitely many singularities (x_j, t_j) , $t_j \leqq \infty$, satisfying the condition

$$\limsup_{t \nearrow t_j} E_R(u(\cdot,t);x_j) \geqq \varepsilon$$

for any R .

A more detailed analysis as in [17] shows that if for $(\bar{x},\bar{t}) \in M \times]0,\infty[$

$$\limsup_{t \nearrow \bar{t}} E_R(u(\cdot,t);\bar{x}) < \varepsilon$$

for some $R > 0$, then u is regular in a neighborhood of $(\bar{x},\bar{t})$.

iii) As $t_m \to \infty$ suitably, by Lemma 4.3

$$\int_M |\partial_t u|^2 dM \Big|_{t=t_m} \to 0 \quad .$$

Suppose that $T = \infty$ is non-singular in the sense that

$$\limsup_{t \to \infty} \left(\sup_{x \in M} E_R(u(\cdot,t),x) \right) < \varepsilon$$

for some $R > 0$.

Then Lemma 4.2 (applied to $v(x,t) \equiv u(x,t_m)$) assures that at $t = t_m$

$$\int_M |\nabla^2 u|^2 dM \leqq c \int_M |\nabla u|^4 dM + c \int_M |\partial_t u|^2 dM \leqq$$

$$\leqq \frac{1}{2} \int_M |\nabla^2 u|^2 dM + c\, E(u_o) + o(1)$$

where $o(1) \to 0$ $(m \to \infty)$. By Rellich's theorem we may assume that $u(\cdot,t_m) \to u_\infty$ strongly in $H^{1,p}(M,N)$, for any $p < \infty$. But then by (2.1)

$$- \Delta_M u(\cdot,t_m) \to {}^N\Gamma(u_\infty)(\nabla u_\infty,\nabla u_\infty)_M = - \Delta_M u_\infty$$

in L^2 . Thus, $u_\infty \in H^{2,2}(M,N)$ is harmonic (and hence regular). Moreover, $u(\cdot,t_m) \to u_\infty$ strongly in $H^{2,2}(M,N)$.

Again, a local analysis shows that also if $T = \infty$ is singular in the sense that at points $\{x_1,\ldots,x_k\}$

$$\limsup_{t\to\infty} E_R(u(\cdot,t),x_j) \geqq \varepsilon$$

for all $R > 0$, then for suitable numbers $t_m \to \infty$ the family $\{u(\cdot,t_m)\}$ will be equi-bounded locally in $H^{2,2}$ on $M\setminus\{x_1,\ldots,x_k\}$, and hence accumulate in $H^{2,2}_{loc}(M\setminus\{x_1,\ldots,x_k\},N)$ at a harmonic map $u_\infty : M\setminus\{x_1,\ldots,x_k\} \to N$. Since $E(u_\infty) \leqq E(u_o)$, by [12, Theorem 3.6] u_∞ may be extended to a smooth harmonic map $u_\infty : M \to N$.

iv) If at $(\bar{x},\bar{t})$ for all $R > 0$ there holds

$$\limsup_{t\nearrow\bar{t}} E_R(u(\cdot,t),\bar{x}) \geqq \varepsilon$$

we can rescale $u \to u^{(m)}(x,t) \equiv u(x_m+r_m x,t_m+r_m^2 t)$ such that

$$\limsup_{t\nearrow 0} \sup_{r_m|x|<R_o} E_1(u^{(m)}(\cdot,t),x) \leqq E_1(u^{(m)}(\cdot,0),0) = \frac{\varepsilon}{2} ,$$

while by Lemma 4.3 and absolute continuity of the Lebesgue integral

$$\int_{-1}^{o} \int_{B_{R_o/r_m}(0)} |\partial_t u^{(m)}|^2 dxdt \to 0 \quad (m \to \infty) .$$

Note that $u^{(m)}$ solves an equation like (2.1) on $B_{R_o/r_m}(0) \times [-1,0]$. Hence a local estimate like Lemma 4.1 is valid and we obtain uniform local a-priori bounds for $u^{(m)}$. A reasoning as in iii) now shows that $u^{(m)}(\cdot,\tau_m) =: u_m \to \tilde{u}$ strongly locally in $H^{2,2}_{loc}(\mathbb{R}^2;N)$ for some sequence $\tau_m \in [-1,0]$, where $\tilde{u}$ is harmonic and has energy $\frac{\varepsilon}{2} \leqq E(\tilde{u}) \leqq E(u_o)$. By [12, Theorem 3.6] again, $\tilde{u}$ may be extended to a smooth, non-constant harmonic map $\overline{u} : S^2 \cong \overline{\mathbb{R}^2} \to N$.

This essentially proves Theorem 3.1.

We now proceed to give further applications of this method to surfaces of prescribed constant mean curvature with free boundaries on a supporting surface in $\mathbb{R}^3$.

5. Applications to H-surfaces with free boundaries

Let S be a smooth embedded compact hypersurface in $\mathbb{R}^3$. An H-surface supported by S is a surface X of mean curvature H which meets S orthogonally along its boundary. Suppose X is of the type of the disc

$$B = \{w \in \mathbb{R}^2 \,|\, |w| < 1\} ,$$

then we may introduce isothermal coordinates $w = (w^1,w^2)$, $X(w) = (X^1(w),X^2(w),X^3(w))$ over B on X . In these coordinates the following relations hold:

$$\Delta X = 2H\, \partial_1 X \wedge \partial_2 X , \tag{5.1}$$

$$|\partial_1 X|^2 - |\partial_2 X|^2 = 0 = \partial_1 X \cdot \partial_2 X , \tag{5.2}$$

$$X|_{\partial B} : \partial B \to S , \tag{5.3}$$

(5.4) $\quad \frac{\partial}{\partial n} X(w) \perp T_{X(w)}S \ , \ \forall \ w \in \partial B \ .$

Here, "$\wedge$" is the exterior product in $\mathbb{R}^3$, "$\cdot$" denotes scalar product, n is the unit exterior normal on ∂B, "$\perp$" means orthogonal; and T_pS denotes the tangent space to S at p.

In particular, if $H = 0$ a solution X to (5.1) - (5.4) is a (parametric) minimal surface supported by S.

Minimal surfaces supported by compact hypersurfaces or Schwarz' chains consisting of a collection of Jordan arcs and plane segments have been studied as early as 1816. The famous Gergonne problem was solved by H.A. Schwarz in 1872 [14]. For smooth supporting surfaces of positive genus Courant [1] proved the existence of minimal surfaces minimizing Dirichlet's integral

$$D(X) = \frac{1}{2} \int_B |\nabla x|^2 dw$$

among surfaces X with boundary $X(\partial B)$ in any given homology class of curves on S.

Courant's results were generalized to H-surfaces by Hildebrandt [6] under the condition that $|H|R \leqq 1$ for some $R > 0$ such that $S \subset B_R(0) \subset \mathbb{R}^3$.

If $S \cong S^2$ Courant's approach fails since S^2 does not support any non-trivial 1-dimensional cycle.

In fact, if S is strictly convex (the boundary of a strictly convex body in $\mathbb{R}^3$) it is easy to see that the only relative minimizers of D among surfaces X with boundary $X(\partial B) \subset S$ are the constants $X(w) \equiv p \in S$.

In 1984, by adapting the Sacks-Uhlenbeck approximation method, the author was able to establish the following existence result [16]:

<u>Theorem 5.1:</u> Suppose S is diffeomorphic to the standard sphere S^2 in $\mathbb{R}^3$. Then S supports a non-constant minimal surface.

Simultaneously, Smyth [15] obtained existence results for the tetrahedron. His proof, however, cannot be extended to more complicated polyhedral boundaries or smooth supporting surfaces.

Later, Grüter and Jost [4], [9] improved Theorem 5.1 for <u>convex surfaces</u> S and established the existence of smoothly <u>embedded</u> minimal discs supported by S.

Recently, the author succeded in extending Theorem 5.1 to surfaces of prescribed constant mean curvature [18]:

<u>Theorem 5.2:</u> Suppose $S \subset B_R(0) \subset \mathbb{R}^3$ is diffeomorphic to S^2. Then for almost every (in the sense of Lebesgue measure) $H \in \mathbb{R}$ satisfying the condition $|H|R < 1$ there exists an H-surface supported by S.

By absence of suitable a-priori bounds of Dirichlet's integral for solutions to (5.1) - (5.4) in the case of non-vanishing H Theorem 5.2 could not be derived by the Sacks-Uhlenbeck method.

Instead, the heat flow methods outlined in Section 3 proved applicable - as we will explain. (For simplicity we restrict ourselves to the case $H = 0$.)

Remark that by regularity of S there exists a δ-neighborhood $U_\delta(s)$ of S such that any point $p \in U_\delta(S)$ has a unique nearest neighbor $\pi(p) \in S$:

$$|p - \pi(p) = \inf_{q \in S}(p-q) \quad .$$

This defines the reflection $\mathcal{R} : U_\delta(S) \to U_\delta(S)$, $\mathcal{R}(p) = 2\pi(p) - p$ in S . If $S \in C^m$, $m \geqq 1$, then $\mathcal{R} \in C^{m-1}$.

Reflecting X in S we obtain a surface

$$\tilde{X}(w) = \begin{cases} X(w) , & w \in B \\ \mathcal{R}(X(\frac{w}{|w|^2})) , & w \notin B , \ X(\frac{w}{|w|^2}) \in U_\delta(S) , \end{cases}$$

defined in a domain $\mathcal{D}(\tilde{X}) \supset \bar{B}$.

Note that since $\mathcal{R}$ is involutory there holds

$$\mathcal{R}(\tilde{X}(w)) = X(\frac{w}{|w|^2}) , \quad \forall\, w \in \mathcal{D}(\tilde{X})\setminus B .$$

Inserting into (5.1) we see that $\tilde{X}$ satisfies an equation like (1.1) with a metric

$$\tilde{g}_{ik}(p) = \frac{\partial}{\partial p^i} \mathcal{R}(p) \cdot \frac{\partial}{\partial p^k} \mathcal{R}(p)$$

on $\mathcal{D}(\tilde{X})\setminus B$ carrying the metric $\tilde{\gamma}_{\alpha\beta}$ induced by the reflection $w \to \frac{w}{|w|^2}$.

If now we let

$$g_{ik}(p,w) = \begin{cases} \delta_{ik} , & w \in B \\ \tilde{g}_{ik}(p,w) , & \text{else} \end{cases}$$

$$\gamma_{\alpha\beta}(w) = \begin{cases} \delta_{\alpha\beta} , & w \in B \\ \tilde{\gamma}_{\alpha\beta}(w) , & \text{else} \end{cases}$$

$\tilde{X}$ may be viewed as a (generalized) harmonic map $\tilde{X} : (\mathcal{D}(\tilde{X}),\gamma) \to (\mathbb{R}^3,g)$.

Note that g , γ are Lipschitz continuous in $w = re^{i\phi}$ and even smooth in angular direction ϕ and as functions of p . Hence the Christoffel symbols Γ related to g will be smooth in p , bounded and measurable in $w \in \mathcal{D}(\tilde{X})$.

Similarly, a solution to the evolution problem[1)]

$$(5.5) \qquad \partial_t X - \Delta X = 0 \quad \text{in} \quad B \times [0,T]$$

$$(5.6) \qquad X(\partial B \times [0,T]) \subset S$$

$$(5.7) \qquad \frac{\partial}{\partial n} X(w,t) \perp T_{X(w,t)} S \ , \quad \forall \ (w,t) \in \partial B \times]0,T]$$

$$(5.8) \qquad X\big|_{t=0} = X_o$$

with $D(X_o) < \infty$, $X_o(\partial B) \subset S$, can be extended to a solution $\tilde{X} : \mathcal{D}(\tilde{X}) \supset \overline{B} \times [0,T] \to R^3$ of an evolution problem

$$(5.9) \qquad \partial_t \tilde{X} - \Delta_\gamma \tilde{X} - {}^g\Gamma(\tilde{X})(\nabla\tilde{X},\nabla\tilde{X}) = 0$$

$$(5.10) \qquad \tilde{X}\big|_{t=0} = \tilde{X}_o \equiv \begin{cases} X_o(w) & \text{on } B \\ R(X_o(\frac{w}{|w|^2})) & \text{else} \end{cases}$$

for a harmonic map from a domain in $(\mathbb{R}^2,\gamma)$ into $(\mathbb{R}^3,g)$.

A-priori bounds for (5.5) - (5.8) may now be derived from (5.9), (5.10) in the same way as outlined for (2.1), (2.2) in Section 4. We rely on the following observations:

---

1) This model of the motion of soap films with free boundaries has the best mathematical properties. Physically, the model corresponds to the assumption that boundary (adhesive) forces act instantaneously to create a vertical contact angle between the soap film and the supporting surface. Surface tension then governs the response "in the large".

Observation 5.3: For any smooth solution X to (5.5) - (5.8)

$$\int_0^T \int_B |\partial_t X|^2 dw dt + D(X(\cdot,T)) \leqq D(X_o) .$$

Proof: Multiply (5.5) by $\partial_t X$ and integrate by parts using that $\partial_t X \cdot \frac{\partial}{\partial n} X \equiv 0$ as a consequence of (5.6), (5.7).

q.e.d.

Observation 5.4: Suppose (as we may) that $\mathcal{D}(\tilde{X}) \subset B_2(0) \times [0,T]$. Then any pointwise or integral estimate for X on $B \times [0,T]$ entrains a corresponding estimate for $\tilde{X}$ on $\mathcal{D}(\tilde{X})$ and vice versa.

Observation 5.5: The analysis of Section 4 can be localized to the set

$$\{(w,t) \in \mathcal{D}(\tilde{X}) \mid w \in B \quad \text{or} \quad \tilde{X}(w,t) \in U_{\delta/2}(S)\}$$

by means of the following cut-off function:

$$\tilde{\varphi}(w,t) = \begin{cases} 1 , & w \in B \\ \psi(\mathrm{dist}(\tilde{X}(w,t),S) , & \text{else} \end{cases}$$

where $\psi \in C_o^\infty$ equals 1 in a neighborhood of 0 and vanishes for arguments $\geqq \delta/2$.

Note that $|\nabla\psi| \leqq c\ |\nabla\tilde{X}|$, etc. and partial derivatives of ψ will contribute "error terms" of the same order as have to be handled in Lemma 4.1.

Lemma 4.1 now conveys to our flow (5.9), (5.10), and we obtain

<u>Theorem 5.6:</u> For any $X_o \in H^{1,2}(B;\mathbb{R}^3)$ with $X_o(\partial B) \subset S$ there exists a unique distribution solution $X : B \times [0,\infty[\to \mathbb{R}^3$ of (5.5) - (5.8) which is regular on $\bar{B} \times]0,\infty[$ with exception of finitely many points $\{(w_j,t_j)\}$, $1 \leq j \leq k$, and unique in this class.

X assumes its initial value X_o continuously in $H^{1,2}$.

At a singularity $(\bar{w},\bar{t})$ a smooth, non-constant minimal surface $\bar{X}$ supported by S separates in the sense that for suitable $r_m \to 0$, $t_m \to \bar{t}$, $w_m \to \bar{w}$ after a possible rotation of coordinates

$$X(w_m+r_m w, t_m) \to \tilde{X} \qquad (m \to \infty)$$

strongly in $H^{2,2}$ on relatively compact subdomains of the upper half-plane $\mathbb{R}^2_+$.

$\tilde{X}$ has finite Dirichlet integral and may be extended to a smooth, non-constant minimal surface $\bar{X} : B \cong \overline{\mathbb{R}^2_+} \to \mathbb{R}^3$.

As $t \to \infty$, $X(\cdot,t)$ converges weakly in $H^{1,2}$ to a minimal surface X_∞ supported by S . If all $t_j < \infty$, $1 \leq j \leq k$, $X(\cdot,t) \to X_\infty$ even strongly in $H^{2,2}$.

We caution the reader that inspite of "good" a-priori estimates local existence for (5.5) - (5.8) cannot be established easily. Indead, it is a consequence of a rather delicate fixed point argument, [18, Lemma 3.16].

Now we can give a <u>proof of Theorem 5.1</u> as follows: Let S be diffeomorphic S^2 , and let

$$C(S) = \{X \in H^{1,2}(B;\mathbb{R}^3) \mid X(\partial B) \subset S\}$$

be the class of $H^{1,2}$-surfaces with boundary on S .

If for some $X_o \in C(S)$ the flow X through X_o becomes singular at some $(\bar{w},\bar{t})$, by Theorem 5.6 there exists a smooth non-constant minimal surface supported by S .

Otherwise, the flow induces a continuous deformation of $C(S)$. Now argue indirectly: Suppose S supports only constant (trivial) solutions to (5.1) - (5.4). Then the flow defines a homotopy equivalence of $C(S)$ with the space of constant maps $X \equiv p \in S$. The space $C(S)$ contracts to the space $C_o(S) = \{X \in C(S) | \Delta X = 0\}$. This latter space (by harmonic extension) is topologically equivalent to the space of closed curves $H^{\frac{1}{2},2}(\partial B;S)$ on S which relative to the constants supports non-trivial 1-dimensional cycles. In particular $C(S)$ cannot be homotopically equivalent to the space of constant maps. The contradiction proves Theorem 5.1.

q.e.d.

<u>Open Problem:</u> Taking account of the $O(2)$-action on Dirichlet's integral induced by rotation $\phi \mapsto \phi + \phi_o$ and reflection $\phi \mapsto 2\pi - \phi$ of the angular coordinates of points $w = re^{i\phi} \in B$ and the rich topological structure of the quotient space of the space of curves on S by this $O(2)$-action, cp. [7], [11], we expect higher multiplicity results for minimal surfaces supported by a surface $S \cong S^2$. However, to this moment only partial results are known [9], [15].
A possible approach to this question would be to prove global regularity of the flow (5.5) - (5.8) (at least for convex supporting surfaces) and thus to make problem (5.1) - (5.4) accessible by Ljusternik-Schnirelman or Morse theory.

<u>Proof of Theorem 5.2:</u> As stated earlier, by lack of suitable a-priori bounds for H-surfaces, $H \neq 0$, Theorem 5.2 is much harder to prove. Employing also variations of the parameter H , however, we do succeed in finding $H^{1,2}$-bounds for the H-surface flow for certain values of H corresponding to the points of differentiability of a monotone function and obtain Theorem 5.2. For details of the proof we refer the interested reader to [18].

References

[1] Courant, R.: Dirichlet's principle, conformal mapping and minimal surfaces, Interscience, New York (1950)

[2] Eells, J. - Sampson, J.H.: Harmonic mappings of Riemannian manifolds, Am. J. Math. 86 (1964), 190-160

[3] Eells, J. - Wood, J.C.: Restrictions on harmonic maps of surfaces, Topology 15 (1976), 263-266

[4] Grüter, M. - Jost, J.: On embedded minimal discs in convex bodies, preprint

[5] Gulliver, R.D. - Osserman, R. - Royden, H.L.: A theory of branched immersions of surfaces, Am J. Math. 95 (1973), 750-812

[6] Hildebrandt, S.: Randwertprobleme für Flächen mit vorgeschriebener mittlerer Krümmung und Anwendungen auf die Kapillaritätstheorie II, Arch. Rat. Mech. Anal. 39 (1970), 275-293

[7] Klingenberg, W.: Lectures on closed geodesics, Springer, Grundlehren 230, Berlin-Heidelberg-New York (1978)

[8] Jost, J.: Ein Existenzbeweis für harmonische Abbildungen, die ein Dirichletproblem lösen, mittels der Methode des Wärmeflusses, manusc. math. 38 (1982), 129-130

[9] Jost, J.: Existence results for embedded minimal surfaces of controlled topological type, preprint

[10] Ladyženskaya, O.A. - Solonikov, V.A. - Ural'ceva, N.N.: Linear and quasilinear equations of parabolic type, AMS Transl. Math. Monogr. 23, Providence (1968)

[11] Ljusternik, L.- Schnirelman, L.: Existence de trois géodésiques fermées sur toute surface de genre 0 , C.R. Acad. Sci. Paris 188 (1929), 534-536

[12] Sacks, J. - Uhlenbeck, K.: The existence of minimal immersions of 2-speres, Ann. Math. 113 (1981), 1-24

[13] Schoen, R. - Uhlenbeck, K.: Boundary regularity and miscellaneous results on harmonic maps, J. Diff. Geom. 18 (1983), 253-268

[14] Schwarz, H.A.: Gesammelte Mathematische Abhandlungen I, Springer, Berlin (1890)

[15] Smyth, B.: Stationary minimal surfaces with boundary on a simplex, Inv. Math. 76 (1984), 411-420

[16] Struwe, M.: On a free boundary problem for minimal surfaces, Inv. Math. 75 (1984), 547-560

[17] Struwe, M.: On the evolution of harmonic mappings of Riemannian surfaces, Comm. Math. Helv. 60 (1985), 558-581

[18] Struwe, M.: The existence of surfaces of constant mean curvature with free boundaries, preprint

[19] von Wahl, W.: Verhalten der Lösungen parabolischer Gleichungen für $t \to \infty$ mit Lösbarkeit im Grossen, Nachr. Akad. Wiss. Göttingen, 5 (1981)

INVERSE BOUNDARY VALUE PROBLEMS

John Sylvester* Gunther Uhlmann**
Duke University University of Washington

In this lecture we shall consider the problem of determining the spatially dependent conductivity of a body Ω from steady state direct current measurements at the boundary. This problem sometimes referred to by the names impedance computed tomography or electrical tomography. It is related to the so called magnetotelluric problem of geophysics, in which one attempts to infer the conductivity of the earth at depth from surface measurements.

§1. ($n = 1$).

The problem has its simplest form in the one dimensional case, which we proceed to discuss.

Let $\Omega = [0, 1]$ represent a one dimensional body (a wire) and let $\rho(x)$ be the resistivity as a function of position: the total resistance of the segment of wire in the interval $[a, b]$ is given by

$$R_{[a,b]} = \int_a^b \rho(x)dx \tag{1.1}$$

and (1) may be taken as the definition of $\rho(x)$.

We define the conductivity $\gamma(x)$ by

$$\gamma(x) = 1/\rho(x).$$

If we set

$$u(x) = \text{voltage potential at } x$$

then

$$\gamma(x)u'(x) = \text{current flux at } x.$$

The assumption that there are no displacement currents (that charge cannot accumulate in the wire) leads to our differential equation

$$(\gamma(x)u'(x))' = 0. \tag{1.2}$$

* Supported by NSF grant DMS-8600797
** Supported by NSF grant DMS-8601118 and an Alfred P. Sloan Research Fellowship

Our task is to obtain information about γ from all possible measurements of the voltage and current flux at the boundary of Ω, that is, all 4-tuples of numbers

$$\{u(0),\ \gamma(0)u'(0),\ u(1),\ \gamma(1)u'(1)\} \tag{1.3}$$

corresponding to solutions of (1.2). We think of each 4-tuple as a measurement or an experiment. We begin our analysis by deciding how many independent measurements we may make.

Because (1.2) is linear and $u(x) \equiv$ constant is a solution for any γ, we see that no information is lost if we take

$$u(0) = 0. \tag{1.4a}$$

As the equation is homogeneous, we may assume also that

$$u(1) = 1 \tag{1.4b}$$

Finally, integrating the equation once yields

$$\gamma(0)u'(0) = \gamma(1)u'(1). \tag{1.4c}$$

As a consequence of (1.4a) – (1.4c) we see that there is in fact only one independent boundary measurement; all the information about $\gamma(x)$ is contained in one measurement, the current flux due to a unit potential difference across the wire (i.e., $\gamma(0)u'(0)$). If we integrate (1.2) directly we obtain, in this case

$$\gamma(0)u'(0) = \int_0^1 \frac{dx}{\gamma(x)} = R_{[0,1]}. \tag{1.5}$$

We conclude that from measurements at the boundary we may only determine the average value of $\frac{1}{\gamma}$, the total resistance of the wire.

§2. ($n \geq 2$).

In dimensions ≥ 2, we shall restrict ourselves to the case where the conductivity is isotropic, i.e., the conductivity at a point x is independent of direction. In this case we take Ω to be a bounded domain in $\mathbb{R}^n$ and let

$$u(x) = \text{the voltage potential}$$

then

$$\gamma(x)\nabla u(x) = \text{the current flux.}$$

The no displacement current assumption yields the equation

$$L_\gamma u = \nabla \cdot \gamma \nabla u = 0 \quad \text{in} \quad \Omega \tag{2.1}$$

and the boundary measurements are

$$u \big|_{\partial\Omega} = f \tag{2.2}$$

$$\gamma \frac{\partial u}{\partial \nu} \bigg|_{\partial\Omega} = \Lambda_\gamma f \tag{2.3}$$

where we have used (2.3) to define the voltage to current map (Dirichlet to Neumann map) Λ_γ: u is the unique solution to (2.1), (2.2) and $\frac{\partial u}{\partial \nu}$ is the normal derivative to $\partial\Omega$.

Λ_γ is a positive self adjoint map on $L^2(\partial\Omega)$ (with domain $H^1(\partial\Omega)$). The corresponding quadratic form is the Dirichlet integral

$$Q_\gamma(f) = \int_\Omega \gamma |\nabla u|^2 = \int_{\partial\Omega} u\gamma \frac{\partial u}{\partial \nu} = (f, \Lambda_\gamma f)_{L^2(\partial\Omega)}. \tag{2.4}$$

We may interpret $Q_\gamma(f)$ as the power expended to maintain the voltage potential f on $\partial\Omega$. Knowledge of the quadratic form Q_γ on $L^2(\partial\Omega)$ (or the map Λ_γ) is equivalent to knowledge of all possible boundary measurements — a particular measurement is just a single pair $(f, \Lambda_\gamma f)$. We may therefore restate our problem: How much information about γ may be obtained from knowledge of $\Lambda_\gamma(Q_\gamma)$? We shall see that the situation here ($n \geq 2$) is much different from the one dimensional case; γ is, in fact, completely determined by Λ_γ. To see that this is plausible, we make a crude counting argument. We are given the map Λ_γ, which is an integral operator with kernel $K(x, y)$ for $(x, y) \in \partial\Omega \times \partial\Omega$; this is a function of $2(n-1)$ variables ($\dim \partial\Omega = n - 1$). γ itself is a function of n variables, so that it is plausible to determine γ from Λ_j whenever

$$2(n-1) \geq n \quad \text{or} \quad n \geq 2.$$

§3. Results.

Special cases of the inverse problem for γ have been studied for over 50 years, going back to Langer [1933]. Most of these cases require γ to depend only on a single real variable, and use separation of variables techniques to reduce the problem to one involving ordinary differential equations only.

The case where γ is a function of n variables was first considered by Calderón [1980]. He showed that the linearized inverse problem at $\gamma \equiv 1$ could be solved (we shall describe his techniques in the next section).

Kohn and Vogelius [1984] next showed that the boundary values of γ and its normal derivatives were determined by boundary measurements. They proved

Theorem 3.1. *Let $\gamma_1 (i = 1, 2)$ be in $L^\infty(\Omega)$ with a positive lower bound. Let $x_0 \in \partial\Omega$ and let B be a neighborhood of x_0 relative to $\overline{\Omega}$. Suppose that*

(3.1a) $$\gamma_i \in C^\infty(B) \qquad i = 0, 1$$

(3.1b) $$Q_{\gamma_0}(\varphi) = Q_{\gamma_1}(\varphi) \quad \forall \varphi \in H^{1/2}(\partial\Omega) \text{ with supp } \varphi \subset B \cap \partial\Omega$$

then

$$\left(\frac{\partial}{\partial x}\right)^\alpha \gamma_0(x_0) = \left(\frac{\partial}{\partial x}\right)^\alpha \gamma_1(x_0)$$

where

$$\left(\frac{\partial}{\partial x}\right)^\alpha \text{ denotes } \left(\frac{\partial}{\partial x_1}\right)^{\alpha_1} \cdots \left(\frac{\partial}{\partial x_n}\right)^{\alpha_n}.$$

As a corollary to the theorem, we see that a real analytic γ is determined by its boundary values. Kohn and Vogelius [1985] have extended this result to include piecewise analytic γ.

In the case where γ is only smooth, the authors, Sylvester and Uhlmann [1986], [1986b] have proved the following results:

Theorem 3.2. ($n = 2$). *There exists $\varepsilon(\Omega, \alpha) > 0$, such that, if $\gamma_i(x)$ $(i = 0, 1)$ belong to $C^\infty(\overline{\Omega})$ with a positive lower bound α and satisfy*

(3.2a) $$\|1 - \gamma_i\|_{C^3(\overline{\Omega})} \leq \varepsilon \quad (i = 0, 1)$$

(3.2b) $$Q_0(\varphi) = Q_1(\varphi) \quad \forall \varphi \in H^{1/2}(\partial\Omega)$$

then

$$\gamma_0(x) = \gamma_1(x) \quad \forall x \in \Omega.$$

Theorem 3.3. ($n \geq 3$). *Let γ_i $(i = 0, 1)$ belong to $C^\infty(\overline{\Omega})$ with a positive lower bound. Suppose that*

(3.3a) $$Q_0(\varphi) = Q_1(\varphi) \quad \forall \varphi \in H^{1/2}(\partial\Omega)$$

then

$$\gamma_0(x) = \gamma_1(x) \quad \forall x \in \Omega.$$

We remark that all of these theorems deal with uniqueness, we know of no general results dealing with continuous dependence or reconstruction, although some results are known in special cases (Kohn and Vogelius [1985], Friedman and Gustafsson [1986]).

§4. Methods.

We give a brief sketch of the proofs of Theorems 3.1 – 3.3.

Sketch of proof of Theorem 3.1.

Kohn and Vogelius [1984] used energy arguments to prove Theorem 3.1. The proof we outline here uses the techniques of microlocal analysis. Assume for the moment that γ has been defined on all of $\mathbb{R}^n$ so that the differential operator $L_\gamma = \nabla \cdot \gamma \nabla$ has a fundamental solution $E_\gamma(x, y)$. It is possible to use E_γ to construct Λ_γ. We first solve the equation

$$\nabla \cdot \gamma \nabla w = g(\beta)\delta_{\partial\Omega} \text{ in } \mathbb{R}^n$$

where $g(\beta) \in C^\infty(\partial\Omega)$ and $\delta_{\partial\Omega}$ is the distribution which acts on $\varphi \in C^\infty(\mathbb{R}^n)$ by

$$\langle \delta_{\partial\Omega}, \varphi \rangle = \int_{\partial\Omega} \varphi(\beta) dS_\beta.$$

We then define the operators P_{12} and P_{22} mapping $C^\infty(\partial\Omega)$ into $C^\infty(\partial\Omega)$ by

$$\begin{aligned} P_{12}g &= \lim_{x \to x_* \in \partial\Omega} w = \lim_{x \to x_*} E_\gamma(g\delta_{\partial\Omega}) \\ P_{22}g &= \lim_{x \to x_* \in \partial\Omega} \frac{\partial w}{\partial \nu} = \lim_{x \to x_*} \frac{\partial}{\partial \nu} E_\gamma(g\delta_{\partial\Omega}) \end{aligned} \tag{4.1}$$

P_{12} and P_{22} are components of the Calderón projector (a.k.a. projection onto the Cauchy data). It turns out that

$$\Lambda_\gamma = P_{22}(P_{12})^{-1} \tag{4.2}$$

and also that all the operators E_γ, P_{12}, P_{22}, and Λ_γ are classical pseudodifferential operators. This means that they have a symbol which as an asymptotic expansion in the cotangent variables, i.e., the symbol of Λ_γ has the form

$$\lambda_\gamma(x, \xi) = \lambda_\gamma^1(x, \xi) + \lambda_\gamma^0(x, \xi) + \ldots + \lambda_\gamma^{-k} + \ldots \tag{4.3}$$

where λ_γ^{-k} is homogeneous of degree $-k$ in the ξ variable. The expansion can be computed explicitly from a similar expansion for E_γ, along with (4.1) and (4.2). It turns out that the jth normal derivative $\frac{\partial^j \gamma}{\partial \nu^j}$ appears for the first time in λ_γ^{1-j} and appears there linearly (with nonzero coefficient). Hence if two γ's give rise to the same Λ_γ, they have the same λ_γ^{1-k} for each k, and, therefore, by induction, the same $\frac{\partial^j \gamma}{\partial \nu^j}$.

It is worth noting that (4.3) can be used to see that the boundary values of γ, and its tangential derivatives do depend on continuously on Λ_γ in an appropriate topology.

Sketch of proof of Theorem 3.3.

The proof of Theorem 3.3 uses the linearization of the inverse problem introduced by Calderón [1980]. We begin with this computation.

We shall denote by $Q_\gamma(\cdot,\cdot)$ the bilinear form obtained from the quadratic form Q_γ by polarization. We fix f and g in $H^{1/2}(\partial\Omega)$ so that

$$Q_\gamma(f,g) = \int_\Omega \nabla u \cdot \nabla v \tag{4.4}$$

where

$$\begin{aligned} L_\gamma u &= 0 \qquad & L_\gamma v &= 0 \\ u\big|_{\partial\Omega} &= f \qquad & v\big|_{\partial\Omega} &= g\,. \end{aligned} \tag{4.5}$$

We set

$$\gamma = (1-t)\gamma_0 + t\gamma_1 \tag{4.6}$$

and compute

$$\begin{aligned} \frac{d}{dt}Q_\gamma(f,g) &= \int_\Omega \Big\{\dot\gamma\nabla u\cdot\nabla v + \gamma(\nabla\dot u\cdot\nabla v + \nabla\dot v\cdot\nabla u)\Big\} \\ &= \int_\Omega \dot\gamma\nabla u\cdot\nabla v + \int_{\partial\Omega}\gamma\Big(\dot u\frac{\partial v}{\partial\nu} + \dot v\frac{\partial u}{\partial\nu}\Big)dS \\ \dot Q_\gamma(f,g) &= \int_\Omega \dot\gamma\nabla u\cdot\nabla v + \int_{\partial\Omega}\dot f\Lambda_\gamma g + \dot g\Lambda_\gamma f \end{aligned} \tag{4.7}$$

where we have integrated by parts and used (4.5). Now f and g are independent of the parameter t so that the second term on the right hand side of (4.7) is zero.

$$\dot Q_\gamma(f,g) = \int_\Omega \dot\gamma\nabla u\cdot\nabla v. \tag{4.8}$$

If we further choose $\gamma_0(x)\equiv 1$ we have (4.8) for all harmonic functions u and v with boundary values f and g, respectively. Calderón chose

$$\begin{aligned} u &= e^{x\cdot\rho_1} \quad \text{or} \quad f = e^{x\cdot\rho_1}\big|_{\partial\Omega} \\ v &= e^{x\cdot\rho_2} \quad \text{or} \quad g = e^{x\cdot\rho_2}\big|_{\partial\Omega} \end{aligned} \tag{4.9}$$

where we insist that each $\rho_i \in \mathbb{C}^n$ satisfies

$$\rho\cdot\rho = 0 \tag{4.10}$$

in order that u and v be harmonic. (4.10) implies that

$$\rho = k + im; \quad k, m \in \mathbb{R}^n$$

such that

$$k \cdot m = 0$$

$$|k| = |m|.$$

Making the choices

$$\begin{aligned} \rho_1 &= k + im \\ \rho_2 &= -k + im \end{aligned} \tag{4.11}$$

(4.8) becomes

$$\begin{aligned} \dot{Q}_\gamma(f,g) &= 2|m|^2 \int_\Omega e^{2ix\cdot m}\dot{\gamma} \\ &= 2|m|^2(\dot{\gamma}\chi_\Omega)^\wedge(2m) \end{aligned} \tag{4.12}$$

where $\wedge$ denotes the Fourier transform and χ_Ω the indicator function for the region Ω. We see that by appropriate choices of f and g we may obtain from $\dot{Q}_\gamma$, essentially the Fourier transform of $\dot{\gamma}$, and hence $\dot{\gamma}$ itself by Fourier inversion. We have thus solved the linearized problem at $\gamma_0 \equiv 1$.

We should remark that the range of the linearized operator (i.e., the set of quadratic forms p such that $p = \dot{Q}_\gamma$ for some $\dot{\gamma}$) is not onto in any reasonable topology; it is therefore not possible to use the implicit function theorem to conclude local uniqueness or existence for the inverse problem.

Our first step in adapting the previous computation to the full problem will be to consider the linearized problem at arbitrary γ, not just at $\gamma \equiv 1$. In order to do this we must find solutions to (4.5) analogous to the exponentials in (4.9). This is contained in the

Lemma 4.1. *Suppose ρ satisfies (4.10) and*

$$\frac{\|q\|_{L^\infty}}{|\rho|} < 1; \quad q = \frac{\Delta(\gamma^{1/2})}{\gamma^{1/2}} \tag{4.13}$$

then there exist solutions to

$$L_\gamma u = 0 \quad in \quad \Omega$$

of the form

$$u(x,\rho) = e^{x\cdot\rho}\gamma^{-1/2}(1 + \Psi_\gamma(x,\rho)) \tag{4.14}$$

where

$$\|\Psi\|_{L^2(\Omega)} \le \frac{k}{|\rho|}\|q\|_{L^2(\Omega)}. \tag{4.15}$$

Moreover, if $\Lambda_{\gamma_0} = \Lambda_{\gamma_1}$; we may choose

(4.16)
$$u_0 \big|_{\partial\Omega} = u_1 \big|_{\partial\Omega}$$

The lemma asserts that for large $|\rho|$ there are special solutions which are small perturbations of $\gamma^{-1/2}e^{x\cdot\rho}$. The lemma does not allow us to prescribe the boundary values of these solutions, but (4.16) says that if γ_0 and γ_1 have the same voltage to current map, these special solutions have the same boundary data.

We now begin the proof of the theorems. Suppose that

$$0 = Q_{\gamma_0}(f,g) - Q_{\gamma_1}(f,g) \quad \forall f,g$$

then

$$0 = Q_{\gamma_0}\Big(u_0(\rho_1)\big|_{\partial\Omega}, u_0(\rho_2)\big|_{\partial\Omega}\Big) - Q_{\gamma_1}\Big(u_1(\rho_1)\big|_{\partial\Omega}, u_1(\rho_2)\big|_{\partial\Omega}\Big)$$

where we have used (4.16). We have, by the fundamental theorem of calculus and recalling (4.6)

$$0 = \int_0^1 dt\{Q_\gamma(u_t(\rho_1), u_t(\rho_2))\}$$

$$0 = \int_0^1 dt\Big\{\int_\Omega \dot\gamma \nabla u_t(\rho_1)\cdot\nabla u_t(\rho_2)\Big\} + \int_0^1 dt\Big\{\int_{\partial\Omega}(\gamma\frac{\partial u(\rho_1)}{\partial\nu}\dot u(\rho_2) + \frac{\partial u}{\partial\nu}(\rho_2)\dot u(\rho_1))\Big\}$$

just as in (4.7). It turns out (see Sylvester-Uhlmann [1986] or [1986b]) that the second term is in fact zero so that we have

$$0 = \int_0^1 dt\int_\Omega \dot\gamma\nabla u_t(\rho_1)\cdot\nabla u_t(\rho_2).$$

If we use the identity

$$L_\gamma(vw) = L_\gamma vw + L_\gamma wv + 2\gamma\nabla w\cdot\nabla v$$

we obtain

$$0 = \int_0^1 dt\int_\Omega \frac{\dot\gamma}{\gamma}L_\gamma(u_t(\rho_1)u_t(\rho_2)).$$

Integrating by parts (according to Theorem 3.1 there are no boundary terms) gives

(4.17)
$$0 = \int_0^1 dt\int_\Omega L_\gamma\Big(\frac{\dot\gamma}{\gamma}\Big)u_t(\rho_1)u_t(\rho_2).$$

We shall choose (for $n \geq 3$)

(4.18)
$$\rho_1 = k + i(m - w)$$
$$\rho_2 = -k + i(m + w)$$

where

(4.19)
$$k\cdot m = k\cdot w = m\cdot w = 0$$
$$|k| = |m - w| = |m + w|$$

and we shall let m be fixed and arbitrary while we send $|w|$ to infinity.

The estimate (4.15) allows us to conclude that

$$\int_0^1 dt \frac{1}{\gamma} L_\gamma\left(\frac{\dot{\gamma}}{\gamma}\right) = 0$$

or

$$0 = \int_0^1 dt\{\Delta(\ln \gamma) + \tfrac{1}{2}|\nabla(\ln \gamma)|^2\}$$

so that

$$\Delta(\ln \gamma_0 - \ln \gamma_1) + \tfrac{1}{2}(\nabla(\ln \gamma_0) + \nabla \ln (\gamma_1)) \cdot \nabla(\ln \gamma_0 - \ln \gamma_1) = 0 \tag{4.20}$$

and by Theorem 3.1

$$(\ln \gamma_0 - \ln \gamma_1)\,|_{\partial\Omega} = 0. \tag{4.21}$$

Hence, from (4.20) and (4.21) we conclude that

$$\ln \gamma_0 - \ln \gamma_1 = 0$$

and Theorem 3.3 is proved.

We remark that the dimension $n \geq 3$ was necessary to choose ρ_1 and ρ_2 in (4.18) and (4.19). For $n = 2$, we are left with the choice (4.11); the resulting computation yields only an estimate on $\hat{\gamma}(m)$ for large m; the assumption (3.29) is necessary to control $\hat{\gamma}(m)$ for m small and to prove Theorem 3.2.

References.

Calderón, A. P. [1980] On an inverse boundary value problem, *Seminar on Numerical Analysis and its Applications to Continuum Physics, Soc. Brasileira de Matemática*, Rio de Janeiro, 1980, 65–73.

Friedman, A. and Gustafsson, B. [1986] Identification of the conductivity coefficient in an elliptic equation, to appear.

Kohn, R. and Vogelius, M. [1984] Determining conductivity by boundary measurements, *CPAM* **37** (1984), 289–298.

Kohn, R. and Vogelius, M. [1985] Determining conductivity by boundary measurements II, *CPAM* **38** (1985), 643–667.

Langer, R. E. [1933] An inverse problem in differential equations, *Bull. Amer. Math. Soc.* **39** (1933), 814–820.

Sylvester, J. and Uhlmann, G. [1986] A uniqueness theorem for an inverse boundary value problem in electrical prospection, *CPAM* **39** (1986), 91–112.

Sylvester, J. and Uhlmann, G. [1986b] A global uniqueness theorem for an inverse boundary value problem, *Annals of Math.*, to appear.

Scattering on the line – an overview

by R. Beals, * P. Deift ** and C. Tomei***

The Schrodinger operator on the line with real potential in Schwartz class is well understood. In particular, the asymptotic properties of solutions of the eigenvalue problem are well known, as well as techniques to recover the potential itself from such information (the inverse problem for the Schrodinger equation [DT]). Inverse data turns out to be very convenient. For example, one can use it to write down explicitly wave operators and the functional calculus for Schrodinger operators. Or one can take the inverse data as a change of variables [GGKM] to linearize certain evolutions, such as the Korteweg-deVries equation.

In this paper, we describe very superficially some aspects of the problem of defining and handling inverse data for arbitrary self-adjoint linear differential operators on the line, with coefficients in Schwartz class.

Detailed statements and proofs will be presented elsewhere. The immediate precursors of this report are [B], [BC], [DT] and [DTT], in which some of these results appeared for the first time.

The direct problem

Consider the eigenvalue problem

$$Lu(x) = z^n u(x), \quad x \in \mathbf{R} \tag{*}$$

* Research supported in part under NSF Grant MCS8104234.

** Research supported in part under NSF Grant MCS8301662 and ONR Grant N00014-76-C-0439.

*** Research supported in part by CNPq, Brazil.

for the differential operator

$$Lu = D^n u - (q_{n-2}(x)D^{n-2}u + \ldots + q_1(x)Du + q_0(x)u),$$
$$Du = \frac{1}{i}u'.$$

For $n = 2$, the Jost functions f and g given by

$$-f'' - q_0 f = z^2 f, \quad f(x,z) \sim e^{izx}, \quad x \to \infty,$$
$$-g'' - q_0 g = z^2 g, \quad g(x,z) \sim e^{-izx}, \quad x \to -\infty,$$
$$\text{Im } z \geq 0,$$

form, for almost all z, a basis for the space of solutions of (*). The Jost functions are solutions of Volterra integral equations, from which one obtains their analytic dependence on the parameter z on the upper half plane, continuous extension (in z) to the real line, and asymptotic behavior (for z large or near zero, and for large real x). For n larger than 2, the situation is more complicated. We still want to obtain solutions to (*) described by their asymptotic behavior in x. Let $n = 3$ for notational convenience – the arguments extend easily to arbitrary order. To fix ideas, let z be a positive real number, and $\alpha = \exp(2\pi i/3)$. We expect to find solutions of (*) F, G and H asymptotic to the free solutions $\exp(i\alpha^2 zx)$, $\exp(izx)$ and $exp(i\alpha zx)$, respectively, for very negative x. Such requirement does not determine G uniquely: we can always add a multiple of F to a candidate G without changing its asymptotic behavior at $-\infty$. Clearly, there is a one-dimensional vector of solutions growing (at most) like (a multiple of) $\exp(izx)$, etc. We have to describe additional asymptotic requirements to obtain well defined solutions. For an arbitrary complex z, we order the numbers $i\alpha^j z$ according to decrasing real part and obtain increasing vector spaces of solutions with asymptotic behavior at $-\infty$ bounded by (multiples of) $\exp(i\alpha^j zx)$. Let Σ be the set of lines in the z-plane in which two of the numbers above have the same real part and consider $z \in \mathbf{C} \setminus \Sigma$. For such z, there is also a sequence of subspaces of solutions of (*) with asymptotic behavior prescribed in a similar fashion at $+\infty$. Those facts are well known, but the usual construction [CL] does not obtain solutions prescribed by asymptotic properties which are meromorphic on maximal domains of the z variable.

In order to prescribe (up to a multiple) a solution of (*), we require the solution to be bounded (up to a multiple) by the same exponential factor on both sides of the real line, as the subspces of solutions with such bounds at $-\infty$ and $+\infty$ have dimensions adding to 4, and hence, have an intersection of dimension 1, generically. We obtain such special solutions normalized at $-\infty$ by performing the steps below.

1. Convert the scalar differential equation into a first order system, in the usual fashion:

$$\begin{aligned} v &= (v_1, v_2, v_3)^T, \\ Dv &= J_z v + qv, \\ J_z &= \begin{pmatrix} 0 & 1 & 0 \\ 0 & 0 & 1 \\ z^3 & 0 & 0 \end{pmatrix}, \quad q = \begin{pmatrix} 0 & 0 & 0 \\ 0 & 0 & 0 \\ q_0 & q_1 & 0 \end{pmatrix}. \end{aligned} \tag{**}$$

Denote by F^-, G^- and H^- the solutions of (**) bounded respectively by multiples of the free solutions

$$e^{i\alpha_1 zx} \begin{pmatrix} 1 \\ \alpha_1 z \\ (\alpha_1 z)^2 \end{pmatrix}, \quad e^{i\alpha_2 zx} \begin{pmatrix} 1 \\ \alpha_2 z \\ (\alpha_2 z)^2 \end{pmatrix}, \quad e^{i\alpha_3 zx} \begin{pmatrix} 1 \\ \alpha_3 z \\ (\alpha_3 z)^2 \end{pmatrix},$$

and asymptotic to the free solutions at $-\infty$. Similarly, denote by F^+, G^+ and H^+ the solutions of (**) with the same exponential bounds normalized at $+\infty$. Set

$$f^{\pm} = F^{\pm} e^{-\alpha_1 zx}, \quad g^{\pm} = G^{\pm} e^{-\alpha_2 zx}, \quad h^{\pm} = H^{\pm} e^{-\alpha_3 zx}.$$

These functions satisfy the differential equations

$$\begin{aligned} Dm^{\pm} &= J_z m^{\pm} - (\alpha_j z) m^{\pm} + q m^{\pm}, \\ &m^{\pm} \text{ bounded}, \\ \lim_{x \to \pm\infty} m^{\pm} &= \begin{pmatrix} i \\ \alpha_j z \\ (\alpha_j z)^2 \end{pmatrix}. \end{aligned} \tag{***}$$

Clearly, all we need is to show the existence of solutions to (***), depending meromorphically on z in each open sector of $\mathbf{C} \setminus \Sigma$.

2. Solutions of (***) are functions taking values on three dimensional vectors. It is not difficult to write the differential equations satisfied by f^-, $f^- \wedge g^-$ and $f^- \wedge g^- \wedge h^-$, and to see that they give rise to Volterra integral equations with integrable kernel, for z in the closure of each sector. There are similar equations for $h^+, h^+ \wedge g^+$ and $h^+ \wedge g^+ \wedge f^+$. So, the existence of solutions for such equations is standard.

3. We still have to identify the solutions of the Volterra equations above as exterior product of solutions of (***). This follows, essentially, from a general fact in exterior algebra (a k-exterior product w is of the form $f \wedge y$, for a vector f if and only if $f \wedge w = 0$) and the uniqueness of solutions for Volterra equations. Without loss, we can take the solutions of the Volterra equation to be of the form $\tilde{f}^-, \tilde{f}^- \wedge \tilde{g}^-$ and $\tilde{f}^- \wedge \tilde{g}^- \wedge \tilde{h}^-$, and $\tilde{h}^+, \tilde{h}^+ \wedge \tilde{g}^+$ and $\tilde{h}^+ \wedge \tilde{g}^+ \wedge \tilde{f}^+$, where $\tilde{f}^\pm, \tilde{g}^\pm$ and $\tilde{h}^\pm$ are orthogonal and normalized at $\pm\infty$.Notice two additional facts. The triple products are Wronskians of solutions, and hence constant in x, as the original scalar equation (*) has no second order coefficient. Moreover, $\tilde{f}^- = f^-$ and $\tilde{h}^+ = h^+$.

4. The problem now reduces to linear algebra. Let M be the matrix with columns given by f^-, g^- and h^-, $\tilde{M}^\pm$ the matrix with columns $\tilde{f}^\pm$, $\tilde{g}^\pm$ and $\tilde{h}^\pm$. From the asymptotic properties of the columns,

$$M = \widetilde{M}^- U \quad \textit{and} \quad M = \widetilde{M}^+ L$$

where L and U are (unknown) lower and upper triangular matrices and the diagonal elements of U are equal to 1. So, to obtain M, it suffices to obtain U or L, the factors in the lower-upper triangular decomposition of the known matrix $(\tilde{M}^+)^{-1}\tilde{M}^-$. It is not hard to check that the columns of M are indeed the solutions of (***) with the required properties. The construction can possibly break down on a discrete set of values z for which the matrix $(\tilde{M}^+)^{-1}\tilde{M}^-$ does not admit a lower-upper factorization.

Thus, we have solutions F^-, G^- and H^- of (**) which behave like free solutions at $-\infty$ and are bounded by multiples of free solutions on the whole line. Moreover, those functions are meromorphic in $\mathbf{C} \setminus \Sigma$ and extend continuously (with a possible exception of a discrete set of points) to the boundaries of the sectors in $\mathbf{C} \setminus \Sigma$. These functions play as central a role in the inverse problem for the higher order scalar equation as Jost solutions do in the Schrodinger case.

For $n = 2$, the factorization in the last step of the above construction breaks down exactly when z is an eigenvalue of (*). In general, the factorization breaks down when a column of either $\tilde{M}^+$ or $\tilde{M}^-$ is asymptotic to two different exponential factors on both sides of the line. In higher order, such singularities are not necessarily L^2-eigenvalues.

The next step is similar to the case $n = 2$: for $z \in \Sigma$, there should be two distinct families of solutions of (*), obtained by continuity from solutions on both sides of the ray. Define the continuous inverse data to be the matrix expressing one set of solutions as a linear combination of the other. Notice that, if we follow the construction above for the $n = 2$ case, we will not obtain as inverse data the usual (unitary) scattering matrix. The standard scattering matrix relates two different bases of solutions for the second order eigenvalue problem than the ones we defined.

Also following the case $n = 2$, we should prescribe "norming constants" to the singularities of the different solutions of (*) for $z \in \mathbf{C} \setminus \Sigma$, but we give no details. We will assume that there are no such singularities.

The inverse problem

For $n = 3$, the construction given above for the inverse data defined $6 \times 3 \times 3$ functions on half-lines, certainly many more objects than the set of two coefficients (on a line) of the scalar equation (*). Our next task is to search for a minimal set of functions equivalent to prescribing the inverse data. One obvious reduction comes from the fact that (*) is invariant under the change from z to αz. This makes the inverse data on alternate rays equal. Moreover, on each ray, some linear algebra computations show that the transition matrix between both bases of solutions breaks in 2×2 blocks, corresponding to solutions with exponential factors of equal amplitude. There are two additional algebraic facts that we do not prove. If we order the basis of solutions in each side of the ray as shown below, the transition matrices admit a special (lower-upper triangular) factorization, indicated in the figure by the special form of the matrix entries. The four functions on the two half-lines provide the minimal set of variables for which we were searching.

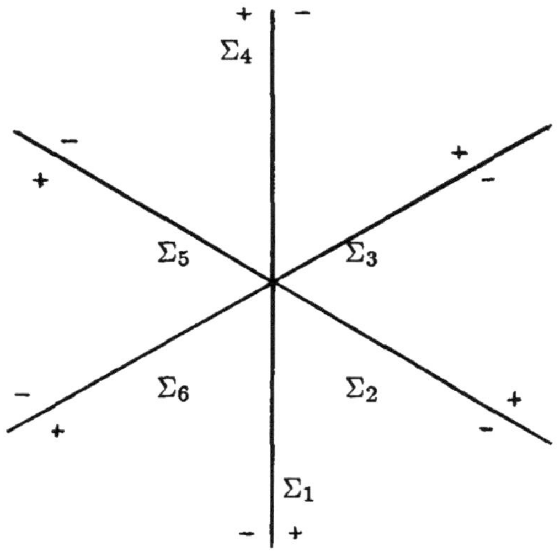

$$\text{On } \Sigma_1, \quad (f^+h^+g^+) = (f^-g^-h^-)\begin{pmatrix} 1 & 0 & 0 \\ 0 & 1+ab & a \\ 0 & b & 1 \end{pmatrix}.$$

$$\text{On } \Sigma_2, \quad (g^+f^+h^+) = (f^-g^-h^-)\begin{pmatrix} 1+cd & c & 0 \\ d & 1 & 0 \\ 0 & 0 & 1 \end{pmatrix}.$$

Figure 1

Moreover, if (*) gives rise to a self-adjoint operator on the line, there are relations among these variables. For $n = 3$,

$$c = \alpha\bar{b}, \qquad \bar{a} = \alpha d.$$

The inverse problem consists in retrieving the coefficients of the scalar equation (*) from the minimal set of variables. In a nutshell, we have to solve a Riemann-Hilbert problem for each fixed x: we are searching for functions which are analytic in sectors, satisfy certain multiplicative jump conditions and have a prescribed behavior for large z. Some things are special about this problem: the underlying geometry is rather unusual, and the corresponding integral equation is not of the form *identity* + *compact* (as in the Fadeev-Marchenko equation, to which the Riemann-Hilbert problem indeed reduces in the Schrodinger case). Actually, both issues go hand in hand: the integral equation is of the form *invertible* + *compact*, so Fredholm theory is still available, and to understand its invertible part, one has to study carefully the behavior of the basis of solutions and of the inverse variables for z near the origin. A critical observation is the fact that the inverse variables arising from a generic set of potentials in Schwartz space admit asymptotic expansions at the origin with coefficients taking universal values. Unfortunately, the free problem is not in this class. This is to be compared with the Schrodinger equation, for which, generically, the reflection coefficient at 0 equals -1, but the free equation is reflectionless. We skip any further comment on such issues. Instead, we provide some details on another important point of the inverse problem: to show that the Fredholm operator equivalent to the Riemann-Hilbert problem has trivial kernel.

We state (and sketch the proof of) a vanishing lemma of many possible uses:

1) to show that the map from pontentials to inverse variables is injective,

2) to show that the integral equation equivalent to the Riemann-Hilbert problem has trivial kernel,

3) to show that the potentials obtained from the integral equation indeed have the prescribed inverse variables,

4) to show that one can use the inverse variables to solve some evolution equations, as indicated in the introduction.

First, denote by f, g and h the first coordinates of the functions f^-, g^- and h^-. We will consider the Riemann-Hilbert problem for the functions f, g and h, with transition matrices given by obvious modifications of the original transition matrices. For large z, F^-, G^- and H^- behave asymptotically like free solutions. So, for fixed x, f, g and h go to 1 at ∞. It is not hard to show that the difference A, B and C of two possible sets of values for f, g and h, would decay like $1/z$ at ∞ and also satisfy the same jump conditions on Σ. The vanishing lemma simply says that such A, B *and* C *have to be equal to zero*, if (*) is self-adjoint.

The proof is as follows.

First, notice that $A(z)$ and $\bar{C}(\bar{z})$ is analytic across the ray Σ_1. By Cauchy's theorem,

$$\int_{\Sigma_0 \nearrow \searrow \Sigma_2} A(z)\overline{C}(\overline{z})dz = 0.$$

From the transition matrices, we have

$$\begin{aligned}
A^+ &= \frac{c}{1+cd}B^+ + \frac{1}{1+cd}B^-, \\
A^- &= \frac{1}{1+cd}B^+ - \frac{d}{1+cd}B^-, \\
C^+ &= bB^+ + B^-, \\
C^- &= B^+ - aB^-,
\end{aligned}$$

where the signs indicate the side from which the limit was taken to reach the ray, as in Figure 1.

3. Substituting in the above expression, we get

$$0=\int_{\Sigma_6}\left(\frac{c}{1+cd}B^+(z)+\frac{1}{1+cd}B^-(z)\right)\left(\overline{B}^+(\overline{z})-\overline{a}\overline{B}^-(\overline{z})\right)dz$$

$$+\int_{\Sigma_2}\left(\frac{1}{1+cd}B^+(z)-\frac{d}{1+cd}B^-(z)\right)\left(\bar{b}\bar{B}^+(\bar{z})+\bar{B}^-(\bar{z})\right)dz$$

$$=\int_{\Sigma_6}B^-(z)\bar{B}^+(\bar{z})dz+\int_{\Sigma_2}B^+(z)\bar{B}^-(\bar{z})dz,$$

from self adjointness and invariance of $A, B,$ and C

under multiplication by α,

$$=\int_{\mathbf{R}-}|B(z)|^2dz+\int_{\mathbf{R}+}|B(z)|^2dz, \quad \text{by Cauchy.}$$

So, $B\equiv 0$ and the result follows.

A similar proof works in the case where there is discrete inverse data. The many applications of Cauchy's Theorem collect residues at poles of A, B and C, but, due to symmetry, the total contribution is again a positive number, provided the poles of B avoid certain regions of the z-plane. One can see that such a restriction is indeed forced for self-adjoint problems: a pole of B essentially corresponds to a value of z for which g^- grows at $+\infty$ not as fast as it does generically. So, for z in certain sectors, such g^- would be an L^2-eigenfunction of (*), contradicting self-adjointness.

The above proof does not apply directly for even n, and an additional step is needed, but we give no details.

The choice of inverse variables above is clearly not the only way to parametrize the set of scalar differential equations. One remarkable property of such variables, however, is that they linearize the Gelfand-Dikii evolutions (of which the Korteweg-deVries and the Boussinesq equation are examples). As a final remark, we notice that we were only able to solve the inverse problem for a generic set of inverse data, even under the hypothesis of self-adjointness of (*). However, it is very easy to check when the Gelfand-Dikii evolutions leave such sets – actually, most of the technical conditions required in our solution of the inverse problem are trivially invariant under such flows.

References

[B] Beals, R., The Inverse Problem for Ordinary Differential Operators on the Line, Amer. J. of Math., 107, 281-366 (1985).

[BC] Beals, R. and Coifman, R., Scattering and Inverse Scattering for First Order Systems, Comm. Pure Appl. Math. XXXVII, 39-90 (1984).

[CL] Coddington, E. and Levinson, N., Theory of Ordinary Differential Equations, McGraw Hill, 1955, p. 104 (Exercise 29, Chapter 3).

[DT] Deift, P. and Trubowitz, E., Inverse Scattering on the Line, Comm. Pure Appl. Math. XXXII, 121-251 (1979).

[DTT] Deift, P., Tomei, C. and Trubowitz, E., Inverse Scattering and the Boussinesq Equation, Comm. Pure Appl. Math. XXXV, 567-628 (1982).

[GGKM] Gardner, C., Greene, J., Kruskal, M. and Miura, R., Method for Solving the Korteweg-deVries Equation, Phys. Rev. Lett. 1095-1097 (1967).

Richard Beals
Department of Mathematics
Yale University
Box 2155, Yale Station
New Haven, CT, 06520, USA

Percy Deift
Courant Institute
251 Mercer Street
New York, NY 10012, USA

Carlos Tomei
Departamento de Matemática
PUC/RJ
Rua Marquês de S. Vicente, 225
Rio de Janeiro, 22453, Brasil

MICROLOCAL COHOMOLOGY IN HYPO-ANALYTIC STRUCTURES

F. Treves

Rutgers University
New Brunswick, N.J. 08903

CONTENTS

INTRODUCTION

Hypo-analytic structures have been defined in [BCT] (see also [T]). One of the simplest hypo-analytic structures is that of $\mathbb{C}^m \times \mathbb{R}^n$ in which the hypo-analytic functions are the functions that are holomorphic with respect to $z = (z^1,\dots,z^m) \in \mathbb{C}^m$ and are independent of $t = (t^1,\dots,t^n) \in \mathbb{R}^n$. These are the functions annihilated by the vector fields $\partial/\partial\bar{z}^j$, $\partial/\partial t^k$ $(1 \le j \le m,\ 1 \le k \le n)$. Locally any smooth hypo-analytic manifold $\mathfrak{M}$ is isomorphic to a submanifold $\Sigma \subset \mathbb{C}^m \times \mathbb{R}^n$ (for a suitable choice of m and n) whose intersection Σ_t with every slice $\mathbb{C}^m \times \{t\} \cong \mathbb{C}^m$ is an m-dimensional totally real submanifold (what we call a maximally real) of $\mathbb{C}^m$. The hypo-analytic functions on Σ are the restrictions to Σ of the holomorphic functions of z. The cotangent structure bundle $T' \subset \mathbb{C}T^*\Sigma$ is spanned by the pull-backs to Σ of $dz^1,\dots,dz^m$. The tangent structure bundle $\mathcal{V} \subset \mathbb{C}T\Sigma$, i.e., the orthogonal of T' for the duality between tangent and cotangent vectors, is spanned by the tangent vectors to Σ that are linear combinations of the $\partial/\partial\bar{z}^j$ and of the $\partial/\partial t^k$. Any function (or distribution) in Σ which is annihilated by any C^∞ section of $\mathcal{V}$ or, which is the same, whose differential is a section of T', is called a "solution". The article [BCT] was devoted to studying the extension properties of solutions, from Σ to wedges with their edge on Σ, in particular to the definition of their hypo-analytic wave-front set and to its delimitation by the Fourier-Bros-Iagolnitzer transform - henceforth abbreviated as the FBI transform.

The present work develops the analogous theory for closed p-forms of degree $p \ge 1$ and of class C^∞. Here "closed" is meant in the sense of the differential complex naturally associated with the hypo-analytic structure under study. Actually the objects we extend are not really differential forms. Rather, they are equivalence classes of differential forms. When studied in Σ these classes have unique representatives of the kind

$$(1)\qquad \sum_{q+r=p}\ \sum_{|J|=q}\ \sum_{|K|=r} f_{J,K}(x,y,t)\,d\bar{z}^J \wedge dt^K$$

(J, K are ordered multi-indices). But of course such a representative will depend on the choice of the embedding into $\mathbb{C}^m \times \mathbb{R}^n$. A change of embedding might add to it a differential p-form involving not only $d\bar{z}^j$'s and dt^k's but also some dz^i's. One might ask if any representative of the class of (1) extends as a $(\bar\partial_z+d_t)$-closed form to a certain wedge. Actually we should also be at liberty to add coboundaries, i.e., exact form (in the complex over Σ); and thus the correct notion is that of local extension of the <u>cohomology class</u> of the form (1). All this is defined in detail in Ch. II. The hypo-analytic wave-front set or <u>microsupport</u> of a cohomology class is defined <u>a la Sato</u> (as was done in [BCT] for closed zero forms, i.e. solutions).

Where we think we add something to the Sato scheme is by showing (in Ch. III) that the FBI transform can be used to delimit the microsupport of a cohomology class. We show that a point γ in $T^*\Sigma$ belongs to the microsupport if and only if there is a representative of that class, of the kind

$$\sum_{|K|=p} f_K(x,y,t)dt^K, \tag{2}$$

endowed with the property that the FBI transform of every coefficient f_K decays exponentially in a conic complex neighborhood of γ (Ch. III, Theorems 5.1, 5.2). The basis for such a result is the vanishing of the $\bar\partial$-cohomology in germs of wedges in $\mathbb{C}^m$ whose edge is a maximally real submanifold of $\mathbb{C}^m$ (proved in Ch. I).

An important species of hypo-analytic structures is that of the <u>Cauchy-Riemann</u> (abbreviated to CR) structures. Rather than embedding them locally into $\mathbb{C}^m\times\mathbb{R}^n$ it is more natural to use local embeddings in $\mathbb{C}^n\times\mathbb{C}^d$, of the kind $(x,y,s)\to(z,w)$ where

$$\begin{aligned} z^j &= x^j+iy^j \qquad (j=1,\dots,n),\\ w^k &= s^k + i\varphi^k(x,y,s) \qquad (k=1,\dots,d). \end{aligned} \tag{3}$$

The reader should be alerted to the fact that the role played, in the embedding into $\mathbb{C}^m\times\mathbb{R}^n$ of the non CR hypo-analytic structures, by the second factor $\mathbb{R}^n$ is played now by z-space $\mathbb{C}^n$; the role of the first factor, $\mathbb{C}^m$, is played by w-space $\mathbb{C}^d$. The ambient differential complex in $\mathbb{C}^{n+d}$ is $\bar\partial=\bar\partial_z+\bar\partial_w$ and thus one is interested, here, in extending cohomology classes for the tangential CR complex to $\bar\partial$-cohomology classes. Needless to say the specific adaptations of the general definitions of Ch. II are routine (see Ch. IV). The great advantage of the special CR embedding is the simplification in the formula for the FBI transform: rather than integration over maximally real submanifolds it is possible and advantageous, in the CR case, to integrate over <u>minimal noncharacteristic</u> submanifolds (see Ch. IV). In practice, in the local "representation" (3), it means to integrate with respect to s alone, and view (x,y) as parameters. This cuts drastically the number of variables of integration. For instance, when the CR structure under study is of the hypersurface type there is a single variable s (i.e., $d=1$) and the FBI transform is a line integral. This leads to the concept of mini-FBI transform (in this terminology the FBI transform carried out over a maximally real submanifold should be called maxi-FBI).

In the last couple of years M.S. Baouendi and the author have come to realize the usefulness of the mini-FBI transform in the study of the holomorphic extension of CR functions (see [BRT]). The present work, and its follow up currently in preparation, will show that the mini-FBI transform is no less useful in the study of the microlocal CR cohomology.

CHAPTER I

VANISHING OF THE $\bar{\partial}$-COHOMOLOGY IN GERMS OF WEDGES

1. SMOOTH FORMS IN WEDGES

The complex coordinates in $\mathbb{C}^m$ will be denoted by $z^j = x^j + \sqrt{-1}y^j$ ($j=1,\dots,m$). We shall write $z = (z^1,\dots,z^m)$. $x = \operatorname{Re} z$, $y = \operatorname{Im} z$, etc. Our basic datum will be a C^∞ map

$$\phi = (\phi^1,\dots,\phi^m) \colon \mathbb{R}^m \to \mathbb{R}^m,$$

having the following properties:

$$(1.1) \qquad \phi(0) = 0, \quad D\phi(0) = 0,$$

where $D\phi$ is the differential of the map ϕ. The equation

$$(1.2) \qquad y = \phi(x)$$

defines an m-dimensional totally real (henceforth abbreviated as "maximally real") submanifold X of $\mathbb{C}^m$. The tangent space to X at 0 is the real space $\mathbb{R}^m$.

We shall be concerned with functions and forms defined in a wedge

$$\mathfrak{w}_\delta(U,\Gamma) = \{z = x+\sqrt{-1}y \in \mathbb{C}^m;\ x \in U,\ |y| < \delta,\ y-\phi(x) \in \Gamma\}.$$

We have used the following notation: U is an open neighborhood of the origin in $\mathbb{R}^m$ (generally a ball centered at 0), δ a number > 0 (generally small), Γ is an open and <u>strictly convex</u> cone in $\mathbb{R}^m$ (all cones considered here will have their vertex at 0 and be non-empty). We shall always assume that the neighborhood U is sufficiently small that

$$(1.3) \qquad |\phi(x)| < \delta, \quad \forall\, x \in U.$$

This implies that, whatever $x \in U$, the point $t = y-\phi(x)$ can converge to zero while staying in Γ or, which is the same, $x+\sqrt{-1}y$ can converge to $x + \sqrt{-1}\phi(x)$ while staying in $\mathfrak{w}_\delta(U,\Gamma)$. By the edge of the wedge $\mathfrak{w}_\delta(U,\Gamma)$ we shall mean the set $\{x+\sqrt{-1}y \in \mathbb{C}^m;\ x\in U,\ y = \phi(x)\}$.

If Ω is an open subset of $\mathbb{R}^n$ ($n \geq 1$ arbitrary) we shall denote by $\mathcal{B}^\infty(\Omega)$ the space of C^∞ functions in Ω whose derivatives of any order are bounded in Ω. The topology of $\mathcal{B}^\infty(\Omega)$ will be its natural one: the topology of uniform convergence in Ω of the functions and of each one of their derivatives. When the set Ω is bounded, which will always be the case for us, then $\mathcal{B}^\infty(\Omega)$ is identical to the space of smooth functions in Ω whose derivatives of any order extend as continuous functions to the closure $\mathcal{Cl}\,\Omega$ of Ω. Indeed, the gradient of any function $f \in \mathcal{B}^\infty(\Omega)$ is uniformly bounded in Ω, and, as a consequence, f is uniformly continuous and extends as a continuous function to $\mathcal{Cl}\,\Omega$. The same is true of all the derivatives of f, since the differential operators with constant coefficients define endomorphisms of $\mathcal{B}^\infty(\Omega)$.

We shall use the standard multi-index notation for differential forms in Ω. Thus

$$(1.4) \qquad f = \sum_{|I|=p,\,|J|=q} f_{I,J}\, dz^I \wedge d\bar{z}^J$$

will be a typical (p,q)-form. As usual I and J are ordered multi-indices, of length $|I|$ and $|J|$ respectively.

We shall denote by $\mathcal{C}^{p,q}$ the space of (p,q)-forms (1.4) whose coefficients belong to $\mathcal{B}^\infty(\mathfrak{W}_\delta(U,\Gamma))$ (we set $\mathcal{C}^{0,0} = \mathcal{B}^\infty(\mathfrak{W}_\delta(U,\Gamma))$). The boundary values of these coefficients on the edge of the wedge $\mathfrak{W}_\delta(U,\Gamma)$ are well-defined, and so is therefore the pull-back to the edge of any form that belongs to $\mathcal{C}^{p,q}$.

The Cauchy-Riemann operator in $\mathbb{C}^m$ defines the differential complex

$$(1.5) \qquad \bar{\partial}: \mathcal{C}^{p,q} \to \mathcal{C}^{p,q+1}.$$

In particular, $\bar{\partial}^2 = 0$.

In the reasoning that follows it is convenient to "flatten" the edge of our wedge $\mathfrak{W}_\delta(U,\Gamma)$. This requires, however, that we modify the complex structure of the ambient space $\mathbb{C}^m$. Recall that ϕ is defined in the whole of R^m. The change of variables

$$(1.6) \qquad t^j = y^j - \phi^j(x), \qquad j=1,\dots,m,$$

determines a diffeomorphism $(x,y) \to (x,t)$ of $\mathbb{R}^{2m}$ onto itself which maps the wedge $\mathfrak{W}_\delta(U,\Gamma)$ onto the set

$$\tilde{\mathfrak{W}}_\delta(U,\Gamma) = \{(x,t) \in \mathbb{R}^{2m};\ x \in U,\ t \in \Gamma,\ |t+\phi(x)| < \delta\}.$$

The edge of $\tilde{\mathfrak{W}}_\delta(U,\Gamma)$ is the set $U\times\{0\}$; it is indeed flat.

The push-forward, under the above diffeomorphism, of the Cauchy-Riemann vector field $\partial/\partial \bar{z}^j$ is the vector field

$$1/2[\partial/\partial x^j + \sqrt{-1} \sum_{k=1}^{m} (\delta_j^k + \sqrt{-1}\, \partial\phi^k/\partial x^j)\partial/\partial t^k] \tag{1.7}$$

(δ_j^k: Kronecker's index). By multiplying the system of vector fields (1.7) by the matrix $(2/\sqrt{-1})(I+\sqrt{-1}\, D\phi)^{-1}$ we obtain a system of pairwise commuting vector fields

$$L_j = \partial/\partial t^j - \sqrt{-1}\, M_j, \qquad j=1,\dots,m, \tag{1.8}$$

where the M_j are smooth vector fields in x-space. Another way of looking at this is to note that, if we define the functions

$$Z^j(x) = x^j + \sqrt{-1}\phi^j(x) + \sqrt{-1}t^j \qquad (j=1,\dots,m),$$

then the vector fields L_j, M_k are defined by the orthonormality relations

$$L_j Z^k = 0,\; L_j t^k = \delta_j^k,\; M_j Z^k = \delta_j^k,\; M_j t^k = 0 \qquad (j,k=1,\dots,m). \tag{1.9}$$

Next consider the differential forms in $\tilde{\mathfrak{W}}_\delta(U,\Gamma)$,

$$u = \sum_{|I|=p,\,|J|=q} u_{I,J}(x,t)\, dZ^I \wedge dt^J. \tag{1.10}$$

We shall say that $u \in \mathfrak{F}^{p,q}$ if $u_{I,J} \in \mathcal{B}^\infty(\tilde{\mathfrak{W}}_\delta(U,\Gamma))$ for every pair of multi-indices I, J, $|I| = p$, $|J| = q$. The vector fields L_j of (1.9) define, in an obvious manner, a differential complex

$$L\colon \mathfrak{F}^{p,q} \to \mathfrak{F}^{p,q+1}. \tag{1.11}$$

We have

$$Lu = \sum_{|I|=p,\,|J|=q} \sum_{j=1}^{m} L_j u_{I,J}(x,t)\, dt^j \wedge dZ^I \wedge dt^J. \tag{1.12}$$

Of course, $L^2 = 0$.

Define the differential operator M exactly in the same manner as we have just defined L, except that M_j must be substitued for L_j. We may write

$$L = d_t - \sqrt{-1}\, M. \tag{1.13}$$

From the properties $(d_t - \sqrt{-1}M)^2 = M^2 = 0$ one derives

$$Md_t + d_t M = 0. \tag{1.14}$$

In vector/matrix notation we may write

$$dt = (I-\sqrt{-1}D\phi)dz - (I+\sqrt{-1}D\phi)d\bar{z}.$$

This implies that the pull-back of differential forms under the diffeomorphism $(x,y) \to (x,t)$ defines a map

$$\mathfrak{F}^{p,q} \to \bigoplus_{k=0}^{m-p} \mathcal{E}^{p+k,q}.$$

If we follow up this pull-back map by the natural projection

$$\bigoplus_{k=0}^{m-p} \mathcal{E}^{p+k,q} \to \mathcal{E}^{p,q},$$

we get an isomorphism of the differential complex (1.11) onto the differential complex (1.5).

We are now in a position to prove

Proposition 1.1. *Let* U *be an open ball centered at the origin in* $\mathbb{R}^m$ *and assume* $0 \le q \le m-1$. *Given any form* $f \in \mathcal{E}^{p,q+1}$ *such that* $\bar{\partial} f = 0$ *there is a form* $u \in \mathcal{E}^{p,q}$ *which vanishes at the edge and is such that* $\bar{\partial} u - f$ *vanishes to infinite order at the edge.*

Proof: Thanks to the remarks that precede the statement it suffices to prove the analogous result for $\mathfrak{F}$ replacing $\mathcal{E}$:

(1.15) $\forall\, f \in \mathfrak{F}^{p,q+1}$ *such that* $Lf = 0$, $\exists\, u \in \mathfrak{F}^{p,q}$ *such that* $u = 0$ *at* $t = 0$, *and such that* $Lu - f$ *vanishes to infinite order at* $t = 0$.

Consider then the Taylor expansion of f with respect to t, $\sum_{\nu=0}^{+\infty} f_\nu$, where the coefficients of the form f_ν (of bidegree $(p,q+1)$, in the sense of (1.10)) are homogeneous polynomials of degree ν with respect to t whose coefficients are functions of x which belong to $\mathcal{B}(U)$. The expression of the differential operator M shows that L acts on the formal series $\sum f_\nu$ and annihilates it. Thus we must have, for $\nu = 0,1,\ldots,$

(1.16) $$d_t f_\nu = \sqrt{-1}\, M f_{\nu-1}.$$

(We abide by the rule that $f_\nu \equiv 0$ if $\nu < 0$.)

Let us first seek a *formal* solution of $Lu = \sum_{\nu\ge 0} f_\nu$ of the kind $u = \sum_{\nu\ge 1} u_\nu$, where the coefficients of the form u_ν, whose bidegree is (p,q), are homogeneous polynomials of degree ν with respect to t with coefficients in $\mathcal{B}^\infty(U)$. We must have

(1.17) $$d_t u_\nu = f_{\nu-1} + \sqrt{-1}\, M u_{\nu-1}.$$

If $\nu \ge 1$ assume that the forms $u_{\nu'}$ have been determined for all

$\nu' < \nu$ recalling that $u_o \equiv 0$. In order that (1.17) have a solution it is necessary and sufficient that the right-hand side be d_t-closed. Indeed,

$$d_t f_{\nu-1} - \sqrt{-1}\, M d_t u_{\nu-1} = d_t f_{\nu-1} - \sqrt{-1}\, M(f_{\nu-2} + \sqrt{-1}\, M u_{\nu-2}) = 0,$$

by virtue of (1.14), (1.16) and of the fact that $M^2 = 0$. We can then solve (1.17) by means of one of the standard inverses of d_t (acting on homogeneous polynomials in t).

Select at random a C^∞ function χ on the real line, $0 \le \chi \le 1$ everywhere, $\chi(\tau)=1$ for $\tau < 1$ and $\chi(\tau) = 0$ for $\tau > 2$. Let $u_{\nu;I,J}(x,t)$ denote the coefficient of $dZ^I \wedge dt^J$ in u_ν. By making use of the fact that the coefficients of $u_{\nu|I,J}$, viewed as a polynomial with respect to t, belong to $\mathcal{B}^\infty(U)$, we can apply induction on ν in choosing increasingly large positive numbers $R_\nu \to +\infty$ so as to insure that the series

$$\sum_{\nu=1}^{+\infty} \chi(R_\nu |t|) u_\nu$$

converges in $\mathfrak{F}^{p,q}$, to an element which we call u and which satisfies the condition in (1.15). ∎

Remark 1.1. The (p,q)-form u resulting from the proof of Prop. 1.1 is defined in the whole tube $U + \sqrt{-1}\mathbb{R}^m$. It has compact support with respect to y. Given any integer $N \ge 0$ the absolute value of every derivative of every coefficient of $f - \bar\partial u$ is bounded by a const. $|y-\phi(x)|^N$. ∎

2. CONVEXIFICATION

We shall make repeated use of the following elementary observation (whose proof is left to the reader):

Proposition 2.1. _Let H be a biholomorphism of an open neighborhood $\mathcal{O}$ of $\mathfrak{w}_\delta(U,\Gamma)$_ (see Sect. 1) _onto an open subset of $\mathbb{C}^m$, having the property that $H(0) = 0$ and $\partial H(0) = $ Identity. Then there exist an open neighborhood $U' \subset U$ of the origin in $\mathbb{R}^m$, an open convex cone $\Gamma' \subset \Gamma$ and a number δ', $0 < \delta' \le \delta$, such that $\mathfrak{w}_{\delta'}(U',\Gamma') \subset \mathfrak{w}_\delta(U,\Gamma)$._

We begin by carrying out a rotation so that, possibly after replacing the cone Γ by a thinner one, we may assume that Γ is defined by the inequality $|t'| < 2t^m$. Here $t = (t^1,\dots,t^m)$ is the variable in $\mathbb{R}^m$, and we use the notation $t' = (t^1,\dots,t^{m-1})$.

Next we carry out a biholomorphic transformation $z \to z_\#$ of $\mathbb{C}^m$ (near the origin) defined by equations

$$(2.1) \qquad z_{\#}^{j} = z^{j} + \sum_{k,\ell=1}^{m} a_{k,\ell}^{j} z^{k} z^{\ell}, \qquad j=1,\dots,m.$$

The coefficients $a_{k,\ell}^{j}$ are chosen in such a way that, if we put $\operatorname{Im} z = \phi(x)$, then all partial derivatives with respect to x, of order ≤ 2, of $\operatorname{Im} z_{\#}^{j}$ $(j=1,\dots,m-1)$ vanish at the origin, while the Hessian of $\operatorname{Im} z_{\#}^{m}$ with respect to x is equal, at the origin, to the $m \times m$ identity matrix. When we put $\operatorname{Im} z = \phi(x)$ the map $x \to x_{\#} = \operatorname{Re} z_{\#}$ is a diffeomorphism of U (possibly contracted about 0) onto an open neighborhood $U_{\#}$ of 0 in $\mathbb{R}^{m}$. The properties of $\operatorname{Im} z_{\#}$ in relation to x that we have just described remain valid with respect to $x_{\#}$.

From Prop. 2.1 it follows that the image of the wedge $\mathfrak{w}_{\delta}(U,\Gamma)$ under the local biholomorphism $z \to z_{\#}$ will contain a similar wedge, with U replaced by a small open neighborhood of 0 in $\mathbb{R}^{m}$, δ by a small number >0 and Γ by a thinner convex, open cone (with same axis $t' = 0$). Actually we shall omit primes and subscripts $\#$, and simply assume that our original wedge $\mathfrak{w}_{\delta}(U,\Gamma)$ is already in the desired form. That is, we shall assume right from the start that

$$(2.2) \qquad \begin{aligned} \phi^{j}(x) &= O(|x|^{3}), \qquad j=1,\dots,m-1, \\ \phi^{m}(x) &= |x|^{2} + O(|x|^{3}). \end{aligned}$$

We also assume that

$$(2.3) \qquad U = \{x \in \mathbb{R}^{m};\ |x| < r\}, \qquad \Gamma = \{t \in \mathbb{R}^{m};\ |t'| < 2t^{m}\},$$

where r is a small number >0.

<u>Proposition 2.2</u>. <u>If</u> (2.2) <u>and</u> (2.3) <u>are satisfied and if</u> r <u>is sufficiently small then the set</u> $\mathfrak{w}_{\delta}(U,\Gamma)$ <u>is convex</u>.

<u>Proof</u>: By definition $\mathfrak{w}_{\delta}(U,\Gamma)$ is equal to the intersection of the set $\mathfrak{w}(U,\Gamma) = \{z \in \mathbb{C}^{m};\ x \in U,\ y-\phi(x) \in \Gamma\}$ with the convex tube $\{z \in \mathbb{C}^{m};\ |y| < \delta\}$ it suffices to show that $\mathfrak{w}(U,\Gamma)$ is convex. Consider two points of $\mathfrak{w}(U,\Gamma)$, z_i $(i=1,2)$, and a number λ, $0 < \lambda < 1$. We may write

$$z_i = x_i + \sqrt{-1}\phi(x_i) + \sqrt{-1}t_i, \quad \text{with} \quad t_i \in \Gamma \qquad (i=1,2).$$

Set then $z = \lambda z_1 + (1-\lambda)z_2$. We have

$$z = \lambda x_1 + (1-\lambda)x_2 + \sqrt{-1}\ \phi[\lambda x_1+(1-\lambda)x_2] + \sqrt{-1}t,$$

where

$$t = \lambda t_1 + (1-\lambda)t_2 + \lambda\phi(x_1) + (1-\lambda)\phi(x_2) - \phi[\lambda x_1+(1-\lambda)x_2].$$

Since Γ is convex, and since therefore $\lambda t_1 + (1-\lambda)t_2 \in \Gamma$, it suffices to prove that $\lambda\phi(x_1) + (1-\lambda)\phi(x_2) - \phi[\lambda x_1+(1-\lambda)x_2] \in \Gamma$. It is readily seen that

$$\lambda\phi(x_1) + (1-\lambda)\phi(x_2) - \phi[\lambda x_1+(1-\lambda)x_2] =$$
$$= \lambda(1-\lambda)\{|x_1-x_2|^2\}\{e_m + O(|x_1|+|x_2|)\},$$

where $e_m = (0,\dots,0,1)$. Since e_m is the unit vector of the central axis of Γ our assertion follows at once. ■

We shall now insert, inside $\mathfrak{W}_\delta(U,\Gamma)$, a <u>strictly convex</u> open set Ω which will contain a smaller wedge $\mathfrak{W}_{\delta'}(U',\Gamma')$ and whose boundary $\partial\Omega$ will have the following property: the intersection $\dot\Omega = \partial\Omega \cap \mathfrak{W}_\delta(U,\Gamma)$ will be defined by an equation $\rho(x,y) = 0$, with $\rho \in C^\infty(\mathfrak{W}_\delta(U,\Gamma))$, and $d\rho \neq 0$ at every point of $\dot\Omega$.

First we select a C^∞ function χ on the real line, with the following properties:

(2.4) $\chi(\lambda) = 0$ for $\lambda < 1/16$, $\chi''(\lambda) > 0$ (and therefore also $\chi(\lambda) > 0$, $\chi'(\lambda) > 0$) for $\lambda > 1/16$;

(2.5) $$\chi(\lambda) > 2\lambda^2 \quad \text{for } \lambda > 1/8.$$

We shall use the notation $t = y-\phi(x)$. Let $\varkappa$ be a suitably large number >0. We define

(2.6) $$\rho(x,y) = \varkappa^{-1}\chi(\varkappa|x|^2) + \varkappa(|t|^2+2|x|^2t^m) + |t| - 2t^m,$$

(2.7) $$\Omega = \{z = x + \sqrt{-1}y \in \mathbb{C}^m;\ \rho(x,y) < 0\}.$$

Let us take note right-away of the fact that ρ is a Lipschitz continuous function in the whole of $\mathbb{R}^{2m}$ and that it is a C^∞ function in the region $t \neq 0$.

<u>Proposition 2.3</u>. <u>If</u> $\varkappa$ <u>is large enough, then the following holds true</u>:

(2.8) $\Omega \subset \mathfrak{W}_\delta(U,\Gamma)$;

(2.9) $\Omega \supset \mathfrak{W}_{\delta'}(U',\Gamma')$, <u>with</u> $U' \subset U$, U' <u>an open ball in</u> $\mathbb{R}^m$ <u>centered at the origin</u>, $\Gamma' = \{t \in \mathbb{R}^m;\ |t'| < t^m/2\}$ <u>and</u> $0 < \delta' < \delta$;

(2.10) <u>in</u> $\mathfrak{W}_\delta(U,\Gamma)$ $\rho = 0$ <u>entails</u> $|\nabla_y\rho| \geq \frac{1}{5}\varkappa[y^m-\phi^m(x)]$.

<u>Proof of (2.8)</u>: The inequality $\rho < 0$ implies

(2.11) $$|t| + \varkappa(|t|^2+2|x|^2t^m) < 2t^m.$$

Recalling that $t^m > 0$ in $\mathfrak{W}_\delta(U,\Gamma)$ we derive from (2.11):

(2.12) $$|x|^2 < 1/\varkappa, \quad |t| < 1/\varkappa.$$

We also have $|t| < 2t^m$, which entails

(2.13) $$|t'| < \sqrt{3}\, t^m.$$

This shows that, if $\varkappa$ is large enough, (2.8) will be true. ■

Proof of (2.9): We suppose that $\varkappa$ has been chosen. Let the radius r' of the ball U' be so small that $\varkappa r'^2 < 1/16$, which implies $\chi(\varkappa|x|^2) \equiv 0$ in U'. In the notation $s = \varkappa t$, we see that $\varkappa\rho(x,y) < |s| + |s|^2 - 15s^m/8$ if $x \in U'$. Suppose $|t'| < t^m/2$. Then $|s| < \sqrt{5}s^m/2$ and $4\varkappa\rho(x,y) < 5(s^m)^2 - 3s^m$ for those same points (x,y). We can further decrease r', and choose δ' so small that $x \in U'$ and $|y| < \delta'$ imply $s^m < 3/5$, and thus $\rho < 0$. ■

Proof of (2.10): We have:

(2.14) $$\nabla_x \rho = 2[\chi'(|x|^2) + 2\varkappa]x - \nabla_t\rho \cdot \phi_x.$$

By virtue of (2.2) we derive from (2.14):

$$|\nabla_x \rho| \geq 4\varkappa|x| - C|x||\nabla_y \rho|,$$

where C is a suitably large positive constant. We take advantage of (2.12):

$$|\nabla_x \rho| \geq 4\varkappa|x| - C\varkappa^{-1/2}|\nabla_y \rho|.$$

As a consequence, provided that $C\varkappa^{-1/2}$ be small enough, we obtain

(2.15) $$|\nabla\rho| \geq 2\varkappa|x| + |\nabla_y\rho|/2.$$

If $|x| \geq 1/4\sqrt{\varkappa}$ we obtain $|\nabla\rho| \geq \sqrt{\varkappa}/2$. In the remainder of the proof we assume $\sqrt{\varkappa}|x| < 1/4$, which implies $\chi(\varkappa|x|^2) = 0$. In this case the equation $\rho=0$ reads

(2.16) $$|t| + \varkappa|t|^2 = 2(1-\varkappa|x|^2)t^m.$$

We have:

(2.17) $$\nabla_{y'}\rho = (1/|t| + 2\varkappa)t',$$

(2.18) $$\partial\rho/\partial y^m = (1/|t| + 2\varkappa)t^m - 2(1-\varkappa|x|^2).$$

If we combine (2.16) and (2.18) we get

(2.19) $$\partial\rho/\partial y^m = (1/|t| + 2\varkappa)t^m - (1+\varkappa|t|)|t|/t^m.$$

If $\mu = |t'|/t^m \geq \varkappa t^m/4$ then, by (2.17), $|\nabla_{y'}\rho| \geq |t'|/|t| \geq \varkappa t^m/5$. If $\mu < \varkappa t^m/4$ ($\leq 1/4$ by (2.12)) then, by (2.19), $\partial\rho/\partial y^m \geq$ $\geq \varkappa t^m(1-\mu^2) - \mu \geq \varkappa t^m/2$. By choosing $\varkappa$ large enough we can achieve that (2.10) be true. ■

<u>Proposition 2.4. If $\varkappa$ is sufficiently large then the Hessian of</u> ρ <u>is uniformly definite-positive in the closure of Ω in $\mathfrak{W}_\delta(U,\Gamma)$.</u>

<u>Proof</u>: Call H_χ the Hessian of $\varkappa^{-1}\chi(\varkappa|x|^2)$. If $\zeta = (\xi,\eta)$ is any vector in $\mathbb{R}^{2m}$ we have

$$\zeta \cdot H_\chi \zeta = 2\chi'(\varkappa|x|^2)|\xi|^2 + 4\varkappa\chi''(\varkappa|x|^2)|x\cdot\xi|^2.$$

We derive from (2.4) that $\lambda^2\chi''(\lambda) \geq \lambda\chi'(\lambda) \geq \chi(\lambda) > 2\lambda^2$ if $\lambda > 1/8$, whence

(2.20) $$\zeta \cdot H_\chi \zeta \geq 4\varkappa|x|^2|\xi|^2 + 8\varkappa|x\cdot\xi|^2 \quad \text{if} \quad \varkappa|x|^2 > 1/8.$$

Next call H_1 the Hessian of the function $|t|$, where $t = y-\phi(x)$. Below ϕ_{xx} denotes the second derivative of ϕ with respect to x; ϕ_{xx} will be viewed as an $m\times m$ matrix with entries that are m-vectors. Call θ the vector of $\mathbb{R}^m$ with components $\sum_k [\delta_{jk}\eta^k - (\partial\phi^k/\partial x^j)\xi^k]$, $j=1,\dots,m$. A straightforward calculation shows that

$$\zeta \cdot H_1 \zeta = |t|^{-1}[|\theta|^2 - |t\cdot\theta|^2/|t|^2] - (\xi\cdot\phi_{xx}\xi)\cdot t/|t|.$$

If we avail ourselves, once again, of (2.2), i.e., of the fact that $\phi(x) = |x|^2 e_m + O(|x|^3)$, we obtain

$$\zeta \cdot H_1 \zeta \geq -(2+C|x|)|\xi|^2.$$

If we then take (2.12) into account, we see that there is $C > 0$ such that

(2.21) $$\zeta \cdot H_1 \zeta \geq -(2+C/\sqrt{\varkappa})|\xi|^2.$$

On the other hand,

(2.22) $$\rho(x,y) - \varkappa^{-1}\chi(\varkappa|x|^2) - |t| =$$
$$= \rho_o(x,y) + O(\varkappa|x|^3|y|, \varkappa|x|^5, |x|^3),$$

where

$$\rho_o(x,y) = \varkappa|y-|x|^2 e_m|^2 - 2(1-\varkappa|x|^2)(y^m-|x|^2)$$
$$= \varkappa|y|^2 - 2y^m + 2|x|^2 - \varkappa|x|^4.$$

Call H_o the Hessian of ρ_o. We have

(2.23) $$\zeta \cdot H_o \zeta = 2\varkappa|\eta|^2 + 4|\xi|^2 - 4\varkappa(|x|^2|\xi|^2 + 2|x\cdot\xi|^2).$$

Call H the Hessian of ρ. Combining (2.20),...,(2.23) yields, when $\varkappa|x|^2 > 1/8$,

(2.24) $$\zeta \cdot H \zeta \geq 2\varkappa|\eta|^2 + (2-C/\sqrt{\varkappa})|\xi|^2.$$

But if $\varkappa|x|^2 \leq 1/8$ we derive from (2.23)

$$\zeta\cdot H_o\zeta \geq 2\varkappa|\eta|^2 + 5|\xi|^2/2,$$

which, when combined with (2.21), yields (2.24) also in this case. It is then obvious that we can choose $\varkappa$ large enough so as to obtain

$$\zeta\cdot H(x,y)\zeta \geq 2\varkappa|\eta|^2 + |\xi|^2, \quad \forall\ (x,y) \in \Omega, \tag{2.25}$$

whereby one reaches the desired conclusion. ■

Corollary 2.5. The open subset Ω of $\mathfrak{W}_\delta(U,\Gamma)$ is convex.

Proof: By Prop. 2.2 we know that $\mathfrak{W}_\delta(U,\Gamma)$ is convex, and by Propositions 2.3 and 2.4 that every point of $C\ell\ \Omega \cap \mathfrak{W}_\delta(U,\Gamma)$ has an open neighborhood in $\mathbb{R}^{2m}$ whose intersection with $C\ell\ \Omega \cap W_\delta(U,\Gamma)$ is convex. On the other hand Ω is clearly connected. This entails that $C\ell\ \Omega \cap W_\delta(U,\Gamma)$, and therefore also Ω, is convex. ■

Consider then the following function in $\mathfrak{W}_\delta(U,\Gamma)$,

$$\Lambda(x,y,x_*,y_*) = (x-x_*)\cdot\nabla_x\rho(x_*,y_*) + (y-y_*)\cdot\nabla_y\rho(x_*,y_*).$$

Proposition 2.6. Provided $\varkappa$ is large enough, we have, for all pairs of points $z = (x,y) \in C\ell\ \Omega$, $z_* = (x_*,y_*) \in \dot{\Omega}$ $(= \partial\Omega \cap \mathfrak{W}_\delta(U,\Gamma))$,

$$\Lambda(x,y,x_*,y_*) \leq \rho(x,y) - |z-z_*|^2/2. \tag{2.26}$$

Proof: By Taylor expansion of order two of ρ about z_* we get:

$$\Lambda(x,y,x_*,y_*) = \rho(x,y) - \rho(x_*,y_*) - R(x,y,x_*,y_*) \leq -R(x,y,x_*,y_*),$$

where R is the remainder of order two:

$$R(x,y,x_*,y_*) =$$

$$\sum_{|\alpha|=2} (x-x_*)^\alpha \int_0^1 (1-\lambda)[(\partial/\partial x)^\alpha\rho](\lambda x+(1-\lambda)x_*,\lambda y+(1-\lambda)y_*)d\lambda\ +$$

$$\sum_{|\alpha|=2} (y-y_*)^\alpha \int_0^1 (1-\lambda)[(\partial/\partial y)^\alpha\rho](\lambda x+(1-\lambda)x_*,\lambda y+(1-\lambda)y_*)d\lambda.$$

By Cor. 2.5 we know that the point $\lambda z + (1-\lambda)z_*$ belongs to $C\ell\ \Omega \cap \mathfrak{W}_\delta(U,\Gamma)$. We may therefore apply (2.25) after substituting $(\lambda x+(1-\lambda)x_*,\lambda y+(1-\lambda)y_*)$ for (x,y), whence (2.26). ■

Proposition 2.7. Given any pair of m-tuples $\alpha,\beta \in \mathbb{Z}_+^m$ such that $|\alpha+\beta| \geq 1$, there is a constant $C_{\alpha,\beta} > 0$ such that, in the wedge $\mathfrak{W}_\delta(U,\Gamma)$,

$$|(\partial/\partial x)^\alpha(\partial/\partial y)^\beta\rho| \leq C_{\alpha,\beta}|y-\phi(x)|^{1-|\alpha+\beta|}. \tag{2.27}$$

Evident, by the definition (2.6) of ρ.

In the remainder of this chapter we shall assume that the constant $\varkappa$ has been chosen so large that the conclusions in every one of the statements in the present section are valid.

3. INTEGRAL REPRESENTATION OF SOLUTIONS OF THE CAUCHY-RIEMANN EQUATIONS

Same notation as in the previous sections. We shall often write $\rho(z)$ for the function $\rho(x,y)$ defined in (2.6); Ω is the open subset of $\mathfrak{W}_\delta(U,\Gamma)$ defined by the inequality $\rho < 0$. We shall also use the notation $\partial\rho = (\partial\rho/\partial z^1,\dots,\partial\rho/\partial z^n)$.

We shall make use of the following function

$$S(z,\zeta) = (\zeta - z)\cdot\partial\rho(\zeta),$$

defined for all $z \in \mathbb{C}^m$ (S is a polynomial of degree one with respect to z) and all $\zeta \in \mathfrak{W}_\delta(U,\Gamma)$. Actually we shall be mainly interested in $S(z,\zeta)$ when $z \in \Omega$ and $\zeta \in \dot{\Omega} = \partial\Omega \cap \mathfrak{W}_\delta(U,\Gamma)$. Notice that if Λ is the function in Prop. 2.6, we have

$$\Lambda(x,y,\xi,\eta) = -\mathrm{Re}\, S(z,\zeta), \tag{3.1}$$

where $\xi = \mathrm{Re}\,\zeta$, $\eta = \mathrm{Im}\,\zeta$. Thus, by Prop. 2.6 we have

$$\mathrm{Re}\, S(z,\zeta) \geq \tfrac{1}{2}\,|z-\zeta|^2, \quad \forall\ (z,\zeta) \in \Omega\times\dot{\Omega}. \tag{3.2}$$

<u>Proposition 3.1</u>. <u>To any</u> $\alpha,\beta \in \mathbb{Z}_+$ <u>there is a constant</u> $C_{\alpha,\beta} > 0$ <u>such that</u>

$$|\partial_\zeta^\alpha\, \bar\partial_\zeta^\beta S| \leq C_{\alpha,\beta}|\eta-\phi(\xi)|^{-|\alpha+\beta|}\{|z-\zeta| + |\eta-\phi(\xi)|\}; \tag{3.3}$$

$$|\partial_\zeta^\alpha\, \bar\partial_\zeta^\beta(\partial_z S)| \leq C_{\alpha,\beta}|\eta-\phi(\xi)|^{-|\alpha+\beta|} \tag{3.4}$$

($\xi = \mathrm{Re}\,\zeta$, $\eta = \mathrm{Im}\,\zeta$).

Follows at once from the definition of S and from Prop. 2.7.

We introduce the following $(0,m-1)$-form on $\mathbb{C}^{2m}$,

$$\omega'(\bar\zeta-\bar z) = \sum_{j=1}^{m} (-1)^{j-1}(\bar\zeta^j-\bar z^j)\, d(\bar\zeta^1-\bar z^1)\wedge\dots\wedge \widehat{d(\bar\zeta^j-\bar z^j)}\wedge\dots\wedge d(\bar\zeta^m-\bar z^m),$$

where the hatted factor must be omitted. Then the Bochner-Martinelli operator, relative to the open set Ω, acting on a form $f \in \mathcal{E}^{0,q+1}$ (see Sect. 1) is defined by the formula

$$B_\Omega f(z) = c_m \int_\Omega |z-\zeta|^{-2m}\, f(\zeta)\wedge\omega'(\bar\zeta-\bar z)\wedge d\zeta, \tag{3.5}$$

where $z \in \Omega$ and $c_m = (-1)^{m(m-1)/2}(m-1)!/(2\pi\sqrt{-1})^m$. We are using partial integration (with respect to ζ) in (3.5): we regard the integrand as a differential form in z-space whose coefficients are differential forms in ζ-space, and we integrate these coefficients over Ω. The only nonvanishing contributions come from those coefficients that have degree equal to $2m$. Since the total degree of the form

$$f(\zeta) \wedge \omega'(\bar{\zeta}-\bar{z}) \wedge d\zeta$$

is equal to $q+2m$, the result of the integration in (3.5) is a $(0,q)$-form in z-space. The Bochner-Martinelli kernel behaves like $|z-\zeta|^{1-2m}$ on the diagonal and therefore the integral (3.5) is a kind of convolution of $\chi_\Omega f$ (χ_Ω: characteristic function of Ω) with a differential form whose coefficients are smooth in the complement of the origin. It follows easily from this that, whatever the form $f \in \mathcal{E}^{0,q+1}$, the coefficients of the $(0,q)$-form $B_\Omega f$ are C^∞ functions in Ω.

Next consider the map $(z,\zeta,\lambda) \to \tau$ from $\Omega \times \Omega \times [0,1]$ into $\mathbb{C}^m$ given by

$$(3.6) \qquad \tau = \lambda|\zeta-z|^{-2}(\bar{\zeta}-\bar{z}) + (1-\lambda)S(z,\zeta)^{-1}\partial\rho(\zeta).$$

We shall make use of the differential form

$$\bar{\omega}'(\tau(z,\zeta,\lambda)) = \sum_{j=1}^{m} (-1)^{j-1} \tau^j \bar{\vartheta}\tau^1 \wedge \ldots \wedge \widehat{\bar{\vartheta}\tau^j} \wedge \ldots \wedge \bar{\vartheta}\tau^m,$$

where $\bar{\vartheta}$ is an abbreviation for $\bar{\partial}_{z,\zeta} + d_\lambda$. It is clear that the coefficients of $\bar{\omega}'(\tau(z,\zeta,\lambda))$ are smooth functions in $\Omega \times \dot{\Omega} \times [0,1]$. Below, what will also matter to us is their behaviour as z and ζ are near the edge of $\mathfrak{W}_\delta(U,\Gamma)$. For the moment we shall be content with remarking that (3.2) entails

$$|\tau^j| + |\partial\tau^j/\partial\lambda| \leq \text{const.}|z-\zeta|^{-2},$$

while, according to Prop. 3.1, the absolute values of the coefficients of $d\tau^j$ are $\leq \text{const.}|z-\zeta|^{-4}|\eta-\phi(\xi)|^{-1}$; as a consequence, those of the coefficients of $\bar{\omega}'(\tau(z,\zeta,\lambda))$ are $\leq \text{const.}|z-\zeta|^{4(1-m)}|\eta-\phi(\xi)|^{2-m}$.

We shall denote by $\mathcal{B}_0^\infty(\Omega)$ the subspace of $\mathcal{B}^\infty(\Omega)$ consisting of the functions in Ω that vanish to infinite order at the edge, i.e., as $|y-\phi(x)| \to 0$. We denote by $\mathcal{E}_0^{p,q}(\Omega)$ the space of (p,q)-forms in Ω whose coefficients belong to $\mathcal{B}_0^\infty(\Omega)$.

The behaviour of the coefficients of $\bar{\omega}'(\tau(z,\zeta,\lambda))$ as ζ approaches the edge allows us to define a linear operator R_Ω acting on $\mathcal{E}_0^{0,q+1}(\Omega)$, by the following formula:

$$(3.8) \qquad R_\Omega f(z) = c_m \int_{\Omega\times[0,1]} f(\zeta) \wedge \bar{\omega}'(\tau(z,\zeta,\lambda)) \wedge d\zeta.$$

In (3.8) we use partial integration with respect to (ζ,λ). We view the integrand, here $f(\zeta) \wedge \bar{\omega}'(\tau(z,\zeta,\lambda)) \wedge d\zeta$, as a differential form in z-space Ω whose coefficients are differential forms in (ζ,λ)-space $\dot{\Omega}\times[0,1]$. We integrate the latter. The only nonzero contributions to the integral in (3.8) come from terms that contain $d\lambda$, and the coefficient of $d\lambda$ in $f(\zeta) \wedge \bar{\omega}'(\tau(z,\zeta,\lambda))$ is a $(0,m+q-1)$-form in (z,ζ)-space. It follows that the result of the integration with respect to (ζ,λ) in (3.8) is a $(0,q)$-form in Ω (regarded as an open subset of z-space), whose coefficients are obviously C^∞ function in Ω.

We introduce the notation

(3.9) $$T_\Omega^{q+1} = (-1)^{q+1}(B_\Omega - R_\Omega);$$

T_Ω^{q+1} acts on forms $f \in \mathcal{E}_o^{0,q+1}(\Omega)$. The following homotopy formula, valid for such forms f, is proved in [H-L1] (see Cor. 1.12.2; [H-L1] uses an orientation of $\mathbb{C}^m$ which differs from ours, whence the different signs):

(3.10) $$f = \bar\partial T_\Omega^{q+1} f + T_\Omega^{q+2}\bar\partial f.$$

In particular,

(3.11) $$\text{If } \bar\partial f = 0, \text{ then } \bar\partial T_\Omega^{q+1} f = f.$$

Now, if f is $(p,q+1)$-form with $p > 0$, it can be written as

(3.12) $$f = \sum_{|J|=p} f_J dz^J,$$

where for each J, $|J| = p$, f_J is a $(0,q+1)$-form. Suppose $f \in \mathcal{E}_o^{p,q+1}(\Omega)$ or, which is the same, $f_J \in \mathcal{E}_o^{0,q+1}(\Omega)$, $\forall J$, $|J| = p$. Then we can define

(3.13) $$T_\Omega^{p,q+1} f = \sum_{|J|=p} T_\Omega^{q+1} f_J \, dz^J.$$

We introduce now the wedge $\mathfrak{W}_{\delta'}(U',\Gamma') \subset \Omega$ of Property (2.9). We recall that $U' = \{x \in \mathbb{R}^m;\ |x| < r'\}$, $\Gamma' = \{t \in \mathbb{R}^m;\ |t'| < t^m/2\}$ and that $|y| < \delta'$ if $(x,y) \in \mathfrak{W}_\delta(U',\Gamma')$.

<u>Proposition 3.2</u>. <u>If the numbers</u> r' <u>and</u> δ' <u>are sufficiently small</u>, <u>then, given any form</u> $f \in \mathcal{E}_o^{p,q+1}(\Omega)$, <u>the restrictions to the wedge</u> $\mathfrak{W}_{\delta'}(U',\Gamma')$ <u>of the coefficients of</u> $T_\Omega^{p,q+1} f$ <u>belong to</u> $\mathcal{B}^\infty(\mathfrak{W}_{\delta'}(U',\Gamma'))$.

<u>Proof</u>: It suffices to consider the case $p=0$. Let $\partial_\nu \ldots \partial_1$ denote a product of partial derivatives $\partial/\partial x^j$ or $\partial/\partial y^k$ $(j,k=1,\ldots,m)$, acting on differential forms coefficientwise. Let us represent the Bochner-Martinelli operator as a convolution cum exterior product, as in (3.5):

$$\mathcal{B}_\Omega f = B * (\chi_\Omega f)$$

where χ_Ω is the characteristic function of the open set Ω and

$$B = c_m |z|^{-2m} \omega'(-\bar z) \wedge dz.$$

By induction on $\nu \geq 1$ we see that

$$(3.14) \qquad \partial_\nu \ldots \partial_1 B_\Omega f = B_\Omega \partial_\nu \ldots \partial_1 f + + \sum_{\iota=1}^{\nu} (\partial_{\nu+1} \ldots \partial_{\iota+1} B) * [(\partial_\iota \chi_\Omega)\partial_{\iota-1} \ldots \partial_o f],$$

with the agreement that $\partial_o = \partial_{\nu+1} =$ (dentity.

By the elementary Hölder estimates we know that the coefficients of the form $B_\Omega \partial_\nu \ldots \partial_1 f$ belong to $L^\infty(\Omega)$. We must prove the following assertion:

(3.15) <u>the coefficients of the forms</u> $(\partial_{\nu+1} \cdots \partial_{\iota+1} B) * [(\partial_\iota \chi_\Omega)\partial_{\iota-1} \cdots \partial_o f]$ <u>belong to</u> $L^\infty(\mathfrak{w}_{\delta'}(U',\Gamma'))$.

The crucial fact, in this connection, is that the forms $(\partial_{\nu+1} \cdots \partial_{\iota+1} B) * [(\partial_\iota \chi_\Omega)\partial_{\iota-1} \ldots \partial_o f]$ are given by integrals over $\dot{\Omega}$. Now, by hypothesis given any inter $K > 0$ there is a constant $C_K > 0$ such that

$$(3.16) \qquad |f_J(\zeta)| \leq C_K |t(\zeta)|^K, \quad \forall\, \zeta \in \dot{\Omega}, \quad J, \ |J| = q,$$

where we have used the notation $t(\zeta) = \eta - \phi(\xi)$ [likewise, below we shall use the notation $t(z) = y - \phi(x)$]. On the other hand, when evaluated at the point $\zeta - z$, the absolute values of the coefficients of the form $\partial_{\nu+1} \ldots \partial_{\iota+1} B$ are bounded by const.$|z-\zeta|^{-N}$ for some integer N depending on m and ν. In order to prove (3.15) it will therefore suffice to show that there is a constant $c > 0$ such that

$$(3.17) \qquad |z-\zeta| \geq c|t(\zeta)|, \quad \forall\, z \in \mathfrak{w}_{\delta'}(U',\Gamma'), \quad \zeta \in \dot{\Omega}.$$

The same reasoning proves that

(3.18) <u>the coefficients of the form</u> $R_\Omega f$ <u>belong to</u> $\mathcal{B}^\infty(\mathfrak{w}_{\delta'}(U',\Gamma'))$.

Indeed, from the definition of $S(z,\zeta)$ and from that of the differential form $\bar{\omega}'(\tau(z,\zeta,\lambda))$ it follows at once that, if $c(z,\zeta,\lambda)$ is any coefficient of this form, to any pair of m-tuples $\alpha, \beta \in \mathbb{Z}_+^m$ there is an integer $N \geq 0$ and a constant $C > 0$ (both depending on α and β) such that

$$(3.19) \qquad |\partial_z^\alpha \bar{\partial}_z^\beta c(z,\zeta,\lambda)| \leq C|t(\zeta)|^{2-m}|z-\zeta|^{-N}.$$

It is clear that (3.18) will follow from the conjunction of (3.16), (3.17) and (3.19).

The remainder of the proof is devoted to showing that (3.17) is valid. Let $z = x + \sqrt{-1}y$, $\zeta = \xi + \sqrt{-1}\eta$ be arbitrary points of $\mathfrak{w}_{\delta'}(U',\Gamma')$ and Ω respectively. Suppose first that $|\xi| > 2r'$. In this case,

$$(3.20) \qquad |z-\zeta| > c,$$

with $c = r'$. Suppose next that $|\eta| > 2\delta'$; then (3.20) is still valid, this time with $c = \delta'$. Henceforth we assume $|\xi| \le 2r'$, $|\eta| \le 2\delta'$. Possibly decreasing r', we may assume that $\varkappa r'^2 < 2^{-6}$ which by (2.4) implies $\chi(\varkappa|\xi|^2) = 0$. As a consequence we have

$$(3.21) \qquad |t(\zeta)| + \varkappa|t(\zeta)|^2 = 2(1-\varkappa|\xi|^2)t^m(\zeta).$$

There is a constant $C > 0$ such that

$$(3.22) \qquad |t(z)-t(\zeta)| \le C|z-\zeta|.$$

If r' and δ' are sufficiently small we shall have $\varkappa|t(\zeta)| \le 2^{-6}$, whence, by (3.21),

$$|t(\zeta)| \ge 2(12/13)t^m(\zeta) \quad (>0),$$

and therefore,

$$(3.23) \qquad |t'(\zeta)| \ge 3t^m(\zeta)/2.$$

We are using the notation $t' = (t^1,\ldots,t^{m-1})$. We have

$$|t(z)-t(\zeta)| \ge \big||t(z)|-|t(\zeta)|\big| \ge |t(\zeta)||\sin\omega|,$$

where ω is the planar angle between the ray (originating at 0) through the point $(|t'(z)|, t^m(z))$ and the ray through the point $(|t'(\zeta)|, t^m(\zeta))$. By virtue of (3.23) and of the fact that $|t'(z)| \le t^m(z)/2$, we have $|\sin\omega| \ge c_o > 0$ with c_o independent of z and ζ. Combining this with (3.22) implies (3.17). ■

We may now combine Prop. 1.1 with the results obtained so far in the present section:

Theorem 3.3. *If the positive numbers r' and δ' are sufficiently small, then, to any $(p,q+1)$-form f in $\mathfrak{w}_\delta(U,\Gamma)$ whose coefficients belong to $\mathcal{B}^\infty(\mathfrak{w}_\delta(U,\Gamma))$ and which is such that $\bar\partial f = 0$, there is a (p,q)-form u in $\mathfrak{w}_{\delta'}(U',\Gamma')$ whose coefficients belong to $\mathcal{B}^\infty(\mathfrak{w}_{\delta'}(U',\Gamma'))$ and which is such that $\bar\partial u = f$ in $\mathfrak{w}_{\delta'}(U',\Gamma')$.*

Actually we shall need a "version with parameters" of Th. 3.3: this simply means that ϕ will not only be a function of $x \in \mathbb{R}^m$ but will also depend, always in C^∞ fashion, on the variable point in $\mathbb{R}^n$ (or rather, in some open neighborhood V of the origin in $\mathbb{R}^n$). Let us here denote by s the variable point in $\mathbb{R}^n$ (in Ch. II, it will be denoted by t). Tye preceding reasoning remains valid if we replace everywhere $\phi(x)$ by $\phi(x,s)$; the wedge $\mathfrak{w}_\delta(U,\Gamma)$ and the domain Ω will now depend (smoothly) on s, which we may indicate by writing $\mathfrak{w}_\delta^s(U,\Gamma)$ and Ω^s. Differentiation with respect to s is then permitted under the integral signs in $B_{\Omega^s}f$ and in $R_{\Omega^s}f$ (see (3.5) and (3.8)) and the reasoning in the preceding pages applies without modification. Let us introduce the notation

(3.24) $$\mathfrak{w}_\delta(U\times V,\Gamma) = \{(z,s) \in \mathbb{C}^m \times V;\ z \in \mathfrak{w}_\delta^s(U,\Gamma)\}.$$

The conclusion in Th. 3.3 can then be strengthened:

Theorem 3.4. There exist an open neighborhood U' of the origin in $\mathbb{R}^m$, an open convex cone $\Gamma' \subset \Gamma$ in $\mathbb{R}^m$, a number δ', $0 < \delta' \le \delta$, and also an open neighborhood $V' \subset V$ of the origin in s-space $\mathbb{R}^n$ such that, if the coefficients of the $\bar{\partial}$-closed form f belong to $\mathcal{B}^\infty(\mathfrak{w}_\delta(U\times V,\Gamma))$, then there is a solution u to the equation $\bar{\partial}u = f$ whose coefficients belong to $\mathcal{B}^\infty(\mathfrak{w}_{\delta'}(U'\times V',\Gamma'))$.

Remark 3.5. Inspection of the preceding argument shows that we can choose any ray r_o in the cone Γ and transform it into the positive half of the y^m-axis. There is then always a local biholomorphism of $\mathbb{C}^m$ which enables us to put ϕ in the form (2.2). We underline the fact that the cone Γ' in Theorems 3.3, 3.4 contains that same ray r_o. ■

CHAPTER II

MICROLOCAL COHOMOLOGY IN A HYPO-ANALYTIC STRUCTURE

1. LOCAL HYPO-ANALYTIC STRUCTURES. THE ASSOCIATED DIFFERENTIAL COMPLEX

Throughout this chapter $\mathfrak{M}$ will denote a C^∞ manifold of dimension ≥ 1, m will denote an integer such that $1 \leq m \leq \dim \mathfrak{M}$. We write $n = \dim \mathfrak{M} - m$. Our viewpoint will always be local: the analysis will take place in the neighborhood of a point of $\mathfrak{M}$ to which we shall systematically refer as the origin and which we denote by 0. Strictly speaking $\mathfrak{M}$ should be viewed as the germ of C^∞ manifold at 0.

Let us call $\mathfrak{Z}^m(\mathfrak{M})$ the collection of all pairs (U,Z) consisting of an open neighborhood of the origin in $\mathfrak{M}$, U, and of a C^∞ map

$$Z = (Z^1,\dots,Z^m): U \to \mathbb{C}^m$$

having the property that

(1.1) $dZ^1,\dots,dZ^m$ <u>are $\mathbb{C}$-linearly independent at every point of</u> U.

We shall define an equivalence relation

$$(U,Z) \approx (U',Z') \tag{1.2}$$

among elements of $\mathfrak{Z}^m(\mathfrak{M})$ by the following condition:

(1.3) there exist an open neighborhood $V \subset U \cap U'$ of 0 in $\mathfrak{M}$ and a biholomorphic map F of an open neighborhood of $Z(V)$ in $\mathbb{C}^m$ onto one of $Z'(V)$, such that we have $Z' = F \circ Z$ in V.

Definition 1.1. By a local hypo-analytic structure in $\mathfrak{M}$ about the point 0 we shall mean an equivalence class for the relation (1.2).

A local hypo-analytic structure in $\mathfrak{M}$ is the germ at the origin of a hypo-analytic structure on $\mathfrak{M}$ as defined in [BCT]. The basic datum in the present work will be such a germ $\mathfrak{G}$ of a hypo-analytic structure.

To any pair $(U,Z) \in \mathfrak{G}$ we shall refer as a hypo-analytic chart.

Consider such a hypo-analytic chart (U,Z). The differentials dZ^i, $i=1,\ldots,m$, span a vector subbundle of the complex cotangent bundle $\mathbb{C}T^*\mathfrak{M}|_U$ over U. The fibre dimension (over $\mathbb{C}$) of this subbundle is evidently equal to m, by (1.1). If $(U',Z') \approx (U,Z)$ the subbundles of $\mathbb{C}T^*\mathfrak{M}|_U$ and $\mathbb{C}T^*\mathfrak{M}|_{U'}$ defined, respectively, by (U,Z) and by (U',Z'), coincide over the subneighborhood V in (4.3). It follows that the hypo-analytic structure $\mathfrak{G}$ defines the germ at 0 of a vector subbundle of $\mathbb{C}T^*\mathfrak{M}$ which we shall always denote by T' and to which we shall refer as the cotangent structure bundle underlying the hypo-analytic structure $\mathfrak{G}$ (and most of the time, simply as the cotangent structure bundle).

The hypo-analytic chart (U,Z) also defines a vector subbundle of the complex tangent bundle $\mathbb{C}T\mathfrak{M}|_U$, specifically the subbundle whose sections are the vector fields L such that $LZ^i = 0$, $\forall\ i=1,\ldots,m$. In other words, it is the orthogonal, for the duality between tangent and cotangent vectors, of the vector subbundle of $\mathbb{C}T^*\mathfrak{M}|_U$ associated above with the chart (U,Z). If (1.3) holds the subbundles of $\mathbb{C}T\mathfrak{M}|_U$ and $\mathbb{C}T\mathfrak{M}|_{U'}$, defined respectively by (U,Z) and (U',Z'), are equal over V. Thus the local hypo-analytic structure $\mathfrak{G}$ defines the germ at 0 of a vector subbundle of $\mathbb{C}T\mathfrak{M}$ which we shall denote by $\mathcal{V}$. Note that $\mathcal{V} = T'^{\perp}$ - in the obvious sense, and that $\mathcal{V}$ is involutive: $[\mathcal{V},\mathcal{V}] \subset \mathcal{V}$, i.e., the commutation bracket of any two smooth sections of $\mathcal{V}$ is also a section of $\mathcal{V}$. We shall refer to $\mathcal{V}$ as the tangent structure bundle underlying the hypo-analytic structure $\mathfrak{G}$.

We shall also say that T' and/or $\mathcal{V}$ define the locally integrable structure underlying $\mathfrak{G}$. Actually this is the germ of a locally integrable structure - on the germ of $\mathfrak{M}$ at the point 0. We point out that different hypo-analytic structures may have the same underlying locally integrable structure.

For the sake of simplicity we shall from now on refer as vector bundles to what, strictly speaking, will be the germs at the origin of vector bundles over $\mathfrak{M}$. Thus let us denote by $T'^{p,q}$ ($p,q \in \mathbb{Z}_+$) the complex vector bundle whose smooth sections are sums of exterior products

$$\psi^1 \wedge \ldots \wedge \psi^{p+q}$$

in which the ψ^i are (the germs at 0 of) smooth one-forms, at least p of which are sections of T'. Obviously we have $T'^{p+1,q-1} \subset T'^{p,q}$, which allows us to introduce the quotient vector bundle

$$(1.4) \qquad \Lambda^{p,q} = T'^{p,q}/T'^{p+1,q-1}.$$

Of course $\Lambda^{0,0}$ is the germ at the origin of the trivial bundle $\mathfrak{M}\times\mathbb{C}$. Note also that whatever p and q,

$$(1.5) \qquad \Lambda^{p,0} = T'^{p,0}, \quad \Lambda^{m,q} = T'^{m,q},$$

and that $\Lambda^{p,q} = 0$ if either $p > m$ or $q > n$.

In the sequel $C^\infty(\Lambda^{p,q})$ will denote the space of germs of smooth sections of $\Lambda^{p,q}$. We simply write C^∞ for $C^\infty(\Lambda^{0,0})$.

Let (U,Z) be a hypo-analytic chart. Since any C^∞ section ψ of T' is a linear combination, with C^∞ coefficients, of (the germs of) $dZ^1,\ldots,dZ^m$, we see that $d\psi$ is a section of $T'^{1,1}$, a property that we abbreviate by writing

$$(1.6) \qquad dT' \subset T'^{1,1}$$

In a similar notation this entails

$$(1.7) \qquad dT'^{p,q} \subset T'^{p,q+1}.$$

As a consequence of (1.7) the exterior derivatives induces a linear differential operator

$$(1.8) \qquad d'^{p,q}: C^\infty(\Lambda^{p,q}) \to C^\infty(\Lambda^{p,q+1}).$$

Of course, we have

$$(1.9) \qquad d'^{p,q+1} \circ d'^{p,q} = 0,$$

and thus, for each $p = 0,1,\ldots,m$, the sequence of operators (1.8), as $q = 0,1,\ldots$, defines a differential complex. We refer to it as <u>the p-th differential complex associated to the local hypo-analytic structure</u> $\mathfrak{G}$ (actually, it is associated to the underlying locally integrable structure, i.e., it is entirely determined by the structure bundle T').

If we take (1.5) into account we see that $d'^{m,q}$ is equal to the exterior derivative d, acting on the smooth sections of $T'^{m,q}$.

We shall say that a section $f \in C^\infty(\Lambda^{p,q})$ is a <u>cocycle</u> if $d'^{p,q}f = 0$ and that it is a <u>coboundary</u> if there is a section

$u \in C^\infty(\Lambda^{p,q-1})$ such that $d'^{\,p,q-1}u = f$ (which, of course, presumes $q \geq 1$). The cohomology spaces of the differential complex (1.8) are defined in the usual manner:

$$H'^{\,p,0} = \text{Ker } d'^{\,p,0}, \tag{1.10}$$

$$H'^{\,p,q} = \text{Ker } d'^{\,p,q}/\text{Im } d'^{\,p,q-1} \text{ if } q > 0. \tag{1.11}$$

When there is no risk of confusion we shall write d' rather than $d'^{\,p,q}$.

Presently we wish to give a more "concrete" representation of the objects that we have just described. This is achieved through the choice of suitable local coordinates in neighborhood of the origin in $\mathfrak{M}$. We shall start from a hypo-analytic chart (U,Z). We are allowed to make biholomorphic substitutions of the Z^i's, in particular $\mathbb{C}$-affine transformations. We are also allowed to contract the neighborhood U about the origin. This enables us to assume that the Z^i have a stronger property than (1.1), specifically that the differentials $d(\text{Re } Z^1),\ldots,d(\text{Re } Z^m)$ are linearly independent at each point of U and, as a matter of fact, that the functions $x^i = \text{Re } Z^i$ are part of a coordinate system in U, in which the remaining coordinates are denoted by $t^1,\ldots,t^n$. We shall also assume that these coordinates, as well as the functions Z^i, all vanish at the origin. Thus we may hypothetize that

$$Z^i = x^i + \sqrt{-1}\phi^i(x,t), \quad i=1,\ldots,m. \tag{1.12}$$

We shall write $\phi = (\phi^1,\ldots,\phi^m)\colon U \to \mathbb{R}^m$. We have $\phi(0) = 0$. Possibly after an additional contraction of U about 0 we may assume that the Jacobian matrix $Z_x = \{\partial Z^h/\partial x^i\}_{1\leq h,i\leq m}$ is nonsingular at every point of U. After substituting $Z_x(0,0)^{-1}Z(x,t)$ for $Z(x,t)$ we may even assume that

$$d_x\phi^i\big|_0 = 0, \quad i=1,\ldots,m. \tag{1.13}$$

The differentials dZ^i, dt^j $(1 \leq i \leq m,\ 1 \leq j \leq n)$ are $\mathbb{C}$-linearly independent or, which amounts to the same, they span $\mathbb{C}T^*\mathfrak{M}|_U$. We may therefore define the vector fields L_j, M_i in U by the "orthonormality" relations

$$L_jZ^h = 0,\quad L_jt^k = \delta^k_j,\quad M_hZ^i = \delta^i_h,\quad M_ht^j = 0, \qquad 1 \leq h,i \leq m,\quad 1 \leq j,k \leq n. \tag{1.14}$$

These vector fields span $\mathbb{C}T\mathfrak{M}|_U$. Moreover, they commute:

$$[L_j,L_k] = [L_j,M_h] = [M_h,M_i] = 0, \qquad 1 \leq h,i \leq m,\quad 1 \leq j,k \leq n. \tag{1.15}$$

The germs at 0 of the vector fields $L_1,\dots,L_n$ span the tangent strucutre bundle $\mathcal{V}$.

It is not difficult to obtain explicit formulas for these vector fields. Denote by $\{\mu_i^h\}_{1\le h,i\le m}$ the inverse of the Jacobian matrix Z_x. Then

(1.16) $$M_h = \sum_{i=1}^{m} \mu_h^i\, \partial/\partial x^i, \quad h=1,\dots,m;$$

(1.17) $$L_j = \partial/\partial t^j - \sqrt{-1} \sum_{i=1}^{m} (\partial\phi^i/\partial t^j) M_i, \quad j=1,\dots,n.$$

If u is a, say smooth, function in U we have

(1.18) $$du = \sum_{i=1}^{m} M_i u\, dZ^i + \sum_{j=1}^{n} L_j u\, dt^j.$$

By the same token, any smooth section $\dot{f}$ of $\Lambda^{p,q}$ has a unique representative of the kind

(1.19) $$f = \sum_{|I|=p} \sum_{|J|=q} f_{I,J}(x,t)\, dZ^I \wedge dt^J.$$

We are using the standard notation for differential forms; thus $I = (i_1,\dots,i_p)$ with $1 \le i_1 <\dots< i_p \le m$, $dZ^I = dZ^{i_1} \wedge\dots\wedge dZ^{i_p}$; $J = (j_1,\dots,j_q)$ with $1 \le j_1 <\dots< j_q \le n$, $dt^J = dt^{j_1} \wedge\dots\wedge dt^{j_q}$. We shall refer to the pair (p,q) as the bidegree of the form f.

We shall refer to (1.19) as the <u>standard representative</u> of the section $\dot{f}$ of $\Lambda^{p,q}$.

By virtue of (1.18) the standard representative of $d'\dot{f}$ is the (germ of) differential form

(1.20) $$Lf = \sum_{|I|=p} \sum_{|J|=q} \sum_{j=1}^{n} L_j f_{I,J}\, dt^j \wedge dZ^I \wedge dt^J.$$

If we avail ourselves of the expressions (1.17) we see that

(1.21) $$Lf = d_t f - \sqrt{-1} \sum_{i=1}^{m} (d_t\phi^i) \wedge M_i f,$$

where M_i acts coefficientwise.

These representations, of $\dot{f}$ by the differential form f and of d' by the differential operator L, are not invariant: they depend on the choice of the coordinates t^j.

We shall extend the cocycle/coboundary terminology to these representations: we shall say that a form (1.19) is a cocycle or, equivalently, that it is L-<u>closed</u> if $Lf = 0$; that it is a coboundary or, equivalently, that it is L-<u>exact</u> if there is a form u of bidegree $(p,q-1)$ such that $Lu = f$. Of course, a cohomology class $[f] \in H'^{p,q}$ will be represented by a coset of cocycles whose pairwise differences are coboundaries.

To say that f is a cocycle is to say that, for any pair of multi-indices I, K, $|I| = p$, $|K| = q+1$, we have

(1.22) $$\sum_{j,J} \varepsilon(j,J) L_j f_{I,J} = 0,$$

where the summation is carried out over all pairs (j,J) consisting on an integer $j \in [1,\ldots,m]$ and of an ordered multi-index $J = (j_1,\ldots,j_q)$ such that $j \notin J$ and such that K is a permutation of $(j,j_1,\ldots,j_q)$. The coefficient $\varepsilon(j,J)$ is equal to +1 if that permutation is even, and to -1 if it is odd.

To say that f is a coboundary is to say, that there is a standard differential form u of bidegree $(p,q-1)$ such that, for any pair of multi-indices I, J, $|I| = p$, $|J| = q$, we have, in the same notation as that used in (1.22),

(1.23) $$\sum_{j\in J} \varepsilon(j,J\setminus\{j\}) L_j u_{I,J\setminus\{j\}} = f_{I,J}.$$

Remark 1.2. A cocycle of bidegree $(p,0)$ is a p-form

(1.24) $$f = \sum_{|I|=p} h_I(x,t)\,dZ^I$$

whose coefficients h_I are solutions of the homogeneous equations

(1.25) $$L_j h = 0, \quad j=1,\ldots,n.$$

Throughout the present article we refer to any distribution that satisfies (1.25) as a solution. ∎

Remark 1.3. The expression (1.19) show that there is a natural isomorphism

(1.26) $$\Lambda^{p,q} \cong \Lambda^{p,0} \otimes \Lambda^{0,q}.$$

Indeed, the form (1.19) can be decomposed, in a unique manner, as

(1.27) $$f = \sum_{|I|=p} dZ^I \wedge f_I,$$

where the f_I are forms of bidegree $(0,q)$. The p-forms dZ^I "span $\Lambda^{p,0}$". We have, by (1.20),

(1.28) $$Lf = (-1)^p \sum_{|I|=p} dZ^I \wedge Lf_I.$$

Formula (1.28) shows that, in order that f be L-closed (resp., L-exact) it is necessary and sufficient that the same property be true of every form f_I, $|I| = p$.

This remark enables us to limit ourselves, in much of the forthcoming reasoning, to the case $p = 0$. ∎

Remark 1.4. When $p = m$ the form (1.19) has the expression

(1.29) $$f = dZ \wedge \sum_{|I|=p} f_J(x,t)dt^J,$$

where we have used the notation (also used throughout the sequel)

(1.30) $$dZ = dZ^1 \wedge \ldots \wedge dZ^m.$$

Indeed, $\Lambda^{m,0}$ is a (complex) line bundle, spanned by (the class of the germ of) the m-form dZ.

When $p = m$, the differential operator L is equal to the exterior derivative acting on sections of $T'^{m,q}$ (cf. (1.5)). ∎

2. EMBEDDING INTO $\mathbb{C}^m \times \mathbb{R}^n$. EXTENSION TO WEDGES OF SMOOTH SECTIONS OF THE BUNDLES $\Lambda^{p,q}$

Let z^i $(i=1,\ldots,m)$ denote the complex coordinates in $\mathbb{C}^m$ and t^j $(j=1,\ldots,n)$ the real coordinates in $\mathbb{R}^n$. What we shall call the <u>standard hypo-analytic structure on</u> $\mathbb{C}^m \times \mathbb{R}^n$ is the hypo-analytic structure defined by the functions $z^1,\ldots,z^m$. It is the pull-back of the complex structure on $\mathbb{C}^m$ under the first coordinate projection $(z,t) \to z$. The tangent structure bundle over $\mathbb{C}^m \times \mathbb{R}^n$ is spanned by the vector fields $\partial/\partial \bar{z}^i$, $\partial/\partial t^j$ $(1 \le i \le m,\ 1 \le j \le n)$. We make no distinction between the standard hypo-analytic structure on $\mathbb{C}^m \times \mathbb{R}^n$ (which is <u>globally</u> defined), and its germ at the origin, which is the <u>standard local hypo-analytic structure on</u> $\mathbb{C}^m \times \mathbb{R}^n$.

In Section 1 of the present chapter the dimension of the base manifold was $m+n$. Presently the base manifold is $\mathbb{C}^m \times \mathbb{R}^n \cong \mathbb{R}^{2m+n}$; and the fibre dimension (over $\mathbb{C}$) of the tangent structure bundle is equal to $m+n$, not to n as in Section 1; that of the cotangent structure bundle is equal to m, both now and in Section 1. If we compare the functions $z^i = x^i + \sqrt{-1}y^i$ to the functions Z^i in (1.12) we see that the role of the variables t^j of Section 1 is now played by the variables t^j together with the variables y^i $(1 \le i \le m,\ 1 \le j \le n)$.

According to the terminology of Section 1 the standard forms of bidegree (p,q) in $\mathbb{C}^m \times \mathbb{R}^n$ are the differential forms of the kind

(2.1) $$\sum_{|I|=p} \sum_{|J|+|K|=q} f_{I,J,K}(x,y,t)\, dz^I \wedge dy^J \wedge dt^K.$$

Since $dy^i = \sqrt{-1}(d\bar{z}^i - dz^i)/2$ we see that the section $\dot{f}$ of the vector bundle $\Lambda^{p,q}$ over $\mathbb{C}^m \times \mathbb{R}^n$ represented by (2.1) has also a unique representative of the following kind:

(2.2) $$\sum_{|I|=p} \sum_{|J|+|K|=q} f_{I,J,K}(x,y,t)\, dz^I \wedge d\bar{z}^J \wedge dt^K.$$

Henceforth we refer to (2.2) as the standard representative of $\dot{f}$: in $\mathbb{C}^m \times \mathbb{R}^n$ we shall not make use of the forms (2.1).

Having agreed to this we see that the analogue of the differential operator L of (1.20), (1.21), is nothing else but the operator $\bar{\partial}_z + d_t$. This suggests that we decompose further the forms of bidegree (p,q), into sums of forms of "tridegree" (p,q',q''), with $q'+q'' = q$, i.e., forms of the following kind:

$$(2.3) \qquad \sum_{|I|=p} \sum_{|J|=q'} \sum_{|E|=q''} f_{I,J,K}(x,y,t)\, dz^I \wedge d\bar{z}^J \wedge dt^K .$$

We now return to the local hypo-analytic structure $\mathfrak{A}$ on the manifold $\mathfrak{M}$ of Section 1. We assume that U is the domain of the real coordinates x^i, t^j and that Z is given by (1.12). More generally, we shall assume that all the properties with which the hypo-analytic chart (U,Z) has been endowed in Section 1 continue to be valid.

The C^∞ map from U into $\mathbb{C}^m \times \mathbb{R}^n$,

$$(2.4) \qquad (x,t) \to (Z(x,t),t)$$

is a diffeomorphism of U onto an (m+n)-dimensional submanifold $\tilde{U}$ of $\mathbb{C}^m \times \mathbb{R}^n$. The submanifold $\tilde{U}$ is defined by the equations

$$(2.5) \qquad y^i = \phi^i(x,t), \quad i=1,\ldots,m.$$

We shall denote by $\tilde{\mathfrak{M}}$ the germ at the origin of the submanifold $\tilde{U}$. (In practice we shall view $\tilde{\mathfrak{M}}$ as a true manifold in which $\tilde{U}$ is an open set). From $\mathbb{C}^m \times \mathbb{R}^n$ $\tilde{\mathfrak{M}}$ inherits a local hypo-analytic structure: that defined by the restriction to $\tilde{\mathfrak{M}}$ of the (germs of) functions z^i, $i=1,\ldots,m$. The (germ at the origin of the) map (2.4) is a hypo-analytic isomorphism of (the germ of) $\mathfrak{M}$, equipped with the local hypo-analytic structure $\mathfrak{A}$, onto $\tilde{\mathfrak{M}}$, equipped with the structure inherited from $\mathbb{C}^m \times \mathbb{R}^n$. We shall sometimes refer to $\tilde{\mathfrak{M}}$ as an <u>embedding</u> (or as a <u>realization</u>) of the (germ of) hypo-analytic manifold $(\mathfrak{M},\mathfrak{A})$.

Different choices of the hypo-analytic chart (U,Z) and of the coordinates t^j yield different realizations. Nonetheless it is sometimes convenient to carry out the analysis on $\tilde{\mathfrak{M}}$ rather than on $\mathfrak{M}$.

Next we define what we shall be calling a wedge in $\mathbb{C}^m \times \mathbb{R}^n$. It is a generalization of the concept introduced in Section 1, Ch. I, taking into account the presence of the variables t^j. Let U denote here an open neighborhood of the origin in $\mathfrak{M}$ which we identify to an open neighborhood of the origin in $\mathbb{R}^{m+n}$ by means of the coordinates x^i, t^j. We write

$$(2.6) \qquad \tilde{\mathfrak{W}}_\delta(U,\Gamma) = \{(z,t) \in \mathbb{C}^m \times \mathbb{R}^n;\ \exists\ (x,t) \in U,\ v \in \Gamma \text{ such that } z = Z(x,t) + \sqrt{-1}v, \text{ and } |y| < \delta\}.$$

Here, as in Ch. I, Γ is an open convex cone in $\mathbb{R}^m$ and δ is a number >0. <u>It will always be tacitly understood that</u>

$$|\mathrm{Im}\, Z(x,t)| < \delta \quad \text{for all} \quad (x,t) \in U \quad (\text{cf. } (1.3), \text{ Ch. I}).$$

As we contract U about 0 and let δ go to zero the sets $\tilde{\mathfrak{w}}_\delta(U,\Gamma)$ define a germ of set $\tilde{\mathfrak{w}}(\mathfrak{M},\Gamma)$, to which we shall refer as <u>a germ of wedge in</u> $\mathbb{C}^m\times\mathbb{R}^n$ <u>with edge</u> $\tilde{\mathfrak{M}}$. We shall also say that $\tilde{\mathfrak{w}}_\delta(U,\Gamma)$ represents (or is a representative of) the germ of wedge $\tilde{\mathfrak{w}}(\mathfrak{M},\Gamma)$.

In order to avoid any confusion we shall denote by $\tilde{\Lambda}^{p,q}$ the analogue, for $\mathbb{C}^m\times\mathbb{R}^n$, of the vector bundle $\Lambda^{p,q}$. (In general the tildas indicate that the objects are attached to $\mathbb{C}^m\times\mathbb{R}^n$, whereas the absence of tildas means that they are attached to $\mathfrak{M}$.) $\tilde{\Lambda}^{p,q}$ is the bundle whose sections have representatives of the kind (2.2) (here, the word "representative" refers to the fact that we are neglecting all $(p+q)$-forms that involve more than p factors dz^i). Then we define $\mathcal{B}^\infty(\tilde{\mathfrak{w}}(\mathfrak{M},\Gamma);\, \tilde{\Lambda}^{p,q})$ as the space of (germs of) smooth sections $\tilde{f}$ of $\tilde{\Lambda}^{p,q}$ which have, for sufficiently small neighborhoods U of 0 and numbers $\delta > 0$, a standard representative $\tilde{f}$ (i.e., a representative of the kind (2.2)) in the wedge $\tilde{\mathfrak{w}}_\delta(U,\Gamma)$, endowed with the following property:

(2.7) <u>the coefficients</u> $\tilde{f}_{I,J,K}$ <u>of</u> $\tilde{f}$ <u>belong to</u> $\mathcal{B}^\infty(\tilde{\mathfrak{w}}_\delta(U,\Gamma))$ (see Ch. I, Sect. 1).

We may now introduce the differential complex

$$\bar{\partial}_z + d_t\colon\ \mathcal{B}^\infty(\tilde{\mathfrak{w}}_\delta(U,\Gamma);\tilde{\Lambda}^{p,q}) \to \mathcal{B}^\infty(\tilde{\mathfrak{w}}_\delta(U,\Gamma);\tilde{\Lambda}^{p,q+1}),\quad q = 0,1,\ldots,m+n-1. \tag{2.8}$$

By letting U contract to 0 and $\delta \to +0$, we can also define the complex

$$\bar{\partial}_z + d_t\colon\ \mathcal{B}^\infty(\tilde{\mathfrak{w}}(\mathfrak{M},\Gamma);\tilde{\Lambda}^{p,q}) \to \mathcal{B}^\infty(\tilde{\mathfrak{w}}(\mathfrak{M},\Gamma);\tilde{\Lambda}^{p,q+1}),\quad q = 0,\ldots,m+n-1. \tag{2.9}$$

If (2.7) holds the pull-back of $\tilde{f}$ to $\tilde{U}$ is well defined. Notice that one can write, in $\mathfrak{M}$,

$$d\bar{z}^i = dz^i = \sqrt{-1}\phi_x^i\cdot(dz+d\bar{z}) - 2\sqrt{-1}d_t\phi^i\cdot dt, \qquad i=1,\ldots,m.$$

In vector/matrix notation this can be rweritten as

$$d\bar{z} = (I+\sqrt{-1}\phi_x)^{-1}(I-\sqrt{-1}\phi_x)dz - 2\sqrt{-1}(I+\sqrt{-1}\phi_x)^{-1}(d_t\phi\cdot dt). \tag{2.10}$$

It obviously entails that the pull-back of $\tilde{f}$ to $\tilde{U}$ is equal to a standard form $\tilde{f}_o$ on $\tilde{U}$ of bidegree (p,q) (i.e, of the kind (1.19) in which the coefficients must be viewed as functions in $\tilde{U}$, not in

U) <u>modulo sums of standard forms on</u> $\tilde{U}$ <u>of bidegree</u> $(p+k,q-k)$ with $k \geq 1$. Let then $\tilde{\tilde{f}}$ be any differential form in $\tilde{\mathfrak{W}}_\delta(U,\Gamma)$ obtained by adding to $\tilde{f}$ a sum of standard forms of bidegree $(p+k,q-k)$ with $k \geq 1$, all of which are also endowed with property (2.8). Then $\tilde{\tilde{f}}$ is also a representative, albeit a nonstandard one, of the section $\dot{\tilde{f}}$. Then the pull-back of $\tilde{\tilde{f}}$ to $\tilde{U}$ will also be congruent to the same standard form $\tilde{f}_o$ modulo sums of standard forms on $\tilde{U}$ of bidegree $(p+k,q-k)$ with $k \geq 1$.

Let then f be the pull-back to U of the form $\tilde{f}_o$ under the map (2.4); f is a standard form of the kind (1.19). It defines a section $\dot{f}$ of $\Lambda^{p,q}$ which clearly belongs to $\mathcal{B}^\infty(U;\Lambda^{p,q})$ (i.e., its coefficients belong to $\mathcal{B}^\infty(U)$ when we identify U to an open subset of R^{m+n} by means of the coordinates x^i, t^j). By going from the section $\dot{f}$ to its germ at 0 we define a linear map

$$\mathcal{B}^\infty(\tilde{\mathfrak{W}}_\delta(U,\Gamma);\tilde{\Lambda}^{p,q}) \to C^\infty(\Lambda^{p,q}). \tag{2.11}$$

The map (2.11) is certainly not injective for it annihilates any form $\tilde{f}$ in the wedege $\tilde{\mathfrak{W}}_\delta(U,\Gamma)$ whose coefficients vanish on the edge $\tilde{U}$.

By letting U contract to 0 and $\delta > 0$ go to zero, (2.11) defines a linear map

$$\mathcal{B}^\infty(\tilde{\mathfrak{W}}(\mathfrak{M},\Gamma);\Lambda^{p,q}) \to C^\infty(\Lambda^{p,q}), \tag{2.12}$$

to which we shall refer as the pull-back to $\mathfrak{M}$.

We recall that the hypo-analytic structure of $\tilde{U}$ is the one inherited from $\mathbb{C}^m \times \mathbb{R}^n$ (i.e., the structure defined by the restrictions to $\tilde{U}$ of the functions $z^1,\dots,z^m$) and that the hypo-analytic structure of U is the pull-back of that of $\tilde{U}$ under the map (2.4). This implies at once

<u>Proposition 2.1</u>. <u>The pull-back to</u> $\mathfrak{M}$ <u>defines a map of the differential complex</u> (2.9) <u>into the differential complex</u> (1.8).

<u>Remark 2.2</u>. The reader may wish to check the assertion in Prop. 2.1 directly. Let $f(x,y,t) \in \mathcal{B}^\infty(\tilde{\mathfrak{W}}_\delta(U,\Gamma))$. It follows from (2.10) that the pull-back to the edge $\tilde{U}$ of the one-form

$$\bar{\partial}_z f + d_t f = \sum_{i=1}^{m} (\partial f/\partial \bar{z}^i) d\bar{z}^i + \sum_{j=1}^{n} (\partial f/\partial t^j) dt^j$$

is congruent, modulo a linear combination of $dz^1,\dots,dz^n$, to the form

$$\sum_{j=1}^{m} \tilde{L}_j f \, dt^j,$$

where

$$\tilde{L}_j f = \partial f/\partial t^j - 2\sqrt{-1} \sum_{h,i=1}^{m} (\partial \phi^h/\partial t^j)\mu_h^i \partial f/\partial \bar{z}^i \tag{2.13}$$

is a function in $\tilde{U}$. The μ_h^i have the same meaning as in (1.16) except that they are viewed as functions in $\tilde{U}$ rather than in U. Observe then that, by the chain rule,

$$(\partial/\partial x^h)f(x,\phi(x,t),t) = \{\partial f/\partial x^h - 2\sqrt{-1}\sum_{i=1}^{m}(\partial\phi^i/\partial x^h)(x,t)\partial f/\partial \bar{z}^i + \sqrt{-1}\sum_{i=1}^{m}(\partial\phi^i/\partial x^h)(x,t)\partial f/\partial x^i\}\Big|_{y=\phi(x,t)},$$

$$(\partial/\partial t^j)f(x,\phi(x,t),t) = \{\partial f/\partial t^j - 2\sqrt{-1}\sum_{i=1}^{m}(\partial\phi^i/\partial t^j)(x,t)\partial f/\partial \bar{z}^i + \sqrt{-1}\sum_{i=1}^{m}(\partial\phi^i/\partial t^j)(x,t)\partial f/\partial x^i\}\Big|_{y=\phi(x,t)}.$$

Call f_o the restriction of f on $\tilde{U}$. In vector/matrix notation what precedes can be summarized as follows:

$$f_{o_x} = f_x + i\phi_x f_x - 2i\phi_x f_{\bar{z}}, \qquad f_{o_t} = f_t + i\phi_t f_x - 2i\phi_t f_{\bar{z}},$$

$(i = \sqrt{-1})$ whence

$$f_{o_t} - i\phi_t\mu f_{o_x} = f_t - 2i\phi_t(1-i\mu\phi_x)f_{\bar{z}}.$$

But $1 - i\mu\phi_x = \mu$. We conclude that the pull-back to U of $\tilde{L}_j f$ is equal to $L_j f$ (see (1.16), (1.17)). From there the assertion follows at once. ∎

We recall that a section $\dot{\tilde{f}} \in \mathcal{B}^\infty(\tilde{\mathfrak{w}}(\mathfrak{m},\Gamma);\tilde{\Lambda}^{p,q})$ is a cocycle if it is $(\bar{\partial}_z+d_t)$-closed.

<u>Definition 2.3. We shall say that a cocycle</u> $\dot{f} \in C^\infty(\Lambda^{p,q})$ <u>extends to the wedge</u> $\tilde{\mathfrak{w}}(\mathfrak{m},\Gamma)$ <u>if there is a cocycle</u> $\dot{\tilde{f}} \in \mathcal{B}^\infty(\tilde{\mathfrak{w}}(\mathfrak{m},\Gamma);\tilde{\Lambda}^{p,q})$ <u>whose pull-back to</u> $\mathfrak{m}$ <u>is equal to</u> $\dot{f}$.

As already pointed out the extension cannot possibly be unique, i.e., the map (2.12) is not injective - not even on cocycles. The following, however, can be said:

<u>Proposition 2.4. Suppose that</u> $\dot{\tilde{f}} \in \mathcal{B}^\infty(\tilde{\mathfrak{w}}(\mathfrak{m},\Gamma);\tilde{\Lambda}^{p,q})$ <u>is a cocycle and that its standard representative</u> $\tilde{f}$ (of the kind (2.2)) <u>has the following property</u>:

(2.14) $$\tilde{f}_{I,J,K} \equiv 0 \quad \text{whenever} \quad J \neq \phi.$$

<u>Then, if the pull-back of</u> $\dot{\tilde{f}}$ <u>to</u> $\mathfrak{m}$ <u>vanishes identically, we must necessarily have</u> $\dot{\tilde{f}} \equiv 0$.

Condition (2.14) means that we have

(2.15) $$\tilde{f} = \sum_{|I|=p}\sum_{|K|=q} f_{I,K}(x,y,t)\, dz^I \wedge dt^K.$$

<u>Proof</u>: It is clear that if (2.15) holds then the pull-back to $\tilde{U}$ of

the standard form $\tilde{f}$ is simply obtained by putting $y = \phi(x)$ in the coefficients of $\tilde{f}$ (we use $dz^1,\dots,dz^m$, $dt^1,\dots,dt^n$ as the basis in $\mathbb{C}T^*\tilde{U}$). On the other hand, to say that $\hat{f}$ is a cocycle means exactly that $\tilde{f}$ is $(\bar{\partial}_z+d_t)$-closed. Since $\tilde{f}$ is $\bar{\partial}_z$-closed its coefficients $\tilde{f}_{I,0,K} \in \mathcal{B}^\infty(\tilde{\mathfrak{w}}_\delta(U,\Gamma))$ are holomorphic in the wedge $\tilde{\mathfrak{w}}_\delta(U,\Gamma)$; but then they must vanish identically if their boundary values on the edge do. ∎

<u>Theorem 2.5. Given any open convex cone</u> Γ <u>in</u> $\mathbb{R}^m$ <u>and any ray</u> $r_o \subset \Gamma$ <u>there is another open convex cone</u> $\Gamma' \subset \Gamma$ <u>such that</u> $r_o \subset \Gamma'$ <u>and such that the following is true:</u>

<u>Each cocycle that belongs to</u> $\mathcal{B}^\infty(\tilde{\mathfrak{w}}(\mathfrak{m},\Gamma);\tilde{\Lambda}^{p,q})$ <u>is cohomologous in</u> $\tilde{\mathfrak{w}}(\mathfrak{m},\Gamma')$ <u>to a cocycle</u> $\dot{\tilde{f}} \in \mathcal{B}^\infty(\tilde{\mathfrak{w}}(\mathfrak{m},\Gamma');\tilde{\Lambda}^{p,q})$ <u>that has a standard representative</u> $\tilde{f}$ <u>endowed with property</u> (2.14).

<u>Proof</u>: Of course we shall deal with representatives of germs, of cohomology classes and of sections of $\tilde{\Lambda}^{p,q}$ over germs of wedges $\tilde{\mathfrak{w}}_\delta(U,\Gamma)$. A moment of thought will convince the reader that it suffices to treat the case $p = 0$. Noting that any zero-form is a function that is holomorphic with respect to z and constant with respect to t, we see that the result is true when $q = 0$. Thus let us consider, in a wedge $\tilde{\mathfrak{w}}_\delta(U,\Gamma)$, a standard form of bidegree $(0,q)$ with $q > 0$, of the kind (2.2). We rewrite it as

$$\tilde{f} = \sum_{q'=0}^{q} \sum_{|K|=q-q'} \tilde{f}_K \wedge dt^K, \tag{2.16}$$

where, for each K, $|K| = q-q'$,

$$\tilde{f}_K = \sum_{|J|=q'} \tilde{f}_{J,K}(x,y,t)d\bar{z}^J. \tag{2.17}$$

We shall deal with forms $\tilde{f}$ such that

(2.18) <u>every coefficient of the form belongs to</u> $\mathcal{B}^\infty(\tilde{\mathfrak{w}}_\delta(U,\Gamma))$.

Express now the fact that $\tilde{f}$ is a cocycle, i.e., that $\bar{\partial}_z\tilde{f} = -d_t\tilde{f}$:

$$\begin{aligned} &\bar{\partial}_z\tilde{f}_K = 0 \quad \text{if} \quad K = \phi\,; \\ &\bar{\partial}_z f_K = -\sum_{j\in K} \epsilon(j,K\backslash\{j\})(\partial/\partial t^j)\tilde{f}_{K\backslash\{j\}} \quad \text{if} \quad K \neq \phi\,. \end{aligned} \tag{2.19}$$

We shall apply Th. 3.4 and Remark 3.5, Ch. I: after contracting U about the origin, Γ about the ray r_o and after decreasing δ we can find a $(0,q-1)$-form u_ϕ that has also property (2.18) and satisfies in $\tilde{\mathfrak{w}}_\delta(U,\Gamma)$,

$$\bar{\partial}_z u_\phi = \tilde{f}_\phi\,. \tag{2.20}$$

Consider now the form $f^{(1)} = f - (\bar\partial_z + d_t)u_\phi$; it is trivially cohomologous to f; in particular, it is $(\bar\partial_z + d_t)$-closed. It has the property, however, that the form $f_\phi^{(1)}$, the analogue of f_ϕ, vanishes identically. If $q = 1$ this means that $f^{(1)}$ has property (2.14). Suppose $q \geq 2$ and consider then the set of equations (2.19) with $f^{(1)}$ substituted for f. We see that

$$(2.21) \qquad \bar\partial_z f_K^{(1)} = 0 \quad \text{if} \quad |K| = 1.$$

By applying once again Th. 3.4, Ch. I, and after some further contraction of $\tilde{\mathfrak{w}}_\delta(U,\Gamma)$, we can find n (q-2)-forms u_K that satisfy the equations in $\mathfrak{w}_\delta(U,\Gamma)$,

$$(2.22) \qquad \bar\partial_z u_K = f_K^{(1)}, \quad \forall\, K, \quad |K| = 1,$$

and have property (2.18). We define then

$$f^{(2)} = f^{(1)} - (\bar\partial_z + d_t)\{\sum_{|K|=1} u_K \wedge dt^K\}.$$

If $q \geq 3$ we repeat the same argument; eventually we end up with a form $f^{(q)}$ in $\tilde{\mathfrak{w}}_\delta(U,\Gamma)$ (contracted a finite number of times) that has the properties (2.14) and (2.18) and is cohomologous to the original form f. Each contraction of the cone Γ is carried out about the ray r_o. ∎

Let us denote by $H^{p,q}(\tilde{\mathfrak{w}}(\mathfrak{m},\Gamma))$ the cohomology spaces of the differential complex (2.8). Let Γ and Γ' be as in Th. 2.5. The natural restriction mapping

$$(2.23) \qquad \mathcal{B}^\infty(\tilde{\mathfrak{w}}(\mathfrak{m},\Gamma);\tilde\Lambda^{p,q}) \to \mathcal{B}^\infty(\tilde{\mathfrak{w}}(\mathfrak{m},\Gamma');\tilde\Lambda^{p,q})$$

induces a natural "restriction" mapping

$$(2.24) \qquad H^{p,q}(\tilde{\mathfrak{w}}(\mathfrak{m},\Gamma)) \to H^{p,q}(\tilde{\mathfrak{w}}(\mathfrak{m},\Gamma')).$$

<u>Corollary 2.6</u>. <u>The restriction mapping</u> (2.24) <u>annihilates</u> $H^{p,q}(\tilde{\mathfrak{w}}(\mathfrak{m},\Gamma))$ <u>for</u> $q > n$.

It is clear that one could define the sheaves on the sphere S^{m-1}, $r_o \to \mathcal{B}^\infty(\tilde{\mathfrak{w}}(\mathfrak{m},r_o);\tilde\Lambda^{p,q})$ and $r_o \to H^{p,q}(\tilde{\mathfrak{w}}(\mathfrak{m},r_o))$: their respective stalks are the inductive limits of the spaces $\mathcal{B}^\infty(\tilde{\mathfrak{w}}(\mathfrak{m},\Gamma);\tilde\Lambda^{p,q})$ and $H^{p,q}(\tilde{\mathfrak{w}}(\mathfrak{m},\Gamma))$ as Γ ranges over the collection of all open and convex cones that contain r_o. In the language of these sheaves Cor. 2.6 states that $H^{p,q}(\tilde{\mathfrak{w}}(\mathfrak{m},\cdot)) = 0$ whenever $q > n$.

Some useful precision can be added to Th. 2.5 by looking at the representatives of the germs of sections:

<u>Proposition 2.7</u>. <u>Suppose the section</u> $\dot f \in \mathcal{C}^\infty(U;\Lambda^{p,q})$ $(q \geq 1)$ <u>extends to the wedge</u> $\tilde{\mathfrak{w}}_\delta(U,\Gamma)$. <u>There is an open neighborhood</u> $U' \subset U$

<u>of</u> O, <u>an open convex cone</u> $\Gamma' \subset \Gamma$ <u>containing a given ray</u> r_o <u>and a number</u> δ', $0 < \delta' < \delta$, <u>such that the following is true:</u>

<u>Let</u> f <u>be the standard representative of</u> $\dot{f}$ <u>in</u> U. <u>There is a standard</u> C^∞ <u>form of bidegree</u> (p,q-1) <u>in</u> U', g, <u>such that</u> f-Lg <u>is the pull-back to</u> U', <u>under the map</u> (2.4), <u>of a standard cocycle</u> $\tilde{f} \in \mathcal{B}^\infty(\tilde{\mathfrak{W}}_{\delta'}(U',\Gamma');\tilde{\Lambda}^{p,q})$ <u>which has property</u> (2.14).

<u>Proof:</u> Let $\tilde{f} \in \mathcal{B}^\infty(\tilde{\mathfrak{W}}_{\delta}(U,\Gamma);\tilde{\Lambda}^{p,q})$ be a cocycle of the kind (2.2) whose pull-back first to the edge $\tilde{U}$ and then to U via the map (2.4), is equal to the standard representative f of $\dot{f}$ modulo forms of bidegree $(p+\ell,p-\ell)$ with $1 \le \ell \le q$. According to Th. 2.5, if U', Γ' and δ' are as indicated in the statement, there is a cocycle $f_1 \in \mathcal{B}^\infty(\mathfrak{W}_{\delta'}(U',\Gamma');\tilde{\Lambda}^{p,q})$, endowed with property (2.14), and a (p,q-1)-form $\tilde{g}_1 \in \mathcal{B}^\infty(\mathfrak{W}_{\delta'}(U',\Gamma');\tilde{\Lambda}^{p,q-1})$ such that $\tilde{f}-\tilde{f}_1 = (\bar{\partial}_z+d_t)g_1$ in $\tilde{\mathfrak{W}}_{\delta'}(U',\Gamma')$. Because of (2.14) the pull-back f_1 of $\tilde{f}_1$ to U' is a standard cocycle. Call g_1 the pull-back of $\tilde{g}_1$ to U'. We have:

$$f - f_1 = Lg_1 + h, \tag{2.25}$$

where we can write

$$g_1 = \sum_{\ell=0}^{q-1} g_{(\ell)}, \qquad h = \sum_{\ell=1}^{q} h_{(\ell)}, \tag{2.26}$$

where $q_{(\ell)} \in C^\infty(U';\Lambda^{p+\ell,q-\ell-1})$, $h_{(\ell)} \in C^\infty(U';\Lambda^{p+\ell,q-\ell})$. Since the bidegree of $Lg_{(\ell)}$ is equal to $(p+\ell,q-\ell)$ we must have

$$Lg_{(\ell)} = -h_{(\ell)}, \qquad \ell = 1,\dots,q-1. \tag{2.27}$$

But then (2.25) entails $h_{(q)} \equiv 0$ and

$$f - Lg_{(0)} = f_1$$

whence the assertion if we take $g = g_{(0)}$. ■

3. TRANSFORMATION OF WEDGES UNDER ISOMORPHISMS

We continue to deal with the (germ of a) C^∞ manifold $\mathfrak{M}$ equipped with a local hypo-analytic structure $\mathcal{A}$. As before the cotangent structure bundle of $(\mathfrak{M},\mathcal{A})$ will be denoted by T' while the tangent structure bundle will be denoted by $\mathcal{V}$ (see Section 1).

We must now introduce the <u>characteristic set</u> T^o of the hypo-analytic structure $\mathcal{A}$; $T^o = T' \cap T^*\mathfrak{M}$. In general T^o is not a vector bundle over $\mathfrak{M}$, the dimension of its fibres might vary from point to point. It is the set of pairs $(\wp,\omega) \in T^*\mathfrak{M}$ such that $\langle \omega, \mathrm{Re}\, L|_\wp \rangle = 0$ whenever the section L of $\mathcal{V}$ over some neighborhood of the point $\wp \in \mathfrak{M}$.

Let then (U,Z) be a hypo-analytic chart in the hypo-analytic manifold $(\mathfrak{M},\mathfrak{A})$ (see Def. 1.1 and following remarks); the chart is "centered" at the point 0, called <u>the origin</u>. We shall assume that U is the domain of real coordinates x^i, t^j $(1 \le i \le m,\ 1 \le j \le n)$ all equal to zero at 0.

We shall reason under the hypotheses (1.12) and (1.13). It is also convenient to assume that the Jacobian matrix of Z with respect to x, $Z_x = \partial Z/\partial x$, is nonsingular at every point of U. Over the open set U the structure bundle T' is spanned by $dZ^1,\dots,dZ^m$.

In order that a covector $\zeta\cdot dZ$ of $T'_\wp$ $(\wp \in U)$ belong to the fibre $T^o_\wp$ is is necessary and sufficient that there exist $\xi \in \mathbb{R}_m$, $\tau \in \mathbb{R}_n$ such that

(3.1) $$\zeta\cdot dZ = \xi\cdot dx + \tau\cdot dt$$

at $\wp$. Taking into account the expressions (1.12) we see that (3.1) means that

$$\xi = (I+\sqrt{-1}\,{}^t\phi_x)\zeta, \qquad \tau = \sqrt{-1}\,{}^t\phi_t\zeta,$$

where ${}^t\phi_x$ (resp., ${}^t\phi_t$) denote the transpose of the differential ϕ_x (resp. ϕ_t) and I is the identity map of $\mathbb{C}^m$. In other words

$$\zeta = (I+\sqrt{-1}\,{}^t\phi_x)^{-1}\xi,$$

and

$$\tau = \sqrt{-1}\,{}^t\phi_t(I-\sqrt{-1}\,{}^t\phi_x)(I-{}^t\phi_x^2)^{-1}\xi.$$

It is convenient to introduce the notation

$$\theta = (I + {}^t\phi_x^2)^{-1}\xi,$$

and to write

(3.2) $$\zeta = (I - \sqrt{-1}\,{}^t\phi_x)\theta,$$

(3.3) $$\tau = {}^t\phi_t\,{}^t\phi_x\theta, \qquad {}^t\phi_t\theta = 0.$$

It follows from (1.13) that we have, at the origin (i.e., when $\wp = 0$)

(3.4) $$\zeta = \xi, \qquad \tau = 0, \qquad {}^t\phi_t\xi = 0.$$

It is often convenient to identify the fibre of T^o at 0 to a linear subspace of $\mathbb{R}^m$. Keep in mind, however, that such an identification does depend on the choice of the hypo-analytic chart and of the coordinates.

<u>Definition 3.1</u>. <u>We shall say that an open, convex and nonempty cone $\Gamma \subset \mathbb{R}^m$ (with vertex at the origin) is definite-negative with respect</u>

<u>to a characteristic cotangent vector</u> ξ^o <u>to</u> $\mathfrak{m}$ <u>at</u> 0 <u>if, for some</u> $c_o > 0$,

(3.5) $$\xi^o \cdot v < -c_o|v|, \quad \forall\, v \in \Gamma.$$

We wish to explore the effect, on wedges $\tilde{\mathfrak{w}}_\delta(U,\Gamma)$ and on germs of wedges $\tilde{\mathfrak{w}}(\mathfrak{m},\Gamma)$ defined by a cone Γ which is definite-negative with respect to some covector ξ^o, of a change of hypo-analytic chart (U,Z) and a concomitant change of local coordinates.

Thus let $(U_\#, Z_\#)$ be another hypo-analytic chart, and $x_\#^i$, $t_\#^j$ $(1 \le i \le m,\ 1 \le j \le n)$ local coordinates in the neighborhood of 0, $U_\#$. We assume that the $x_\#^i$, the $t_\#^j$ and the map $Z_\#$ all vanish at 0. The manner in which we have defined our wedges compels us to assume also that, whatever $i = 1,\dots,m$,

(3.6) $$Z_\#^i = x_\#^i + \sqrt{-1}\phi_\#^i(x_\#, t_\#),$$

(3.7) $$\phi_\#^i\big|_0 = 0, \qquad d_{x_\#}\phi_\#^i\big|_0 = 0.$$

Recall that there is a biholomorphic map H of an open neighborhood of 0 in $\mathbb{C}^m$ onto another such neighborhood, such that $Z_\# = H \circ Z$. Therefore

(3.8) $$x_\# = \operatorname{Re} H(Z(x,t)), \qquad t_\# = g(x,t),$$

where g is a C^∞ map an open neighborhood of 0 in $\mathbb{R}^{m+n}$, into $\mathbb{R}^n$. Of course, $H(0) = 0$, $g(0,0) = 0$ and $D(x_\#, t_\#)/D(x,t) \ne 0$.

Lemma 3.2. <u>There is a linear map</u> $K: \mathbb{R}^m \to \mathbb{R}^n$ <u>such that if we define</u>

(3.9) $$G(x,y,t) = g(x,t) - K[y - \phi(x,t)],$$

<u>then</u>

(3.10) $$\det G_t(0,0,0) \ne 0.$$

Note that putting $x = 0$, $t = 0$ in (3.9) yields $G(0,0,0) = 0$.

<u>Proof</u>: According to (3.9), $G_t = g_t + K\phi_t$. Call A the transpose of $\phi_t\big|_0: \mathbb{R}^n \to \mathbb{R}^m$, B that of $g_t\big|_0: \mathbb{R}^n \to \mathbb{R}^n$. We have

$$2\partial x_\#/\partial t = i[H'(x+i\phi(x,t)) - \bar{H}'(x-i\phi(x,t))]\phi_t(x,t) \qquad (i = \sqrt{-1}),$$

whence

$$\partial x_\#/\partial t\big|_0 = C\,{}^tA: \mathbb{R}^n \to \mathbb{R}^m, \quad \text{with } C = \operatorname{Im} H'(0).$$

We know that the Jacobian matrix of the change of variables $(x,t) \to (x_\#, t_\#)$ is nonsingular. This demands that the map

$$\mathbb{R}_m \times \mathbb{R}_n \ni (\xi, \tau) \to A({}^tC\xi) + B\tau \in \mathbb{R}_n$$

be surjective. But then the map $(\xi,\tau) \to A\xi + B\tau$ must also be. If we take $E = \mathbb{R}^m$, $F = \mathbb{R}^n$ and the linear map K in Lemma 3.2 to be the transpose of the linear map K' below, we see that Lemma 3.2 is a consequence of the following:

Lemma 3.3. *Let* E, F *be two finite-dimensional real vector spaces and* $A: E \to F$, $B: F \to F$ *be two linear maps. If* $F = A(E) + B(F)$ *then there is a linear map* $K': F \to E$ *such that* $AK' + B$ *is an automorphism of* F.

Proof of Lemma 3.3. Let F_o be a linear subspace of F such that $F = F_o \oplus \operatorname{Ker} B$ and let F_1 be a linear subspace of $A(F)$ such that $F = B(F) \oplus F_1$. We can select a linear subspace E_1 of E such that $A|_{E_1}$ is a bijection of E_1 onto F_1. This implies $\dim E_1 = \dim F_1 = \dim \operatorname{Ker} B$. Let K_o be any linear bijection of $\operatorname{Ker} B$ onto E_1 and define K' as being equal to 0 on F_o and to K_o on $\operatorname{Ker} B$. An arbitrary element of F has the form $y = By_o + y_1$ with $y_o \in F_o$ and $y_1 \in F_1$. There is a unique element $x_1 \in E_1$ such that $y_1 = Ax_1$ and there is a unique element $y' \in \operatorname{Ker} B$ such that $x_1 = K'y'$. We obtain, thus, $y = AK'y' + By_o = (AK' + B)(y' + y_o)$. ■

Remark 3.4. It is clear that the map K' in Lemma 3.3 and therefore also the map G in Lemma 3.2 are not unique. ■

Call $\tilde{U} \subset \mathbb{C}^m \times \mathbb{R}^n$ the image of U under the map $(x,t) \to (Z(x,t),t)$ and $\tilde{U}_\#$ the analogue for the objects with subscript $_\#$. The map $(x,t) \to (x_\#, t_\#)$ defined in (3.8) can be transfered, by means of the two maps $(x,t) \to (Z(x,t),t)$ and $(x_\#,t_\#) \to (Z_\#(x_\#,t_\#),t_\#)$, as a diffeomorphism $\mathcal{J}$ of some open neighborhood of 0 in $\tilde{U}$ onto one in $\tilde{U}_\#$. Lemma 3.2 has the following consequence:

Corollary 3.5. *Let* H *be the biholomorphism introduced above and* G *the map* (3.9). *Then the map*

$$(z,t) \to (H(z), G(x,y,t)) \qquad (z = x + \sqrt{-1}y), \tag{3.11}$$

is a diffeomorphism of an open neighborhood of the origin in $\mathbb{C}^m \times \mathbb{R}^n$ *onto another such neighborhood. Its restriction to some open neighborhood of* 0 *in* $\tilde{U}$ *is equal to* $\mathcal{J}$.

Remark 3.6. It is important to note that a diffeomorphism such as (3.11) defines an automorphism of the germ of $\mathbb{C}^m \times \mathbb{R}^n$ at 0 equipped with its standard hypo-analytic structure, which can be defined by the coordinate system $(z^1, \ldots, z^m)$, as well as by the biholomorphism $H(z)$. ■

Let $\zeta \cdot dZ = \xi \cdot dx$ be any nonzero characteristic cotangent vector to $\mathfrak{m}$ at 0. It can also be represented by $\zeta_{\#} \cdot dZ_{\#} = \xi_{\#} \cdot dx_{\#}$. In passing note that because of our hypotheses (3.6), (3.7), the analogue of (3.4) is valid with subscripts #. We have $\zeta_{\#} \cdot dZ_{\#} = \zeta_{\#} \cdot H_Z(0)dZ = {}^tH_Z(0)\xi_{\#} \cdot dx$, whence

$$\xi = {}^tH_Z(0)\xi_{\#}, \tag{3.12}$$

which shows, among other things, that ${}^tH_Z(0)\xi_{\#}$ is a real covector.

Assume now that U and δ are small enough that (3.11) defines a diffeomorphism of $\tilde{\mathfrak{w}}_{\delta}(U,\Gamma)$ onto an open subset of $\mathbb{C}^m \times R^n$.

Proposition 3.7. Let the cone Γ be definite-negative (Def. 3.1) with respect to a characteristic cotangent vector ξ^o to $\mathfrak{m}$ at 0. If the neighborhood $U_{\#}$ and the number $\delta_{\#} > 0$ are sufficiently small the image of $\tilde{\mathfrak{w}}_{\delta}(U,\Gamma)$ under the diffeomorphism (3.11) contains a wedge $\tilde{\mathfrak{w}}_{\delta_{\#}}(U_{\#},\Gamma_{\#})$ defined by means of a cone $\Gamma_{\#}$ which is definite-negative with respect to $\xi^o_{\#} = {}^tH_Z(0)^{-1}\xi^o$.

Proof: We contend that to every $(x,t,v) \in U \times \mathbb{R}^m$ sufficiently close to the origin there is a unique point $(x',t',v_{\#}) \in U \times \mathbb{R}^m$ such that

$$\begin{aligned} &H(Z(x',t')) + \sqrt{-1}v_{\#} = H(Z(x,t) + \sqrt{-1}v), \\ &g(x',t') = G(x,\phi(x,t) + v,t). \end{aligned} \tag{3.13}$$

Indeed, the equations (3.13) have the solution

$$x' = x, \quad t' = t, \quad v_{\#} = 0 \ \text{when} \ v = 0. \tag{3.14}$$

On the other hand, by our hypotheses, the Jacobian determinant of the left-hand sides with respect to $(x',t',v_{\#})$ does not vanish, provided, of course, $(x',t',v_{\#})$ is close enough to the origin. It suffices then to apply the implicit function theorem.

Let us differentiate with respect to v both sides of the first equation (3.13) and "freeze" the result at the origin. For the sake of brevity we shall use the notation

$$A = \phi_t\big|_0, \quad B = g_x\big|_0, \quad C = g_t\big|_0, \quad M = H_Z(0),$$
$$X = (\partial x'/\partial v)\big|_0, \quad \theta = (\partial t'/\partial v)\big|_0 .$$

Observe that (1.13) implies

$$(\partial/\partial v)\phi(x',t')\big|_0 = A\theta. \tag{3.15}$$

We may write:

$$\operatorname{Im}\{M(I+i(X+iA\theta))\} = 0. \tag{3.16}$$

Since

$$v_{\#} = \mathrm{Im}\{H(Z(x,t)+iv) - H(Z(x',t'))\},$$

it follows from (3.14) that, for some constant $C > 0$, independent of v,

(3.17) $$|v_{\#}| \le C|v|.$$

Direct computation shows that

$$\partial v_{\#}/\partial v|_0 = \mathrm{Re}\, M - \mathrm{Im}\, M(X+iA\theta),$$

which is to say,

(3.18) $$\partial v_{\#}/\partial v|_0 = \mathrm{Re}\{M(I + i(X+iA\theta))\}.$$

If we take (3.16) into account, (3.18) can be rewritten as

(3.19) $$\partial v_{\#}/\partial v|_0 = M(I + i(X+iA\theta)).$$

Observe that

$$\xi^o_{\#}\cdot Mv = {}^tH_z(0)\xi^o_{\#}\cdot v = \xi^o\cdot v.$$

On the other hand, by virtue of (3.4),

$$\xi^o_{\#}\cdot M(X+iA\theta)v = \xi^o\cdot Xv + i\,{}^t(\phi_t|_0)\xi^o\cdot\theta v = \xi^o\cdot Xv.$$

Thus,

$$\xi^o_{\#}\cdot(\partial v_{\#}/\partial v|_0)v = \xi^o\cdot(v+iXv).$$

But since the left-hand side is real we must have

(3.20) $$\xi^o_{\#}\cdot(\partial v_{\#}/\partial v|_0)v = \xi^o\cdot v.$$

We derive that there is a constant $C_1 > 0$ independent of v such that

$$\xi^o_{\#}\cdot v_{\#} \le \xi^o\cdot v + C_1|v|^2,$$

and therefore, by virtue of (3.5),

$$\xi^o_{\#}\cdot v_{\#} < -c_o|v|/2,$$

provided $C_1|v| < c_o/2$. If we combine this with (3.17) we get

(3.21) $$\xi^o_{\#}\cdot v_{\#} < -c_{\#}|v_{\#}|,$$

where $c_{\#} = c_o/2C$.

We rephrase the first equation (3.13) by setting

(3.22) $$z_{\#} = H(Z(x',t')).$$

We obtain

(3.23) $$z_{\#}+iv_{\#} = H(Z(x,t)+iv).$$

Set then

(3.24) $$t_{\#} = G(x,\phi(x,t)+v,t).$$

This shows that $(z_{\#}+iv_{\#},t_{\#})$ is the image of $(Z(x,t)+iv,t)$ under the map (3.11). By the second equation (3.13) we also get

(3.25) $$t_{\#} = g(x',t').$$

If now we combine (3.22) and (3.25) we reach the conclusion that

(3.26) $$z_{\#} = Z_{\#}(x_{\#},t_{\#})$$

(cf. (3.6)), i.e., $(z_{\#},t_{\#})$ belongs to $\tilde{U}_{\#}$ (assuming, as we may, that $(x_{\#},t_{\#}) \in U_{\#}$). Finally, the inequality (3.21) obviously defines a cone $\Gamma_{\#}$ which is definite-negative with respect to $\xi^{o}_{\#}$.

In summary we have shown that if we are willing to contract U about 0, the image under the map (3.11) of $\tilde{\mathfrak{w}}_{\delta}(U,\Gamma)$ is contained in $\tilde{\mathfrak{w}}_{\delta_{\#}}(U_{\#},\Gamma_{\#})$ where $\delta_{\#} = C\delta$ (C is the constant in (3.17)). But the same argument applies to the inverse of the map (3.11): if $U_{\#}$, $\delta_{\#}$ and $c_o/c_{o\#}$ are suitably small then the image of $\tilde{\mathfrak{w}}_{\delta_{\#}}(U_{\#},\Gamma_{\#})$ under the inverse of the map (3.11) is contained in $\tilde{\mathfrak{w}}_{\delta}(U,\Gamma)$ or, equivalently, the image of $\tilde{\mathfrak{w}}_{\delta}(U,\Gamma)$ under the map (3.11) contains $\tilde{\mathfrak{w}}_{\delta_{\#}}(U_{\#},\Gamma_{\#})$. ■

It is convenient to re-interpret Prop. 3.7 in the language of germs (of sets and of maps). Recall that the wedge $\tilde{\mathfrak{w}}(\mathfrak{m},\Gamma)$ is defined by means of the hypo-analytic chart (U,Z) and of the coordinates t^{j}. Call $\tilde{\mathfrak{w}}_{\#}(\mathfrak{m},\Gamma)$ the analogue, defined by the chart $(U_{\#},Z_{\#})$ and the coordinates t^{j}_{*}. It is also time to take into account the fact that $\xi^{o}_{\#} = {}^{t}H_{z}(0)^{-1}\xi^{o}$ represents, in the latter chart, the same characteristic covector that ξ^{o} represents in the chart (U,Z). Below we denote by $\bar{\omega}$ the covector in question.

<u>Corollary 3.8</u>. <u>Let the cone</u> Γ <u>be definite-negative with respect to the characteristic covector</u> $\bar{\omega}$ <u>at</u> 0. <u>The image of</u> $\tilde{\mathfrak{w}}(\mathfrak{m},\Gamma)$ <u>under the germ of the isomorphism</u> (3.11) <u>contains a germ of wedge</u> $\tilde{\mathfrak{w}}_{\#}(\mathfrak{m},\Gamma_{\#})$ <u>defined by means of a cone</u> $\Gamma_{\#}$ <u>which is also definite-negative with respect to</u> $\bar{\omega}$.

4. THE MICROSUPPORT OF A COHOMOLOGY CLASS

Consider a germ of wedge $\tilde{\mathfrak{w}}(\mathfrak{m},\Gamma) \subset \mathbb{C}^{m}\times\mathbb{R}^{n}$ as before. It follows from Prop. 2.1 that the pull-back map (2.12) induces a map

(4.1) $$H^{p,q}(\tilde{\mathfrak{w}}(\mathfrak{m},\Gamma)) \to H'^{\,p,q}$$

to which we shall also refer as the <u>pull-back to</u> $\mathfrak{m}$. We recall that the $H'^{p,q}$ are the cohomology spaces of the complex d' on $\mathfrak{M}$, (1.8), and that the $H^{p,q}(\tilde{\mathfrak{W}}(\mathfrak{M},\Gamma))$ are those of the complex $\bar{\partial}_z+d_t$ on $\tilde{\mathfrak{W}}(\mathfrak{M},\Gamma)$, (2.9).

Definition 4.1. We shall say that a cohomology class $[h] \in H'^{p,q}$ extends to the wedge $\tilde{\mathfrak{W}}(\mathfrak{M},\Gamma)$ if $[h]$ belongs to the image of the pull-back map (4.1).

It is equivalent to say that $[h] \in H'^{p,q}$ extends to $\tilde{\mathfrak{W}}(\mathfrak{M},\Gamma)$ or that there is a cocycle $\dot{h} \in \mathcal{C}^{\infty}(\Lambda^{p,q})$ in the class $[h]$ which extends to the wedge $\tilde{\mathfrak{W}}(\mathfrak{M},\Gamma)$ (Def. 2.3).

Def. 4.1 lacks invariance, since the wedge $\tilde{\mathfrak{W}}(\mathfrak{M},\Gamma)$ is defined by means of the hypo-analytic chart (U,Z) and of the coordinates t^j. But let $\bar{\omega}$ be a characteristic cotangent vector to $\mathfrak{M}$ at the origin and consider the following property:

(4.2) There is an open convex cone $\Gamma \subset \mathbb{R}^m$ which is definite-negative with respect to $\bar{\omega}$ (Def. 3.1) and such that $[h]$ extends to $\tilde{\mathfrak{W}}(\mathfrak{M},\Gamma)$.

Let $\dot{\tilde{f}} \in \mathcal{B}^{\infty}(\tilde{\mathfrak{W}}(\mathfrak{M},\Gamma);\tilde{\Lambda}^{p,q})$ be a cocycle whose pull-back $\dot{f} \in \mathcal{C}^{\infty}(\Lambda^{p,q})$ to $\mathfrak{M}$ is a cocycle in the class $[h]$. Let $\tilde{f} \in \mathcal{B}^{\infty}(\tilde{\mathfrak{W}}_{\delta}(U,\Gamma);\tilde{\Lambda}^{p,q})$ be the standard representative of $\dot{\tilde{f}}$ in $\tilde{\mathfrak{W}}_{\delta}(U,\Gamma)$ (we are assuming that U and δ are suitably small). The pull-back of $\tilde{f}$ to the edge $\tilde{U}$ is a form of bidegree (p,q) whose pull-back to U under the diffeomorphism $(x,t) \to (Z(x,t),t)$ we shall call f_o. This form f_o need not be standard; but its class $\dot{f}_o$ in $\mathcal{C}^{\infty}(U;\Lambda^{p,q})$ has a standard representative which we denote by f. Of course f is L-closed. The germ of $\dot{f}_o$ at the origin is the cocycle $\dot{f} \in \mathcal{C}^{\infty}(\Lambda^{p,q})$ so denoted at the beginning of this paragraph.

Consider then a new hypo-analytic chart $(U_{\#},Z_{\#})$ and new coordinates $x_{\#}^j$, $t_{\#}^j$ (satisfying (3.6) and (3.7)). Let the map $(x,t) \to (x_{\#},t_{\#})$ be defined by (3.8). Cor. 3.5 tells us that it is "induced" by the diffeomorphism (3.11). We apply Prop. 3.7. After choosing $U_{\#}$, $\Gamma_{\#}$ and $\delta_{\#}$ as required by Prop. 3.7 we can push forward, via (3.11), the differential form $\tilde{f}$ from $\tilde{\mathfrak{W}}_{\delta}(U,\Gamma)$ to $\tilde{\mathfrak{W}}_{\delta_{\#}}(U_{\#},\Gamma_{\#})$; call $\tilde{f}_{\#}$ the resulting differential form (which has bidegree (p,q)). The original form $\tilde{f}$ is $(\bar{\partial}_z+d_t)$-closed and so is the form $\tilde{f}_{\#}$, due to the fact that the diffeomorphism (3.11) is a local automorphism of the standard hypo-analytic structure on $\mathbb{C}^m \times \mathbb{R}^n$.

The pull-back of $\tilde{f}_{\#}$ to $U_{\#}$ under the diffeomorphism $(x_{\#},t_{\#}) \to (Z_{\#}(x_{\#},t_{\#}),t_{\#})$ is equal to the form f_o since it could have been obtained simply by making the change of coordinates (3.8).

We reach the following conclusion:

(4.3) <u>The cohomology class</u> $[h]$ <u>extends to the wedge</u> $\tilde{\mathfrak{W}}_{\#}(\mathfrak{M},\Gamma_{\#})$.

We recall (Prop. 3.7, also Cor. 3.8) that <u>the cone</u> $\Gamma_{\#}$ <u>is definite-negative with respect to the characteristic covector</u> $\bar{\omega}$. Thus, in a sense, Property (4.2) is independent of the choice of the hypo-analytic chart (U,Z) and of the coordinates x^i, t^j.

<u>Definition 4.2.</u> <u>Let</u> $\bar{\omega}$ <u>be a characteristic cotangent vector to</u> $\mathfrak{M}$ <u>at the origin.</u> <u>We shall say that the cohomology class</u> $[h] \in H'^{p,q}$ <u>vanishes at</u> $\bar{\omega}$ <u>if there is a finite family of cohomology classes</u> $[h_\iota] \in H'^{p,q}$ $(1 \le \iota \le J)$ <u>such that</u>

$$[h] = \sum_{\iota=1}^{J} [h_\iota], \tag{4.4}$$

<u>and such that every cohomology class</u> $[h_\iota] \in H'^{p,q}$ <u>has Property</u> (4.2).

<u>The set of characteristic covectors to</u> $\mathfrak{M}$ <u>at the origin, at which the class</u> $[h]$ <u>does not vanish, will be called the microsupport of</u> $[h]$ <u>and denoted by</u> $\mu\mathrm{supp}\,[h]$.

If a cone Γ is definite-negative with respect to $\bar{\omega}$ it is also definite-negative with respect to all characteristic covectors in some conic neighborhood of $\bar{\omega}$ in T_0^o. It follows from this that the <u>microsupport of a cohomology class is a closed conic subset of</u> $T_0^o \backslash 0$.

The reader will notice that all the definitions introduced here are relative to the origin. Thus only the fibre at 0 of the characteristic set T^o enters in what precedes, and perhaps what we have chosen to call "microsupport" ought to be called "microsupport at the origin". But recall that the concept of a cohomology class, defined as it is by means of the germs at 0 of sections of $\Lambda^{p,q}$, is itself tied to the central point. In dealing with a "true" hypo-analytic manifold $\mathfrak{M}$ (rather than with the germ of one at 0) one can repeat at each point what we have done at 0. This leads straightaway to a system of sheaves and to sheaf-theoretical notions whose definition is cumbersome and yet rather routine.

<u>Example 4.3.</u> Consider the case $q = 0$, i.e., $[h] \in H'^{p,0}$ and for the sake of simplicity, take also $p = 0$. There is a single cocycle $\dot{h}$ in the class $[h]$ since there are no 0-coboundaries $\neq 0$; $\dot{h}$ is the germ at 0 of a solution h in some open neighborhood U of 0 in $\mathfrak{M}$. To say that $[h]$ extends to $\tilde{\mathfrak{W}}(\mathfrak{M},\Gamma)$ is the same as saying that, if U and $\delta > 0$ are small enough, there is a 0-cocycle $\tilde{h} \in \mathcal{B}^\infty(\tilde{\mathfrak{W}}_\delta(U,\Gamma))$ whose restriction to the edge $\tilde{U}$ of $\tilde{\mathfrak{W}}_\delta(U,\Gamma)$ is

equal to $\tilde{h}_o$, the push-forward of h under the map $(x,t) \to (Z(x,t),t)$. But to say that $(\bar{\partial}_z + d_t)\tilde{h} = 0$ is simply to say that the function $\tilde{h}$ is independent of t and that it is a holomorphic function of z. In summary $h = \tilde{h}_o \circ Z$ and $\tilde{h}_o$ extends holomorphically to the wedge $\tilde{\mathfrak{W}}_\delta(U,\Gamma)$.

With $[h] \in H'^{0,0}$ as before suppose that (4.4) holds. It means exactly that the point $(0,\bar{\omega}) \in T^o$ does not belong to the hypo-analytic wave-front set ([BCT], Def. 1.2, Ch. II) of the solution h (whose germ at the origin is the cocycle in the class [h]). ■

CHAPTER III

THE FBI TRANSFORM IN A HYPO-ANALYTIC STRUCTURE

1. THE LIFT OF THE HYPO-ANALYTIC STRUCTURE TO THE COTANGENT STRUCTURE BUNDLE

Let $(\mathfrak{M},\mathfrak{A})$ be the (germ of) hypo-analytic manifold we have been dealing with until now; and let T' be its cotangent structure bundle. We recall that T' is a complex vector subbundle of $\mathbb{C}T^*\mathfrak{M}$; if (U,Z) is a hypo-analytic chart, T' is spanned over U by $dZ^1,\ldots,dZ^m$.

The local hypo-analytic structure $\mathfrak{A}$ on $\mathfrak{M}$ defines, in a natural fashion, a local hypo-analytic structure $\mathfrak{A}'$ on T': view T' as a germ of vector bundle over the germ of manifold $\mathfrak{M}$ at the point $0 \in \mathfrak{M}$; as defining hypo-analytic charts take the charts $(T'|_U, Z, \zeta)$ where U is an open neighborhood of 0 in $\mathfrak{M}$, $Z = (Z^1,\ldots,Z^m)\colon U \to \mathbb{C}^m$ such that $dZ^1,\ldots,dZ^m$ form a basis of T' over U, and $\zeta_1,\ldots,\zeta_m$ are the complex coordinates with respect to this basis, in the fibres of T' at point of U. We write $\zeta = (\zeta_1,\ldots,\zeta_m)$ and (Z,ζ) is a (hypo-analytic) map $T'|_U \to \mathbb{C}^m \times \mathbb{C}_m$.

Let $(U_\#, Z_\#)$ be a different hypo-analytic chart in the base $\mathfrak{M}$ and let F be a biholomorphic map of an open neighborhood of $Z(U \cap U_\#)$ in $\mathbb{C}^m$ onto one of $Z_\#(U \cap U_\#)$ such that $Z_\# = F \circ Z$ in $U \cap U_\#$. Then

$$\zeta^\# \cdot dZ_\# = \sum_{i=1}^{m} \zeta_i^\# dZ_\#^i = \sum_{i,j=1}^{m} \zeta_i^\# [(\partial F^i/\partial z^j) \circ Z]\, dZ^j,$$

and thus the change of hypo-analytic chart in T' is expressed by

(1.1) $$Z_{\#} = F \circ Z, \qquad \zeta^{\#} = ({}^tF_z \circ Z)^{-1}\zeta.$$

The map $(z,\zeta) \to (F(z), {}^tF_z(z)^{-1}\zeta)$ is clearly a biholomorphism from an open neighborhood in $\mathbb{C}^m \times \mathbb{C}_m$ of the image of $T'|_{U \cap U_{\#}}$ under the map (Z,ζ) onto one of the image of the same set under the map $(Z_{\#},\zeta^{\#})$.

Definition 1.1. <u>We shall refer to the local hypo-analytic structure $\mathcal{G}'$ on the cotangent structure bundle T' as the lift to T' of the local hypo-analytic structure $\mathcal{G}$ on $\mathfrak{M}$.</u>

The hypo-analytic structure $\mathcal{G}'$ induces, on each fibre of T', its natural complex structure (which is inherited from the complex structure of the fibre of $\mathbb{C}T^*\mathfrak{M}$ at the same point). If we identify $\mathfrak{M}$ to the zero section of T' then $\mathcal{G}'$ induces on $\mathfrak{M}$ the hypo-analytic structure $\mathcal{G}$.

The dimension of the C^∞ manifold T' is equal to $3m+n$. Let us denote by ${}^2T'$ the structure bundle of the hypo-analytic structure $\mathcal{G}'$. It is the vector subbundle of $\mathbb{C}T^*T'$ spanned, over an arbitrary local chart (U,Z) of $\mathfrak{M}$, by $dZ^1,\dots,dZ^m$, $d\zeta_1,\dots,d\zeta_m$. Thus it has fibre dimension (over $\mathbb{C}$) equal to $2m$. We shall denote by ${}^2\mathcal{V}$ its orthogonal for the duality between tangent and cotangent vectors on T'; the fibre dimension of ${}^2\mathcal{V}$ is equal to $m+n$. Notice that the Cauchy-Riemann vector fields along the fibres of T' are section of ${}^2\mathcal{V}$. They span a vector subbundle of ${}^2\mathcal{V}$ of fibre dimension m.

Let (U,Z) be a hypo-analytic chart in $\mathfrak{M}$ and, as before, let ζ_j denote the complex coordinates, with respect to the basis $dZ^1,\dots,dZ^m$, in the fibres of T' at points of U. Suppose now that U is the domain of local coordinates x^i, t^j $(i=1,\dots,m,\ j=1,\dots,n)$ and that (1.12) and (1.13), Ch. II, hold true. Moreover, let us make the following hypothesis

(1.2) <u>The Jacobian matrix $Z_x = (\partial Z^i/\partial x^j)_{1\le i,j\le m}$ is nonsingular at every point of U.</u>

We shall take U in the product form:

(1.3) $$U = U_1 \times U_2,$$

with U_1 (resp. U_2) an open neighborhood of the origin in x-space $\mathbb{R}^m$ (resp., t-space $\mathbb{R}^n$).

The map (2.4), Ch. II, defines an embedding of U into $\mathbb{C}^m \times \mathbb{R}^n$. We can now associate with the hypo-analytic chart (U,Z) an embedding of $T'|_U$ into $\mathbb{C}^m \times \mathbb{R}^n \times \mathbb{C}_m$, namely the map

$$((x,t),\zeta\cdot dZ|_{(x,t)}) \to (Z(x,t),t,\zeta). \tag{1.4}$$

The image of $T'|_U$ under (1.4) is equal to $\tilde{U}\times\mathbb{C}_m$. If we are willing to identify $\mathbb{C}^m$ to its dual, $\mathbb{C}_m$, and if we write (z,ζ,t) instead of $((z,t),\zeta)$, then (1.4) is an embedding of $T'|_U$ into $\mathbb{C}^{2m}\times\mathbb{R}^n$, the analogue for $T'|_U$ of the map (2.4), Ch. II.

We recall that a submanifold X of the hypo-analytic manifold $(\mathfrak{M},\mathfrak{A})$ is said to be <u>maximally real</u> if the pull-back to X of cotangent vectors to $\mathfrak{M}$ at points of X induces an isomorphism of $T'|_X$ onto $\mathbb{C}T^*X$. This is equivalent to saying that, if $(U_\#,Z_\#)$ is any hypo-analytic chart, the pull-backs of $dZ^1_\#,\dots,dZ^m_\#$ to $X\cap U_\#$ are $\mathbb{C}$-linearly independent.

We return to the chart (U,Z). For each $t_o \in U_2$ denote by X_t the submanifold of $\mathfrak{M}$ contained in U and defined by the equation $t = t_o$; X_{t_o} is a maximally real submanifold of $\mathfrak{M}$. The map (2.4), Ch. II, $(x,t) \to (Z(x,t),t)$, transforms X_{t_o} into a maximally real submanifold $\tilde{X}_{t_o}$ of $\tilde{U}$, the image of U. If we regard $\tilde{X}_{t_o}$ as a submanifold of $\mathbb{C}^{m_o}$ (via the map $(z,t) \to z$) it is a totally real submanifold of dimension m. Note that now the natural diffeomorphism of X_t onto $\tilde{X}_t$ is the map $x \to Z(x,t)$.

The pull-back to X_t of cotangent vectors to $\mathfrak{M}$ defines the natural map $\mathbb{C}T^*\mathfrak{M}|_{X_t} \to \mathbb{C}T^*X_t$; it induces an isomorphism $T'|_{X_t} \cong \mathbb{C}T^*X_t$. In the basis $dx^1,\dots,dx^m$, the isomorphism under consideration is expressed by the map

$$(x,\zeta\cdot dZ) \to (x,{}^tZ_x(x,t)\zeta\cdot dx). \tag{1.5}$$

We have a natural isomorphism of $T'|_{X_t}$ onto $\tilde{X}_t\times\mathbb{C}_m$, namely $(x,\zeta\cdot dZ) \to (z,\zeta)$ with $z = Z(x,t)$. As t sweeps U_2 this defines an isomorphism of $T'|_U$ onto $\tilde{U}\times\mathbb{C}_m$ which is nothing but the map (1.4).

We are now going to pay particular attention to a special family of maximally real submanifolds of $T'|_U$. Let $t \in U_2$ arbitrary.

<u>Notation 1.2. We shall denote by</u> $\mathbb{R}T'_t$ <u>the preimage of the real cotangent bundle of</u> X_t, T^*X_t, <u>under the isomorphism</u> (1.5).

Thus $\mathbb{R}T'_t$ is a real vector subbundle of $T'|_{X_t}$; $\dim \mathbb{R}T'_t = 2m$. Since the pull-backs to $\mathbb{R}T'_t$ of the dZ^i and $d\zeta_j$ $(1 \le i,j \le m)$ are $\mathbb{C}$-linearly independent we see that, for each $t \in U_2$, $\mathbb{R}T'_t$ is a maximally real submanifold of T' (T' is equipped with the hypo-analytic structure $\mathcal{a}'$). As t sweeps U_2 the union of the $\mathbb{R}T'_t$ makes up a smooth submanifold of $T'|_U$ which we shall denote by $\mathcal{R}_U$.

Suppose that we use, in the fibres of T^*X_t, the (real) coordinates $\xi_1,\dots,\xi_m$ with respect to the basis $dx^1,\dots,dx^m$. Then the natural isomorphism of T^*X_t onto $\mathbb{R}T'_t$ is the map

$$(1.6)\qquad (x,\xi\cdot dx) \to (x,\zeta\cdot dZ), \qquad \zeta = {}^tZ_x(x,t)^{-1}\xi.$$

If we follow up this map with the diffeomorphism $(x,\zeta\cdot dZ) \to (Z(x,t),\zeta)$ of $T'|_{X_t}$ onto $\tilde{X}_t\times\mathbb{C}_m$, T^*X_t is transformed into a subset $\tilde{\mathbb{R}}T'_t$ of $\tilde{X}_t\times\mathbb{C}_m$. As t ranges over U_2 the sets $\tilde{\mathbb{R}}T'_t$ make up the image of $\mathcal{R}_U$ under the map (1.4); it is a subset of $\tilde{U}\times\mathbb{C}_m$ which we shall denote by $\tilde{\mathcal{R}}_U$.

In the same set-up we want to describe the natural generators of the tangent structure bundle $\mathcal{V}^2$ over T'. First of all we have the "vertical" Cauchy-Riemann vector fields $\partial/\partial\bar{\zeta}_i$ $(1 \le i \le m)$. In order to get a linear basis of $\mathcal{V}^2$ over $T'|_U$ we adjoin to these the vector fields $\mathcal{L}_j$ defined by the following relations (cf. (1.14), Ch. II):

$$(1.7)\qquad \mathcal{L}_jZ^h = \mathcal{L}_j\zeta_i = \mathcal{L}_j\bar{\zeta}_i = 0, \qquad \mathcal{L}_jt^k = \delta^k_j$$
$$(1 \le h,i \le m, \quad 1 \le j,k \le n).$$

The basis $dZ^1,\dots,dZ^m$ of T' over U enables us to define the "horizontal" submanifolds of $T'|_U$, namely the subsets U_ζ consisting of the points $((x,t),\zeta\cdot dZ)$ with $\zeta \in \mathbb{C}_m$ fixed. Clearly, for each ζ, the base projection $((x,t),\zeta\cdot dZ) \to (x,t)$ is a diffeomorphism of U_ζ onto U. We may therefore take the pull-back of the vector field L_j to U_ζ; this is precisely $\mathcal{L}_j$.

It is also possible to define the analogues of the vector fields M_i (see (1.14), Ch. II). There are $2m$ of these; the anti-Cauchy-Riemann vector fields $\partial/\partial\zeta_i$ $(1 \le i \le m)$ are m of them. The m remaining ones are the vector fields $\mathfrak{m}_i$ defined by the relations

$$(1.8)\qquad \mathfrak{m}_iZ^h = \delta^h_i, \quad \mathfrak{m}_i\zeta_h = \mathfrak{m}_i\bar{\zeta}_h = \mathfrak{m}_it^j = 0,$$
$$(1 \le i,h \le m, \quad 1 \le j \le n).$$

For each i, at points of U_ζ, $\mathfrak{m}_i$ is equal to the lift of M_i via the inverse of the base projection. The vector fields $\mathcal{L}_j$, $\mathfrak{m}_i$, $\partial/\partial\zeta_k$, $\partial/\partial\bar{\zeta}_\ell$ $(j=1,\dots,n, i,k,\ell = 1,\dots,m)$ commute pairwise.

<u>Links with symplectic geometry</u>

Actually a certain amount of invariance can be added to the

picture by looking at it from the viewpoint of <u>holomorphic symplectic geometry</u> on the complexified cotangent bundle. Let σ denote the <u>tautological form</u> on $\mathbb{C}T'^*\mathfrak{M}$: the value of σ at a point $(\rho,\theta) \in \mathbb{C}T^*\mathfrak{M}$ is the pull-back, under the base projection $\pi: \mathbb{C}T^*\mathfrak{M} \to \mathfrak{M}$, of the cotangent vector θ to $\mathfrak{M}$ at the point ρ. Let us refer to it as the <u>first fundamental form</u> on T'; the <u>second fundamental form</u> will be its exterior derivative $d\sigma$. The two-form $d\sigma$ defines a skew-symmetric bilinear functional on each complexified tangent space to $\mathbb{C}T^*\mathfrak{M}$. Let $(U,x^1,\ldots,x^m,t^1,\ldots,t^n)$ be a coordinate patch in $\mathfrak{M}$, and call ξ_i, τ_j the <u>complex</u> coordinates in the fibres of $\mathbb{C}T^*\mathfrak{M}$, with respect to the linear basis $dx^1,\ldots,dx^m$, $dt^1,\ldots,dt^n$. We have, in $\mathbb{C}T^*\mathfrak{M}|_U$,

$$\sigma = \sum_{i=1}^{m} \xi_i dx^i + \sum_{j=1}^{n} \tau_j dt^j,$$

$$d\sigma = \sum_{i=1}^{m} d\xi_i \wedge dx^i + \sum_{j=1}^{n} d\tau_j \wedge dt^j.$$

Regard $\mathbb{C}T^*\mathfrak{M}$ as a C^∞ manifold and let $\mathbb{C}T(\mathbb{C}T^*\mathfrak{M})$ denote its complexified tangent bundle. Denote by $\bar{D}$ the subbundle of $\mathbb{C}T(\mathbb{C}T^*\mathfrak{M})$ whose sections are the Cauch-Riemann vector fields tangent to the fibres of $\mathbb{C}T^*\mathfrak{M}$. Over the coordinate patch $(U,x^i,t^j)_{1\le i\le m,1\le j\le n}$ these are the linear combinations of the vector fields $\partial/\partial\bar{\xi}_i$ and $\partial/\partial\bar{\tau}_j$. It is clear that $\bar{D}$ <u>is the orthogonal of</u> $\mathbb{C}T(\mathbb{C}T^*\mathfrak{M})$ <u>for the bilinear form</u> $d\sigma$.

We introduce the following bundle over the manifold $CT^*\mathfrak{M}$:

$$\text{Hol}\,\mathbb{C}T^*\mathfrak{M} = \mathbb{C}T(\mathbb{C}T^*\mathfrak{M})/\bar{D}. \tag{1.9}$$

The sections of Hol $\mathbb{C}T^*\mathfrak{M}$ are not vector fields on $\mathbb{C}T^*\mathfrak{M}$, rather they are equivalence classes of such vector fields modulo vertical Cauchy-Riemann vector fields. Since $\bar{D}$ is the orthogonal of $\mathbb{C}T(\mathbb{C}T^*\mathfrak{M})$ for the second fundamental form the latter defines a <u>complex symplectic form</u> on Hol $\mathbb{C}T^*\mathfrak{M}$, i.e., a nondegenerated skew symmetric complex bilinear functional on each complexified tangent space to $\mathbb{C}T^*\mathfrak{M}$, which we denote also by $d\sigma$.

The vector bundle dual to Hol $\mathbb{C}T^*\mathfrak{M}$ can be identified to the orthogonal of $\bar{D}$ (for the duality between tangent and cotangent vectors to $\mathbb{C}T^*\mathfrak{M}$); we denote it by $\text{Hol}^*\mathbb{C}T^*\mathfrak{M}$. It is generated by the differentials of the C^∞ functions in $\mathbb{C}T^*\mathfrak{M}$ which are holomorphic with respect to the fibre variables (ξ_i and τ_j in the coordinate patch used earlier). The form $d\sigma$ defines an isomorphism of Hol $\mathbb{C}T^*\mathfrak{M}$ onto $\text{Hol}^*\mathbb{C}T^*\mathfrak{M}$, by means of which we can transfer $d\sigma$ itself to $\text{Hol}^*\mathbb{C}T^*\mathfrak{M}$; we shall denote by $d\sigma^*$ the transfer. Let f,

g be two arbitrary smooth functions in an open subset $\mathcal{O}$ of $\mathbb{C}T^*\mathfrak{M}$, both holomorphic with respect to the fibre variables. The function in Ω, $d\sigma^*(df,dg)$, can be called the <u>holomorphic Poisson bracket</u> of f and g; in the sequel we denote it by $\{f,g\}$. There are unique sections of Hol $\mathbb{C}T^*\mathfrak{M}$ over $\mathcal{O}$, $\mathcal{H}_f$ and $\mathcal{H}_g$, such that

(1.10) $$\{f,g\} = d\sigma(\mathcal{H}_f,\mathcal{H}_g).$$

The vector fields $\mathcal{H}_f$ and $\mathcal{H}_g$ are the <u>holomorphic Hamiltonian fields</u> of f and g respectively. The standard real theory extends routinely, with the adjective "holomorphic" added wherever it makes sense. Thus

(1.11) $$\{f,g\} = \mathcal{H}_f g, \qquad [\mathcal{H}_f,\mathcal{H}_g] = \mathcal{H}_{\{f,g\}} .$$

Let v be a smooth vector field in an open subset U of $\mathfrak{M}$. Its symbol $\sigma(v)$ is the function in $\mathbb{C}T^*\mathfrak{M}|_U$ which to any point $(\wp,\bar{\omega})$ assigns the complex number $\langle\bar{\omega}, v|_\wp\rangle$. In passing note that we can write $\sigma(v) = \langle\sigma,v\rangle$ if we interpret the right-hand side in the following manner: the value of $\langle\sigma,v\rangle$ at the point $(\wp,\bar{\omega})$ is the value of the cotangent vector to $\mathbb{C}T^*\mathfrak{M}$, $\sigma|_\wp$, on any tangent vector to $\mathbb{C}T^*\mathfrak{M}$ at $\wp$ whose base projection is equal to $v|_\wp$ (indeed, σ annihilates all vertical tangent vectors). The function $\sigma(v)$ is smooth in $\mathbb{C}T^*\mathfrak{M}|_U$ and is holomorphic with respect to the fibre variables. The Hamiltonian field of $\sigma(v)$ is a well defined section of Hol $\mathbb{C}T^*\mathfrak{M}$ which we shall call the Hamiltonian lift of v and denote by $\mathcal{H}_v$. If w is another C^∞ vector field and if f is a C^∞ function in U we have

(1.12) $$\sigma([v,w]) = \{\sigma(v),\sigma(w)\};$$

(1.13) $$[\mathcal{H}_v,\mathcal{H}_w] = \mathcal{H}_{[v,w]};$$

(1.14) $$\mathcal{H}_{fv} = (f\circ\pi)\mathcal{H}_v.$$

The cotangent structure bundle T' of the hypo-analytic manifold $(\mathfrak{M},\mathfrak{A})$ is the set of common zeros in $\mathbb{C}T^*\mathfrak{M}$ of all the functions $\sigma(L)$ as L ranges over the space of smooth sections of the tangent structure bundle $\mathcal{V}$. For the sake of simplicity we shall also denote by σ and $d\sigma$ the pull-backs to T' of the first and second fundamental forms respectively.

If we recall that σ is the tautological form we see that its pull-back to T' is a section of the pull-back of T' to T' under the base projection. It is orthogonal, not only to the vertical tangent vectors, i.e., the vectors tangent to the fibres of T', but also to any vector field Λ whose base projection L is a section of $\mathfrak{V}$. Let $\alpha^1,\ldots,\alpha^m$ be smooth sections of the vector bundle T' over some open subset Ω of $\mathfrak{M}$, spanning $T'|_\Omega$. Denote by $\zeta_1,\ldots,\zeta_m$ the complex coordinates in the fibres of T' at points of Ω with respect to the basis $\alpha^1,\ldots,\alpha^m$. Call α^i the pull-back of α^i to $T'|_\Omega$ under the base projection. We have

$$\sigma = \sum_{i=1}^{m} \zeta_i \alpha^i. \tag{1.14}$$

Let now $M_1,\ldots,M_m$ be m smooth vector fields in Ω such that

$$\langle \alpha^h, M_i \rangle = \delta_i^h \quad (h,i=1,\ldots,m). \tag{1.15}$$

We see that

$$\zeta_i = \sigma(M_i). \tag{1.16}$$

On the other hand there are m^2 smooth one-forms γ_i^h in Ω such that

$$d\alpha^h = \sum_{i=1}^{m} \gamma_i^h \alpha^i. \tag{1.17}$$

Consequently,

$$d\sigma = \sum_{i=1}^{m} \lambda_i \wedge \alpha^i, \tag{1.18}$$

where

$$\lambda_i = d\zeta_i + \sum_{h=1}^{m} \zeta_h \gamma_i^h. \tag{1.19}$$

Taking the exterior derivatives of both sides in (1.18) shows that there are m^2 one-forms ρ_{hi} such that

$$d\lambda_i - \sum_{h=1}^{m} \gamma_i^h \wedge \lambda_h = \sum_{h=1}^{m} \rho_{hi} \wedge \alpha^h. \tag{1.20}$$

The formulas (1.18) and (1.20) show that the one-forms α^i, λ_h span a vector subbundle of $\mathbb{C}T^*(T')$ over $T'|_\Omega$ which is closed in the sense of Cartan. Let $\beta^1,\ldots,\beta^m$ be another basis of T' over Ω and write

$$\beta^i = \sum_{h=1}^{m} f_h^i \alpha^h,$$

where the f_h^i are smooth functions and $\det(f_h^i) \neq 0$ in Ω. Let μ_i $(i=1,\ldots,m)$ be one-forms in $T'|_\Omega$ such that

$$d\sigma = \sum_{i=1}^{m} \mu_i \wedge \beta^i.$$

We have

$$\sum_{i=1}^{m} (\lambda_i - f_i^h \mu_h) \wedge \alpha^i = 0$$

which implies

$$\lambda_i = \sum_{h=1}^{m} f_i^h \mu_h \text{ modulo linear combinations of } \alpha^1,\ldots,\alpha^m.$$

This proves that the vector bundle spanned by the β^i and the μ_h is equal to the one spanned by the α^i and the λ_h.

Finally let (U,Z) be a hypo-analytic chart in $\mathfrak{M}$. We can apply what precedes to $\alpha^i = dZ^i$; in this case, we can take $\gamma_i^h = 0$ in (1.17); thus $\lambda_h = d\zeta_h$. We reach the conclusion that the vector bundle over T' spanned, over the open subset Ω, by the α^i and the λ_h is nothing else but ${}^2T'|_\Omega$.

The natural injection $\mathbb{C}T(T') \subset \mathbb{C}T(\mathbb{C}T^*\mathfrak{M})$ induces a natural map

$$\mathbb{C}T(T') \to \mathbb{C}T(\mathbb{C}T^*\mathfrak{M})/\bar{D};$$

call Hol T' the image of this map. If v is a smooth vector field in an open subset Ω of $\mathfrak{M}$ its Hamiltonian lift H_v will not, in general, be tangent to T'. If it is tangent to T' it is a section of Hol T'. For certain kinds of vector fields this will be the case:

<u>Proposition 1.3</u>. <u>Let</u> v <u>be a smooth vector field in an open subset</u> Ω <u>of</u> $\mathfrak{M}$ <u>whose commutator with any section of</u> $\mathcal{V}$ <u>is a section of</u> $\mathcal{V}$. <u>Then the Hamiltonian lift</u> $\mathcal{H}_v$ <u>of a smooth section</u> L <u>of</u> $\mathcal{V}$ <u>over</u> Ω <u>is a section of</u> Hol T' <u>over</u> $T'|_\Omega$.

<u>Proof</u>: Let $\wp$ be a point of Ω and let $L_1,\ldots,L_n$ be n smooth vector fields in an open neighborhood $U \subset \Omega$ of $\wp$, which span $\mathcal{V}$ over U and commute pairwise. The equations $\sigma(L_j) = 0$, $j=1,\ldots,n$, define $T'|_U$ in $\mathbb{C}T^*\mathfrak{M}|_U$. Since $[v,L_j] = 0$ we also have $\mathcal{H}_v\sigma(L_j)=0$, whence the assertion. ∎

<u>Corollary 1.4</u>. <u>The Hamiltonian lift of a section of</u> $\mathcal{V}$ <u>is a section of</u> Hol T'.

Let (U,Z) be a hypo-analytic chart in $\mathfrak{M}$. As before assume that U is the domain of local coordinates $x^1,\ldots,x^m$, $t^1,\ldots,t^n$, and that (1.12) and (1.13), Ch. II, as well as (1.2) in the present section, hold true. Let L_j, M_i be the vector fields in U associated with the Z^i and the t^j (see (1.14), Ch. II). In $T'|_U$ we have

$$(1.21) \qquad \sigma = \sum_{i=1}^{m} \zeta_i dZ^i, \quad d\sigma = \sum_{i=1}^{m} d\zeta_i \wedge dZ^i, \quad \zeta_i = \sigma(M_i).$$

Introduce also the vector fields $\mathcal{L}_j$ and $\mathfrak{M}_i$ defined in (1.7) and (1.8). It is readily checked that

$$(1.22) \qquad \mathcal{L}_j = \mathcal{H}_{L_j}, \quad \mathfrak{M}_i = \mathcal{H}_{M_i} \quad (i=1,\ldots,m,\ j=1,\ldots,n).$$

By Prop. 1.3 we know that the Hamiltonian lifts $\mathcal{H}_{M_i}$ are sections of Hol T′ over $T'|_U$.

The hypo-analytic structure $\mathcal{Q}'$ on T' induces a hypo-analytic structure on $\mathcal{R}_U$ which we shall denote by $\mathcal{Q}'_U$. It is defined by the function $Z^h(x,t)$, $\sigma(M_h)(x,t,\xi)$. Indeed, $\sigma(M_h)$ is equal to the restriction of the complex coordinate ζ_h to the fibres of $\mathbb{R}T'_t$ or, which is the same, on these fibres, $\zeta = {}^tZ_x(x,t)^{-1}\xi = (\sigma(M_1),\ldots,\sigma(M_m))$. We recall that $\dim \mathcal{R}_U = 3m+n$. The fibre dimension of the cotangent structure bundle of $\mathcal{Q}'_U$ (which is the pull-back to $\mathcal{R}_U$ of the bundle ${}^2T'$) is equal to $2m$. Its orthogonal, the tangent structure bundle underlying $\mathcal{Q}'_U$ is spanned by the vector fields $\mathcal{H}_{L_j}$ and $\mathcal{H}_{M_i}$ where $\zeta = {}^tZ_x(x,t)^{-1}\xi$.

2. THE FBI TRANSFORM

We continue to reason in the hypo-analytic chart (U,Z); U is the domain of local coordinates x^i, t^j $(1 \le i \le m,\ 1 \le j \le n)$. We assume that (1.12), (1.13), Ch. II, hold, as well as (1.2) of the present chapter. We take U in the product form (1.3).

Let u be any distribution in U whose support with respect to x is compact, i.e., there is a compact subset K of U_1 such that $u \equiv 0$ in $(U_1\backslash K)\times U_2$. For any (z,ζ) in the set

$$z \in \mathbb{C}^m, \quad \zeta \in \mathbb{C}_m; \qquad |\mathrm{Im}\,\zeta| < |\mathrm{Re}\,\zeta|, \tag{2.1}$$

we consider the "integral"

$$\mathfrak{F}(u;z,\zeta,t) = \int_{U_1} e^{i\zeta\cdot(z-Z(x,t))-\langle\zeta\rangle\langle z-Z(x,t)\rangle^2}\, u(x,t)\,dZ(x,t), \tag{2.2}$$

where we use the following notation: $\langle z\rangle = \{(z^1)^2+\ldots+(z^m)^2\}^{1/2}$ $(z \in \mathbb{C}^m$ or $z \in \mathbb{C}_m)$; of course, $i = \sqrt{-1}$;

$$dZ = dZ^1 \wedge\ldots\wedge dZ^m = \det Z_x(x,t)\,dx^1 \wedge\ldots\wedge dx^m. \tag{2.3}$$

Actually z, ζ, t play the role of parameters and the integral is a duality bracket in U_1, between the compactly supported zero-current $u(t,x)$ and the C^∞ m-form $e^{i\zeta\cdot(z-Z(x,t))-\langle\zeta\rangle\langle z-Z(x,t)\rangle^2}\,dZ$.

We shall make use of the maximally real submanifolds X_t of $\mathfrak{M}$ defined in Section 1. The right-hand side in (2.2) can be regarded as an integral over X_t of $\mathfrak{M}$ or, equivalently, as one over its image under the map $x \to Z(x,t)$, the submanifold $\tilde{X}_t$ of $\tilde{U} \subset \mathbb{C}^m\times\mathbb{R}^n$ (in fact all these integrals are duality brackets between currents and test-forms). Call $\tilde{u}$ the push-forward of the distribution u under the map $(x,t) \to (Z(x,t),t)$. Then

(2.4) $$\mathfrak{F}(u;z,\zeta,t) = \int_{\tilde{X}_t} e^{i\zeta\cdot(z-\tilde{z})-\langle\zeta\rangle\langle z-\tilde{z}\rangle^2}\,\tilde{u}\,d\tilde{z}.$$

Observe that $\mathfrak{F}(u;z,\zeta,t)$ is a holomorphic function of (z,ζ) in the set (2.1) valued in the space of distributions of t in U_2.

<u>Definition 2.1.</u> $\mathfrak{F}(u;z,\zeta,t)$ <u>will be called the</u> Fourier-Bros-Iagolnitzer (in short, FBI) <u>transform of the distribution</u> u.

Introduce now the vector fields L_j and M_i defined in (1.14), Ch. II. A straightforward integration by parts shows that

(2.5) $$\mathfrak{F}(L_j u;t,z,\zeta) = (\partial/\partial t^j)\mathfrak{F}(u;t,z,\zeta), \qquad j=1,\dots,n;$$

(2.6) $$\mathfrak{F}(M_j u;t,z,\zeta) = (\partial/\partial z^i)\mathfrak{F}(u;t,z,\zeta), \qquad i=1,\dots,m.$$

Let now f be a current in U of the kind

(2.7) $$f = \sum_{|I|=p}\ \sum_{|J|=q} f_{I,J}(x,t)\,dZ^I \wedge dt^J,$$

whose coefficients $f_{I,J}$ all have compact support with respect to x.

<u>Definition 2.2.</u> <u>We shall call FBI transform of</u> f <u>the current in</u> $\mathbb{C}^m \times U_2$,

(2.8) $$\mathfrak{F}(f;z,\zeta,t) = \sum_{|I|=p}\ \sum_{|J|=q} \mathfrak{F}(f_{I,J};z,\zeta,t)\,dz^I \wedge dt^J.$$

In (2.8) we regard ζ as a parameter, varying in the set

(2.9) $$\zeta \in \mathbb{C}_m; \qquad |\mathrm{Im}\,\zeta| < |\mathrm{Re}\,\zeta|.$$

It follows at once from (2.5) applied to the coerfficients $f_{I,J}$, that

(2.10) $$\mathfrak{F}(Lf;z,\zeta,t) = d_t\mathfrak{F}(u;z,\zeta,t).$$

Henceforth we shall reason under a more restrictive hypothesis about the map $\phi: U \to \mathbb{R}^m$. We shall assume that

(2.11) $$\frac{\partial^2\phi^i}{\partial x^j \partial x^k}(0,0) = 0, \qquad \forall\ i,j,k = 1,\dots,m.$$

Actually (2.11) is not a restriction on the hypo-analytic structure, only on the hypo-analytic chart (U,Z). For observe that (2.11) can always be achieved if we replace Z^i by

$$Z^i - \frac{i}{2}\sum_{j,k=1}^{m} \frac{\partial^2\phi^i}{\partial x^j \partial x^k}(0,0)Z^jZ^k$$

for each $i=1,\dots,m$.

In the next statement we make use of the real vector bundle

$\tilde{\mathbb{R}}T'_t$ over $\tilde{X}_t$ introduced in Section 1 (see Notation 1.2 and following remarks):

<u>Proposition 2.3.</u> <u>If</u> (2.11) <u>holds then, to every</u> ε, $0 < \varepsilon < 1$, <u>there is</u> $\delta > 0$ <u>such that, if</u> diam $U < \delta$ <u>we have</u>

$$(2.12)\qquad -\mathrm{Re}\{i\zeta\cdot(z-\tilde{z})-\langle\zeta\rangle\langle z-\tilde{z}\rangle^2\} \geq (1-\varepsilon)|z-\tilde{z}|^2|\zeta|,$$
$$\forall\ (z,\zeta) \in \tilde{\mathbb{R}}T'_t,\quad \tilde{z} \in \tilde{X}_t,\quad t \in U_2.$$

<u>Proof</u>: Thus suppose $z = Z(x,t)$, $\zeta = {}^tZ_x(x,t)^{-1}\xi$, $\tilde{z} = Z(x',t)$ for some $x,x' \in U_1$, $t \in U_2$ and $\xi \in \mathbb{R}_m$ (cf. (1.6)). Then

$$-\mathrm{Re}\ i\zeta\cdot(z-\tilde{z}) = \xi\cdot\mathrm{Im}\{Z_x(x,t)^{-1}[Z(x,t)-Z(x',t)]\}.$$

But if (2.11) holds,

$$(2.13)\qquad |Z_x(x,t)^{-1}[Z(x,t)-Z(x',t)] - (x-x')| \leq \text{const.}(\mathrm{diam}\ U)|x-x'|^2,$$

and therefore

$$(2.14)\qquad \mathrm{Re}\ i\zeta\cdot(z-\tilde{z}) \leq \text{const.}(\mathrm{diam}\ U)|x-x'|^2.$$

On the other hand (1.13), Ch. II, means that

$$(2.15)\qquad \|Z_x(x,t)-I\| \leq \text{const.}(\mathrm{diam}\ U),$$

whence

$$(2.16)\qquad |Z(x,t)-Z(x',t)-(x-x')| \leq \text{const.}(\mathrm{diam}\ U)|x-x'|,$$

$$(2.17)\qquad ||\langle\zeta\rangle|/|\xi|-1| + ||\zeta|/|\xi|-1| \leq \text{const.}(\mathrm{diam}\ U).$$

By combining all this we obtain that

$$|\mathrm{Re}\{i\zeta\cdot(z-\tilde{z})-\langle\zeta\rangle\langle z-\tilde{z}\rangle^2\} + |z-\tilde{z}|^2|\zeta|| \leq$$
$$\leq \text{const.}(\mathrm{diam}\ U)|z-\tilde{z}|^2|\zeta|,$$

whereby we reach the desired conclusion. ∎

Henceforth we shall reason under the hypothesis that (2.12) <u>holds with a number</u> $\varepsilon < 1$.

Recall that the vector fields M_i and L_j (see (1.14), Ch. II) span $\mathbb{C}T\mathcal{M}$ over U and commute pairwise. It follows that, given any compactly supported distribution u in U, there is a finite family of compactly supported continuous functions in U, $u_{\alpha,\beta}$ ($\alpha \in \mathbb{Z}^m_+$, $\beta \in \mathbb{Z}^n_+$, $|\alpha|+|\beta| \leq k$), such that

$$(2.18)\qquad u = \sum_{|\alpha|+|\beta|\leq k} M^\alpha L^\beta u_{\alpha,\beta}$$

($M^\alpha = M_1^{\alpha_1}\cdots M_m^{\alpha_m}$, $L^\beta = L_1^{\beta_1}\cdots L_n^{\beta_n}$). Applying then (2.5) and (2.6) yields

(2.19) $$\mathfrak{F}(u;z,\zeta,t) = \sum_{|\alpha|+|\beta|\le k} \partial_z^\alpha \partial_t^\beta \mathfrak{F}(u_{\alpha,\beta};z,\zeta,t).$$

We shall now use the following consequence of (2.12):

(2.20) To every $\alpha \in \mathbb{Z}_+^m$ there is a constant $C_\alpha > 0$ such that, for all $(z,\zeta) \in \widetilde{\mathbb{R}T}'_t$ and all $\tilde{z} \in \tilde{X}_t$,
$$|\partial_z^\alpha e^{i\zeta\cdot(z-\tilde{z})-\langle\zeta\rangle\langle z-\tilde{z}\rangle^2}| \le C_\alpha(1+|\zeta|)^{|\alpha|} e^{-(1-\epsilon)|\zeta||z-\tilde{z}|^2/2}.$$

Us derive from this that, for each $\beta \in \mathbb{Z}_+^n$, $|\beta| \le k$,

(2.21) $$(1+|\zeta|)^{-k+|\beta|} \sum_{|\alpha|=k-|\beta|} \partial_z^\alpha \mathfrak{F}(u_{\alpha,\beta};z,\zeta,t)$$

is a continuous and bounded function of (z,ζ,t) in the manifold
$$\tilde{\mathfrak{R}}_U = \{(z,\zeta,t) \in \mathbb{C}^m\times\mathbb{C}_m\times\mathbb{R}^n;\ \ (z,\zeta) \in \widetilde{\mathbb{R}T}'_t,\ \ t \in U_2\}.$$

Conclusion:

Proposition 2.4. Suppose that (2.12) holds with some ϵ, $0 < \epsilon < 1$. Let u be a compactly supported distribution in U. There is an integer $k \ge 0$ and a finite family of continuous and bounded functions of (z,ζ,t) in the manifold $\tilde{\mathfrak{R}}_U$, g_β $(\beta \in \mathbb{Z}_+^n,\ |\beta| \le \ell)$, such that, in $\tilde{\mathfrak{R}}_U$,

(2.22) $$\mathfrak{F}(u;z,\zeta,t) = (1+|\zeta|)^k \sum_{|\beta|\le\ell} \partial_t^\beta g_\beta(z,\zeta,t).$$

Remark 2.5. Assume that a compact set $K \subset U_2$ contains the t-projection of supp u. Then K also contains the t-projection of supp $\mathfrak{F}(u;z,\zeta,t)$. Whatever $\epsilon > 0$, it can be arranged that the t-projection of the support of every g_β be contained in the neighborhood of order ϵ of K. ∎

If the (compactly supported) distribution u is a continuous function of z (resp., t) then the integer k (resp. ℓ) in (2.22) can be taken to be zero. If we increase the hypotheses of regularity of u with respect to either z or t (or to both) those integers k or/and ℓ can be taken to be negative. At the limit, when $\ell = +\infty$, the same kind of argument as above leads to the next statement:

Proposition 2.6. Suppose that (2.12) holds with some ϵ, $0 < \epsilon < 1$. Let u be a C^∞ function with compact support in U. Then $\mathfrak{F}(u;z,\zeta,t)$ is a C^∞ function of $t \in U_2$ valued in the space of holomorphic functions of (z,ζ) in the set (2.1). Its restriction to the manifold $\tilde{\mathfrak{R}}_U$ has the following property:

Given any $\beta \in \mathbb{Z}_+^n$ and any integer $k \ge 0$ there is a constant $C_{\beta,k} > 0$ such that

(2.23) $$|\partial_t^\beta \mathfrak{F}(u;z,\zeta,t)| \le C_{\beta,k}(1+|\zeta|)^{-k}.$$

We must now define the FBI transform of an arbitrary distribution (i.e., whose support need not project into a compact subset of x-space) - or rather, the FBI transform of the germ at the origin of a distribution in $\mathfrak{M}$. The definition is based on the following

Lemma 2.7. Suppose that (2.12) holds with some ε, $0 < \varepsilon < 1$. To every open neighborhood of 0 in x-space, $V_1 \subset U_1$, there is an open neighborhood of 0 in t-space, $V_2 \subset U_2$, an open neighborhood $\mathfrak{G}$ of 0 in $\mathbb{C}^m$, and open cone Γ in $\mathbb{C}_m$ which contains $\mathbb{R}_m \backslash \{0\}$ and a number $R > 0$ such that the following is true:

Given any distribution u in U whose support is compact and contained in $(U_1 \backslash V_1) \times U_2$ and any function $\chi \in C_c^\infty(V_2)$, there is a constant $C > 0$ such that

(2.24) $$\left|\int \mathfrak{F}(u;z,\zeta,t)\chi(t)dt\right| \le C\, e^{-|\zeta|/R}, \quad \forall\ (z,\zeta) \in \mathfrak{G}\times\Gamma.$$

Remark 2.8. If the compactly supported distribution u is a continuous function of t with values in $\mathcal{D}'(U_1)$ its FBI transform $\mathfrak{F}(u;z,\zeta,t)$ is a continuous function of (z,ζ,t), holomorphic with respect to (z,ζ), in the product of the set (2.1) with U_2. If furthermore $u \equiv 0$ when $x \in V_1$ the inequality (2.24) can be replaced by the following one:

(2.25) $$|\mathfrak{F}(u;z,\zeta,t)| \le C\, e^{-|\zeta|/R}, \quad \forall\ (z,\zeta) \in \mathfrak{G}\times\Gamma, \quad t \in V_2.$$

Proof of Lemma 2.7: By (2.19) we have

$$(1+|\zeta|)^{-k}\int \mathfrak{F}(u;z,\zeta,t)\chi(t)dt =$$
$$= \sum_{|\alpha|+|\beta|\le k} \partial_z^\alpha \int \mathfrak{F}(u_{\alpha,\beta};z,\zeta,t)(-\partial_t)^\beta\chi(t)dt.$$

This formula, combined with the Cauchy inequalities in z-space, shows that it suffices to prove the assertion when the compactly supported continuous functions $u_{\alpha,\beta}$ are substituted for the distribution u (recall that they can be selected in such a way that their support be contained in the neighborhood of order $\delta > 0$ of supp u, with δ as small as needed). We shall therefore assume that $u \in \mathbb{C}_o^0(U)$ and that $u(x,t) \equiv 0$ if $x \in V_1$, and prove that (2.25) holds.

Denote by $\tilde{u}$ the push-forward of u from U to $\tilde{U}$ under the map $(x,t) \to (Z(x,t),t)$; $\tilde{u} \in C_o^\infty(\tilde{U})$ and $\tilde{u}(z,t) = 0$ whenever z belongs to a suitably small neighborhood $\tilde{V}$ of 0 in $\mathbb{C}^m$. We derive from (2.12) there is a number $c_o > 0$ such that

(2.26) $$\mathrm{Re}\{i\zeta\cdot(z-\tilde{z}) - \langle\zeta\rangle\langle z-\tilde{z}\rangle^2\}/|\zeta| \le -c_o$$
$$\forall\ (z,\zeta) \in \widetilde{\mathbb{R}T}'_t, \quad \tilde{z} \in {}^o\tilde{X}_t \cap (\mathbb{C}^m \backslash \tilde{V}),$$

whatever $t \in U_2$, provided $|z|$ is small enough. Take $z = Z(0,t)$, $\zeta = {}^tZ_x(0,t)^{-1}\xi$ for some $\xi \in \mathbb{R}_m$; according to (2.26) there is a constant $C > 0$ such that

$$\mathrm{Re}\{i\xi\cdot\tilde{z} + |\xi|\langle\tilde{z}\rangle^2\}/|\xi| \geq c_0 - C|t|.$$

If $R > 2/c_0$ and if we restrict the variation of t to an open neighborhood $V_2 \subset U_2$ of 0 sufficiently small we obtain that

$$\mathrm{Re}\{i\zeta\cdot(z-\tilde{z}) - \langle\zeta\rangle\langle z-\tilde{z}\rangle^2\}/|\zeta| < -1/R \tag{2.27}$$
$$\forall\ t \in U_2,\quad \tilde{z} \in \tilde{X}_t \cap (\mathbb{C}^m\backslash\tilde{V}),$$

when $z = 0$ and $\zeta = \xi \in \mathbb{R}_m\backslash\{0\}$. But of course this will remain true if z varies in a sufficiently small neighborhood of 0 in $\mathbb{C}^m$ and ζ in a sufficiently thin cone in $\mathbb{C}_m\backslash\{0\}$ which contains $\mathbb{R}_m\backslash\{0\}$. Taking (2.27) into account in the integral at the righ in (2.4) yields what we want. ∎

Let $f_1(z,\zeta,t)$, $f_2(z,\zeta,t)$ be two holomorphic functions of (z,ζ) in the set (2.1) valued in the space of distributions of t in U_2.

Definition 2.9. We shall say that f_1 and f_2 are equivalent and we shall write $f_1 \approx f_2$ if there is an open neighborhood of 0 in t-space, $V_2 \subset U_2$, one, $\mathcal{O}$, in $\mathbb{C}^m$, an open cone Γ in $\mathbb{C}_m$ containing $\mathbb{R}_m\backslash\{0\}$ and constants $C, R > 0$ such that, given any function $\chi \in C_c^\infty(V_2)$, we have

$$\left|\int[f_1(z,\zeta,t) - f_2(z,\zeta,t)]\chi(t)dt\right| \leq C\,e^{-|\zeta|/R}, \tag{2.28}$$
$$\forall\ z \in \mathcal{O},\quad \zeta \in \Gamma.$$

Lemma 2.7 entails that, if u_1 and u_2 are any two compactly supported distributions in U which are equal in some open neighborhood of 0 contained in U, then $\mathfrak{F}(u_1;z,\zeta,t) \approx \mathfrak{F}(u_2;z,\zeta,t)$. This allows us to introduce the following

Definition 2.10. Let u_0 be a distribution in some open neighborhood U_0 of 0 in $\mathfrak{M}$. By the FBI transform of u_0 at the origin, in the hypo-analytic chart (U,Z), we shall mean the equivalence class, for the equivalence relation in Def. 2.9, of the FBI transform $\mathfrak{F}(u;z,\zeta,t)$ of any compactly supported distribution u in U equal to u_0 in some neighborhood of 0 contained in $U_0 \cap U$.

We shall denote by $\mathfrak{F}(u_0;z,\zeta,t)$ the FBI transform of the distribution u_0.

3. INVERSION OF THE FBI TRANSFORM

We continue to use the concepts and notation of Section 2, and to reason under hypothesis (2.12) with $0 < \epsilon < 1$. We shall also assume that $\operatorname{diam} U$ is sufficiently small that

$$\operatorname{Re}\langle {}^tZ_x(x,t)^{-1}\xi\rangle^2 \geq |\xi|^2/2, \qquad \forall\ (x,t) \in U,\ \xi \in \mathbb{R}_m. \tag{3.1}$$

We look first at a compactly suported distribution u in $U = U_1 \times U_2$. Here we regard $\mathfrak{F}(u;z,\zeta,t)$ as a distribution on the manifold $\tilde{\mathcal{R}}_U$. Let us take note, in passing, that it has a pull-back to $\mathcal{R}_U$,

$$\mathfrak{F}(u;Z(x,t),{}^tZ(x,t)^{-1}\xi,t) \qquad ((x,t) \in U,\ \xi \in \mathbb{R}_m).$$

Let then K denote a compact subset of U_1 and $\tilde{K}_t$ the image of K under the map $x \to Z(x,t)$. Call then $\mathcal{K}_t$ the portion of $\tilde{\mathbb{R}}T'_t$ which lies over $\tilde{K}_t$ (or $\tilde{X}_t$). Let δ be a number >0. We shall study the integral

$$\mathfrak{F}^\delta_K(u;z,t) = (4\pi^3)^{-m/2}\int\!\!\int_{\mathcal{K}_t} e^{i\zeta\cdot(z-z')-\langle\zeta\rangle\langle z-z'\rangle^2-\delta\langle\zeta\rangle^2}\,\mathfrak{F}(u;z',\zeta,t)\,\langle\zeta\rangle^{m/2}\Delta(z-z',\zeta)\,dz'\,d\zeta,$$

where $\Delta(z,\zeta)$ is the Jacobian determinant of the map $\zeta \to \zeta + i\langle\zeta\rangle i\langle\zeta\rangle z$. One can view the integration in $\mathfrak{F}^\delta_K$ as being performed with respect to x' over K and ξ over $\mathbb{R}_m$; then $z' = Z(x',t)$, $\zeta = {}^tZ_x(x',t)^{-1}\xi$.

<u>The integral $\mathfrak{F}^\delta_K$ is an entire holomorphic function of z in $\mathbb{C}^m$ valued in the space of distributions of t in U_2.</u>

Let $\mathcal{O}$ be an open subset of $\mathbb{C}^m$, Ω_2 an open subset of U_2. We shall denote by $\mathcal{H}(\mathcal{O};\mathcal{D}'(\Omega_2))$ the space of holomorphic maps $\mathcal{O} \to \mathcal{D}'(\Omega_2)$. It is equipped with its natural topology, that of uniform convergence on the compact subsets of $\mathcal{O}$ (it is then a <u>Montel space</u>). The space $\mathcal{H}(\mathcal{O};\mathcal{D}'(\Omega_2))$ can be naturally identified to the space $\mathcal{D}'(\Omega_2;\mathcal{H}(\mathcal{O}))$ of distributions of t in Ω_2 valued in the space of holomorphic functions in $\mathcal{O}$. More symmetrically,

$$\mathcal{H}(\mathcal{O};\mathcal{D}'(\Omega_2)) \cong \mathcal{D}'(\Omega_2;\mathcal{H}(\mathcal{O})) \cong \mathcal{H}(\mathcal{O})\,\hat{\otimes}\,\mathcal{D}'(\Omega_2),$$

where $\hat{\otimes}$ stands for the completion of the topological tensor product (in the present set-up it does not matter which tensor product topology we select, the π or the ϵ one). What is important for us to note is that, if Ω_1 is an open subset of U_1 and if $\mathcal{O}$ contains the image of $\Omega_1\times\Omega_2$ under the map $(x,t) \to Z(x,t)$, then, given any $\tilde{h}(z,t) \in \mathcal{H}(\mathcal{O};\mathcal{D}'(\Omega_2))$, we have the right to form $\tilde{h}(Z(x,t),t)$. It is an element of $C^\infty(\Omega_1;\mathcal{D}'(\Omega_2))$; its definition is obvious when $\tilde{h}(z,t)$ is a finite sum of products $\tilde{h}_1(z)\tilde{h}_2(t)$ with $\tilde{h}_1 \in \mathcal{H}(\mathcal{O})$, $\tilde{h}_2 \in \mathcal{D}'(\Omega_2)$; from there the general definition follows by approximation.

Lemma 3.1. Suppose that the origin does not lie in the compact set K. There is an open neighborhood $\mathcal{O}$ of the origin in $\mathbb{C}^m$ and one in t-space, $V_2 \subset U_2$, such that, if u is any compactly supported distribution in U, then, as $\delta \to +0$, the integral $\mathfrak{F}_K^\delta(u;z,t)$ converges in $\mathcal{H}(\mathcal{O};\mathcal{D}'(V_2))$.

Proof: The hypothesis allows us to select an open neighborhood $\tilde{V}$ of 0 in C^m such that $\tilde{V} \cap \tilde{K}_t = \emptyset$ whatever $t \in U_2$. We derive from (2.12) that the analogue of (2.26) is valid here also, provided $z \in \tilde{X}_t$ is sufficiently close to the origin:

(3.2) $$\mathrm{Re}\{i\zeta\cdot(z-\tilde{z}) - \langle\zeta\rangle\langle z-\tilde{z}\rangle^2\}/|\zeta| \le -c_o, \quad \forall\ (\tilde{z},\zeta) \in \mathcal{H}_t.$$

The same reasoning as in the proof of Lemma 2.7 shows that the same inequality is valid, with a reduced constant $c_o > 0$, when t remains in a sufficiently small open neighborhood of 0, $V_2 \subset U_2$, and when $z = 0$, therefore also when z stays in a sufficiently small open neighborhood of 0 in $\mathbb{C}^m$, $\mathcal{O}$. Let us then integrate $\mathfrak{F}_K^\delta$ against a test-form $\chi(t)dt$ with $\chi \in C_c^\infty(V_2)$. It is immediately clear that the result converges uniformly (as a function of z in $\mathcal{O}$) when $\delta \to +0$. ∎

Integration by parts shows that the difference

(3.3) $$\partial_z^\alpha\partial_t^\beta\mathfrak{F}_K^\delta(u;z,t) -$$
$$(4\pi^3)^{-m/2}\iint_{\mathcal{H}_t} e^{i\zeta\cdot(z-z')-\langle\zeta\rangle\langle z-z'\rangle^2-\delta\langle\zeta\rangle^2}\partial_{z'}^\alpha\partial_t^\beta\mathfrak{F}(u;z',\zeta,t)\langle\zeta\rangle^{m/2}\Delta(z-z',\zeta)dz'd\zeta$$

is a sum of integrals with respect to (x,ξ) carried out over the set $\partial K\times\mathbb{R}_m$. By the same argument used in proving Lemma 3.1 we can prove

Lemma 3.2. Suppose that the compact subset K of U_1 is a neighborhood of 0. Then there is an open neighborhood of 0 in $\mathbb{C}^m$, $\mathcal{O}$, and one in $\mathbb{R}^n$, $V_2 \subset U_2$, such that, whatever the compactly supported distribution u in U and the multiplets $\alpha \in \mathbb{Z}_+^m$, $\beta \in \mathbb{Z}_+^n$, the difference (3.3) converges in $\mathcal{H}(\mathcal{O};\mathcal{D}'(V_2))$.

We can now prove the "inversion formula" for the FBI transform:

Theorem 3.3. Suppose that (2.12) holds for some ϵ, $0 < \epsilon < 1/2$. There is an open neighborhood of 0 in $\mathbb{R}^m$, $V_1 \subset U_1$, one in $\mathbb{R}^n$, $V_2 \subset U_2$, and one in $\mathbb{C}^m$, $\mathcal{O}$, such that the image of $V_1\times V_2$ under the map $(x,t) \to Z(x,t)$ is contained in $\mathcal{O}$, and that the following holds:

(3.4) *To each compactly supported distribution* u *in* U *there is a function* $\tilde{h} \in \mathcal{H}(\mathcal{O};\mathcal{D}'(V_2))$ *such that the difference* $u(x,t) - \mathcal{F}_K^\delta(u;Z(x,t),t)$ *converges in* $\mathcal{D}'(V_1 \times V_2)$ *to* $\tilde{h}(Z(x,t),t)$.

Proof: Thanks to Lemma 3.1 we can modify the compact neighborhood $K \subset U$ of 0 as we wish; here we shall assume that it is a ball centered at 0. We refer the reader to the representation (2.18). We derive from Lemma 3.2 and from (2.19) that it suffices to prove that

$$\mathcal{F}_K^\delta(u;Z(x,t),t) - \sum_{|\alpha|+|\beta|\le k} (\partial_z^\alpha \partial_t^\beta \mathcal{F}_K^\delta)(u_{\alpha,\beta};Z(x,t),t)$$

converges in $\mathcal{D}'(V_1 \times V_2)$ to a function of the kind $\tilde{h}(Z(x,t),t)$. On the other hand note that

$$(\partial_z^\alpha \partial_t^\beta \mathcal{F}_K^\delta)(u_{\alpha,\beta};Z(x,t),t) = M^\alpha L^\beta [\mathcal{F}_K^\delta(u_{\alpha,\beta};Z(x,t),t)].$$

This shows that it suffices to prove (3.4) when $u_{\alpha,\beta}$ replaces u; in other words we may assume that u is a compactly supported continuous function in U.

We may then write

$$(3.5) \qquad \mathcal{F}_K^\delta(u;z,t) = (4\pi^3)^{-m/2} \iint_{\mathcal{H}_t} \int_{\tilde{X}_t} e^{i\zeta\cdot(z-\tilde{z})-\delta\langle\zeta\rangle^2} \cdot e^{-\langle\zeta\rangle(\langle z-z'\rangle^2+\langle z'-\tilde{z}\rangle^2)} \tilde{u}(\tilde{z},t)\langle\zeta\rangle^{m/2}\Delta(z-z',\zeta)\,d\tilde{z}\,dz'\,d\zeta.$$

As usual we have denoted by $\tilde{u}$ the push-forward of the function u to $\tilde{U}$ under the map $(x,t) \to (Z(x,t),t)$. It is clear that we may deform the domain of integration with respect to ζ from the image of $\mathbb{R}_m$ under the map $\xi \to {}^tZ_x(x',t)^{-1}\xi$ to the image of $\mathbb{R}_m$ under the map $\xi \to {}^tZ_x(x,t)^{-1}\xi$ for an arbitrary $x \in U_1$. It is convenient, at this juncture, to extend the map $\phi\colon U \to \mathbb{R}^m$ as a C^∞ map from $\mathbb{R}^m \times \mathbb{R}^m$ into $\mathbb{R}^m$ vanishing outside a compact neighborhood of $C\ell\, U$. Set then

$$\mathcal{F}^\delta(u;z,t) = (4\pi^3)^{-m/2} \iint_{(x',\xi)\in\mathbb{R}^m\times\mathbb{R}_m} \int_{\tilde{z}\in\tilde{X}_t} e^{i\zeta\cdot(z-\tilde{z})-\delta\langle\zeta\rangle^2} \cdot e^{-\langle\zeta\rangle(\langle z-z'\rangle^2+\langle z'-\tilde{z}\rangle^2)} \tilde{u}(\tilde{z},t)\langle\zeta\rangle^{m/2}\Delta(z-z',\zeta)\,d\tilde{z}\,dz'\,d\zeta,$$

where $z' = Z(x',t) = x'+i\phi(x',t)$, $\zeta = {}^tZ_x(x,t)^{-1}\xi$. By the same argument that allowed us to prove Lemmas 3.1 and 3.2, we see easily that

$$\mathcal{F}_K^\delta(u;z,t) - \mathcal{F}^\delta(u;z,t)$$

converges in the space $\mathcal{H}(\mathcal{O};\mathcal{D}'(V_2))$ provided the neighborhoods $\mathcal{O}$ and V_2 are small enough. In $\mathcal{F}^\delta(u;z,t)$ we can also deform the domain of integration with respect to z', from the image of $\mathbb{R}^m$ under the map $x' \to Z(x',t)$, to $\mathbb{R}^m$ itself. It is seen at once that

$$(3.6)\qquad (2\langle\zeta\rangle/\pi)^{m/2}\int e^{-\langle\zeta\rangle(\langle z-z'\rangle^2+\langle z'-\tilde{z}\rangle^2)}\Delta(z-z',\zeta)dz' = e^{-\langle\zeta\rangle\langle z-\tilde{z}\rangle^2/2}\,\Delta(\tfrac{1}{2}(z-\tilde{z}),\zeta),$$

Theorem 3.1 will be proved if we show that the integral

$$(3.7)\qquad (2\pi)^{-m}\int_{\xi\in\mathbb{R}_m}\int_{\tilde{z}\in\tilde{X}_t} e^{i\zeta\cdot(z-\tilde{z})-\langle\zeta\rangle\langle z-\tilde{z}\rangle^2/2-\delta\langle\zeta\rangle^2}\,\tilde{u}(\tilde{z},t)\Delta(\tfrac{1}{2}(z-\tilde{z}),\zeta)d\tilde{z}d\zeta,$$

in which $z = Z(x,t)$ and $\zeta = {}^tZ_x(x,t)^{-1}\xi$, converges uniformly to $u(x,t)$ in U. This is where the validity of (2.12) for some $\varepsilon < 1/2$ is important: for it implies that, in (3.7),

$$\mathrm{Re}\{i\zeta\cdot(z-\tilde{z}) - \tfrac{1}{2}\langle\zeta\rangle\langle z-\tilde{z}\rangle^2\} \le 0,$$

and, as a consequence, that the limit of (3.7) when $\delta \to +0$, is equal to the limit of the similar integral

$$(3.8)\qquad (2\pi)^{-m}\int_{\xi\in\mathbb{R}_m}\int_{\tilde{z}\in\tilde{X}_t} e^{i\theta\cdot(z-\tilde{z})-\delta\langle\theta\rangle^2}\,\tilde{u}(\tilde{z},t)d\tilde{z}d\theta,$$

where $\theta = \zeta + i\langle\zeta\rangle(z-\tilde{z})$ and $z = Z(x,t)$, $\zeta = {}^tZ_x(x,t)^{-1}\xi$ as before. That the integral (3.8) converges to $u(x,t)$ is well-known and easy to verify. ∎

Corollary 3.4. *Let V_1 and V_2 be as in Th. 3.3. Given any compactly supported distribution u in U, when $\delta \to +0$, $\mathfrak{F}_K^\delta(u;Z(x,t),t)$ converges in $\mathcal{D}'(V_1\times V_2)$.*

Definition 3.5. *We shall say that two distributions u_1 and u_2 in $U = U_1\times U_2$ are equivalent and we shall write $u_1 \approx u_2$ if there are open neighborhoods of the origin $V_1 \subset U_1$, $V_2 \subset U_2$ and $\mathcal{O} \subset \mathbb{C}^m$ such that the image of $V_1\times V_2$ under the map $(x,t) \to Z(x,t)$ is contained in $\mathcal{O}$ and if there is a function $\tilde{h} \in \mathcal{H}(\mathcal{O};\mathcal{D}'(V_2))$ such that $u_1-u_2 = \tilde{h}(Z(x,t),t)$ in $V_1\times V_2$.*

The following result relates Def. 3.5 to Definitions 2.9 and 2.10:

Theorem 3.6. *In order that a distribution in $\mathfrak{m}$ be equivalent to zero it is necessary and sufficient that its FBI transform be equivalent to zero.*

Proof: Let $\mathfrak{F}(u;z,\zeta,t)$ be equivalent to zero, i.e., be exponentially decaying as ζ goes to ∞ within a cone in $\mathbb{C}_m$ which contains $\mathbb{R}_m\backslash\{0\}$. We can choose the compact neighborhood $K \subset U$ and the open neighborhood $\mathcal{O} \subset \mathbb{C}^m$ sufficiently small that, when $z \in \mathcal{O}$, $z' = Z(x',t)$,

$\zeta = {}^tZ(x',t)^{-1}\xi$ for $(x',t) \in K$ and $\xi \in \mathbb{R}_m$, the growth of the absolute value of the factor $e^{i\zeta\cdot(z-z')-\langle\zeta\rangle\langle z-z'\rangle^2}$ is completely compensated by the decay of $\mathfrak{F}(u;z',\zeta,t)$, so that the integrals $\mathfrak{F}_K^\delta(u;z,t)$ converge, as $\delta \to +0$ and in the appropriate sense, to a function $\tilde{h}(z,t) \in \mathcal{H}(\mathcal{O};\mathcal{D}'(U_2))$. The sufficiency of the condition follows then at once from Th. 3.3.

We shall not give here the proof of the necessity as it is a duplication of the proof of Th. 4.1, Ch. I, in [BCT]. Moreover we shall prove a more precise result in the next section. ∎

Theorem 3.6 entails that the equivalence class (in the sense of Def. 3.5) of the limit, as $\delta \to +0$, of the "integrals" $\mathfrak{F}_K^\delta(u;Z(x,t),t)$ is equal to that of $u(x,t)$. This is the sense in which those integrals provide us with the inversion of the FBI transform.

Theorem 3.6 shows that the FBI transform (Def. 2.10) defines an injection of the space of equivalence classes (in the sense of Def. 3.5) of distribtuions in $\mathfrak{M}$, into the space of equivalence classes (in the sense of Def. 2.9) of holomorphic functions of (z,ζ) in the set (2.1) valued in $\mathcal{D}'(U_2)$. It would not be exceedingly difficult to characterize, by means of a Paley-Wiener type theorem, the precise subspace of such classes of functions which is the range of that injection: the holomorphic functions are those that can be substituted for $\mathfrak{F}(u;z,\zeta,t)$ in the integrals $\mathfrak{F}_K^\delta$ without impairing the convergence of $\mathfrak{F}_K^\delta$ in the distribution sense, in some neighborhood of 0. Denote by $\mathfrak{F}\mathcal{D}'$ the range in question; and denote by $\mathfrak{F}\mathcal{D}'(\Lambda^{p,q})$ the space of differential forms of bidegree (p,q),

$$F = \sum_{|I|=p,\,|J|=q} F_{I,J}(z,\zeta,t)\, dz^I \wedge dt^J,$$

whose coefficients belong to $\mathfrak{F}\mathcal{D}'$. We can form the complex

$$d_t : \mathfrak{F}\mathcal{D}'(\Lambda^{p,q}) \to \mathfrak{F}\mathcal{D}'(\Lambda^{p,q+1}), \quad q=0,1,\dots . \tag{3.9}$$

On the other hand we may define the quotient space of the space $\mathcal{D}'$ of distributions in $\mathfrak{M}$ modulo the equivalence relation of Def. 3.5. Denote it by $\dot{\mathcal{D}}'$, and denote by $\dot{\mathcal{D}}'(\Lambda^{p,q})$ the corresponding space of differential forms of bidegree (p,q) (whose representatives have expressions of the kind (1.19), Ch. II). Then we have the differential complex (derived from the distribution analogue of the complex (1.8), Ch. II):

$$d' : \dot{\mathcal{D}}'(\Lambda^{p,q}) \to \dot{\mathcal{D}}(\Lambda^{p,q+1}), \quad q=0,1,\dots . \tag{3.10}$$

Formula (2.5) implies that <u>the FBI transform is an isomorphism</u>

of the differential complex (3.10) onto the differential complex (3.9).

Theorem 3.6 has a C^∞ analogue which can also be regarded as a converse to Prop. 2.6. Notice that the following property only depends on the equivalence class (in the sense of Def. 3.5) of the compactly supported distribution u in U:

(3.11) There is an open neighborhood $U' \subset U$ of the origin such that $\mathfrak{F}(u;z,\zeta,t)$ is a C^∞ function in $\tilde{\mathfrak{R}}_U$, and that, given any $\beta \in \mathbb{Z}^n_+$ and any integer $k \geq 0$, there is a constant $C_{\beta,k} > 0$ such that (2.23) holds for all $(z,\zeta,t) \in \tilde{\mathfrak{R}}_{U'}$.

Theorem 3.7. In order that a distribution in $\mathfrak{M}$ be a C^∞ function in some open neighborhood of the origin it is necessary and sufficient that its FBI transform have property (3.11).

The necessity is of course stated in Prop. 2.6. The sufficiency follows easily from Theorem 3.3 and from differentiation under the integral sign in the integrals $\mathfrak{F}^\delta_K(u;Z(x,t),t)$. We leave the details to the reader.

4. HOLOMORPHIC EXTENSION OF DISTRIBUTIONS TO WEDGES AND EXPONENTIAL DECAY OF THEIR FBI TRANSFORM

We continue to reason in the hypo-analytic chart (U,Z). From now on we restrict the kind of distributions in $U = U_1 \times U_2$ we deal with: they will belong to the space $C^0(U_2;\mathcal{D}'(U_1))$ of continuous functions of $t \in U_2$ valued in the space of distributions of x, $\mathcal{D}'(U_1)$. This will help simplify the statements, without compromising our current aim.

In the present section we make the connection between holomorphic extension to wedges $\tilde{\mathfrak{W}}_\delta(U,\Gamma)$ (see (2.6), Ch. II) of distributions in $\tilde{U}$ and the exponential decay, as $\zeta \to \infty$ along certain cones, of the FBI transform of their pull-backs to U.

We shall denote by $\mathcal{S}(\tilde{\mathfrak{W}}_\delta(U,\Gamma))$ the space of continuous functions $\tilde{u}$ in the open subset $\tilde{\mathfrak{W}}_\delta(U,\Gamma)$ of $\mathbb{C}^m \times \mathbb{R}^n$ which are holomorphic with respect to z and have the following property:

(4.1) Given any open neighborhood U' of 0 whose closure is compact and contained in U and any open convex cone Γ', there are numbers r, C > 0 such that

(4.2) $$|\tilde{u}(z,t)| \leq C[\operatorname{dist}((z,t),\tilde{U})]^{-r}, \quad \forall\ (z,t) \in \tilde{\mathfrak{W}}_\delta(U',\Gamma').$$

Lemma 4.1. Suppose that (4.1) holds. Let K_o be any compact subset of $\Gamma \cap S^{m-1}$. Then, whatever $\chi \in C_c^\infty(U_1)$, the function in $U_2 \times K_o \times [0,\delta[$,

$$(t,v,\lambda) \to \int \tilde{u}(Z(x,t)+i\lambda v,t)\chi(x)dx$$

is continuous.

Proof: Set $\chi(x,t) = \chi(x)/\det Z_x(x,t)$ and, as usual, $dZ = (\det Z_x)dx$. Integrantion by parts shows that

$$\int (\partial\tilde{u}/\partial z^i)(Z(x,t)+i\lambda v,t)]\chi(x,t)dZ = \int M_i[\tilde{u}(Z(x,t)+i\lambda v,t)]\chi(x,t)dZ =$$
$$- \int \tilde{u}(Z(x,t)+i\lambda v,t)M_i\chi(x,t)dZ.$$

On the other hand,

$$\partial_\lambda \int \tilde{u}(Z(x,t)+i\lambda v,t)\chi(x,t)dZ = \int (v\cdot\partial_z\tilde{u})(Z(x,t)+i\lambda v,t)\chi(x,t)dZ$$

where $v\cdot\partial_z = v^1\partial/\partial z^1 + \ldots + v^m\partial/\partial z^m$. We reach the conclusion that

$$\partial_\lambda^k \int \tilde{u}(Z(x,t)+i\lambda v,t)\chi(x,t)dZ =$$
$$= \int \tilde{u}(Z(x,t)+i\lambda v,t)(-v\cdot M)^k\chi(x,t)dZ \quad (k=0,1,\ldots).$$

We conclude from (4.1) and from this that, to every compact subset K_2 of U_2 and to every integer $k \geq 0$, there is a constant $C_{K_2,k} > 0$ such that

$$\left|\partial_\lambda^k \int \tilde{u}(Z(x,t)+i\lambda v,t)\chi(x,t)dZ\right| \leq C_{K_2,k}\lambda^{-r},$$
$$\forall\, v \in K_o, \quad 0 < \lambda < \delta, \quad t \in K_2$$

This entails what we wanted. ∎

Thanks to Lemma 4.1 we can take the limit, as $\lambda \to +0$, of the integrals

$$(4.3) \qquad \int \tilde{u}(Z(x,t)+i\lambda v,t)\chi(x)dx.$$

Lemma 4.2. Let $\tilde{u}$ belong to $\mathcal{S}(\tilde{\mathfrak{W}}_\delta(U,\Gamma))$. The limit, as $\lambda \to +0$, of the integral (4.3) is independent of the vector v.

Proof: We have

$$\partial_v \int \tilde{u}(Z(x,t)+i\lambda v,t)\chi(x,t)dZ =$$
$$= -i\lambda \int \tilde{u}(Z(x,t)+i\lambda v,t)M\chi(x,t)dZ$$

where $M = (M_1,\ldots,M_m)$. But it follows from Lemma 4.1 that

$$\int \tilde{u}(Z(x,t)+i\lambda v,t)M\chi(x,t)dZ$$

is a continuous function of $\lambda \in [0,\delta[$. We conclude that

$$\lim_{\lambda\to+0} \partial_v \int \tilde{u}(Z(x,t)+i\lambda v,t)\chi(x,t)dZ = 0,$$

whence the result. ∎

Actually, in order to prove the main theorems of the present section, it is convenient to modify slightly our definition of a wedge. What we shall call a wedge with edge $\tilde{U}$ in the remainder of the present section will be a set of the following kind:

(4.4) $\hat{\mathfrak{W}}_\delta(U,\Gamma) = \{(z,t) \in \mathbb{C}^m \times \mathbb{R}^n;\ \exists\ (x,t) \in U,\quad v \in \Gamma$ such that $z = Z(x,t) + \sqrt{-1}Z_x(x,t)v,$ and $|y| < \delta\}.$

Here, as before, Γ is a convex cone in $\mathbb{R}^m$ and δ is a number >0. It is always assumed (cf. (1.3), Ch. I) that

$$|\operatorname{Im} Z(x,t)| < \delta, \quad \forall\ (x,t) \in U. \tag{4.5}$$

As we contract U about 0 and let δ go to zero the sets $\hat{\mathfrak{W}}_\delta(U,\Gamma)$ define a germ of set $\hat{\mathfrak{W}}(\mathfrak{m},\Gamma)$. This is what we shall call, now, a <u>germ of wedge</u> in $\mathbb{C}^m \times \mathbb{R}^n$ <u>with edge</u> $\mathfrak{m}$. Concerning the relation of the new wedges with those defined in (2.6), Ch. II, we can say that given any open and convex cone Γ' whose closure in $\mathbb{R}^m \setminus \{0\}$ is contained in Γ there is an open neighborhood $U' \subset U$ of the origin and a number δ', $0 < \delta' \leq \delta$, such that

$$\hat{\mathfrak{W}}_{\delta'}(U',\Gamma') \subset \tilde{\mathfrak{W}}_\delta(U,\Gamma), \qquad \tilde{\mathfrak{W}}_{\delta'}(U',\Gamma') \subset \hat{\mathfrak{W}}_\delta(U,\Gamma). \tag{4.6}$$

As a consequence,

$$\hat{\mathfrak{W}}(\mathfrak{m},\Gamma') \subset \tilde{\mathfrak{W}}(\mathfrak{m},\Gamma), \qquad \tilde{\mathfrak{W}}(\mathfrak{m},\Gamma') \subset \hat{\mathfrak{W}}(\mathfrak{m},\Gamma). \tag{4.7}$$

The analogues of Lemmas 4.1 and 4.2 for a function $\tilde{u} \in \mathcal{S}(\hat{\mathfrak{W}}_\delta(U,\Gamma))$ can be proved by a slight modification of the proofs of those lemmas; we leave the details to the reader. They allow us to define the boundary value of $\tilde{u}$ on the edge $\tilde{U}$. We shall denote by $b\tilde{u}$ its pull-back to U under the map $(x,t) \to (Z(x,t),t)$. It is a continuous function of $t \in U_2$ valued in $\mathcal{D}'(U_1)$,

$$U_2 \ni t \to \{C_c^\infty(U_1) \ni \chi \to \lim_{\lambda\to+0} \int \tilde{u}(Z(x,t)+i\lambda Z_x(x,t)v,t)\chi(x)dx\},$$

where v is an arbitrary vector in Γ. The map $\tilde{u} \to b\tilde{u}$ is a linear injection of $\mathcal{S}(\hat{\mathfrak{W}}_\delta(U,\Gamma))$ into $C^0(U_2;\mathcal{D}'(U_1))$. Indeed, any holomorphic function in a wedge in $\mathbb{C}^m$ which grows slowly at the edge and whose boundary value on the edge is equal to zero must vanish identically. It suffices to apply this for each fixed $t \in U_2$.

We shall adopt the "conic" terminology: if E is a vector bundle over a manifold, a conic subset of E is a subset whose intersection with every fibre of E is either empty or else a cone (with vertex at the origin of that fibre). Belw we apply this in $\mathbb{C}^m \times \mathbb{C}_m \times \mathbb{R}^n$ which we view as a vector bundle over $\mathbb{C}^m \times \mathbb{R}^n$ with fibre $\mathbb{C}_m$.

Let Γ be an open and convex cone in $\mathbb{R}^m$. We recall that the polar of Γ is the closed and convex cone

(4.8) $$\Gamma^o = \{\xi \in R_m;\ \forall\ U \in \Gamma,\ \xi\cdot v \geq 0\}.$$

In the sequel we shall denote by Γ^* the complement of Γ^o,

(4.9) $$\Gamma^* = \{\xi \in R_m;\ \exists\ v \in \Gamma,\ \xi\cdot v < 0\}.$$

Theorem 4.3. Assume that Condition (2.12) is satisfied for some ϵ, $0 < \epsilon < 1$. Let Γ be an open and convex cone in $\mathbb{R}^m$.

Given any open neighborhood $U' \subset U$ of the origin, any open and convex cone Γ' whose closure in $\mathbb{R}^m \setminus \{0\}$ is contained in Γ and any number $\delta > 0$, there is a conic and open neighborhood $\mathcal{C}$ of the set $\{0\} \times \Gamma'^* \times \{0\}$ in $\mathbb{C}^m \times \mathbb{C}_m \times \mathbb{R}^n$ and a constant $R > 0$ such that the following property of a compactly supported distribution u in U:

(4.10) There is a function $\tilde{u} \in \mathcal{S}(\hat{\mathcal{W}}_\delta(U',\Gamma'))$ such that $u = b\tilde{u}$ in U'

implies the following one:

(4.11) There is a constant $C > 0$ such that
$$|\mathfrak{F}(u;z,\zeta,t)| \leq C\, e^{-|\zeta|/R}, \quad \forall\ (z,\zeta,t) \in \mathcal{C}.$$

Proof: It is essentially a repetition of the proof of Theorem 2.1, Ch. II [BCT]. Let $g \in C_c^\infty(U')$ be equal to one in some neighborhood $V = V_1 \times V_2$ of 0. Denote by $\tilde{g}(z,t)$ the function in $\tilde{U}$ such that $g(x,t) = \tilde{g}(Z(x,t),t)$ for all (x,t). Consider the integral

$$\mathfrak{F}(gu;z,\zeta,t) = \int_{\tilde{X}_t} e^{i\zeta\cdot(z-\tilde{z})-\langle\zeta\rangle\langle z-\tilde{z}\rangle^2}\,(\tilde{g}\tilde{u})(\tilde{z},t)\,d\tilde{z}.$$

Here $\tilde{u}(\tilde{z},t)$ is the push-forward of $u(x,t)$ under the map $(x,t) \to (Z(x,t),t)$. It is also the boundary value on the subset $\tilde{X}_t \cap \tilde{U}'$ of the edge $\tilde{U}$ of the function $\tilde{u} \in \mathcal{S}(\hat{\mathcal{W}}_\delta(U',\Gamma))$. Select a function $g_o \in C_c^\infty(V_1)$ such that $0 \leq g_o \leq 1$ everywhere, and $g_o = 1$ in a neighborhood $\mathcal{N}_1$ of 0; call $\tilde{g}_o(z,t)$ the function in $\tilde{U}$ such that $g_o(x) = \tilde{g}_o(Z(x,t),t)$. If t stays in V_2, the hypothesis (4.10) allows us to deform the domain of integration, in the integral $\mathfrak{F}(gu;z,\zeta,t)$, from $\tilde{X}_t$ to the image of $\tilde{X}_t$ under the map $\tilde{z} \to \tilde{z} + i\lambda\tilde{g}_o(\tilde{z},t)\tilde{M}(\tilde{z},t)v$, where $\tilde{M}(\tilde{z},t)$ is the push-forward, from U to $\tilde{U}$, of the matrix-valued function $Z_x(x,t)$ and where v is an arbitrary unit vector in Γ and $0 < \lambda < \delta_o$. The number $\delta_o > 0$ is sufficiently small that

$$|\mathrm{Im}\{Z(x,t)+i\lambda g_o(x)Z_x(x,t)v\}| < \delta, \quad \forall\ (x,t) \in V,\ 0 < \lambda < \delta_o.$$

We must then study

$$\mathcal{D} = -\mathrm{Re}\{i\zeta\cdot(z-\tilde{z}-i\lambda\tilde{g}_o(\tilde{z},t)\tilde{M}(\tilde{z},t)v) - \langle\zeta\rangle\langle z-\tilde{z}-i\lambda\tilde{g}_o(\tilde{z},t)\tilde{M}(\tilde{z},t)v\rangle^2\},$$

where $\tilde{z} \in \tilde{X}_t$. Take $(z,\zeta) \in \widetilde{\mathbb{R}T}'_t$, i.e., $z \in \tilde{X}_t$, $\zeta = {}^t\tilde{M}(\tilde{z},t)^{-1}\xi$ for some $\xi \in R_m$. We have:

$$\mathcal{D} \geq -\mathrm{Re}\{i\zeta\cdot(z-\tilde{z})-\langle\zeta\rangle\langle z-\tilde{z}\rangle^2\} - \lambda\tilde{g}_o(\tilde{z},t)\xi\cdot v - \epsilon_1|\zeta||z-\tilde{z}|^2 - C_{\epsilon_1}\lambda^2\tilde{g}_o(\tilde{z},t)|\xi|,$$

where $\epsilon_1 > 0$ will now be chosen ($C_{\epsilon_1} > 0$ depends on ϵ_1).

Suppose now that $\xi \in \Gamma^*$ and select v so as to have $\xi\cdot v < 0$. By virtue of (2.12) we have

$$\mathcal{D} \geq (1-\epsilon-\epsilon_1)|\zeta||z-\tilde{z}|^2 + (|\xi\cdot v|-C_{\epsilon_1}\lambda|\xi|)\lambda\tilde{g}_o(\tilde{z},t).$$

First we take $\epsilon_1 < (1-\epsilon)/2$. Next we choose $\lambda = |\xi\cdot v|/(2C_{\epsilon_1}|\xi|)$. We take into account (1.2); on supp $\tilde{g}$ we have, for some constant $C > 0$,

$$C\mathcal{D}/|\xi| \geq |z-\tilde{z}|^2 + (|\xi\cdot v|/|\xi|)^2\,\tilde{g}_o(\tilde{z},t).$$

Take $z = Z(0,t)$; we have $\tilde{z} = Z(x,t)$ for some $x \in U_1$ and therefore $C\mathcal{D}/|\xi| \geq |z-\tilde{z}|^2 \geq |x|^2$. If $x \in h_1$ we have $C\mathcal{D}/|\xi| \geq (|\xi\cdot v|/|\xi|)^2$. Thus $\mathcal{D} \geq c(\xi)|\zeta|$ for some $c(\xi) > 0$. After a decrease of $c(\xi)$ this will remain true for all (z,ζ) in a suitable "thin" conic (and open) neighborhood C_t of $(Z(0,t), {}^tZ_x(0,t)^{-1}\xi)$ in $\mathbb{C}^m\times\mathbb{C}_m$. The "conic diameter" of C_t i.e., the diameter of its intersection with $\mathbb{C}^m\times S^{2m-1}$, depends only on the constant $c(\xi)$. It follows from this that, if we decrease further $c(\xi)$ and restrict the variation of t to a sufficiently small open neighborhood $W_2 \subset V_2$ of the origin, the inequality $\mathcal{D} \geq c(\xi)|\zeta|$ will be true in a conic and open neighborhood of $(0,\xi)$ in $\mathbb{C}^m\times\mathbb{C}_m$. Of course we can take $c(\xi)$ to be a continuous function in Γ^*, homogeneous of degree zero. It suffices then to observe that the closure of the cone Γ'^* is contained in the open cone Γ^*, to reach the desired conclusion. ■

Theorem 4.4. *Assume that Condition* (2.12) *is satisfied for some* ϵ, $0 < \epsilon < 1/2$. *Let* Γ *be an open and convex cone in* $\mathbb{R}^m$, C *a conic and open neighborhood of* $\{0\}\times\Gamma^*\times\{0\}$ *in* $\mathbb{C}^m\times\mathbb{C}_m\times\mathbb{R}^n$ *and* R *a constant* >0.

To every open and convex cone Γ' *whose closure in* $\mathbb{R}^m\setminus\{0\}$ *is contained in* Γ *there is an open neighborhood* $U'\subset U$ *of* 0 *and a number* $\delta'> 0$ *such, whatever the compactly supported distribution* $u \in C^0(U_2;\mathcal{D}'(U_1))$, *Property* (4.11) *entails the following property:*

(4.12) <u>There is a function</u> $\tilde{u} \in \mathcal{S}(\hat{\mathfrak{W}}_\delta, (U',\Gamma'))$ <u>such that</u> $u = b\tilde{u}$ <u>in</u> U'.

<u>Proof:</u> It is essentially a repetition of the proof of Th. 2.2, Ch.II [BCT]. Let Γ' be as in the statement; select a third open and convex cone Γ'', such that the closure of Γ'' is contained in Γ and that the closure of Γ' is contained in Γ''. We shall use the fact that the closure of Γ'^* is contained in Γ''^* and there is an open neighborhood $V = V_1 \times V_2 \subset U$ of the origin such that $\mathcal{C}$ contains the image of $V_1 \times \Gamma''^* \times V_2$ under the map $(x,\xi,t) \to (Z(x,t), {}^tZ_x(x,t)^{-1}\xi, t)$. Consider the integral (3.1) where we take K to be a compact neighborhood of 0 in x-space, $K \subset V_1$. Call $J_1^\delta(z,t)$ the portion of that integral in which the integration with respect to ξ is carried out over Γ''^*, $J_2^\delta(z,t)$ the portion in which the integration is carried out over Γ''^o (which is the complement of Γ''^* with respect to $\mathbb{R}^m$). Thanks to (4.11) we derive at once that, when $\delta \to +0$, J_1^δ converges uniformly in $\mathcal{O} \times V_2$ where $\mathcal{O}$ is an open neighborhood of 0 in $\mathbb{C}^m$. Its limit is a continuous function of (z,t) in $\mathcal{O} \times V_2$ which is holomorphic with respect to z.

We concentrate on $J_2^\delta(z,t)$ $(t \in V_2)$. Let us write $z = Z(x,t) + iZ_x(x,t)v$ with $v \in \Gamma'$. We have $\tilde{z} = Z(x',t)$, $\zeta = {}^tZ_x(x',t)^{-1}\xi$ for some $x' \in K \subset V_1$, $\zeta \in \Gamma''^o$. Observe that there is a constant $c_o > 0$ such that

(4.13) $$\xi \cdot v \geq c_o |v|\,|\xi|, \quad \forall\; v \in \Gamma',\; \xi \in \Gamma''^o.$$

In the notation of the proof of Theorem 4.3 we may write

$$\begin{aligned}\mathfrak{D} = -\mathrm{Re}\{i\zeta\cdot(z-\tilde{z}) - \langle\zeta\rangle\langle z-\tilde{z}\rangle^2\} \geq \\ -\mathrm{Re}\{i\zeta\cdot[Z(x,t)-Z(x',t)] - \langle\zeta\rangle\langle Z(x,t)-Z(x',t)\rangle^2]\} \\ +\mathrm{Re}\; \xi\cdot Z_x(x',t)^{-1}Z_x(x,t)v - C|\zeta|\,|v|\,|Z(x,t)-Z(x',t)| - C|\zeta|\,|v|^2.\end{aligned}$$

Recalling that $Z(x,t) = x + i\phi(x,t)$ we have

(4.14) $$\|Z_x(x',t)^{-1}Z_x(x,t) - I\| \leq \mathrm{const.}|x-x'|.$$

By combining (4.13) and (4.14) we see that, given any $\epsilon_1 > 0$ there is $C_{\epsilon_1} > 0$ such that

$$\begin{aligned}\mathfrak{D} \geq -\mathrm{Re}\{i\zeta\cdot[Z(x,t)-Z(x',t)] - \langle\zeta\rangle\langle Z(x,t)-Z(x',t)\rangle^2]\} \\ + c_o|v|\,|\xi| - \epsilon_1|\zeta|\,|x-x'|^2 - C_{\epsilon_1}|v|^2|\xi|.\end{aligned}$$

Observing that $|x-x'| \leq |Z(x,t)-Z(x',t)|$ we apply (2.12) (with $\epsilon < 1/2$):

$$\mathfrak{D} \geq (1-\epsilon-\epsilon_1)|\zeta|\,|Z(x,t)-Z(x',t)|^2 + c_o|v|\,|\xi|/2$$

provided $|v|$ is sufficiently small. We apply Prop. 2.4 with $\ell = 0$ (since u is a continuous function of t with values in $\mathcal{D}'(U_1)$). We derive for all λ, $u > 0$,

$$|J_2^\lambda(Z(x,t)+i{}^tZ_x(x,t)v,t)| \le \text{const.} \int e^{-c|v||\xi|-\lambda|\xi|^2}(1+|\xi|)^k d\xi ;$$

$$|J_2^\lambda(Z(x,t)+i{}^tZ_x(x,t)v,t)-J_2^\mu(Z(x,t)+i{}^tZ_x(x,t)v,t)| \le$$
$$\text{const.} \int e^{-c|v||\xi|}\, |e^{-\lambda|\xi|^2}-e^{-\mu|\xi|^2}|(1+|\xi|)^k d\xi .$$

Thus, provided δ' is sufficiently small,

$$|v|^{k+m}\, J_2^\lambda(Z(x,t)+i{}^tZ_x(x,t)v,t)$$

converges in $\hat{\mathfrak{W}}_{\delta'}(V,\Gamma')$ to a continuous and bounded function u_o. It follows from Theorem 3.3 that, in a sufficiently small open neighborhood of the origin, $U' \subset V$, u is equal to the sum of the boundary value of u_o, bu_o, and of the pull-back (under the map $(x,t) \to (Z(x,t),t)$) of the restriction to $\tilde{U}'$ of a continuous function $\tilde{h}$ in some open neighborhood of $\tilde{U}'$ in $\mathbb{C}^m \times \mathbb{R}^n$ which is holomorphic with respect to z. ∎

If we assume that $u \in C_c^\infty(U)$ and if we apply Prop. 2.6 instead of Prop. 2.4 we obtain, by the same reasoning as in the proof of Th. 4.4, the following result:

<u>Theorem 4.5</u>. <u>Assume that Condition</u> (2.12) <u>is satisfied for some</u> ϵ, $0 < \epsilon < 1/2$. <u>Let</u> Γ <u>be an open and convex cone in</u> $\mathbb{R}^m$, $\mathcal{C}$ <u>a conic and open neighborhood of</u> $\{0\}\times\Gamma^*\times\{0\}$ <u>in</u> $\mathbb{C}^m\times\mathbb{C}_m\times\mathbb{R}^n$ <u>and</u> R <u>a constant</u> >0.

<u>To every open and convex cone</u> Γ' <u>whose closure in</u> $\mathbb{R}^m\backslash\{0\}$ <u>is contained in</u> Γ <u>there is an open neighborhood</u> $U' \subset U$ <u>of the origin and a number</u> $\delta' > 0$ <u>such, whatever the compactly supported</u> C^∞ <u>function</u> u <u>in</u> U, <u>Property</u> (4.11) <u>entails the following property</u>:

(4.15) <u>There is a function</u> $\tilde{u} \in \mathcal{B}^\infty(\hat{\mathfrak{W}}_{\delta'}(U',\Gamma'))$ (see Section 1, Ch. I) <u>which is holomorphic with respet to</u> z <u>and such that</u> $u = b\tilde{u}$ <u>in</u> U'.

That $u = b\tilde{u}$ with $\tilde{u} \in \mathcal{B}^\infty(\hat{\mathfrak{W}}_{\delta'}(U',\Gamma'))$ simply means that u is equal to the pull-back, under the map $(x,t) \to (Z(x,t),t)$, of the restriction to the edge $\tilde{U}'$ of the function $\tilde{u}$ regarded as a C^∞ function in the closure of $\hat{\mathfrak{W}}_{\delta'}(U',\Gamma')$.

From the preceding theorems there follow various versions of the theorem on <u>the edge of the wedge</u>. We shall content ourselves with the following result:

Theorem 4.6. *Assume that Condition* (2.12) *is satisfied for some* ϵ, $0 < \epsilon < 1/2$. *Let* Γ_ι $(\iota = 1, \ldots, \nu)$ *be convex and open cones in* $\mathbb{R}_m \backslash \{0\}$. *Let* $u \in C^0(U_2; \mathcal{D}'(U_1))$ *have compact support as a distribution in* U. *The property*

(4.16) *whatever* $\iota = 1, \ldots, \nu$, *there is a function* $\tilde{u}_\iota \in \mathcal{S}(\tilde{\mathfrak{W}}_\delta(U, \Gamma_\iota))$ *such that* $u = b\tilde{u}_\iota$ *in* U,

entails that

(4.17) *to every open and convex cone* Γ' *whose closure in* $\mathbb{R}^m \backslash \{0\}$ *is contained in the convex hull of* $\Gamma_1 \cup \ldots \cup \Gamma_\nu$ *there is an open neighborhood* $U' \subset U$ *of* 0, *a number* $\delta' > 0$ *and a function* $\tilde{u} \in \mathcal{S}(\tilde{\mathfrak{W}}_{\delta'}(U', \Gamma'))$ *such that* $u = b\tilde{u}$ *in* U'.

Proof: Theorem 4.3 implies that (4.11) holds if we take C to be a conic open neighborhood in $\mathbb{C}^m \times \mathbb{C}_m \times \mathbb{R}^n$ of $\{0\} \times (\Gamma_1^* \cup \ldots \cup \Gamma_\nu^*) \times \{0\}$. But $\Gamma_1^* \cup \ldots \cup \Gamma_\nu^*$ is the complement of the intersection of the polars Γ_ι^0. This intersection is equal to the polar of the convex hull of $\Gamma_1 \cup \ldots \cup \Gamma_\nu$, whence the result by Theorem 4.4. ■

Remark 4.7. If the convex hull of $\Gamma_1 \cup \ldots \cup \Gamma_\nu$ is equal to the whole of $\mathbb{R}^m$ we can conclude, by the same argument as in the proof of Theorem 4.6 and by applying the sufficiency part in Theorem 3.6, that the distribution u is equivalent to zero (Def. 3.5). ■

5. DELIMITATION OF THE MICROSUPPORT BY MEANS OF THE FBI TRANSFORM

In the present section we establish a link between the concepts and results of Ch. II and the theory of the FBI transform. The latter continues to be defined in the hypo-analytic chart (U, Z) used in all the preceding sections. We shall always reason under the hypothesis that (2.12) holds for some ϵ, $0 < \epsilon < 1$.

We need the microlocal analogue of Def. 2.9. Here, however, we limit ourselves to the FBI transforms $\mathfrak{F}(u; z, \zeta, t)$ of distributions $u(x,t)$ in $U = U_1 \times U_2$ which are continuous functions of $t \in U_2$ valued in $\mathcal{D}'(U_1)$.

We say that the FBI transform $\mathfrak{F}(u; z, \zeta, t)$ of such a distribution u *decays exponentially in the direction of* $\xi^0 \in \mathbb{R}_m \backslash \{0\}$ if there is a conic and open neighborhood C of $(0, \xi^0, 0)$ in $\mathbb{C}^m \times \mathbb{C}_m \times \mathbb{R}^n$, a compactly supported distribution v, equal to u in an open neighborhood of the base projection of C, and constants $C, R > 0$ such that

$$(5.1) \qquad |\mathcal{F}(v;z,\zeta,t)| \leq C\, e^{-|\zeta|/R}, \quad \forall\ (z,\zeta,t) \in \mathcal{C}.$$

Lemma 2.7 shows that the property that $\mathcal{F}(u;z,\zeta,t)$ decays exponentially in the direction of ξ_o does not depend on the choice of the compactly supported distribution U equal to u in an open neighborhood of 0.

Theorem 3.6 may now be rephrased as follows: In order that u be equivalent to zero (in the sense of Def. 3.5) it is necessary and sufficient that $\mathcal{F}(u;z,\zeta,t)$ decay exponentially in the direction of every vector in $\mathbb{R}_m\setminus\{0\}$.

In the statement that follow we identify $\mathbb{R}_m$ to the cotangent space to the maximally real submanifold $X_o = \{(x,t) \in U;\ t=0\}$ of $\mathfrak{M}$ and also (thanks to (1.12), (1.13), Ch. II) to the fibre of the bundle $\tilde{\mathcal{R}}_U$ at the origin (see Section 1, present chapter). We recall that there is an injection of the characteristic set T^o over U into $\tilde{\mathcal{R}}_U$ naturally associated with the map $(x,t) \to (Z(x,t),t)$. Thus to any characteristic cotangent vector to $\mathfrak{M}$ at 0, $\bar{\omega}$, we can associate a unique element ξ^o of $\mathbb{R}_m$ identified to the fibre of $\tilde{\mathcal{R}}_U$ at the origin.

<u>Theorem 5.1</u>. <u>Let</u> $\bar{\omega} \neq 0$ <u>be a characteristic cotangent vector to</u> $\mathfrak{M}$ <u>at the origin and let</u> $\xi^o \in \mathbb{R}_m\setminus\{0\}$ <u>be the associated covector</u>. <u>If a cohomology class</u> $[h] \in H'^{p,q}$ <u>vanishes at</u> $\bar{\omega}$ (Def. 4.2, Ch. II) <u>then, possibly after contracting</u> U <u>about</u> 0, <u>there is a standard cocycle representing</u> [h] <u>in the hypo-analytic chart</u> (U,Z),

$$(5.2) \qquad f = \sum_{|I|=p,\,|J|=q} f_{I,J}(x,t)\,dZ^I \wedge dt^J$$

<u>such that</u>

(5.3) $\forall$ I, J, $|I| = p$, $|J| = q$, $\mathcal{F}(f_{I,J};z,\zeta,t)$ <u>decays exponentially in the direction of</u> ξ^o.

<u>Proof</u>: Assume (Def. 4.2, Ch. II) that

$$[h] = \sum_{\iota=1}^{\nu} [h_\iota]$$

where, for each $\iota = 1,\dots,\nu$, there is an open and convex cone $\Gamma_\iota \subset \mathbb{R}^m\setminus\{0\}$ which is definite-negative with respect to $\bar{\omega}$ (Def. 3.1, Ch. II) and is such that the cohomology class $[h_\iota]$ extends to $\tilde{\mathfrak{W}}(\mathfrak{M},\Gamma_\iota)$. In turn the latter means that, if the neighborhood U and the number $\delta > 0$ are small enough, there is a cocycle in the wedge $\tilde{\mathfrak{W}}_\delta(U,\Gamma_\iota)$ (for the natural hypo-analytic structure of $\mathbb{C}^m\times\mathbb{R}^n$),

$$\tilde{f}_\iota = \sum_{|I|=p,\,|J+K|=q} \tilde{f}_{\iota,I,J,K}(z,t)\, dz^I \wedge d\bar{z}^J \wedge dt^K,$$

whose coefficients belong to $\mathcal{B}^\infty(\tilde{\mathfrak{W}}_\delta(U,\Gamma_\iota))$ and whose pull-back to U, under the map $(x,t) \to (Z(x,t),t)$, is a representative of $[h_\iota]$.

We apply Th. 2.5, Ch. II: after contracting U about 0 and Γ_ι about one of its rays, r_ι, and after decreasing δ we may assume that the coefficients $\tilde{f}_{\iota,I,J}$ have property (2.14), Ch. II. In other words we may write

$$\tilde{f}_\iota = \sum_{|I|=p,\,|J|=q} \tilde{f}_{\iota,I,J}(z,t)\, dz^I \wedge dt^J,$$

with $\bar{\partial}\tilde{f}_{\iota,I,J} = 0$ in $\tilde{\mathfrak{W}}_\delta(U,\Gamma_\iota)$ for every I, J. Let then

$$f_\iota = \sum_{|I|=p,\,|J|=q} f_{\iota,I,J}(x,t)\, dZ^I \wedge dt^J,$$

be the pull-back of $\tilde{f}_\iota$ to U. f_ι is a standard representative of $[h_\iota]$ in U. We apply Th. 4.3 to each coefficient $f_{\iota,I,J}$: there is a conic open neighborhood $\mathcal{C}_\iota$ of $\{0\}\times\Gamma_\iota^*\times\{0\}$ in $\mathbb{C}^m\times\mathbb{C}_m\times\mathbb{R}^n$, and constants $C,R > 0$ such that, for all I, J,

(5.4) $$|\mathfrak{F}(f_{\iota,I,J};z,\zeta,t)| \le C\, e^{-|\zeta|/R}, \quad \forall\ (z,\zeta,t) \in \mathcal{C}_\iota .$$

But since Γ_ι is definite-negative with respect to $\bar{\omega}$ we have $\xi^0\in\Gamma_\iota^*$. In order to reach the sought conclusion it suffices to define

$$f = \sum_{\iota=1}^{\nu} f_\iota .$$

■

Throughout the remainder of this section we assume that Condition (2.12) is satisfied for some ϵ, $0 < \epsilon < 1/2$.

Let Γ be an open and convex cone in $\mathbb{R}_m\backslash\{0\}$. We shall make the following hypothesis:

(5.5) <u>There is a standard cocycle</u> f, <u>as in</u> (5.2), <u>which represents the cohomology class</u> [h] <u>in the hypo-analytic chart</u> (U,Z), <u>and which has the following property</u>:

(5.6) $\forall$ I, J, $|I| = p$, $|J| = q$, $\mathfrak{F}(f_{I,J};z,\zeta,t)$ <u>decays expoentially in the direction of every covector</u> $\xi \in \Gamma^*$.

We recall that $\Gamma^* = \{\xi \in \mathbb{R}_m\backslash\{0\};\ \exists\ v \in \Gamma,\ \xi\cdot v < 0\}$.

<u>Theorem 5.2. Under the preceding hypotheses, to any open and convex cone</u> Γ' <u>whose closure in</u> $\mathbb{R}_m\backslash\{0\}$ <u>is contained in</u> Γ <u>there is an open neighborhood</u> $U'\subset U$ <u>of</u> 0, <u>a number</u> $\delta' > 0$ <u>and a cocycle in</u> $\hat{\mathfrak{W}}_{\delta'}(U',\Gamma')$,

$$\tilde{f} = \sum_{|I|=p,|J|=q} \tilde{f}_{I,J}(z,t)dz^I \wedge dt^J,$$

<u>such that, whatever</u> I, J, $|I| = p$, $|J| = q$, <u>the following is valid:</u>

(5.7) <u>The function</u> $\tilde{f}_{I,J} \in \mathcal{B}^\infty(\hat{\mathfrak{W}}_{\delta'}(U',\Gamma'))$ <u>is holomorphic with respect to</u> z <u>and</u> $f_{I,J} = b\tilde{f}_{I,J}$ <u>in</u> U'.

<u>Proof</u>: That there exists a form $\tilde{f}$ of bidegree (p,q) in $\hat{\mathfrak{W}}_{\delta'}(U',\Gamma')$ with Property (5.7) is a direct consequence of Th. 4.5. All we have to do is to show that $\tilde{f}$ is a d_t-closed. Consider the function in $\hat{\mathfrak{W}}_{\delta'}(U',\Gamma')$,

(5.8)
$$\sum_{j,J} \varepsilon(j,J)\partial\tilde{f}_{I,J}/\partial t^j,$$

where the summation convention, on j and J, is the same as that in (1.22), Ch. II. The value of (5.8) on the edge $\tilde{U}'$ is precisely equal to

$$\sum_{j,J} \varepsilon(j,J)L_j[f_{I,J}(Z(x,t),t)],$$

which vanishes since, by hypothesis, f is a cocycle. We conclude that the function (5.8) itself vanishes, which proves that $\tilde{f}$ is a cocycle. ■

<u>Corollary 5.3.</u> <u>Under the hypotheses of Th. 5.2 the cohomology class</u> $[h]$ <u>vanishes at every characteristic cotangent vector</u> $\bar{\omega}$ <u>whose associated covector</u> ξ <u>belongs to</u> Γ^*.

In the terminology of Def. 2.2 one can rephrase property (5.6) by saying that <u>the FBI transform of the standard cocycle</u> f <u>decays exponentially in the direction of every covector</u> $\xi \in \Gamma^*$.

<u>Remark 5.4</u>. One can add some precision to Th. 5.2 by combining it with Prop. 2.7, Ch. II. Suppose that the section $\dot{f} \in C^\infty(U;\Lambda^{p,q})$ $(q \geq 1)$ extends to the wedge $\tilde{\mathfrak{W}}_\delta(U,\Gamma)$. Let f be the standard representative of $\dot{f}$ in U. After contracting U about 0, Γ about one of its rays and after decreasing δ we can find a standard form $g \in C^\infty(U;\Lambda^{p,q-1})$ such that $f - Lg$ is the pull-back to U of a standard cocycle $\tilde{f}$ of bidegree (p,q) in $\tilde{\mathfrak{W}}_\delta(U,\Gamma)$ which has property (2.14), Ch. II.

Apply then (2.5): we reach the conclusion that

(5.9) $\mathfrak{F}(f;z,\zeta,t) - d_t\mathfrak{F}(g;z,\zeta,t)$ <u>decays exponentially in the direction of every covector</u> $\xi \in \Gamma^*$.

Conversely, the existence of a form $g \in C^\infty(U;\Lambda^{p,q-1})$ such that (5.9) holds will yield an extension of $\dot{f}$ in the sense of Th.5.2. ■

CHAPTER IV

MICROLOCAL COHOMOLOGY IN A CR STRUCTURE

1. HYPO-ANALYTIC MINIFIBRATIONS IN CR MANIFOLDS

Throughout the present chapter the hypo-analytic structure $\mathcal{G}$ on the manifold $\mathfrak{M}$ will be a CR structure. This means that the underlying cotangent and tangent structure bundles T' and $\mathcal{V} = T'^{\perp}$ are such that

$$\mathbb{C}T^*\mathfrak{M} = T' + \bar{T}', \qquad \mathcal{V} \cap \bar{\mathcal{V}} = 0. \tag{1.1}$$

A consequence of (1.1) is that $T' \cap \bar{T}'$ is a complex vector subbundle of $\mathbb{C}T^*\mathfrak{M}$; we shall denote by d its fibre dimension (over $\mathbb{C}$). Obviously it is generated by its real part which is the <u>characteristic set</u> T^0: in a CR structure T^0 is a real vector bundle, i.e., it is a vector subbundle of $T^*\mathfrak{M}$; its fibre dimension (over $\mathbb{R}$) is equal to d. By virtue of (1.1) we have $\dim \mathfrak{M} = m+n = 2m-d$ and thus

$$d = m-n. \tag{1.2}$$

We refer to the integer d as the <u>codimension</u> of the CR structure $\mathcal{G}$ on $\mathfrak{M}$. In the present chapter we shall always assume $n \geq 1$ and $d \geq 1$, that is to say, the structure is a "true" CR structure, not a totally real structure (for which $n=0$) nor a complex structure (for which $d=0$).

There exist hypo-analytic charts in the CR manifold $(\mathfrak{M},\mathcal{G})$ that reflect especially well the CR structure. Let (U,Z) be an arbitrary hypo-analytic chart (centered at 0). Since $dZ^1,\dots,dZ^m$ span T' over U $n = m-d$ of these differentials must be $\mathbb{C}$-linerly independent mod $T' \cap \bar{T}'$, which means that $dZ^1,\dots,dZ^n,d\bar{Z}^1,\dots,d\bar{Z}^n$ must be linearly independent. After suitable linear subsitutions (and, possibly, a contraction of U about 0) we can take $\operatorname{Re} Z^i$, $\operatorname{Im} Z^j$ ($i=1,\dots,m$, $j=1,\dots,n$) as coordinates. We shall modify slightly the notation used in the case of general (i.e., not necessarily CR) hypo-analytic structures. We shall systematically write

$$x^i = \operatorname{Re} Z^i, \quad y^i = \operatorname{Im} Z^i \quad (1 \le i \le n), \quad s^k = \operatorname{Re} Z^{n+k} \quad (1 \le k \le d).$$

We shall also write $z^i = x^i + iy^i$ $(i = \sqrt{-1})$, $w^k = Z^{n+k}$ $(1 \le k \le d)$. Possibly after further contraction of U about 0 and $\mathbb{C}$-linear substitution of the w^k, we can assume that, for all $k = 1,\dots,d$,

$$w^k = s^k + i\phi^k(z,\bar{z},s), \tag{1.3}$$

$$\phi^k\big|_0 = 0, \quad d\phi^k\big|_0 = 0. \tag{1.4}$$

Here $\phi = (\phi^1,\dots,\phi^d)$ is a C^∞ map $U \to \mathbb{R}^d$. Note that the Jacobian matrix of $w = (w^1,\dots,w^d)$ with respect to s is nonsingular at every point of U. It will be convenient to take U in the product form

$$U = \Delta \times \mathcal{B}, \tag{1.5}$$

where Δ is an open poldysk in z-space $\mathbb{C}^n$ and $\mathcal{B}$ is an open ball centered at the origin in s-space $\mathbb{R}^d$.

Thus, in the open neighborhood of 0, U, the hypo-analytic structure is defined by the C^∞ map

$$(x,y,s) \to (z,w) \colon U \to \mathbb{C}^n \times \mathbb{C}^d. \tag{1.6}$$

The image of U under the map (1.6) is a <u>generic submanifold</u> $\tilde{U}$ of $\mathbb{C}^n \times \mathbb{C}^d$; the (real) codimension of $\tilde{U}$ in $\mathbb{C}^{n+d}$ is equal to d. The first coordinate projection $(z,w) \to z$ maps $\tilde{U}$ onto the polydisk Δ, and both U and $\tilde{U}$ can be regarded as fibre bundles over Δ. The fibre at a point $z_0 \in \Delta$ in U is the submanifold $\Sigma_{z_0} = \{(x,y,s) \in U;\ x+iy = z_0\}$. In $\tilde{U}$ it is the image $\tilde{\Sigma}_{z_0}$ of Σ_{z_0} under the map

$$s \to w = s + i\phi(z_0,\bar{z}_0,s) \colon \mathcal{B} \to \mathbb{C}^d. \tag{1.7}$$

Observe that $\tilde{\Sigma}_{z_0}$ is a totally real submanifold of $\mathbb{C}^d$ of real dimension d (i.e., a maximally real submanifold of $\mathbb{C}^d$); (1.7) is a diffeomorphism.

Over U the cotangent structure bundle T' is spanned by the differentials dz^i, dw^k. Let $\mu = (\mu^k_\ell)_{1 \le k,\ell \le d}$ denote the inverse of the Jacobian matrix $w_s = \partial w/\partial s$. We may write

$$\sum_{\ell=1}^{d} \mu^k_\ell dw^\ell = ds^k + i\{\sum_{j=1}^{n} A^k_j dz^j + B^k_j d\bar{z}^j\}$$

and define

$$\omega^k = \sum_{k=1}^{d} \mu^k_\ell dw^\ell - i \sum_{j=1}^{n} (A^k_j + \bar{B}^k_j) dz^j.$$

It is clear that $\omega^1, \ldots, \omega^d$ are smooth section of T' over U. Since

$$\omega^k = ds^k - i \sum_{j=1}^{n} (\bar{B}^k_j dz^j - B^k_j d\bar{z}^j), \quad k=1,\ldots,d, \tag{1.8}$$

the forms ω^k are real and thus they generate T^o over U. Notice also that the pull-back of ω^k to Σ_z is equal to that of ds^k. In particular this shows that the pull-back map $T^*\mathfrak{M}|_{\Sigma_z} \to T^*\Sigma_z$ induces a natural isomorphism $T^o|_{\Sigma_z} \cong T^*\Sigma_z$ and therefore that the submanifolds Σ_z are <u>noncharacteristic</u> (a submanifold is noncharacteristic when its conormal bundle does not intersect the characteristic set off the zero section). Actually the submanifold Σ_z are <u>minimal noncharacteristic</u>, which means that there are no noncharacteristic submanifolds of dimension $< d$ $(= \dim \Sigma_z)$.

Notice also that the submanifold Σ_{z_o} is <u>hypo-analytic</u>, in the sense that it is the zero-set of a family of hypo-analytic functions, namely the functions $z^i - z^i_o$ $(i=1,\ldots,m)$. Because of all of these we shall refer to the fibre bundle U over Δ as a <u>hypo-analytic minifibration</u> of $\mathfrak{M}$ at the origin.

The hypo-analytic chart (U,z,w) defines an embedding of $T^o|_U$ into $\mathbb{C}^m \times \mathbb{C}^d \times \mathbb{C}_d$, namely

$$(x,y,s,\sigma\cdot\omega) \to (z,w,{}^t\mu\sigma) \qquad (\sigma\cdot\omega = \sum_{k=1}^{d} \sigma_k \omega^k). \tag{1.9}$$

We recall that ${}^t\mu$ is the <u>contragredient</u> of the Jacobian matrix $\partial w/\partial s$, i.e., ${}^t\mu = {}^t(\partial w/\partial s)^{-1}$. The image of $T^o|_U$ under the map (1.9) is what we shall call $\tilde{\mathfrak{R}}_U$ in the present chapter; $\tilde{\mathfrak{R}}_U$ can be regarded as a real vector bundle over $\tilde{U}$. A moment of thought will show that this is the natural adaptation to CR structures of what we have done in Section 1 of Ch. II for general hypo-analytic structures. According to what was said earlier, the covector $\sigma\cdot\omega$ is the preimage of $\sigma\cdot ds$ (the latter is regarded as a cotangent vector to Σ_z) under the pull-back map $T'|_{\Sigma_z} \to \mathbb{C}T^*\Sigma_z$, which here, however, is merely surjective, never injective since $n \geq 1$. The pull-back map induces a bijection onto $\mathbb{C}T^*\Sigma_z$ of the vector subbundle generated over $\mathbb{C}$ by $T^o|_U$, i.e., $T' \cap \bar{T}'|_U$.

Over U the tangent structure bundle $\mathcal{V}$ is spanned by the vector fields (cf. (1.17), Ch. II)

$$(1.10) \qquad L_j = \partial/\partial \bar{z}^j - i \sum_{k=1}^{d} (\partial\phi^k/\partial\bar{z}^j) M_k, \qquad j=1,\dots,n,$$

where (cf. (1.16), Ch. II)

$$(1.11) \qquad M_k = \sum_{\ell=1}^{d} \mu_k^\ell \, \partial/\partial s^\ell, \qquad k=1,\dots,d.$$

In order to obtain a basis of $\mathbb{C}T\mathfrak{M}$ over U one can adjoin to $L_1,\dots,L_n$, $M_1,\dots,M_d$ the following n vector fields:

$$(1.12) \qquad L'_j = \partial/\partial z^j - i \sum_{k=1}^{d} (\partial\phi^k/\partial z^j) M_k, \qquad j=1,\dots,n.$$

Notice that, for all $i,j = 1,\dots,n$, $k,\ell = 1,\dots,d$,

$$(1.13) \qquad \begin{aligned} L_j z^i &= L_j w^k = 0, & L_j \bar{z}^i &= \delta_j^i, \\ M_k z^i &= M_k \bar{z}^i = 0, & M_k w^\ell &= \delta_k^\ell, \\ L'_j \bar{z}^i &= L'_j w^k = 0, & L'_j z^i &= \delta_j^i. \end{aligned}$$

Thus, if $u \in C^\infty(U)$,

$$(1.14) \qquad du = \sum_{j=1}^{n} L_j u \, d\bar{z}^j + L'_j u \, dz^j + \sum_{k=1}^{d} M_k u \, dw^k.$$

Here any smooth section $\dot{f}$ of $\Lambda^{p,q}$ has a unique representative

$$(1.15) \qquad f = \sum_{|I|+|K|=p} \; \sum_{|J|=q} f_{I,J,K}(x,y,s) dz^I \wedge d\bar{z}^J \wedge dw^K,$$

to which we refer as the <u>standard representative</u> of $\dot{f}$. As before the pair (p,q) will be called the bidegree of the differential form (1.15). The differential operator $d'^{\,p,q}$ (see (1.8), Ch. II) will be represented in U by the operator

$$(1.16) \qquad L: C^\infty(U;\Lambda^{p,q}) \to C^\infty(U;\Lambda^{p,q+1})$$

whose action on the form (1.15) is given by

$$(1.17) \quad Lf = \sum_{|I|+|K|=p} \; \sum_{|J|=q} \; \sum_{j=1}^{n} L_j f_{I,J,K}(x,y,s) d\bar{z}^j \wedge dz^I \wedge d\bar{z}^J \wedge dw^K.$$

We close this section with a few remarks about the lift of a CR structure. The consideration of Section 1, Ch. III, can be modified to better reflect the CR nature of the structure. Here we can equip the characteristic bundle T^0 with a CR structure: Let (U,z,w) be the CR chart studied above. Let us call (ξ,η,σ) the coordinates in the cotangent spaces to $\mathfrak{M}$ at points of U with respect to the basis dx^i, dy^j, ds^k $(1 \le i,j \le n,\; 1 \le k \le d)$. We may define a CR structure on $T^0|_U$ by means of the functions

(1.18) $$z^j = x^j + iy^j,\ w^k\ \text{(given by (1.3))},\ \sigma(M_\ell)$$
$$(j=1,\dots,n,k,\ \ell=1,\dots,d),$$

where $\sigma(M_\ell)$ is the symbol of the vector field M_ℓ defined in (1.11). We have:

(1.19) $$\sigma(M_\ell) = \sum_{k=1}^{d} \mu_\ell^k \sigma_k.$$

We may regard $T^o|_U$ as a submanifold of $T^*\mathfrak{m}|_U$, specifically the submanifold defined by the equations

(1.20) $$\sigma(L_j)(x,y,s,\xi,\eta,\sigma) = 0, \qquad j=1,\dots,n,$$

where $\sigma(L_j)$ is the symbol of the vector field L_j defined in (1.10). Since $\mathcal{V} \cap \bar{\mathcal{V}} = 0$ the equations (1.20) are equivalent to a set of $2n$ independent real equations. We know that the dimension of the manifold T^o is equal to $2(n+d)$. Denote by H_{L_j} the Hamiltonian vector field of the functions $\sigma(L_j)$. Since the L_j and the M_k commute pairwise we have

(1.21) $$H_{L_i}\sigma(L_j) = H_{L_i}\sigma(M_k) = 0,$$
$$i,j=1,\dots,n,\quad k=1,\dots,d.$$

From (1.21) we deduce first that the vector fields H_{L_j} in $T^*\mathfrak{m}|_U$ are tangent to the submanifold $T^o|_U$ and second, that they annihilate all the function (1.18). Thus the tangent structure bundle of the lifted CR structure on T^o is spanned, over $T^o|_U$, by the Hamiltonian vector fields H_{L_j}. Since the base projection of H_{L_j} is equal to L_j we see at once that the vector fields $H_{L_1},\dots,H_{L_n}$ and their complex conjugates are $\mathbb{C}$-linearly independent, which proves that the structure on T^o is indeed a CR structure.

Let us make the link with lift of the CR structure to the cotangent structure bundle T' as described in Section 1, Ch. III. In our present notation the "tautological form" (denoted by $\zeta\cdot dZ$ in Section 1, Ch. III) is equal, over U, to the one-form on $T'|_U$,

$$\omega = \sum_{j=1}^{n} \zeta_j dz^j + \sum_{k=1}^{d} \tau_k dw^k.$$

From the relations (1.13) it follows at once that

(1.22) $$\zeta_j = \sigma(L'_j), \qquad \tau_k = \sigma(M_k).$$

Here the symbols $\sigma(L'_j)$ and $\sigma(M_k)$ are holomorphic functions along the fibres of T' at points of U, depending smoothly on the variable point in U. If we go back to the forms (1.8) we see that we may write

$$(1.23)\qquad L_j = \partial/\partial\bar{z}^j - i\sum_{k=1}^{d} B_j^k\partial/\partial s^k, \qquad j=1,\dots,n.$$

We have (cf. remarks preceding (1.8)):

$$0=\langle \sum_{\ell=1}^{d} \mu_\ell^k dw^\ell, L'_j\rangle = \langle ds^k, L'_j\rangle + iA_j^k,$$

whence

$$(1.24)\qquad L'_j = \partial/\partial z^j - i\sum_{k=1}^{d} A_j^k\partial/\partial s^k, \qquad j=1,\dots,n,$$

and therefore

$$(1.25)\qquad L'_j = \bar{L}_j - i\sum_{k=1}^{d} (A_j^k+\bar{B}_j^k)\partial/\partial s^k.$$

Recalling that $T'\cap\bar{T}'$ is defined by the equations $\sigma(L_j) = \sigma(\bar{L}_j) = 0$, $j = 1,\dots,n$, we reach the conclusion that, in $T'\cap\bar{T}'$,

$$(1.26)\qquad \sigma(L'_j) = -i\sum_{k=1}^{d} (A_j^k+\bar{B}_j^k)\sigma_k.$$

If we combine this with the fact that

$$\sum_{k=1}^{d} \sigma(M_k)dw^k = \sum_{\ell=1}^{d} \sigma_\ell\mu_k^\ell dw^k = \sum_{\ell=1}^{d} \sigma_\ell\omega^\ell + i\sum_{j=1}^{n}\sum_{\ell=1}^{d} \sigma_\ell(A_j^\ell + \bar{B}_j^\ell)dz^j$$

we reach the conclusion that the tautological form is equal, in $T'\cap\bar{T}'$, to the form

$$(1.27)\qquad \omega = \sum_{k=1}^{d} \sigma_k\omega^k.$$

We may also consider the <u>Hamiltonian lifts</u> $\mathcal{H}_{L_j}$ of the vector fields L_j. We recall that $\mathcal{H}_{L_j}$ is a tangent vector field to the manifold T' which is holomorphic along the fibres of T'. It induces naturally a tangent vector field on the submanifold T^o which is nothing else but the Hamiltonian field H_{L_j}. By this we mean that if F is a C^∞ function in some open neighborhood $\mathcal{O}$ in T' of a point $\bar{\omega}$ of T^o and if F is holomorphic with respect to the fibre variables, then the restriction of $\mathcal{H}_{L_j}F$ to $\mathcal{O}\cap T^o$ equals the action of H_{L_j} on the restriction of F to $\mathcal{O}\cap T^o$.

2. EXTENSION OF COHOMOLOGY CLASSES TO WEDGES IN CR STRUCTURES

Let (U,Z) be the CR chart introduced in Section 1. Let Γ be an open and convex cone in $\mathbb{R}^d$; $\Gamma\neq\emptyset$ and has its vertex at 0, like all the cones we deal with. By a wedge we shall mean a set of the kind

$$(2.1)\qquad \tilde{\mathfrak{W}}_\delta(U,\Gamma) = \{(z,w)\in\mathbb{C}^{n+d};\ z\in\Delta,\ \ w = s+it,$$
$$s\in\mathcal{B},\ |t|<\delta,\ t-\phi(z,\bar{z},s)\in\Gamma\}.$$

We always assume $|\phi| < \delta$ in U so that the closure of $\tilde{\mathfrak{W}}_\delta(U,\Gamma)$ contains the edge $\tilde{U} = \{(z,w) \in \mathbb{C}^d;\ z \in \Delta,\ s \in \mathcal{B},\ w = s+i\phi(z,\bar{z},s)\}$. We propose to study the effect on such a wedge of certain changes of the CR chart and of the local coordinates. Let us write

$$z_\# = H_1(z,w), \qquad w_\# = H_2(z,w), \tag{2.2}$$

where $H = (H_1,H_2)$ is a biholomorphism of an open neighborhood of 0, $\mathcal{N}$, in $\mathbb{C}^{n+d}$ onto another such neighborhood, and $H(0,0) = 0$. We require that the hypo-analytic chart $(U_\#,z_\#,w_\#)$, in which $U_\#$ is a suitably small open neighborhood of 0 in $\mathcal{M}$, have the properties analogous to those of the chart (U,z,w).

If we look at the Taylor expansion of order one of H at 0 and express the fact that $dw_\#$ must be real at the origin, and then the fact that the functions $x^i_\# = \mathrm{Re}\, z^i_\#$, $y^j_\# = \mathrm{Im}\, z^j_\#$ and $s^k_\# = \mathrm{Re}\, w^k_\#$ form a coordinate system in $U_\#$ we see at once that we must have

$$\begin{gathered} \det \partial H_1/\partial z|_0 \neq 0, \qquad \det \partial H_2/\partial w|_0 \neq 0 \\ \partial H_2/\partial w|_0 \text{ is real}, \quad \partial H_2/\partial z|_0 = 0. \end{gathered} \tag{2.3}$$

In what follows we assume that the neighborhood U and the number δ are so small that the wedge $\tilde{\mathfrak{W}}_\delta(U,\Gamma)$ is contained in the neighborhood $\mathcal{N}$ of 0 in which the map H is defined.

Proposition 2.1. Let Γ' be any open and convex cone whose closure in $\mathbb{R}^d\backslash\{0\}$ is contained in Γ. Define $\Gamma'_\# = (\partial H_2/\partial w)(0)\Gamma'$.

There is, then, an open neighborhood of 0 in $\mathcal{M}$, $U'_\# \subset U_\#$, and a number $\delta'_\# > 0$ such that the image of $\tilde{\mathfrak{W}}_\delta(U,\Gamma)$ under the biholomorphism H contains $\tilde{\mathfrak{W}}_{\delta'_\#}(U'_\#,\Gamma'_\#)$.

Proof: It is a variant of the proof of Prop. 3.7, Ch. II. We apply the implicit function thorem: to every $(x,y,s,v) \in U\times\mathbb{R}^d$ sufficiently close to 0 there is a unique point $(x',y',s',v_\#) \in U\times\mathbb{R}^d$ such that

$$\begin{gathered} H_1(z',w') = H_1(z,w+iv) \\ H_2(z',w') + iv_\# = H_2(z,w+iv), \end{gathered} \tag{2.4}$$

where $w = s + i\phi(z,\bar{z},s)$, $w' = s' + i\phi(z,\bar{z},s')$. By uniqueness we have

$$z' = z, \quad s' = s, \quad v_\# = 0 \quad \text{when} \quad v = 0. \tag{2.5}$$

We want to prove that, if (x,y,s) remains in a suitably small open neighborhood $U' \subset U$ of the origin and if $v \in \Gamma'$, $|v| < \delta'$ with $\delta' > 0$ suitably small, then $v_\#$ stays in $\Gamma_\# = (\partial H_2/\partial w)(0)\Gamma$ and $|v_\#| < \delta_\#$ with $\delta_\#$ a suitably small number >0 tending to zero with δ'.

Provided U' and δ' are small enough, the point (x',y',s') will remain in U. This shows that the biholomorphism H mpas $\mathfrak{W}_{\delta'}(U',\Gamma')$ into $\mathfrak{W}_{\delta_\#}(U_\#,\Gamma_\#)$ where $U_\#$ is the image of U under the map H. Applying the analogous result to the inverse of the biholomorphism H yields the sought conclusion.

By (2.4) we have

$$iv_\# = H_2(z,w+iv) - H_2(z',w'), \tag{2.6}$$

which, when combined with (2.5), yields $v_\# = M'v$, where $M' = M'(x,y,s,v)$ is a $d\times d$ complex matrix. The result will follow if we show that $M' = \partial H_2/\partial w$ at the origin. Let us differentiate the right-hand side in (2.6) with respect to v and freeze the result at 0. If we take into account the fact that $d\phi|_0 = 0$ we see that the resulting matrix is equal to

$$\{i(\partial H_2/\partial w) - (\partial H_2/\partial z)(\partial z'/\partial v) - (\partial H_2/\partial w)(\partial s'/\partial v)\}|_0.$$

By (2.3) we reach the conclusion that

$$\partial v_\#/\partial v = (\partial H_2/\partial w)[I+i(\partial s'/\partial v)] \quad \text{at the origin.}$$

But since $\partial H_2/\partial v$, $\partial s'/\partial v$ and $\partial v_\#/\partial v$ are all real, we must have

$$\partial v_\#/\partial v = \partial H_2/\partial v, \quad \partial s'/\partial v = 0 \quad \text{at the origin.} \tag{2.7}$$

■

Remark 2.2. We wish to emphasize the contrast between Prop. 3.7, Ch.II, and Prop. 2.1 of the present chapter: In the general hypo-analytic set-up we were able to claim that the image of a wedge under a local automorphism of $\mathbb{C}^m\times\mathbb{R}^n$ (equipped with its standard hypo-analytic structure) contains a thinner wedge, only under the proviso that the cones defining the wedges be definite-negative with respect to a given characteristic covector. No such proviso is needed when one deals with CR structures. The reason for this lies in the fact that the cotangent bundle of the fibres Σ_z of the minifibration (defined by the CR chart (U,z,w)) are isomorphic to the characteristic set T^0 over Σ_z. The situation is quite different in the nonCR case: the cotangent bundle of the fibre X_t, in the fibration of U by these maximally real submanifold (see Section 1, Ch. III), is not in general, isomorphic to $T^0|_{X_t}$; in general the "canonical" image of $T^0|_{X_t}$ is a proper subset of T^*X_t. ■

Once again we regard the map (z,w) as an embedding of U into $\mathbb{C}^m\times\mathbb{C}^d$. The latter space is always equipped with its natural complex structure. It induces on the image $\tilde{U}$ of U the tangential CR structure; (z,w) is then an isomorphism of U onto $\tilde{U}$ for their respective CR structures. As U shrinks to $\{0\}$ the subsets $\tilde{U}$ define the germ of a generic submanifold of $\mathbb{C}^{n+d}$, which we shall denote by $\tilde{\mathfrak{M}}$. Recall, however, that this "realization" of $\mathfrak{M}$ depends on the embedding (z,w).

As U shrinks to $\{0\}$ and $\delta \to +0$ the wedges $\tilde{\mathfrak{W}}_\delta(U,\Gamma)$ define the germ at the origin of a subset of $\mathbb{C}^{n+d}$ which we shall denote $\tilde{\mathfrak{W}}(\mathfrak{M},\Gamma)$, as before: we refer to $\tilde{\mathfrak{W}}(\mathfrak{M},\Gamma)$ as <u>the germ of a wedge with edge</u> $\tilde{\mathfrak{M}}$.

The next item to be discussed is the extension of cocycles in U to cocycles in the wedge $\tilde{\mathfrak{W}}_\delta(U,\Gamma)$. We may as well deal with a "standard" form of bidegree (p,q), f, as in (1.15), whose coefficients are C^∞ functions in U and which, moreover, is a cocycle, i.e., satisfies the analogue of (1.22), Ch. II:

$$\sum_{j,J} \epsilon(j,J) L_j f_{I,J,K} \equiv 0. \tag{2.8}$$

Here (I,K) ranges over the family of all pairs of ordered multi-indices $i_1 <\dots< i_{p'}$, $k_1 <\dots< k_{p''}$, such that $1 \le i_\alpha \le n$, $1 \le k_\gamma \le d$ for all α, γ, and $p'+p'' = p$. The summation is carried out over an equivalence class of pairs (j,J) consisting of an integer $j \in [1,\dots,n]$ and of an ordered multi-index $J = \{j_1 <\dots< j_q;\ \forall\,\gamma,\ 1 \le j_\gamma \le n\}$, which are equal after a permutation; $\epsilon(j,J) = +1$ or -1 depending on whether the permutation that orders the set $\{j,j_1,\dots,j_q\}$ is even or odd.

By a cocycle (of bidegree (p,q)) in $\tilde{\mathfrak{W}}_\delta(U,\Gamma)$ we mean a $\bar\partial$-closed (p,q)-form

$$\tilde{f} = \sum_{|I|+|K|=p}\ \sum_{|J|+|K'|=q} \tilde{f}_{I,J,K,K'}(z,w)\,dz^I \wedge d\bar{z}^J \wedge dw^K \wedge d\bar{w}^{K'}, \tag{2.9}$$

whose coefficients $\tilde{f}_{I,J,K,K'}$ belong to $\mathcal{B}^\infty(\tilde{\mathfrak{W}}_\delta(U,\Gamma))$. Here, of course, $\bar\partial = \bar\partial_z + \bar\partial_w$.

Let a section $\dot{f} \in C^\infty(U;\Lambda^{p,q})$ be d'-closed; its standard representative, f, is a cocycle. We say that $\dot{f}$ <u>extends to the wedge</u> $\tilde{\mathfrak{W}}_\delta(U,\Gamma)$ is there is a cocycle (2.9) whose pull-back to the neighborhood U, under the embedding (z,w) with w as in (1.3), (1.4), is equal to f modulo sums of forms of bidegree $(p+\ell,q-\ell)$ with $\ell = 1,\dots,q$. The cocycle $\tilde{f}$ can be viewed as the standard representative of a section $\dot{\tilde{f}} \in \mathcal{B}^\infty(\tilde{\mathfrak{W}}_\delta(U,\Gamma);\tilde{\Lambda}^{p,q})$, and we may say then that $\dot{\tilde{f}}$ <u>extends</u> $\dot{f}$ to $\tilde{\mathfrak{W}}_\delta(U,\Gamma)$ or, equivalently, that $\dot{f}$ is the <u>pull-back</u> of $\dot{\tilde{f}}$ (on all this cf. Section 2, Ch. II).

<u>Example 2.3</u>. When $p = q = 0$ there is only one representative f of

$\dot{f}$; f is a smooth CR function in U. If $\dot{f}$ extends to the wedge $\tilde{\mathfrak{w}}_\delta(U,\Gamma)$ it means that there is a holomorphic function $\tilde{f} \in \mathcal{B}^\infty(\tilde{\mathfrak{w}}_\delta(U,\Gamma))$ (this is the cocycle (2.9) when $p = q = 0$) whose restriction to $\tilde{U}$ is equal to the push-forward of f under the embedding (z,w). ∎

The extension to wedges of a cocycle of bidegree (p,q), with $q > 0$, cannot be unique. But the following can be said (cf. Prop. 2.4, Ch. II):

<u>Proposition 2.4.</u> <u>Suppose that</u> $\dot{\tilde{f}} \in \mathcal{B}^\infty(\tilde{\mathfrak{w}}_\delta(U,\Gamma);\tilde{\Lambda}^{p,q})$ <u>is</u> $\bar{\partial}$<u>-closed and that its standard representative</u> $\tilde{f}$ (of the kind (2.9)) <u>has the following property</u>:

$$\tilde{f}_{I,J,K,K'} \equiv 0 \quad \text{whenever} \quad K' \neq \emptyset . \tag{2.10}$$

<u>Then, if the pull-back of</u> $\dot{\tilde{f}}$ <u>to</u> $\tilde{U}$ <u>vanishes identically we must necessarily have</u> $\dot{\tilde{f}} \equiv 0$.

<u>Proof</u>: From the fact that $\tilde{f}$ is $\bar{\partial}$-closed and from (2.10) we derive that the coefficients of $\tilde{f}$ are holomorphic with respect to w. Since the pull-backs of these coefficients to the submanifolds $\tilde{\Sigma}_z$ (for $z \in \Delta$ fixed) vanish identically the coefficients themselves must be identically zero in $\tilde{\mathfrak{w}}_\delta(U,\Gamma)$, whence the assertion. ∎

The analogue of Theorem 2.5, Ch. II, is best stated by dealing with germs of section $\dot{\tilde{f}}$ and germs of cocycles $\tilde{f}$ in germ of wedges $\tilde{\mathfrak{w}}(\mathfrak{m},\Gamma)$:

<u>Theorem 2.5.</u> <u>Given any open convex cone</u> Γ <u>in</u> $\mathbb{R}^m$ <u>and any ray</u> $r_o \subset \Gamma$ <u>there is another open convex cone</u> $\Gamma' \subset \Gamma$ <u>containing</u> r_o <u>and such that the following is true</u>:

<u>Each cocycle that belongs to</u> $\mathcal{B}^\infty(\tilde{\mathfrak{w}}(\mathfrak{m},\Gamma);\tilde{\Lambda}^{p,q})$ <u>is cohomologous in</u> $\tilde{\mathfrak{w}}(\mathfrak{m},\Gamma')$ <u>to a cocycle</u> $\dot{\tilde{f}} \in \mathcal{B}^\infty(\tilde{\mathfrak{w}}(\mathfrak{m},\Gamma');\tilde{\Lambda}^{p,q})$ <u>that has a standard representative endowed with property</u> (2.10).

The proof of Theorem 2.5 is identical to that of Theorem 2.5, Ch. II, except that the role of t is played now by $\bar{z}$; in particular, d_t is replaced everywhere by $\bar{\partial}_z$.

There is a result that parallels Prop. 2.7, Ch. II (and has an identical proof):

<u>Proposition 2.6.</u> <u>Suppose that the section</u> $\dot{f} \in C^\infty(U;\Lambda^{p,q})$ $(q \geq 1)$ <u>extends to the wedge</u> $\tilde{\mathfrak{w}}_\delta(U,\Gamma)$. <u>There is an open neighborhood</u> $U' \subset U$ <u>of</u> 0, <u>an open convex cone</u> $\Gamma' \subset \Gamma$ <u>containing a given ray</u> r_o <u>and a number</u> δ', $0 < \delta' < \delta$, <u>such that the following is true</u>:

Let f *be standard representative of* $\dot{f}$ *in* U. *There is a standard* C^∞ *form* g *of bidegree* (p,q-1) *in* U′ *such that* f - Lg *is the pull-back to* U′, *under the embedding* (z,w) *with* w *as in* (1.3), (1.4), *of a standard* $\bar{\partial}$*-closed* (p,q)*-form* $\tilde{f} \in \mathcal{B}^\infty(\tilde{\mathfrak{W}}_\delta, (U',\Gamma'); \tilde{\Lambda}^{p,q})$ (*as in* (2.9)) *which has property* (2.10).

3. THE MINI-FBI TRANSFORM IN A CR MANIFOLD

We continue to reason in the CR chart (U,z,w) of the preceding sections. Let u be a distribution in U such that the projection of supp u under the coordinate map $(x,y,s) \to s$: $\Delta \times \mathcal{B} \to \mathcal{B}$ is contained in a compact subset of $\mathcal{B}$. For any (z,w,τ) in the set

$$z \in \mathbb{C}^n, \quad w \in \mathbb{C}^d, \quad \tau \in \mathbb{C}_d; \quad |\mathrm{Im}\,\tau| < |\mathrm{Re}\,\tau|, \tag{3.1}$$

we consider the "integral" (cf. (2.2), Ch. III):

$$\mathfrak{F}(u;z,w,\tau) = \int_{\mathcal{B}} e^{i\tau\cdot(w-s-i\phi(z,\bar{z},s))-\langle\tau\rangle\langle w-s-i\phi(z,\bar{z},s)\rangle^2} u(x,y,s)\det[I+i\phi_s(z,\bar{z},s)]\,ds. \tag{3.2}$$

We are using a notation analogous to that of Section 2, Ch. III. Here also the integral represents a duality bracket between a zero-current and a differential form (here, of degree d); z, w, τ play the role of parameters. We may regard (3.2) as an integral on the submanifold $\tilde{\Sigma}_z$, the image of $\{z\}\times\mathcal{B}$ under the map $s \to s + i\phi(z,\bar{z},s)$. If $\tilde{u}$ denotes the transfer of u under this map, we have

$$\mathfrak{F}(u;z,w,\tau) = \int_{\mathcal{B}} e^{i\tau\cdot(w-\tilde{w})-\langle\tau\rangle\langle w-\tilde{w}\rangle^2}\,\tilde{u}(z,\tilde{w})\,d\tilde{w}. \tag{3.3}$$

Definition 3.1. $\mathfrak{F}(u;z,w,\tau)$ *will be called the* mini-FBI transform *of the distribution* u.

The motivation for this terminology lies in Def. 2.1, Ch. II (here as there, FBI stands for Fourier-Bros-Iagolnitzer) and in the fact that the integration is performed over the minimal noncharacteristic submanifold Σ_z. The FBI transform of Ch. III could be called the *maxi*-FBI transform since it is carried out over the maximally real submanifold X_t. In a sense the integration in (3.2) is carried out with respect to a minimal number of variables (the variables $s^1,\dots,s^d$). In the important "hypersurface" case, when d = 1, it is a line integral. (When d = 1 $\mathfrak{M}$ is the germ at 0 of a real hypersurface in $\mathbb{C}^{n+1}$.)

Introduce the vector fields L_i, L'_j and M_k of Section 1 (see (1.10) to (1.13)). A straightforward integration by part shows that we have (cf. (2.5), (2.6), Ch. III):

$$(3.4)\qquad \mathfrak{F}(L_j u;z,w,\tau) = (\partial/\partial\bar{z}^j)\mathfrak{F}(u;z,w,\tau),\qquad j=1,\dots,n,$$

$$(3.5)\qquad \mathfrak{F}(L'_j u;z,w,\tau) = (\partial/\partial z^j)\mathfrak{F}(u;z,w,\tau),\qquad j=1,\dots,n,$$

$$(3.6)\qquad \mathfrak{F}(M_k u;z,w,\tau) = (\partial/\partial w^k)\mathfrak{F}(u;z,w,\tau),\qquad k=1,\dots,d.$$

Let then f be a current in U of the kind (1.15) in which the coefficients $f_{I,J,K}$ all have compact support with respect to s.

<u>Definition 3.2. We shall call</u> mini-FBI transform of f <u>the current in</u> $\Delta\times C^d$,

$$(3.7)\quad \mathfrak{F}(f;z,w,\tau) = \sum_{|I|+|K|=p}\ \sum_{|J|=q} \mathfrak{F}(f_{I,J,K};z,w,\tau)dz^I \wedge d\bar{z}^J \wedge dw^k.$$

In (3.7) we view τ as a parameter, varying in the set

$$(3.8)\qquad \tau \in \mathbb{C}_d;\quad |\mathrm{Im}\,\tau| < |\mathrm{Re}\,\tau|.$$

It follows at once from (3.4) applied to the coefficients $f_{I,J,K}$, that

$$(3.9)\qquad \mathfrak{F}(Lf;z,w,\tau) = \bar{\partial}_z\mathfrak{F}(f;z,w,\tau).$$

In analogy with the hypothesis (2.11), Ch. III, we strengthen our requirements on the map $\phi\colon U \to \mathbb{R}^d$. We shall assume that

$$(3.10)\qquad \frac{\partial^2\phi^k}{\partial s^\ell \partial s^{\ell'}}(0,0,0) = 0,\qquad \forall\ k,\ell,\ell' = 1,\dots,d.$$

This can always be achieved if we replace w^k, as given in (1.3), by

$$w^k - \frac{i}{2}\sum_{\ell,\ell'=1} \frac{\partial^2\phi^k}{\partial s^\ell \partial s^{\ell'}}(0,0,0)\ w^\ell w^{\ell'},$$

for each $k = 1,\dots,d$.

Let $\tilde{\mathbb{R}}_U$ be the real vector bundle over $\tilde{U}$ defined in Sect. 1. Denote by $\tilde{\mathbb{R}}T'_z$ the pull-back of $\tilde{\mathbb{R}}_U$ to $\tilde{\Sigma}_z$; $\tilde{\mathbb{R}}T'_z$ is the image of the real cotangent bundle of Σ_z, $T^*\Sigma_z$, under the map

$$(3.11)\qquad (x,y,s,\sigma) \to (z,w,{}^t\mu\sigma) \in \mathbb{C}^{m+d}\times\mathbb{C}_d,$$

where $w = s + i\phi(z,\bar{z},s)$ and $\mu = (\partial w/\partial s)^{-1}$. To see this it suffices to factor the map (1.9) into the pull-back map $T^o|_{\Sigma_z} \to T^*\Sigma_z$ (which transforms ω^k into ds^k for each k) followed up by the map (3.11). We have the exact analogue of Prop. 2.3, Ch. III:

<u>Proposition 3.3. If</u> (3.10) <u>holds then, to every</u> ϵ, $0 < \epsilon < 1$, <u>there is</u> $\delta > 0$ <u>such that, if</u> diam $U < \delta$, <u>we have</u>

$$(3.12)\qquad -\mathrm{Re}\{i\tau\cdot(w-\tilde{w}) - \langle\tau\rangle\langle w-\tilde{w}\rangle^2\} \geq (1-\epsilon)|w-\tilde{w}|^2|\tau|,$$
$$\forall\ z \in \Delta,\quad (w,\tau) \in \tilde{\mathbb{R}}T'_z,\quad \tilde{w} \in \tilde{\Sigma}_z.$$

The proof of Prop. 3.3 is exactly the same as that of Prop. 2.3, Ch. III.

In what follows we shall always assume that (3.12) holds for some ϵ, with $0 < \epsilon < 1$ and sometimes with $\epsilon < 1/2$.

For the sake of simplicity we limit ourselves to dealing with distribtuions that are continuous functions of $(x,y) \in \Delta$ valued in the space of distributions of s in $\mathcal{B}$, $\mathcal{D}'(\mathcal{B})$. We shall denote by $C^0(\Delta;\mathcal{D}'(\mathcal{B}))$ the space of such distributions in $U = \Delta\times\mathcal{B}$; it will be equipped with the topology of uniform convergence on the compact subsets of Δ.

If we assume that the support of $u \in C^0(\Delta;\mathcal{D}'(\mathcal{B}))$ is compact, there is a finite family of compactly supported continuous functions in U, u_α $(\alpha \in \mathbb{Z}^d_+,\ |\alpha| \leq k)$ such that

$$(3.13)\qquad u = \sum_{|\alpha|\leq k} M^\alpha u_\alpha$$

$(M^\alpha = M_1^{\alpha_1}\cdots M_d^{\alpha_d})$. By virtue of (3.6) we have

$$(3.14)\qquad \mathfrak{F}(u;z,w,\tau) = \sum_{|\alpha|\leq k} \partial_w^\alpha \mathfrak{F}(u;z,w,\tau).$$

The analogues of Propositions 2.4, 2.6 and of Lemma 2.7, Ch. III, are valid. We remind the reader that we are reasoning under hypothesis (3.12) (for some ϵ, $0 < \epsilon < 1$).

Proposition 3.4. *Suppose that $u \in C^0(\Delta;\mathcal{D}'(\mathcal{B}))$, as a distribution in U, is compactly supported. Then there is an integer $k \geq 0$ such that $(1+|\tau|)^{-k}\mathfrak{F}(u;z,w,\tau)$ is a bounded and continuous function in $\tilde{\mathcal{R}}_U$.*

Proposition 3.5. *The mini-FBI transform of a function $u \in C^\infty_c(U)$ has the following property:*

(3.15) *Whatever the integer ℓ and the n-tuples α, $\beta \in \mathbb{Z}^n_+$ such that $|\alpha+\beta| \leq \ell$,*

$$(1+|\tau|)^\alpha \partial_x^\alpha \partial_y^\beta \mathfrak{F}(u;z,w,\tau)$$

is a bounded and continuous function in $\tilde{\mathcal{R}}_U$.

Proposition 3.6. *To every open ball $\mathcal{B}' \subset \mathcal{B}$ centered at the origin there is an open neighborhood $\mathcal{O}$ of 0 in $\mathbb{C}^{n+d}$, an open cone Γ in $\mathbb{C}_d$ which contains $\mathbb{R}_d\backslash\{0\}$ and a number $R > 0$ such that the following is true:*

Given any $u \in C^0(\Delta;\mathcal{D}'(\mathcal{B}))$ which, as a distribution in U,

<u>has a compact support contained in</u> $\Delta\times(\mathcal{B}\setminus\mathcal{B}')$, <u>there is a constant</u> $C > 0$ <u>such that</u>

$$(3.16)\qquad |\mathfrak{F}(u;z,w,\tau)| \le C\, e^{-|\tau|/R}, \qquad \forall\ (z,w) \in \mathcal{O},\ \ \tau \in \Gamma.$$

This allows us to introduce the analogues of Def. 2.9 and 2.10, Ch, III; we leave their precise formulation to the reader. From now we shall talk of the mini-FBI transform at the origin of an arbitrary distribution u in the CR manifold $\mathfrak{M}$, in the minifibration defined by the CR chart (U,Z). We shall denote it by $\mathfrak{F}(u;z,w,\tau)$.

We must now describe the inversion of the mini-FBI transform. We shall assume that $\operatorname{diam} U$ is so small that

$$(3.17)\qquad \operatorname{Re}\langle {}^t w_s(z,\bar z,s)^{-1}\sigma\rangle^2 \ge |\sigma|^2/2, \qquad \forall\ (z,s) \in U,\ \ \sigma \in R_d.$$

As before we assume that $u \in C^0(\Delta;\mathcal{D}'(\mathcal{B}))$; we shall also assume, at first, that the support of u, regarded as a distribution in U, is compact (this hypothesis will be removed at the very end). Let K be a compact neighborhood of 0 in $\mathcal{B}$ and denote by $\tilde K_z$ the image of K under the map $s \to s + i\phi(z,\bar z,s)$; then call $\mathcal{K}_z$ the portion of the fibre bundle $\tilde{\mathcal{R}}_U$ which lies above $\tilde K_z$. For any $\delta > 0$ we consider the integral

$$\mathfrak{F}^\delta_K(u;z,w) = (4\pi^3)^{-d/2} \iint_{\mathcal{K}_z} e^{i\tau\cdot(w-w')-\langle\tau\rangle\langle w-w'\rangle^2-\delta\langle\tau\rangle^2}\mathfrak{F}(u;z,w',\tau)\cdot \langle\tau\rangle^{d/2}\Delta(w-w',\tau)\,dw'\,d\tau\,,$$

where $\Delta(w,\tau)$ is the Jacobian determinant of the map $\tau \to \tau+i\langle\tau\rangle w$. One can view the integration in $\mathfrak{F}^\delta_K$ as being performed with respect to s' over K and σ over $\mathbb{R}_d$; then $w' = s' + i\phi(z,\bar z,s')$, $\tau = {}^t\mu(z,\bar z,s')\sigma$. We recall that $\mu(z,\bar z,s) = [I+i\phi_s(z,\bar z,s)]^{-1}$.

<u>The integral</u> $\mathfrak{F}^\delta_K$ <u>is a continuous function of</u> (z,w) <u>in</u> $\Delta\times\mathbb{C}^d$, <u>holomorphic with respect to</u> w.

By duplicating the proof of Theorem 3.3, Ch. III, one can prove the analogue of that result.

<u>Theorem 3.7.</u> <u>Suppose that</u> (3.12) <u>holds for some</u> ε, $0 < \varepsilon < 1/2$. <u>There is an open polydisk</u> $\Delta' \subset \Delta$ <u>in</u> z-<u>space</u> $\mathbb{C}^m$, <u>an open ball</u> $\mathcal{B}' \subset \mathcal{B}$ <u>in</u> s-<u>space</u> $\mathbb{R}^d$ <u>and one</u>, $\mathcal{O}$, <u>in</u> $\mathbb{C}^d$, <u>all centered at the origin, such that the image of</u> $\Delta'\times\mathcal{B}'$ <u>under the map</u> $(z,s) \to s + i\phi(z,\bar z,s)$ <u>is contained in</u> $\mathcal{O}$, <u>and that the following holds:</u>

(3.18) <u>To each</u> $u \in C^0(\Delta;\mathcal{D}'(\mathcal{B}))$ <u>which, when regarded as a distribution in</u> U, <u>is compactly supported, there is a continuous function</u> $\tilde h$ <u>in</u> $\Delta'\times\mathcal{O}$, <u>holomorphic with respect to</u> w, <u>such that the difference</u>

(3.19) $$u(x,t) - \mathfrak{F}_K^{\delta}(u;z,s+i\phi(z,\bar{z},s))$$

converges in $C^0(\Delta';\mathcal{B}')$ to $\tilde{h}(z,s+i\phi(z,\bar{z},s))$.

The proof of Theorem 3.7 duplicates that of Theorem 3.4, Ch. III.

The analogue of Def. 3.5, Ch. III, is the following:

Definition 3.8. We shall say that two distributions u_1, u_2 belonging to $C^0(\Delta;\mathcal{D}'(\mathcal{B}))$ are equivalent and we shall write $u_1 \approx u_2$ if there is an open polydisk $\Delta' \subset \Delta$, an open ball $\mathcal{B}' \subset \mathcal{B}$ and one, $\mathcal{O}$, in $\mathbb{C}^d$, all centered at the origin, such that the image of $\Delta' \times \mathcal{B}'$ under the map $s \to s + i\phi(z,\bar{z},s)$ is contained in $\mathcal{O}$, and if there is a function $\tilde{h} \in C^0(\Delta' \times \mathcal{O})$ such that $u_1 - u_2 = \tilde{h}(z,s+i\phi(z,\bar{z},s))$ in $\Delta' \times \mathcal{B}'$.

We remove now the requirement that the support of u be compact. In accordance with Def. 2.9, Ch. III, we shall say that the mini-FBI transform of u, $\mathfrak{F}(u;z,w,\tau)$, is equivalent to zero if it satisfies (3.16) for a suitable choice of $\mathcal{O}$, Γ, R and C. We have the analogues of Theorems 3.6 and 3.7, Ch. III:

Theorem 3.9. In order that a distribution $u \in C^0(\Delta;\mathcal{D}'(\mathcal{B}))$ be equivalent to zero it is necessary and sufficient that its mini-FBI transform be equivalent to zero.

Theorem 3.10. In order that a distribution $u \in C^\infty(\Delta;\mathcal{D}'(\mathcal{B}))$ be equal to a C^∞ function in some open neighborhood of the origin it is necessary and sufficient that its mini-FBI transform have Property (3.15).

4. HOLOMORPHIC EXTENSION OF DISTRIBUTIONS TO WEDGES AND EXPONENTIAL DECAY OF THEIR MINI-FBI TRANSFORMS

The content of the present section is the CR analogue of that of Section 4, Ch. III. In Theorem 4.1 below we make use of wedges of the following kind (cf. (4.4), Ch. III):

(4.1) $$\hat{\mathcal{W}}_\delta(U,\Gamma) = \{(z,w) \in \mathbb{C}^{n+d};\ z \in \Delta,\ w = s+it,\ s \in \mathcal{B},\ t \in \mathbb{C}^d,\ |t| < \delta,\ \exists\, v \in \Gamma \text{ such that } t = \phi(z,\bar{z},s) + [I+i\phi_s(z,\bar{z},s)]v\}.$$

As before Γ is an open and convex cone in $\mathbb{R}^d$ and δ is a number >0. It is always assumed that the edge $\tilde{U}$ lies entirely in the closure of $\hat{\mathcal{W}}_\delta(U,\Gamma)$, i.e., that $|\phi(z,\bar{z},s)| < \delta$, $\forall\ (z,s) \in U$. The inclusions (4.6) and (4.7), Ch. III, are valid here, once the meaning of the notation has been adapted to the CR situation (cf. (2.1)).

It should perhaps be pointed out that, when v ranges over $\mathbb{R}^d$, the point (z,w), with $w = s+it$ and $t = \phi(z,\bar{z},s) + [I+i\phi_s(z,\bar{z},s)]v$, describes the d-plane normal to $\tilde{U}$ at the point $(z,s+i\phi(z,\bar{z},s))$. Normal is meant here in the sense of the standard Euclidean structure on $\mathbb{C}^{n+d}$. Thus we see that the wedge $\hat{\mathfrak{W}}_\delta(U,\Gamma)$ is the union of the cones that are the natural copies of Γ in the d-planes normal to $\tilde{U}$.

The space $\mathcal{S}(\hat{\mathfrak{W}}_\delta(U,\Gamma))$ in the present set-up has an obvious definition: its elements are the continuous functions in $\hat{\mathfrak{W}}_\delta(U,\Gamma)$ which are holomorphic with respect to w and have the following property:

(4.2) <u>Given any open neighborhood</u> U' <u>of</u> 0 <u>whose closure is compact and contained in</u> U <u>and any open convex cone</u> Γ' <u>whose closure in</u> $\mathbb{R}^d\backslash\{0\}$ <u>is contained in</u> Γ <u>there are numbers</u> r, $C > 0$ <u>such that</u>

(4.3) $$|\tilde{u}(z,w)| \leq C[\mathrm{dist}((z,w),\tilde{U})]^{-r} \qquad \forall\ (z,w) \in \hat{\mathfrak{W}}_\delta(U',\Gamma').$$

The analogues of Lemmas 4.1 and 4.2, Ch. III, are valid here and enable us to define the boundary value $b\tilde{u}$ on the edge $\tilde{U}$ of any function $\tilde{u} \in \mathcal{S}(\hat{\mathfrak{W}}_\delta(U,\Gamma))$. The map $\tilde{u} \to b\tilde{u}$ is an injection of $\mathcal{S}(\hat{\mathfrak{W}}_\delta(U,\Gamma))$ into $C^0(\Delta;\mathcal{D}'(\mathcal{B}))$.

We call Γ^* the complement of the polar Γ^0 of Γ:

$$\Gamma^* = \{\sigma \in \mathbb{R}_d;\ \exists\ v \in \Gamma,\ \ \sigma\cdot v < 0\}.$$

The analogues of Theorems 4.3, 4.4 and 4.5, Ch. III, are valid here. We shall summarize this set of results as follows:

<u>Theorem 4.1.</u> <u>Suppose</u> (2.12) <u>is valid for some</u> ϵ, $0 < \epsilon < 1/2$. <u>The properties</u> (4.4) <u>and</u> (4.6) <u>below, of an arbitrary element</u> u <u>of</u> $C^0(\Delta;\mathcal{D}'(\mathcal{B})) \cap \mathcal{E}'(U)$, <u>are equivalent</u>:

(4.4) <u>Given any open and convex cone</u> Γ' <u>whose closure in</u> $\mathbb{R}_d\backslash\{0\}$ <u>is contained in</u> Γ, <u>there is an open neighborhood</u> $U' \subset U$ <u>of the origin in</u> $\mathfrak{M}$ <u>and a number</u> $\delta' > 0$ <u>such that the following is true</u>:

(4.5) <u>There is a function</u> $\tilde{u} \in \mathcal{S}(\hat{\mathfrak{W}}_{\delta'}(U',\Gamma'))$ <u>such that</u> $u = b\tilde{u}$ <u>in</u> U'.

(4.6) <u>Given any open and convex cone</u> Γ' <u>whose closure in</u> $\mathbb{R}_d\backslash\{0\}$ <u>is contained in</u> Γ <u>there is an open and conic neighborhood</u> $\mathcal{C}'$ <u>of</u> $\{0\}\times\Gamma'^*$ <u>in</u> $\mathbb{C}^{n+d}\times\mathbb{C}_d$ <u>such that, for suitably large constants</u> R', $C > 0$ <u>we have</u>

(4.7) $\qquad |\mathfrak{F}(u;z,w,\tau)| \le C\, e^{-|\tau|/R'}, \qquad \forall\ (z,w,\tau) \in \mathcal{C}'$.

If $u \in C_c^\infty(U)$ the properties (4.4) and (4.6) are each equivalent to the following one:

(4.8) To every open and convex cone Γ' whose closure in $\mathbb{R}_d \backslash \{0\}$ is contained in Γ there is an open neighborhood $U' \subset U$ of the origin in $\mathfrak{M}$ and a number $\delta' > 0$ such that the following is true:

(4.9) There is a function $\tilde{u} \in \mathcal{B}^\infty(\hat{\mathcal{W}}_{\delta'}(U',\Gamma'))$ which is holomorphic with respect to w and such that $u = b\tilde{u}$ in U'.

The notation $u \in C^0(\Delta;\mathcal{D}'(\mathcal{B})) \cap \mathcal{E}'(U)$ means that the support of $u \in C^0(\Delta;\mathcal{D}'(\mathcal{B}))$, regarded as a distribution in U, is compact.

The proof of Theorem 4.1 is the same as those of Theorems 4.3, 4.4, 4.5, Ch. III, combined.

Remark 4.2. Inspection of those proofs shows that, once the assignment of the pair U', δ' to the cone Γ' in Property (4.4) is determined, the conic neighborhood $\mathcal{C}'$ and the constant R' in (4.6) can be chosen independently of the distribution u. Conversely, once the assignment of the pair $\mathcal{C}'$, R' to Γ' is determined, the neighborhood U' and the number δ' can be chosen independently of u. ∎

Concerning the edge of the wedge theorem we shall content ourselves with pointing ou that Theorem 4.6 and Remark 4.7, Ch. III, can be restated within the present context. In fact the only modifications required in their statements consist in substituting the polydisk Δ in z-space $\mathbb{C}^n$ for the neighborhood U_2 in t-space $\mathbb{R}^n$ and d for m.

We conclude by mentioning that all the results about differential forms and cohomology classes of Section 5, Ch. III, have natural analogues for CR structures. The role of z in general hypo-analytic structures is played by w, that of t is played by $\bar{z}$ (constant functions with respect to t are replaced by holomorphic functions with respect to z).

5. THE HYPERSURFACE CASE

In this section we take a closer look at the case where the codimension d of the CR structure $\mathfrak{A}$ on the manifold $\mathfrak{M}$ is equal to one. If we reason in the CR chart (U,z,w) of the preceding sections we see that (z,w) [with $z = x+iy$ and with w defined by (1.3), (1.4)] is a diffeomorphism of U onto the submanifold $\tilde{U}$ of $\mathbb{C}^{n+1}$, whose real codimension is equal to one: $\tilde{\mathfrak{m}}$ is the germ of a real hypersurface in $\mathbb{C}^{n+1}$.

The hypersurface $\tilde{U}$ is defined by the scalar equation

$$\text{Im } w = \phi(z,\bar{z},\text{Re } w). \tag{5.1}$$

Note also, in passing, that there is only one vector field (1.11) in the CR chart (U,z,w), namely $M = (1 + i\phi_s(z,\bar{z},s))^{-1}\partial/\partial s$.

The characteristic set T^o is a real line bundle over $\mathfrak{m}$. The open and convex cones used to define the wedges (either in (2.1) or in (4.1)) will perforce be equal to one of the two half-lines in $\mathbb{R}_1\backslash\{0\}$. Suppose that Γ is the positive half-line $\mathbb{R}_+$. Then (see (2.1))

$$\tilde{\mathfrak{w}}_\delta(U,\mathbb{R}_+) = \{(z,w) \in \mathbb{C}^{n+1};\ z \in \Delta,\ \text{Re } w \in \mathcal{B},\ |\text{Im} w| < \delta,\ \text{Im } w - \phi(z,\bar{z},\text{Re } w) > 0\}.$$

Thus $\tilde{\mathfrak{w}}_\delta(U,\mathbb{R}_+)$ is one side of the hypersurface $\tilde{U}$. Of course the orientation on T^o is a matter of convention, and the side represented by $\tilde{\mathfrak{w}}_\delta(U,\mathbb{R}_+)$ depends on the choice of the defining equation (5.1) for $\tilde{U}$.

If u is a distribution in U, say compactly supported, its mini-FBI transform is given by a line "integral" (cf. (3.2)):

$$\mathfrak{F}(u;z,w,\tau) = \int_{-\infty}^{+\infty} e^{i\tau(w-s-i\phi(z,\bar{z},s)) - \langle\tau\rangle(w-s-i\phi(z,\bar{z},s))^2} u(x,y,s)[1+i\phi_s(z,\bar{z},s)]\,ds. \tag{5.2}$$

[Although the function $\phi(z,\bar{z},s)$ is only defined when z stays in the polydisk Δ and s in the open interval $\mathcal{B}$ we have the right to "integrate" from $-\infty$ to $+\infty$ since supp u is compact.]

The minimal noncharacteristic submanifolds Σ_z and $\tilde{\Sigma}_z$ are real curves. For each $z_o \in \Delta$, $\tilde{\Sigma}_{z_o}$ is the intersection of the hypersurface $\tilde{U}$ with the complex curve $z = z_o$. Here $\tilde{\mathcal{R}}_U$ is a line bundle over $\tilde{U}$ and $\mathfrak{F}_K^\delta(u;z,w)$ is a double integral.

Let us choose Γ to be the positive-half line $\mathbb{R}_+$; then Γ^* is the negative half-line $\mathbb{R}_-$. When $d = 1$ the statement of Theorem 4.1 can be simplified. Let Ω be an open ball centered at 0 in $\mathbb{C}^{n+1}$ whose radius is so small that any open ball $\Omega' \subset \Omega$ also centered at 0 is "cut" by $\tilde{U}$ into exactly two parts, one above $\tilde{U}$, defined by $\text{Im } w > \phi(z,\bar{z},\text{Re } w)$, the other one below $\tilde{U}$, defined by $\text{Im } w < \phi(z,\bar{z},\text{Re } w)$. We shall call U_{Ω}, the pre-image of $\tilde{U} \cap \Omega'$ under the map $(x,y,s) \to (z,s+i\phi(z,\bar{z},s))$ (where, as always, $z = x+iy$).

Theorem 5.1. *Suppose* (3.12) *is valid for some* ε, $0 < \varepsilon < 1/2$. *The following properties of an arbitrary element* u *of* $C^0(\Delta;\mathcal{D}'(\mathcal{B})) \cap \cap \mathcal{E}'(U)$, *are equivalent:*

(5.3) *There is an open ball* $\Omega' \subset \Omega$ *centered at the origin in* $\mathbb{C}^{n+1}$ *and a function* $\tilde{u} \in \mathcal{S}(\Omega'^{+})$ *such that* $u = b\tilde{u}$ *in* $U_{\Omega'}$.

(5.4) *There is an open neighborhood of the origin in* $\mathbb{C}^{n+1}$, $\mathcal{O}$, *and suitably large constants* $R, C > 0$ *such that*

$$|\mathcal{F}(u;z,w,\tau)| \leq C\, e^{-|\tau|/R}, \tag{5.5}$$

$$\forall\ (z,w) \in \mathcal{O},\quad \tau \in \mathbb{C} \ \text{ such that } \ R|\operatorname{Im}\tau| < -\operatorname{Re}\tau .$$

If $u \in C_c^{\infty}(U)$ *the previous properties are equivalent to the following one:*

(5.6) *There is an open ball* $\Omega' \subset \Omega$ *centered at* 0 *in* $\mathbb{C}^{n+1}$ *and a function* $\tilde{u} \in \mathcal{B}^{\infty}(\Omega'^{+})$ *such that* $u = b\tilde{u}$ *in* $U_{\Omega'}$.

The version of Theorem 4.6, Ch. III, that is meaningful here is the following one:

Theorem 5.2. *Same hypothesis as in Theorem 5.1. Let* u *be a distribution in* U *belonging to* $C^0(\Delta;\mathcal{D}'(\mathcal{B}))$. *The following two conditions are equivalent:*

(5.7) *There is an open neighborhood of* 0 *in* $\mathbb{C}^{n+1}$, Ω, *and a continuous function* $\tilde{h}(z,w)$ *in* Ω, *holomorphic with respect to* w *and such that* $u(x,y,s) = \tilde{h}(z, s+i\phi(z,\bar{z},s))$ *in* U_Ω.

(5.8) *There is an open neighborhood of* 0 *in* $\mathbb{C}^{n+1}$, Ω, *and two functions* $\tilde{u}^{+} \in \mathcal{S}(\Omega^{+})$, $\tilde{u}^{-} \in \mathcal{S}(\Omega^{-})$ *such that* $u = b\tilde{u}^{+} = b\tilde{u}^{-}$ *in* U_Ω.

REFERENCES

[BCT] Baouendi, M.S., Chang, C.H. & Treves, F. - Microlocal hypo-analyticity and extension of CR-functions, J. Diff. Geom. 18 (1983), 331-391.

[BRT] Baouendi, M.S., Rothschild, L.P. & Treves, F. - CR structures with group action and extendability of CR functions, Inventiones Math. 82 (1985), 359-396.

[HL] Henkin, G. & Leiterer, J. - Theory of functions on complex manifolds, Monographs in Math., Birkhäuser, Basel-Boston (1984).

[T] Treves, F. - Hypo-analytic structures, Contemporary Math. Vol. 27 (1984), 23-44.

LECTURE NOTES IN MATHEMATICS

Edited by A. Dold and B. Eckmann

Some general remarks on the publication of proceedings of congresses and symposia

Lecture Notes aim to report new developments - quickly, informally and at a high level. The following describes criteria and procedures which apply to proceedings volumes. The editors of a volume are strongly advised to inform contributors about these points at an early stage.

§1. One (or more) expert participant(s) of the meeting should act as the responsible editor(s) of the proceedings. They select the papers which are suitable (cf. §§ 2, 3) for inclusion in the proceedings, and have them individually refereed (as for a journal). It should not be assumed that the published proceedings must reflect conference events faithfully and in their entirety. Contributions to the meeting which are not included in the proceedings can be listed by title. The series editors will normally not interfere with the editing of a particular proceedings volume - except in fairly obvious cases, or on technical matters, such as described in §§ 2, 3. The names of the responsible editors appear on the title page of the volume.

§2. The proceedings should be reasonably homogeneous (concerned with a limited area). For instance, the proceedings of a congress on "Analysis" or "Mathematics in Wonderland" would normally not be sufficiently homogeneous.

One or two longer survey articles on recent developments in the field are often very useful additions to such proceedings - even if they do not correspond to actual lectures at the congress. An extensive introduction on the subject of the congress would be desirable.

§3. The contributions should be of a high mathematical standard and of current interest. Research articles should present new material and not duplicate other papers already published or due to be published. They should contain sufficient information and motivation and they should present proofs, or at least outlines of such, in sufficient detail to enable an expert to complete them. Thus resumes and mere announcements of papers appearing elsewhere cannot be included, although more detailed versions of a contribution may well be published in other places later.

Surveys, if included, should cover a sufficiently broad topic, and should in general not simply review the author's own recent research. In the case of surveys, exceptionally, proofs of results may not be necessary.

"Mathematical Reviews" and "Zentralblatt für Mathematik" require that papers in proceedings volumes carry an explicit statement that they are in final form and that no similar paper has been or is being submitted elsewhere, if these papers are to be considered for a review. Normally, papers that satisfy the criteria of the Lecture Notes in Mathematics series also satisfy this

.../...

requirement, but we would strongly recommend that the contributing authors be asked to give this guarantee explicitly at the beginning or end of their paper. There will occasionally be cases where this does not apply but where, for special reasons, the paper is still acceptable for LNM.

§4. Proceedings should appear soon after the meeeting. The publisher should, therefore, receive the complete manuscript within nine months of the date of the meeting at the latest.

§5. Plans or proposals for proceedings volumes should be sent to one of the editors of the series or to Springer-Verlag Heidelberg. They should give sufficient information on the conference or symposium, and on the proposed proceedings. In particular, they should contain a list of the expected contributions with their prospective length. Abstracts or early versions (drafts) of some of the contributions are very helpful.

§6. Lecture Notes are printed by photo-offset from camera-ready typed copy provided by the editors. For this purpose Springer-Verlag provides editors with technical instructions for the preparation of manuscripts and these should be distributed to all contributing authors. Springer-Verlag can also, on request, supply stationery on which the prescribed typing area is outlined. Some homogeneity in the presentation of the contributions is desirable.

Careful preparation of manuscripts will help keep production time short and ensure a satisfactory appearance of the finished book. The actual production of a Lecture Notes volume normally takes 6 -8 weeks.

Manuscripts should be at least 100 pages long. The final version should include a table of contents and as far as applicable a subject index.

§7. Editors receive a total of 50 free copies of their volume for distribution to the contributing authors, but no royalties. (Unfortunately, no reprints of individual contributions can be supplied.) They are entitled to purchase further copies of their book for their personal use at a discount of 33.3 %, other Springer mathematics books at a discount of 20 % directly from Springer-Verlag. Contributing authors may purchase the volume in which their article appears at a discount of 33.3 %.

Commitment to publish is made by letter of intent rather than by signing a formal contract. Springer-Verlag secures the copyright for each volume.

LECTURE NOTES

ESSENTIALS FOR THE PREPARATION
OF CAMERA-READY MANUSCRIPTS

Springer
Springer-Verlag
Berlin Heidelberg New York
London Paris Tokyo Hong Kong

The preparation of manuscripts which are to be reproduced by photo-offset require special care. Manuscripts which are submitted in technically unsuitable form will be returned to the author for retyping. There is normally no possibility of carrying out further corrections after a manuscript is given to production. Hence it is crucial that the following instructions be adhered to closely. If in doubt, please send us 1 - 2 sample pages for examination.

General. The characters must be uniformly black both within a single character and down the page. Original manuscripts are required: photocopies are acceptable only if they are sharp and without smudges.

On request, Springer-Verlag will supply special paper with the text area outlined. The standard TEXT AREA (OUTPUT SIZE if you are using a 14 point font) is 18 x 26.5 cm (7.5 x 11 inches). This will be scale-reduced to 75% in the printing process. If you are using computer typesetting, please see also the following page.

Make sure the TEXT AREA IS COMPLETELY FILLED. Set the margins so that they precisely match the outline and type right from the top to the bottom line. (Note that the page number will lie outside this area). Lines of text should not end more than three spaces inside or outside the right margin (see example on page 4).

Type on one side of the paper only.

Spacing and Headings (Monographs). Use ONE-AND-A-HALF line spacing in the text. Please leave sufficient space for the title to stand out clearly and do NOT use a new page for the beginning of subdivisons of chapters. Leave THREE LINES blank above and TWO below headings of such subdivisions.

Spacing and Headings (Proceedings). Use ONE-AND-A-HALF line spacing in the text. Do not use a new page for the beginning of subdivisons of a single paper. Leave THREE LINES blank above and TWO below headings of such subdivisions. Make sure headings of equal importance are in the same form.

The first page of each contribution should be prepared in the same way. The title should stand out clearly. We therefore recommend that the editor prepare a sample page and pass it on to the authors together with these instructions. Please take the following as an example. Begin heading 2 cm below upper edge of text area.

MATHEMATICAL STRUCTURE IN QUANTUM FIELD THEORY

John E. Robert
Mathematisches Institut, Universität Heidelberg
Im Neuenheimer Feld 288, D-6900 Heidelberg

Please leave THREE LINES blank below heading and address of the author, then continue with the actual text on the same page.

Footnotes. These should preferable be avoided. If necessary, type them in SINGLE LINE SPACING to finish exactly on the outline, and separate them from the preceding main text by a line.

Symbols. Anything which cannot be typed may be entered by hand in BLACK AND ONLY BLACK ink. (A fine-tipped rapidograph is suitable for this purpose; a good black ball-point will do, but a pencil will not). Do not draw straight lines by hand without a ruler (not even in fractions).

Literature References. These should be placed at the end of each paper or chapter, or at the end of the work, as desired. Type them with single line spacing and start each reference on a new line. Follow "Zentralblatt für Mathematik"/"Mathematical Reviews" for abbreviated titles of mathematical journals and "Bibliographic Guide for Editors and Authors (BGEA)" for chemical, biological, and physics journals. Please ensure that all references are COMPLETE and ACCURATE.

IMPORTANT

Pagination. For typescript, number pages in the upper right-hand corner in LIGHT BLUE OR GREEN PENCIL ONLY. The printers will insert the final page numbers. For computer type, you may insert page numbers (1 cm above outer edge of text area).

It is safer to number pages AFTER the text has been typed and corrected. Page 1 (Arabic) should be THE FIRST PAGE OF THE ACTUAL TEXT. The Roman pagination (table of contents, preface, abstract, acknowledgements, brief introductions, etc.) will be done by Springer-Verlag.

If including running heads, these should be aligned with the inside edge of the text area while the page number is aligned with the outside edge noting that right-hand pages are odd-numbered. Running heads and page numbers appear on the same line. Normally, the running head on the left-hand page is the chapter heading and that on the right-hand page is the section heading. Running heads should not be included in proceedings contributions unless this is being done consistently by all authors.

Corrections. When corrections have to be made, cut the new text to fit and paste it over the old. White correction fluid may also be used.

Never make corrections or insertions in the text by hand.

If the typescript has to be marked for any reason, e.g. for provisional page numbers or to mark corrections for the typist, this can be done VERY FAINTLY with BLUE or GREEN PENCIL but NO OTHER COLOR: these colors do not appear after reproduction.

COMPUTER-TYPESETTING. Further, to the above instructions, please note with respect to your printout that
- the characters should be sharp and sufficiently black;
- it is not strictly necessary to use Springer's special typing paper. Any white paper of reasonable quality is acceptable.

If you are using a significantly different font size, you should modify the output size correspondingly, keeping length to breadth ratio 1 : 0.68, so that scaling down to 10 point font size, yields a text area of 13.5 x 20 cm (5 3/8 x 8 in), e.g.

Differential equations.: use output size 13.5 x 20 cm.

Differential equations.: use output size 16 x 23.5 cm.

Differential equations.: use output size 18 x 26.5 cm.

Interline spacing: 5.5 mm base-to-base for 14 point characters (standard format of 18 x 26.5 cm).
If in any doubt, please send us 1 - 2 sample pages for examination. We will be glad to give advice.

Vol. 1173: H. Delfs, M. Knebusch, Locally Semialgebraic Spaces. XVI, 329 pages. 1985.

Vol. 1174: Categories in Continuum Physics, Buffalo 1982. Seminar. Edited by F.W. Lawvere and S.H. Schanuel. V, 126 pages. 1986.

Vol. 1175: K. Mathiak, Valuations of Skew Fields and Projective Hjelmslev Spaces. VII, 116 pages. 1986.

Vol. 1176: R.R. Bruner, J.P. May, J.E. McClure, M. Steinberger, H_∞ Ring Spectra and their Applications. VII, 388 pages. 1986.

Vol. 1177: Representation Theory I. Finite Dimensional Algebras. Proceedings, 1984. Edited by V. Dlab, P. Gabriel and G. Michler. XV, 340 pages. 1986.

Vol. 1178: Representation Theory II. Groups and Orders. Proceedings, 1984. Edited by V. Dlab, P. Gabriel and G. Michler. XV, 370 pages. 1986.

Vol. 1179: Shi J.-Y. The Kazhdan-Lusztig Cells in Certain Affine Weyl Groups. X, 307 pages. 1986.

Vol. 1180: R. Carmona, H. Kesten, J.B. Walsh, École d'Été de Probabilités de Saint-Flour XIV – 1984. Édité par P.L. Hennequin. X, 438 pages. 1986.

Vol. 1181: Buildings and the Geometry of Diagrams, Como 1984. Seminar. Edited by L. Rosati. VII, 277 pages. 1986.

Vol. 1182: S. Shelah, Around Classification Theory of Models. VII, 279 pages. 1986.

Vol. 1183: Algebra, Algebraic Topology and their Interactions. Proceedings, 1983. Edited by J.-E. Roos. XI, 396 pages. 1986.

Vol. 1184: W. Arendt, A. Grabosch, G. Greiner, U. Groh, H.P. Lotz, U. Moustakas, R. Nagel, F. Neubrander, U. Schlotterbeck, One-parameter Semigroups of Positive Operators. Edited by R. Nagel. X, 460 pages. 1986.

Vol. 1185: Group Theory, Beijing 1984. Proceedings. Edited by Tuan H.F. V, 403 pages. 1986.

Vol. 1186: Lyapunov Exponents. Proceedings, 1984. Edited by L. Arnold and V. Wihstutz. VI, 374 pages. 1986.

Vol. 1187: Y. Diers, Categories of Boolean Sheaves of Simple Algebras. VI, 168 pages. 1986.

Vol. 1188: Fonctions de Plusieurs Variables Complexes V. Séminaire, 1979–85. Edité par François Norguet. VI, 306 pages. 1986.

Vol. 1189: J. Lukeš, J. Malý, L. Zajíček, Fine Topology Methods in Real Analysis and Potential Theory. X, 472 pages. 1986.

Vol. 1190: Optimization and Related Fields. Proceedings, 1984. Edited by R. Conti, E. De Giorgi and F. Giannessi. VIII, 419 pages. 1986.

Vol. 1191: A.R. Its, V.Yu. Novokshenov, The Isomonodromic Deformation Method in the Theory of Painlevé Equations. IV, 313 pages. 1986.

Vol. 1192: Equadiff 6. Proceedings, 1985. Edited by J. Vosmansky and M. Zlámal. XXIII, 404 pages. 1986.

Vol. 1193: Geometrical and Statistical Aspects of Probability in Banach Spaces. Proceedings, 1985. Edited by X. Fernique, B. Heinkel, M.B. Marcus and P.A. Meyer. IV, 128 pages. 1986.

Vol. 1194: Complex Analysis and Algebraic Geometry. Proceedings, 1985. Edited by H. Grauert. VI, 235 pages. 1986.

Vol.1195: J.M. Barbosa, A.G. Colares, Minimal Surfaces in $\mathbb{R}^3$. X, 124 pages. 1986.

Vol. 1196: E. Casas-Alvero, S. Xambó-Descamps, The Enumerative Theory of Conics after Halphen. IX, 130 pages. 1986.

Vol. 1197: Ring Theory. Proceedings, 1985. Edited by F.M.J. van Oystaeyen. V, 231 pages. 1986.

Vol. 1198: Séminaire d'Analyse, P. Lelong – P. Dolbeault – H. Skoda. Seminar 1983/84. X, 260 pages. 1986.

Vol. 1199: Analytic Theory of Continued Fractions II. Proceedings, 1985. Edited by W.J. Thron. VI, 299 pages. 1986.

Vol. 1200: V.D. Milman, G. Schechtman, Asymptotic Theory of Finite Dimensional Normed Spaces. With an Appendix by M. Gromov. VIII, 156 pages. 1986.

Vol. 1201: Curvature and Topology of Riemannian Manifolds. Proceedings, 1985. Edited by K. Shiohama, T. Sakai and T. Sunada. VII, 336 pages. 1986.

Vol. 1202: A. Dür, Möbius Functions, Incidence Algebras and Power Series Representations. XI, 134 pages. 1986.

Vol. 1203: Stochastic Processes and Their Applications. Proceedings, 1985. Edited by K. Itô and T. Hida. VI, 222 pages. 1986.

Vol. 1204: Séminaire de Probabilités XX, 1984/85. Proceedings. Edité par J. Azéma et M. Yor. V, 639 pages. 1986.

Vol. 1205: B.Z. Moroz, Analytic Arithmetic in Algebraic Number Fields. VII, 177 pages. 1986.

Vol. 1206: Probability and Analysis, Varenna (Como) 1985. Seminar. Edited by G. Letta and M. Pratelli. VIII, 280 pages. 1986.

Vol. 1207: P.H. Bérard, Spectral Geometry: Direct and Inverse Problems. With an Appendix by G. Besson. XIII, 272 pages. 1986.

Vol. 1208: S. Kaijser, J.W. Pelletier, Interpolation Functors and Duality. IV, 167 pages. 1986.

Vol. 1209: Differential Geometry, Peñíscola 1985. Proceedings. Edited by A.M. Naveira, A. Ferrández and F. Mascaró. VIII, 306 pages. 1986.

Vol. 1210: Probability Measures on Groups VIII. Proceedings, 1985. Edited by H. Heyer. X, 386 pages. 1986.

Vol. 1211: M.B. Sevryuk, Reversible Systems. V, 319 pages. 1986.

Vol. 1212: Stochastic Spatial Processes. Proceedings, 1984. Edited by P. Tautu. VIII, 311 pages. 1986.

Vol. 1213: L.G. Lewis, Jr., J.P. May, M. Steinberger, Equivariant Stable Homotopy Theory. IX, 538 pages. 1986.

Vol. 1214: Global Analysis – Studies and Applications II. Edited by Yu.G. Borisovich and Yu.E. Gliklikh. V, 275 pages. 1986.

Vol. 1215: Lectures in Probability and Statistics. Edited by G. del Pino and R. Rebolledo. V, 491 pages. 1986.

Vol. 1216: J. Kogan, Bifurcation of Extremals in Optimal Control. VIII, 106 pages. 1986.

Vol. 1217: Transformation Groups. Proceedings, 1985. Edited by S. Jackowski and K. Pawalowski. X, 396 pages. 1986.

Vol. 1218: Schrödinger Operators, Aarhus 1985. Seminar. Edited by E. Balslev. V, 222 pages. 1986.

Vol. 1219: R. Weissauer, Stabile Modulformen und Eisensteinreihen. III, 147 Seiten. 1986.

Vol. 1220: Séminaire d'Algèbre Paul Dubreil et Marie-Paule Malliavin. Proceedings, 1985. Edité par M.-P. Malliavin. IV, 200 pages. 1986.

Vol. 1221: Probability and Banach Spaces. Proceedings, 1985. Edited by J. Bastero and M. San Miguel. XI, 222 pages. 1986.

Vol. 1222: A. Katok, J.-M. Strelcyn, with the collaboration of F. Ledrappier and F. Przytycki, Invariant Manifolds, Entropy and Billiards; Smooth Maps with Singularities. VIII, 283 pages. 1986.

Vol. 1223: Differential Equations in Banach Spaces. Proceedings, 1985. Edited by A. Favini and E. Obrecht. VIII, 299 pages. 1986.

Vol. 1224: Nonlinear Diffusion Problems, Montecatini Terme 1985. Seminar. Edited by A. Fasano and M. Primicerio. VIII, 188 pages. 1986.

Vol. 1225: Inverse Problems, Montecatini Terme 1986. Seminar. Edited by G. Talenti. VIII, 204 pages. 1986.

Vol. 1226: A. Buium, Differential Function Fields and Moduli of Algebraic Varieties. IX, 146 pages. 1986.

Vol. 1227: H. Helson, The Spectral Theorem. VI, 104 pages. 1986.

Vol. 1228: Multigrid Methods II. Proceedings, 1985. Edited by W. Hackbusch and U. Trottenberg. VI, 336 pages. 1986.

Vol. 1229: O. Bratteli, Derivations, Dissipations and Group Actions on C*-algebras. IV, 277 pages. 1986.

Vol. 1230: Numerical Analysis. Proceedings, 1984. Edited by J.-P. Hennart. X, 234 pages. 1986.

Vol. 1231: E.-U. Gekeler, Drinfeld Modular Curves. XIV, 107 pages. 1986.

Vol. 1232: P.C. Schuur, Asymptotic Analysis of Soliton Problems. VIII, 180 pages. 1986.

Vol. 1233: Stability Problems for Stochastic Models. Proceedings, 1985. Edited by V.V. Kalashnikov, B. Penkov and V.M. Zolotarev. VI, 223 pages. 1986.

Vol. 1234: Combinatoire énumérative. Proceedings, 1985. Edité par G. Labelle et P. Leroux. XIV, 387 pages. 1986.

Vol. 1235: Séminaire de Théorie du Potentiel, Paris, No. 8. Directeurs: M. Brelot, G. Choquet et J. Deny. Rédacteurs: F. Hirsch et G. Mokobodzki. III, 209 pages. 1987.

Vol. 1236: Stochastic Partial Differential Equations and Applications. Proceedings, 1985. Edited by G. Da Prato and L. Tubaro. V, 257 pages. 1987.

Vol. 1237: Rational Approximation and its Applications in Mathematics and Physics. Proceedings, 1985. Edited by J. Gilewicz, M. Pindor and W. Siemaszko. XII, 350 pages. 1987.

Vol. 1238: M. Holz, K.-P. Podewski and K. Steffens, Injective Choice Functions. VI, 183 pages. 1987.

Vol. 1239: P. Vojta, Diophantine Approximations and Value Distribution Theory. X, 132 pages. 1987.

Vol. 1240: Number Theory, New York 1984–85. Seminar. Edited by D.V. Chudnovsky, G.V. Chudnovsky, H. Cohn and M.B. Nathanson. V, 324 pages. 1987.

Vol. 1241: L. Gårding, Singularities in Linear Wave Propagation. III, 125 pages. 1987.

Vol. 1242: Functional Analysis II, with Contributions by J. Hoffmann-Jørgensen et al. Edited by S. Kurepa, H. Kraljević and D. Butković. VII, 432 pages. 1987.

Vol. 1243: Non Commutative Harmonic Analysis and Lie Groups. Proceedings, 1985. Edited by J. Carmona, P. Delorme and M. Vergne. V, 309 pages. 1987.

Vol. 1244: W. Müller, Manifolds with Cusps of Rank One. XI, 158 pages. 1987.

Vol. 1245: S. Rallis, L-Functions and the Oscillator Representation. XVI, 239 pages. 1987.

Vol. 1246: Hodge Theory. Proceedings, 1985. Edited by E. Cattani, F. Guillén, A. Kaplan and F. Puerta. VII, 175 pages. 1987.

Vol. 1247: Séminaire de Probabilités XXI. Proceedings. Edité par J. Azéma, P.A. Meyer et M. Yor. IV, 579 pages. 1987.

Vol. 1248: Nonlinear Semigroups, Partial Differential Equations and Attractors. Proceedings, 1985. Edited by T.L. Gill and W.W. Zachary. IX, 185 pages. 1987.

Vol. 1249: I. van den Berg, Nonstandard Asymptotic Analysis. IX, 187 pages. 1987.

Vol. 1250: Stochastic Processes – Mathematics and Physics II. Proceedings 1985. Edited by S. Albeverio, Ph. Blanchard and L. Streit. VI, 359 pages. 1987.

Vol. 1251: Differential Geometric Methods in Mathematical Physics. Proceedings, 1985. Edited by P.L. García and A. Pérez-Rendón. VII, 300 pages. 1987.

Vol. 1252: T. Kaise, Représentations de Weil et GL_2 Algèbres de division et GL_n. VII, 203 pages. 1987.

Vol. 1253: J. Fischer, An Approach to the Selberg Trace Formula via the Selberg Zeta-Function. III, 184 pages. 1987.

Vol. 1254: S. Gelbart, I. Piatetski-Shapiro, S. Rallis. Explicit Constructions of Automorphic L-Functions. VI, 152 pages. 1987.

Vol. 1255: Differential Geometry and Differential Equations. Proceedings, 1985. Edited by C. Gu, M. Berger and R.L. Bryant. XII, 243 pages. 1987.

Vol. 1256: Pseudo-Differential Operators. Proceedings, 1986. Edited by H.O. Cordes, B. Gramsch and H. Widom. X, 479 pages. 1987.

Vol. 1257: X. Wang, On the C*-Algebras of Foliations in the Plane. V, 165 pages. 1987.

Vol. 1258: J. Weidmann, Spectral Theory of Ordinary Differential Operators. VI, 303 pages. 1987.

Vol. 1259: F. Cano Torres, Desingularization Strategies for Three-Dimensional Vector Fields. IX, 189 pages. 1987.

Vol. 1260: N.H. Pavel, Nonlinear Evolution Operators and Semigroups. VI, 285 pages. 1987.

Vol. 1261: H. Abels, Finite Presentability of S-Arithmetic Groups. Compact Presentability of Solvable Groups. VI, 178 pages. 1987.

Vol. 1262: E. Hlawka (Hrsg.), Zahlentheoretische Analysis II. Seminar, 1984–86. V, 158 Seiten. 1987.

Vol. 1263: V.L. Hansen (Ed.), Differential Geometry. Proceedings, 1985. XI, 288 pages. 1987.

Vol. 1264: Wu Wen-tsün, Rational Homotopy Type. VIII, 219 pages. 1987.

Vol. 1265: W. Van Assche, Asymptotics for Orthogonal Polynomials. VI, 201 pages. 1987.

Vol. 1266: F. Ghione, C. Peskine, E. Sernesi (Eds.), Space Curves. Proceedings, 1985. VI, 272 pages. 1987.

Vol. 1267: J. Lindenstrauss, V.D. Milman (Eds.), Geometrical Aspects of Functional Analysis. Seminar. VII, 212 pages. 1987.

Vol. 1268: S.G. Krantz (Ed.), Complex Analysis. Seminar, 1986. VII, 195 pages. 1987.

Vol. 1269: M. Shiota, Nash Manifolds. VI, 223 pages. 1987.

Vol. 1270: C. Carasso, P.-A. Raviart, D. Serre (Eds.), Nonlinear Hyperbolic Problems. Proceedings, 1986. XV, 341 pages. 1987.

Vol. 1271: A.M. Cohen, W.H. Hesselink, W.L.J. van der Kallen, J.R. Strooker (Eds.), Algebraic Groups Utrecht 1986. Proceedings. XII, 284 pages. 1987.

Vol. 1272: M.S. Livšic, L.L. Waksman, Commuting Nonselfadjoint Operators in Hilbert Space. III, 115 pages. 1987.

Vol. 1273: G.-M. Greuel, G. Trautmann (Eds.), Singularities, Representation of Algebras, and Vector Bundles. Proceedings, 1985. XIV, 383 pages. 1987.

Vol. 1274: N. C. Phillips, Equivariant K-Theory and Freeness of Group Actions on C*-Algebras. VIII, 371 pages. 1987.

Vol. 1275: C.A. Berenstein (Ed.), Complex Analysis I. Proceedings, 1985–86. XV, 331 pages. 1987.

Vol. 1276: C.A. Berenstein (Ed.), Complex Analysis II. Proceedings, 1985–86. IX, 320 pages. 1987.

Vol. 1277: C.A. Berenstein (Ed.), Complex Analysis III. Proceedings, 1985–86. X, 350 pages. 1987.

Vol. 1278: S.S. Koh (Ed.), Invariant Theory. Proceedings, 1985. V, 102 pages. 1987.

Vol. 1279: D. Ieşan, Saint-Venant's Problem. VIII, 162 Seiten. 1987.

Vol. 1280: E. Neher, Jordan Triple Systems by the Grid Approach. XII, 193 pages. 1987.

Vol. 1281: O.H. Kegel, F. Menegazzo, G. Zacher (Eds.), Group Theory. Proceedings, 1986. VII, 179 pages. 1987.

Vol. 1282: D.E. Handelman, Positive Polynomials, Convex Integral Polytopes, and a Random Walk Problem. XI, 136 pages. 1987.

Vol. 1283: S. Mardešić, J. Segal (Eds.), Geometric Topology and Shape Theory. Proceedings, 1986. V, 261 pages. 1987.

Vol. 1284: B.H. Matzat, Konstruktive Galoistheorie. X, 286 pages. 1987.

Vol. 1285: I.W. Knowles, Y. Saitō (Eds.), Differential Equations and Mathematical Physics. Proceedings, 1986. XVI, 499 pages. 1987.

Vol. 1286: H.R. Miller, D.C. Ravenel (Eds.), Algebraic Topology. Proceedings, 1986. VII, 341 pages. 1987.

Vol. 1287: E.B. Saff (Ed.), Approximation Theory, Tampa. Proceedings, 1985–1986. V, 228 pages. 1987.

Vol. 1288: Yu. L. Rodin, Generalized Analytic Functions on Riemann Surfaces. V, 128 pages, 1987.

Vol. 1289: Yu. I. Manin (Ed.), K-Theory, Arithmetic and Geometry. Seminar, 1984–1986. V, 399 pages. 1987.

MIX
Papier aus verantwortungsvollen Quellen
Paper from responsible sources
FSC® C105338

If you have any concerns about our products,
you can contact us on
ProductSafety@springernature.com

In case Publisher is established outside the EU,
the EU authorized representative is:
Springer Nature Customer Service Center GmbH
Europaplatz 3, 69115 Heidelberg, Germany

Printed by Libri Plureos GmbH
in Hamburg, Germany